International Joint Conference on Metallurgical and Materials Engineering (JCMME 2018)

MATEC Web of Conferences Volume 277 (2019)

Wellington, New Zealand
10-12 December 2018

Editors:

Willvanto Anggono

ISBN: 978-1-5108-8456-4

Printed from e-media with permission by:

Curran Associates, Inc.
57 Morehouse Lane
Red Hook, NY 12571

Some format issues inherent in the e-media version may also appear in this print version.

No changes have been made to the content of these proceedings. There may be changes to pagination and minor adjustments for aesthetics.

Printed by Curran Associates, Inc. (2019)

For additional information, please contact EDP Sciences – Web of Conferences at the address below.

EDP Sciences – Web of Conferences
17, Avenue du Hoggar
Parc d'Activité de Courtabœuf
BP 112
F-91944 Les Ulis Cedex A
France

Phone: +33 (0) 1 69 18 75 75
Fax: +33 (0) 1 69 28 84 91

contact-edps@webofconferences.org

Additional copies of this publication are available from:

Curran Associates, Inc.
57 Morehouse Lane
Red Hook, NY 12571 USA
Phone: 845-758-0400
Fax: 845-758-2633
Email: curran@proceedings.com
Web: www.proceedings.com

TABLE OF CONTENTS

Preface

We are pleased to introduce the proceedings of the 2018 International Joint Conference on Metallurgical and Materials Engineering (JCMME 2018). The conference was held in Wellington, New Zealand on December 10 to 12, 2018. A major conference theme was related to the recent advances of Metallurgical and Material Engineering. This event was organized by International Academy of Science and Engineering for Development (IASED) as part of their commitment to spread the recent development of researches in engineering and related sciences.

This conference provided many opportunities for delegates to exchange new ideas directly, to build business or research relationships, and to find global partners for future cooperation. Their contributions have supported this conference into an extraordinary one. The results of the conference have significantly contributed to the knowledge in the latest scientific fields, encouraging interaction between researchers and academic improvement in a more established scholastic community.

The internationally acclaimed distinguished professors and distinguished experts delivered keynote speakers to enthusiastic audience, sharing their in-depth experiences of challenges and future trends, which made this conference remarkable. The participants had the privilege of hearing from the main researchers in their fields, as they presented the latest findings from their work.

This conference has provided a great experience for participants from all over the world, with an excellent technical program and social activities. The conference was a huge success with a lot of peer reviewed papers from various nations presented to maintain the promise of contributors on a global scale. The number of delegates of the conference was also highly gratifying, indicating a high level of international enthusiasm in the subject of Metallurgical and Material Engineering.

These proceedings will provide a permanent record of what has been presented and complement world scientists with good references. The results of this conference are an encouragement to stimulate further study and research in all scientific fields. We thank all contributors for their support and participation in making this conference a great success.

With our warmest regards,

Assoc. Prof. Dr. Willyanto Anggono, S.T., M.Sc.
On behalf of The Conference Committee

A new process of smelting laterite by low-temperature reduction and microwave irradiation

Lei Wang, *Peimin* Guo*, and *Lingbing* Kong

China Iron and Steel Research Institute, 76 Xueyuannan Road, Beijing 100081, China

Abstract. A new smelting process for the reduction of laterite with a relatively high yield of Ni has been presented in this paper; particularly, this process has been underwent industrialized experiment in China. This new process was based on the following fundamental research results. (1) Since nickel oxide can be reduced more easily than iron oxide, the nickel content of the produced ferronickel alloy increases significantly when the quantity of coal that is mixed with laterite ore is optimized, leading to a right oxygen potential of gas for reduction reaction. (2) When laterite is reduced at 1150°C, only 75% of the reduced ferronickel product can be magnetically separated from slag, although more than 95% Ni has been reduced from laterite. (3) The key technology is to realize a fast carburization at 1300°C by microwave irradiation, which causes ferronickel fines grow to a large size in the semi-melting state so that they can be separated magnetically with a high efficiency. The first demonstration production line with an annual capacity of 100 kt green balls has been established in China. The Ni content reaches 13.5%, moreover, both P and S contents are also within the expected range.

1 Introduction

Nickel resource is important in the world. About 80% of the total known nickel reserves are present as nickel oxide in the laterite ores [1]. Nevertheless, at the present time about 60% of the total metallic nickel production arises from the sulfide ore deposits [2]. There are a number of reasons for this disparity, the major one is economic and thus there is considerable incentive for the development of more efficient and profitable processes for the extraction of metallic nickel from these laterite ores. According to the Ni and Fe content, Laterite can be classified into three catalogues: limonite, intermediate and saprolite, shown in Table 1. The nickel content is over 1.8% and iron content is less than 25% for saprolite, which is suitable for RK-EF smelting; The nickel content is less than 1.5% and iron content is over 30% for limonite, which is suitable for BF process; The intermediate can be applied for EF or BF [3-5].

BF process needs over much coke and RK-EF needs over much electricity. With the surplus capacity of ferronickel and the unceasingly decreased price of metal nickel, Both

* Corresponding author: guopm@pku.org.cn

BF and RK-EF processes lost their original economic advantages. So these existed processes need to be upgraded or new energy saving processes need to be developed.

Table 1. Distribution and composition of laterite.

Ore bed	Composition/ wt, %				
	Ni	Co	Fe	Cr_2O_3	MgO
Limonite	0.8~1.5	0.1~0.2	40~50	2~5	0.5~5
Intermediate	1.5~1.8	0.02~0.1	25~40	1~2	5~15
Saprolite	1.8~3	0.02~0.1	10~25	1~2	15~35

We have developed a new process of low-temperature reduction combined with microwave smelting. Laterite ores underwent reduction and fresh ferronickel fines underwent gathering and growing up during the semi-molten state. A demonstrate production line showed that the new process is of the features such as low smelting temperature, low energy consumption and low emission. The main results are introduced here.

2 Fundamentals on semi-molten reduction and smelting of laterite

2.1 Composition and phases of saprolite

The composition of saprolite used in Lab is shown in Table 2. Phases of the ore include $Fe_2O_3{\cdot}H_2O$, $(Mg,Fe)_2SiO_4$, $MgSiO_3$ and $Mg_3Si_2O_5(OH)_4$ etc. [6] The ore contains up to about 30~45% water, both as free moisture and also as chemically combined water in hydrated minerals.

Table 2. Composition of saprolite /wt%.

TFe	Ni	SiO_2	Al_2O_3	CaO	MgO	MnO	S	P	Cr	Burning loss
20~25	1.5~2.0	20~30	~1.0	~0.2	15~20	~0.5	~0.02	~0.01	~0.2	30~45

2.2 Thermodynamics of saprolite reduction at low temperature

According to the main phases in saprolite, the reduction reactions are shown in Table 3. Reaction temperature of nickel oxide with carbon is remarkably less than that of iron oxide with carbon, so nickel oxide is reduced easier than iron oxide. As shown in Fig.1, the equilibrium content of $CO/(CO+CO_2)$ is over 70% for obtaining metal iron, if oxidizing atmosphere occurs, the fresh metal iron can be oxidized again. However, the equilibrium content of $CO/(CO+CO_2)$ is very low(less than 5%) for obtaining metal nickel, the fresh metal nickel can not be reoxidized easily . So the selective reduction can be realized by controlling ratio of carbon/oxygen and reactive atmosphere, and higher nickel content of ferronickel can be produced economically.

Table 3. Thermodynamic data of nickel oxide and iron oxide reduced by carbon.

Reaction	Gibbs energy	Starting temperature of reaction at standard condition
$NiO+C=Ni+CO_{(g)}$	$\Delta G°=121470-173T$	t=429°C
$FeO+C=Fe+CO_{(g)}$	$\Delta G°=151880-153T$	t=720°C
$Ni_2SiO_4+2C=2Ni+Si+2CO_{(g)}$	$\Delta G°=254680-352T$	t=450°C
$Fe_2SiO_4+2C=2Fe+SiO_2+2CO_{(g)}$	$\Delta G°=331480-330T$	t=730°C

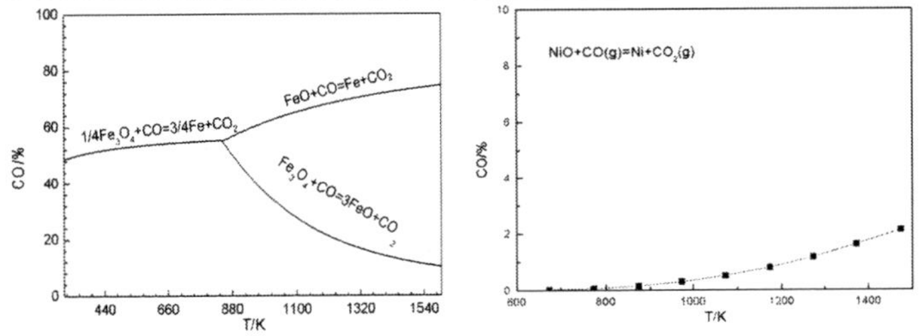

Fig.1. The equilibrium diagrams of iron oxide and nickel oxide reduced by CO-CO_2.

2.3 Reduction and magnetic separation experiment

The composition of chosen saprolite is shown in Table 2. The main reduction device is a vertical heating furnace. Fine coke is used as reducing agent, whose fixing carbon content is 85% and ash content is 12%. Firstly, saprolite and fine coke are refined to -74μm and then mixed in certain ratio. The 10g mixed sample is put in a corundum crucible and then in a vertical heating furnace for reduction. Nitrogen was used as protective atmosphere.

The experimental parameters, such as temperature, time and ratio of carbon/oxygen, are carefully controlled to get higher reduction ratio of nickel. After reduction, the reduced sample is milled and magnetically separated to measure the separation efficiency of nickel.

As shown in Fig.2, at the conditions of 1150°C and C:O=1:1, the reduction ratio of nickel is over 90% after 30min reduction and the final ratio reaches 98%. However, the reduction rate of iron oxide is slow, which reaches 75% after 30min reduction and 90% after 60min. When the ratio of carbon/oxygen decreases from 1:1 to 0.8:1, the reduction rate of nickel oxide is still faster. It reaches 80% after 30min reduction and 95% after 60min. The experiment result shows that the ratio of carbon/oxygen has a weak influence on nickel reduction, but it has a significant influence on iron reduction.

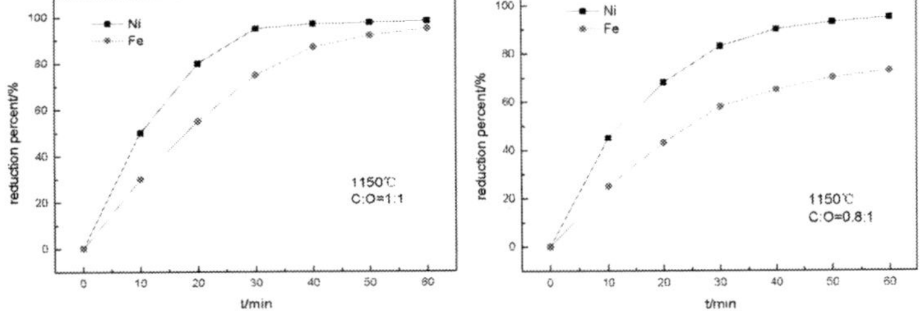

Fig.2. Effects of temperature, ratio of carbon/oxygen on reduction ratios of nickel and iron.

After the reduction at 1150°C, the megascopic ferronickel granules, diffused in the slag, are too small to be seen. Therefore, the mixture of reduced ferronickel fine and slag need to be crushed and magnetically separated. The yield of fine nickel is only 75% in the experiment.

2.4 Effective separation of ferronickel from slag at the semi-molten condition

According to the above results, nickel oxide is reduced easily and fine ferronickel can be obtained at 1150°C, but separation efficiency is unsatisfactory. Slag and ferronickel can be separated at 1500°C~1600°C in BF or RK-EF process, but high energy consumption is required. For saprolite, the amount of slag is much more than that of ferronickel and enthalpy of slag is also more than that of ferronickel, so for 1t ferronickel molten, much power is needed for melting slag.

Our R&D team proposed a new route of efficiently separating ferronickel from slag at the semi-melting condition: the ferronickel granules with low melting point can be formed in slag by fast carburization at 1300°C, then grow up and gather together due to surface tension and gravity. Theoretically, the complete separation of ferronickel and slag would be realized after enough time. In our experiment, the real smelting time is limited, so, a massive volume of ferronickel alloy is rarely found and alloy granules are bound by slag. The final size distribution of ferronickel granules is shown as follows: the granules larger than 5mm are up to 70%, the granules of 1~5mm up to 20%, and the fine particles less than 1mm about 10%.

2.4.1 Melting points of ferronickel and slag

Carbon content can decrease the melting point of ferronickel remarkably. When carbon content is about 4wt%, the melting point of ferronickel will be less than 1200°C. On the contrary, nickel content has a pretty weak effect on the melting point of ferronickel. So the key measure to lower ferronickel melting point is to accelerate carburizing process [7].

Main phase of slag after smelting is magnesium silicate with a small amount of FeO, Al_2O_3, and CaO in slag. As shown in Fig.3, there are two low melting regions in CaO-MgO-SiO_2 system, one is labelled as A, and the other is labelled as B. The composition of A is different from the ferronickel slag, so, we do not take A region into our consideration. Although the melting points are less than 1400°C in B region, a large amount of CaO needs to be replenished for saprolite smelting, leading to more power consumption. So, if 1300℃ is chosen for smelting temperature of ferronickel, the mixture of alloy and slag is actually in semi-molten state.

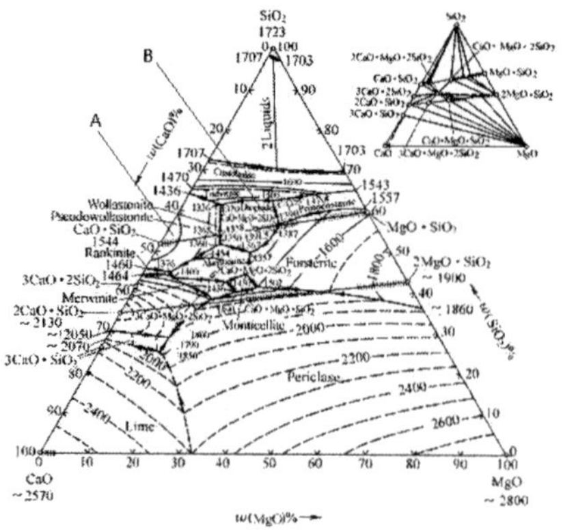

Fig.3. Phase diagram of CaO-MgO-SiO_2 system.

2.4.2 Fast carburization

It is easy to carburize ferronickel alloy at high smelting temperature during BF or EF process. However, the carburizing speed is slow in solid reduction process at 1100°C~1300°C.Solid reaction can be conducted in rotary and flush reactors. In rotary reactor, such as rotary kiln, charging materials rotate with rotary kiln. In flush reactor, such as rotary hearth furnace (RHF) or tunnel kiln (TK), charging materials move with hearth or pallet.

Semi-molten state can be realized so long as flame temperature arrives to 1400°C in rotary kiln and overturning materials can receive the energy from the flame evenly, so carburization can be conducted uniformly. Risk of semi-molten smelting in RK is the sticking problem. When materials move from low-temperature region to high temperature region, it is relatively safe, but semi-molten material moves from high temperature region to low temperature region, semi-molten materials start to cool, and it is possible to freeze in a certain area of the kiln.

For RHF or TK process, heating from material surface to inner is unidirectional. Heat radiation is of surface heating characteristics and the efficiency of heating the inner material mainly depends on heat conducting efficiency. So, the efficiency of this unidirectional heating is pretty low.

We developed a kind of microwave smelting technique to change the unidirectional heating style. The microwave is used to irradiate materials of 1150°C directly. It heats material very fast and is in favour of carburizing reaction. The microwave irradiation heating style include following advantages:

(1)Hot laterite ore can absorb more microwave energy with temperature increase .

(2)Carbon can adsorb much more microwave energy than slag and accelerate carburization reaction.

(3)Microwave changes traditional heating direction from surface to inner of materials. For flush reactor, microwave heats inner of materials directly and realizes heating from inner to surface.

3 New process based on low-temperature reduction and microwave irradiation

3.1 Process description

For decreasing electricity or coke consumption, the ferronickel smelting process based on low-temperature reduction and microwave irradiation has been developed. The first step is to dry laterite ores to decrease water content from 35% to 10%. Waste heat is used for drying laterite ores in the drier. The second step is to refine laterite ores and anthracite coals. The third step is to manufacture green balls with uniform composition by batching, mixing and briquetting. The green balls are dried and conveyed to a long low-temperature reduction kiln, where the maximum reduction temperature is 1200℃.Then, the heated and reduced balls are sent into a microwave equipment, where balls are heated up to 1300℃ and semi-molten smelting is realized to obtain ferronickel granules. The hot mixture of ferronickel and slag are water quenched and crushed. Then, the ferronickel granules are separated from slag magnetically. The fine ferronickel can be recycled as a sort of raw material and fine slag is used as an additive agent of cement. The 300℃ off gas in the reduction kiln is used for drying laterite and enters into cyclone and sack for dust eliminating, then enters CFD desulfurizer for removing sulphur, finally low temperature and clean off gas is discharged into atmosphere.

Technological process is shown in Fig.4.

3.2 The first demonstration line

The first industrialized demonstration production line was installed in 2014. Its capacity is 100 kt green balls /a. Some photographs from the demonstration factory and final product are shown in Fig.5 and Fig.6. The main raw material is saprolite, Ni and Fe Contents of the saprolite are 1.84% and 14.5% respectively. Pulverized coal is used as reducing agent and fuel. Ni, S, P contents are 13.5%, 0.2% and 0.025% respectively in the final ferronickel granules. The selective reduction is realized in the demonstration line.

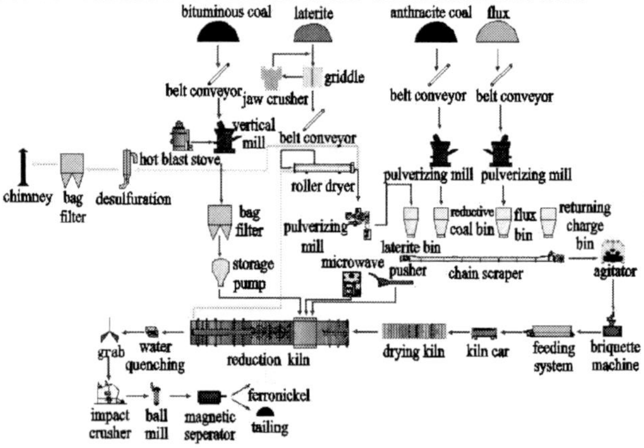

Fig.4. Process diagram of ferronickel smelting by low-temperature reduction and microwave irradiation.

Low-temperature reduction kiln Microwave devices

Ball milling and magnetic separation Environmental devices

Fig.5. Photographs from the demonstration production line.

Fig.6. Morphology of ferronickel product.

4 Conclusions

We invented a new smelting process to reduce laterite in the semi-melted state and set up a demonstration line for industrial production. The main results are as follows:

(1)Selective reduction of Ni from laterite strongly depends on the ratio of carbon to oxygen of the charge, i.e., optimized quantity of coal in the charge could lead to much higher nickel content of produced ferronickel granules.

(2)When laterite is reduced at 1150°C, the yield ratio of nickel is over 95% but the efficiency of magnetic separation of ferronickel from slag is just 75%.

(3)In the new process, the reduction is operated at 1300°C in the semi-melting state. The key technology is to realize a fast carburization by means of powerful microwave irradiation, which helps ferronickel fines diffuse and merge to larger sizes in the semi-melting state for highly efficient separation.

(4)The first demonstration production line with this process has been constructed with an annual capacity of 100 kt green balls. The nickel content of produced ferronickel granules reaches 13.5%. The P and S contents are acceptable.

Acknowledgements

The presented research was financed by Projects from National Natural Science Foundation of China (U1560201).

References

1. Bingxi Geng. Steel & Management, (2008)(6):27~29.
2. Yuting Hu. China Steel, (2007)(10):10~12.
3. Wuzhuang Zhao. World Nonferrous Metals, (2007)(3):5~8.
4. Huanhua He. China Nonferrous Metallurgy, 33(2004)(6):12~16.
5. Jingyou Chen, Juming Tan. Ferro-alloys, (2008)(3):13~16.
6. Rongqiu Peng. Nickel Metallurgy [M]. Changsha: Central South University Press, 2005.
7. Peimin Guo, Pei Zhao. Efficient Utilization of Metallurgical Resources [M]. Beijing: Metallurgical Industry Press, 2012.

Solvent extraction of scandium from chlorination system with EHPAEH

Lan Jiang, *Hongliang* Yang*, and *Yang* Sun

Northeastern University(School of materials and metallurgy Northeastern University)

Abstract. The extraction of scandium from hydrochloric acid media with EHPAEH is investigated. Fe3+ in synthesized solution would reduce the purity of scandium, therefore, N235 is selected as extractant for separating Fe3+ and Sc3+. Experimental results indicated that separation factor βFe/Sc reached 5416.4 with the organic system consisting of 20 vol% trinoctylamine(N235), phase volume ratio(O/A) of 1:1, extraction stage of 3. Then mono(2-ethyl hexyl)-2-ethyl hexyl phosphonate(EHPAEH) is employed to extract Sc3+ from the leaching solution of Fe3+ free. With the organic phase containing 5 vol% EHPAEH, phase volume ratio O/A of 1:5, at pH 0.3, nearly 99% scandium is extracted. Finally, the scandium-loaded organic phase is stripped with 4.5N hydrochloric acid, under phase volume ratio of 1:1 at 65°C.

1 Introduction

Scandium, an important Rare Earth element, is widely used in advanced ceramics, electric lighting source, electronic and other high-tech fields due to its excellent physical and chemical properties. In nature, independent scandium deposits are rare, it is usually associated with other metals and mainly recovered as by-product from residues, tailings and waste liquors in the smelting process of some metals such as rare earths, uranium, titanium, tungsten, aluminium, tantalum and niobium[1,2]. In this experiment, raw material is the residue originated in $TiCl_4$ production by boiling chlorinating method (and the scandium compounds are enriched in it)[3].

Scandium in chlorination residue can be leached with hydrochloric acid. The methods of recovering scandium from synthesized solutions usually include solvent extraction[4-8]and ion exchange[4,9-11].

Solvent extraction with the characteristics of efficiency and economica is widely used in the enrichment and purification of many metals. Plenty of extractants can be used for scandium concentrate and separate it from other metal ions [2]. Acidic organophosphorus and neutral organophosphorus are two common types of extractants. Nearly 100% scandium is extracted from chloride solutions with 20 vol% di-2-ethylhexyl phosphoric acid(D2EHPA, in China used to call P_{204})[5].

In this paper, Fe^{3+} is removed by N_{235} before extracting Sc^{3+} with EHPAEH. Experimental parameters which affecting extraction process, such as concentration of

* Corresponding author: 421589185@qq.com

extractants, phase volume ratio (O/A), temperature and extraction stages are investigated in detail and the optimal experimental conditions are carried out.

2 Experimental

2.1 Aqueous solution

Synthetic leaching solution for this experiment was prepared by dissolving chlorination dust with 1mol/L HCl, liquid to solid ratio (L/S) of 2:1, at 75°C holding for 3h. (The concentration of Sc^{3+} and Fe^{3+} were respectively 238.3mg/L and 3.1g/L). All aqueous solutions were made up using distilled water.

2.2 Organic reagents

Extractant Trinoctylamine (N235) and Mono(2-ethyl hexyl)-2-ethyl hexyl phosphonate (EHPAEH) were of industrial grade. Modifier Sec-Octyl alcohol analytical grade was of analytical grade. Diluent sulfonated kerosene was of industrial grade. All reagents were directly used without further purification.

2.3 Test produces

Extraction and stripping experiments were carried out in a 125 ml separating funnel by contacting the organic and aqueous phases under the given experimental condition including certain pH, composition of organic phase, phase volume ratio O/A and temperature. The funnel was immersed in a temperature controlled water bath to keep temperature constant (and ensure a well mixed model between organic and aqueous solution). When the reaction reached equilibrium, organic phase and aqueous phase was separated, then ion concentrations in aqueous solution was analyzed.

2.4 Test method and calculation formula

Each test, about 10mL aqueous solution was taken for chemical analysis. Inductively Coupled Plasma and Atomic Emission Spectroscopy (ICP-AES) was used to test the concentration of scandium. The model of 722s spectrophotometer was used to test the concentration of Fe^{3+}, and sulfosalicylic acid was served as chromogenic agent. Sartorius PHS-3 pH meter was employed to measure pH value in aqueous solution.

3. Results and discussion

3.1 Removal of Fe^{3+} by Trinoctylamine (N_{235})

The main impurities in the synthetic leaching solutions are iron, aluminum, manganese, calcium, titanium, manganese, etc. Preliminary experiments have shown that the co-extraction rate of Mn^{2+} and Fe^{3+}were relatively high when extracted with acidic organophosphorus. It will pose interference to the purity of scandium. Mn^{2+} can be removed by hydrolysis precipitation method due to the big difference of precipitation pH. In this experiment, the effect of N_{235} concentration, modifier concentration, extraction stage on and separation factor were studied.

3.1.1 The effect of N_{235} concentration

Different concentrations of N_{235} in sulfonated kerosene between 5 vol% and 40 vol% were employed in the preliminary experiment and kept phase volume ratio O/A of 1:1. The results shows that with the increase concentration of N_{235}, the extraction rate of Fe^{3+} gradually increased, while it presents very little effect on the Sc^{3+} with the extraction rate always less than 2 vol%. However, organic phase turns into a turbidity, when concentration exceeded 20 vol%. The extraction rate only increased by 3.31% with N_{235} concentration increasing from 20 vol% to 30 vol% and the change of extraction rate and separation factor are very small after over 40 vol%. The appropriate concentration of N_{235} in sulfonated kerosene is 20 vol% and it was set up in the following experiments.

Table 1. Effect of N_{235} concentration on extraction.

No.	N_{235} concentrations (vol%)	E_{Fe} (%)	E_{Sc} (%)	$\beta_{Fe/Sc}$
1	5	71.61	0.88	284.1
2	10	80.65	1.26	326.6
3	20	85.81	1.34	445.2
4	30	88.71	1.72	449.0
5	40	87.74	1.59	442.9

3.1.2 The effect of modifier concentration

Although N_{235} has a good property on separation of Sc^{3+} and Fe^{3+}, the separation rate of organic phase and aqueous phase is slow and easy to appear a third phase in organic phase. In order to solve this problem, a modifier Sec-Octyl alcohol was added into the organic phase. In this experiments, the concentration of Sec-Octyl alcohol was varied from 0 vol% to 25 vol% with other parameters constant.

Table 2. Effect of modifier concentration on extraction.

No.	Sec-Octyl alcohol concentration (vol%)	E_{Fe} (%)	E_{Sc} (%)	$\beta_{Sc/Fe}$	The third phase
1	0	85.65	1.34	445.2	existent
2	5	85.32	1.34	427.9	non-existent
3	10	85.16	1.30	435.7	non-existent
4	15	85.16	1.26	449.7	non-existent
5	20	80.00	1.28	308.5	non-existent
6	25	71.61	1.20	207.7	non-existent

No difficulty was encountered in phase separation when adding Sec-Octyl alcohol. As shown in Table 2, the change of Fe^{3+} extraction rate is small with Sec-Octyl alcohol less than 15%, but sharply dropped when it exceeded 15%, while the change of Sc^{3+} is almost negligible. This may be due to the formation of complexation between Sec-Octyl alcohol and N_{235} thus reduce the effective concentration of extractant then prevented the combination of Fe^{3+} and N_{235}. Therefore, the concentration of Sec-Octyl alcohol should be no more than 15 vol%.

3.2 Solvent extraction of scandium

EHPAEH is a kind of organophosphorus acids extractant which is similar to DE2HPA. But EHPAEH is easier to be stripped owing to its weaker extraction ability. In addition, it is more suitable for extracting Sc^{3+} from high acidity solution[4]. In this experiment,

EHPAEH is employed to extract Sc^{3+}. The extraction principles are as follows[4]:

① in low acidity system:

$$Sc^{3+}+3HA \rightarrow ScA_3+3H^+ \quad (1)$$

② in high acidity system:

$$Sc^{3+}+nHA \rightarrow Sc(HA)_n^{3+} \quad (2)$$

It can be seen from reaction Eq. (1) and (2) that cation exchange reaction exits in low acidity solution and solvation exits in high acidity solution[4,12].

The effects of experimental parameters, such as aqueous pH, EHPAEH concentration, and phase volume ratio are investigated to determine the optimum operating conditions.

Poor phase separation easily appears in the extraction process when organic phase contained only EHPAEH in sulfonated kerosene. After adding 15 vol% sec-Octyl alcohol as modifier, phase separation improvs significantly.

3.2.1 Effect of pH

In the experiment, pH was regulated in the range of 0~2.0 (under the following conditions) with 5 vol% EHPAEH in sulfonated kerosene, phase volume ratio O/A ratio of 1:1 and the temperature of 25°C).

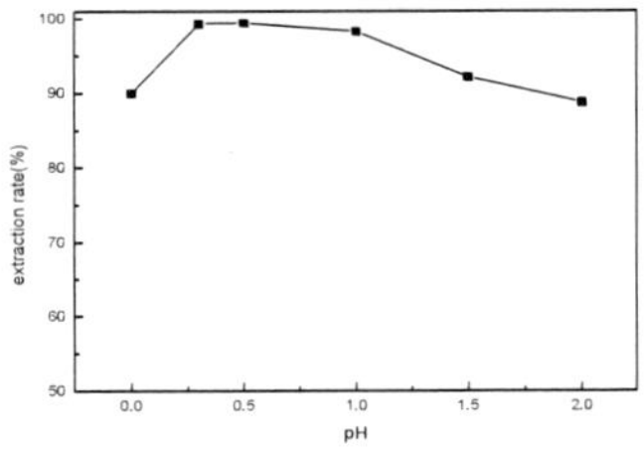

Fig. 1. Effect of solution pH on extraction rate.

It can be seen from Fig.1 that at beginning, with the increase of pH, the extraction rate of Sc^{3+} gradually increased then significantly decreased after pH greater than 0.9. The preferable pH range for scandium extraction is 0.3~1.0. The extraction rate gradually decreased with pH greater than 1.0, possibly caused by impurity irons hydrolysis. The initial pH 0.3 is in the appropriate range, so directly extraction without adjusting pH is feasible.

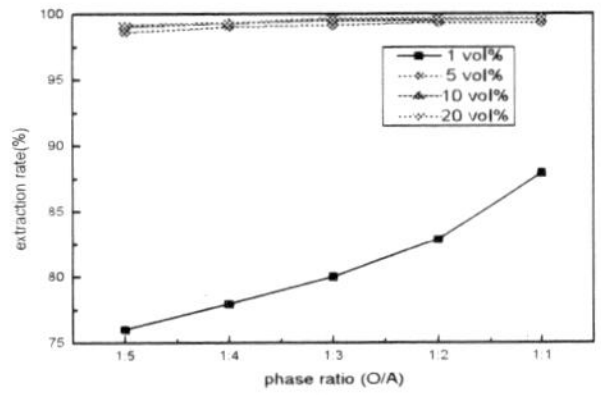

Fig. 2. Effect of phase ratio on extraction rate with different EHPAEH concentration.

3.2.2 Effect of EHPAEH concentration and phase volume ratio

Fig.2 shows the effect of phase volume ratio O/A on the Sc^{3+} extraction rate at different EHPAEH concentrations. Extraction rate increased with both increasing concentrations and increasing volume phase ratio. But lower O/A is beneficial to concentrating scandium in the organic phase[4]. Scandium has more charge and smaller ionic radius which shows the great electron affinity and extractability, therefore, the selectivity of scandium over other metals is increased with low EHPAEH concentration. Since extraction rate doesn't increase obviously with the concentration of EHPAEH over 5 vol% and O/A over 1:5, 5 vol% and 1:5 is the decent concentration and phase volume ratio, respectively.

3.3 Stripping of scandium

3.3.1 Effect of hydrochloric acid concentration and phase volume ratio

Lanthanide element has the characteristic of lanthanide contraction. It means the atomic radius of the sixth periodic element is smaller than the fifth. From lanthanum to lutetium with the increase of nuclear charge, the effective nuclear charge gradually increases, which cause the atomic volume reduced. Consequently the heavy rare earth elements with larger atomic number has more smaller radius. Scandium that located in the forth periodic is similar to heavy rare earth elements in terms of atomic radius. Therefore it requires high hydrochloric acid concentration in the stripping process[4].

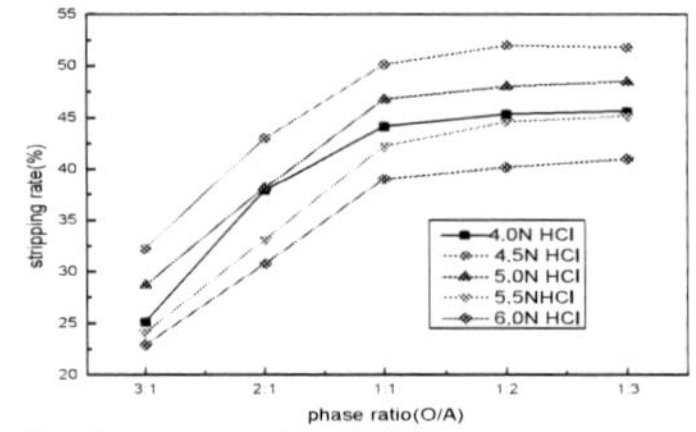

Fig. 3. Effect of phase ratio on stripping rate with different concentration of hydrochloric acid.

The effect of phase volume ratio O/A with different hydrochloric acid concentration on stripping rate of scandium was investigated. Fig.3 shows that at the outset the stripping rate of scandium gradually increased with the increasing acid concentration and it reaches the maximum of 50.13% with 4.5mol/L HCl then begin to drop. Similarly, with the increase of phase volume ratio(O/A), the stripping rate of scandium gradually increased, then stables in a certain range with phase volume ratio over 1:1.

3.3.2 Effect of temperature

Scandium was stripped from scandium-loaded organic phase at different temperatures with the fixed condition of 4.5N HCl, phase ratio O/A of 1:1. Fig.4 shows that at beginning with the increase of temperature, the stripping rate of Sc^{3+} gradually increased, but it only increased by 0.91% (from 80.36% to 81.27%) when temperature increased from 65°C to 85°C. Furthermore, high temperature easily led to the organics. To sum up, 65°C is appropriate.

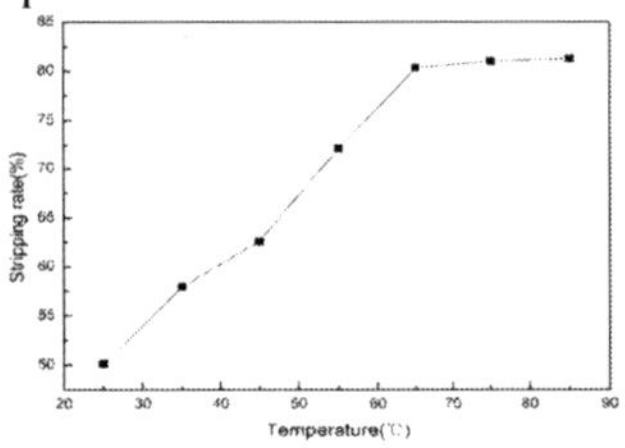

Fig. 4. Effect of temperature on stripping rate.

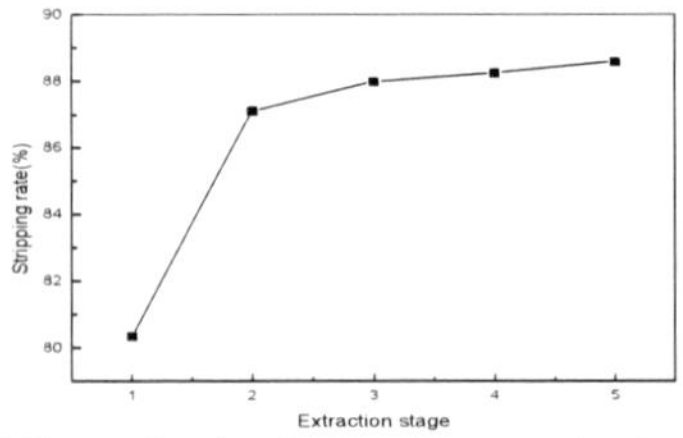

Fig. 5. Effect of stripping stage on stripping rate.

3.3.3 Effect of stripping stage

According to Fig.5, the stripping rate of scandium is improved by increasing stripping stage. After two stages, the stripping rate reached 87.12% then increased slowly. After certain treatment, the organic solvent is reused for extraction.

4 Conclusion

N_{235} demonstrates better selectivity for Fe^{3+} and Sc^{3+}. More than 99% Fe^{3+} in synthetic solution is extracted into organic phase while little scandium is co-extracted. The separation factor $\beta_{Fe/Sc}$ reaches 5416.4 with the organic system consisting of 20 vol% N_{235}+15 vol% Sec-Octyl alcohol+65 vol% sulfonated kerosene, phase volume ratio O/A of 1:1, temperature of 25°C, and extraction stages of three.

EHPAEH has a high scandium extraction ability, even small amount can acquire good results. The extraction rate of Sc^{3+} reaches 98.59% with the organic system consisting of 5 vol% EHPAEH, 15 vol% sec-Octyl alcohol+80 vol% sulfonated kerosene, phase volume ratio (O/A) of 1:5, at pH 0.3.

Finally the Sc extracted complexes is stripped by hydrochloric acid. With increasing of temperature and hydrochloric acid concentration, the stripping rate of Sc^{3+} increases quickly but tend to be smooth finally. Under the conditions of hydrochloric acid concentration 4.5N, temperature of 65°C, stripping stages of two, 87.12% scandium in organic phase returns into aqueous phase.

References

1. Wang W W, Pranolo Y and Cheng C Y. Metallurgical processes for scandium recovery from various resources: A review[J]. Hydrometallurgy, 2011, 108: 100-108.
2. Xu S Q and Li S Q. Review of the extractive metallurgy of scandium in China (1978-1991)[J]. Hydrometallurgy, 1996, 42: 337-343.
3. Zhang Z B and Zhang Z H. Scandium resources and its extraction technology[J]. Yunnan Metallurgy, 2006, 5: 23-25.
4. Rongjun M. Principles on hydrometallurgy[M]. Beijing: Metallurgical Industry Press, 2007.
5. Ditze A and Kongolo K. Recovery of scandium from magnesium, aluminium and iron scrap[J]. Hydrometallurgy, 1997, 44: 179-184.
6. Wu D B, Niu C J, Li D Q and Bai Y. Solvent extraction of scandium(III), yttrium(III), lanthanum(III) and gadolinium(III) using Cyanex 302 in heptane from hydrochloric acid solutions[J]. Journal of Alloys and Compounds, 2004, 374: 442-446.
7. Zhang P W, You S T, Zhang L, Feng S and Hou S H. A solvent extraction process for the preparation of ultrahigh purity scandium oxide[J]. Hydrometallurgy, 1997, 47: 47-56.
8. Kostikova G V， Danilov N A， KrylovG Y S. Extractin of Scandium from Various Media with Triisoamyl Phosphate: 3. Development of Extractive Refining of Scandium[J], Radiochemistry, 2006, 48(5).
9. Liu Z B and Li H X. Metallurgical process for valuable elements recovery from red mud—A review[J]. Hydrometallurgy, 2015, 155:29-43.
10. Srnirnov D I and Molchanova T V. The investigation of sulphuric acid sorption recovery of scandium and uranium from the red mud of alumina production[J]. Hydrometallurgy, 1997, 45: 249-259.
11. Petropulu M O, Lyberopulu T and Parissakis G. Selective separation and determination of scandium from yttrium and lanthanides in red mud by a combined ion exchange/solvent extraction method[J]. Analytica Chimica Acta, 1995, 315: 231-237.
12. Zhou H L, Li D Y, Tian Y J and Chen Y F. Extraction of scandium from red mud by modified activated carbon and kinetics study[J]. Rare Metals, 2008, 3: 223-227.

Online semi-supervised multi-person tracking with gaussian process regression

Baobing Zhang*, *Zhengwen* Huang , *Babak H.* Rahi, *Qicong* Wang, and *Maozhen* Li

Department of Electronic and Computer Engineering, Brunel University London, Uxbridge, UB8 3PH, UK

Abstract. Most existing multi-person tracking approaches are affected by lighting condition, pedestrian pose change abruptly, scale changes, real-time processing to name a few, resulting in detection error, drift and other issues. To cope with this challenge, we propose an enhanced multi-person framework by introducing a new observation model, which adaptively updates fully online to avoid the loss of sample diversity and learning in a semi-supervised manner. We fuse prior information for tracking decision, meanwhile extracted knowledge from current frame is used to assist to make tracking decision, which can be viewed as a transfer learning strategy, and both aspects can ameliorate the tendency to drift. The new approach does not need any calibration or batch processing. Experimental results show that the approach yields comparable or better performance in comparison with the state-of-the-arts, which do calibration or batch processing.

1 Introduction

Tracking-by-detection approach is very popular during recent years [1] [2]. In practice most tracking-by-detection approaches are still limited to special scenarios and affected by occlusion, scale change, real-time processing etc. Moreover harmed by false and missing detection, some methods employ occlusion reasoning to smoothing the trajectories [3]. However, these methods are sensitive to detection error, because they build the trajectory based on two consecutive frames. Thus, during long-term occlusion or abrupt changes in pose, a danger is the tracked target trends to drift.

To deal with this problem, dynamic and observation model are combined used for the tracking problem. Dynamic model take pedestrian behaviour into account, moreover often used for estimating the new location of pedestrian. However, most existing dynamic model use only the previous one state information for predicting and lack of utilizing prior information, when the pedestrian motion change abruptly tend to incorrect estimation. Observation model represents the pedestrian's appearance change, particularly when adapted online take account for the gradual appearance change. Most of the existing observation models gather past appearance information over time, but in fact, these methods lack utilize current state information of the pedestrian's appearance, inevitably lead to drift problem.

* Corresponding author: Baobing.Zhang@brunel.ac.uk

In this paper, we introduce a new observation model to cope with these problems mentioned above in several aspects. First, we fuse prior information for tracking decisions. Second, the observation model is learnt in a semi-supervised manner through using both labelled and unlabelled sample. Third, background information is taking into consideration in the process of observation model updating. Fourth, re-weighting knowledge is used for tracking decision can be viewed as a transfer learning strategy. All the aspects mentioned above tend to alleviate drift. The main contributions of our work are:

- The new observation model, updated adaptively avoids the loss of sample diversity and learnt in a semi-supervised manner.
- We extract re-weighting knowledge from the current pedestrian status information and used for tracking inference, can be viewed as a transfer learning strategy.

We test our method on two multi-person tracking benchmark sequences. Our method achieved promising results better than previously tested state-of-the-arts. The rest of the paper is organized as follows. Section 2 presents the new observation model used for tracking. Section 3 evaluates the performance of the observation model in comparison with a number of typical methods. Section 4 concludes the paper and points out some future work.

2 Tracking with Gaussian process regression

2.1 New Observation model

In this section, we present the process of tracking. At each frame f_t , the current location information of a tracker is stored in a bounding box $X_t = [x_t^i, y_t^i, w_t^i, h_t^i,]$, which is estimated by the observatio T_i n model based on the previous information. Once a tracker is initialized, $\mathcal{I}_t^{T_i} = \{\mathbf{I}_1, \cdots, \mathbf{I}_t\}$ is a set of observation images, we aim to inference the hidden state variable $X_t^{T_i}$ in a Markov model which is given by equation (1).

$$p\left(X_t^{T_i} \middle| \mathcal{I}_t^{T_i}\right) \propto p\left(\mathbf{I}_t \middle| X_t^{T_i}\right) \int p\left(X_t^{T_i} \middle| X_{t-1}^{T_i}\right) p\left(X_{t-1}^{T_i} \middle| \mathcal{I}_{t-1}^{T_i}\right) dX_{t-1}^{T_i} \tag{1}$$

We use Kalman filter's prediction as input to stochastically generate a set of pedestrian candidate location in the current frame, which is $\mathcal{X}_U^{T_i} = \{X_t^{T_i,j}, j = 1, 2, ..., n_U\}$. Tracking results of tracker T_i can be estimated by MAP as shown in equation (2).

$$X_t^{\hat{T}_i} = \underset{X_t^{T_i,j}}{\arg\max}\, \mathrm{P}\left(X_t^{T_i,j} \middle| \mathcal{I}_t^{T_i}\right) \tag{2}$$

For each sample, we introduce an indicator variable $y_j \in \{-1, +1\}$ to indicate positive sample $(y_j = +1)$ or negative $(y_j = -1)$ of $X_t^{T_i,j}$. $\mathcal{X}_U^{T_i}$ is the unlabeled sample set of tracker T_i . From the tracking results up to the (t-1)-th frame for each tracker, we extract n_L labelled training samples with indicator variables, and then we divide n_L labelled training samples into two groups. First is $\mathcal{D}_T$ which consist n_T samples, refers target sample set gathered from the most recent frame, the other is $\mathcal{D}_A$ which consist n_A samples, we called auxiliary sample set which collected every few intervals, and $n_L = n_T + n_A$. Then, the regression function for the indicators of unlabelled samples y_U can be written as equation (3).

$$\mathcal{R} = P\left(y_U = 1 \middle| \mathcal{X}_U, \mathcal{D}_A, \mathcal{D}_T\right) \tag{3}$$

We introduce two real valued latent vectors l_A and l_U, corresponding to the label y_A and y_U respectively. We connect regression and classification by using a sigmoid output model. The Gaussian process model restricted to the auxiliary data and unlabelled data is as shown in equation (4).

$$P(l_A, l_U | \mathcal{X}_A, \mathcal{X}_U, \mathcal{D}_T) \sim \mathcal{N}(\mu, \widetilde{\Delta}^{-1}) \tag{4}$$

2.2 Graph Laplaacians

We construct the prior covariance matrix based on the weighted graph $\mathcal{G} = (V, E)$, which has the node set $V$and edge E, corresponding to all samples in the way similar with [4]. We explore the manifold structure of all samples. Furthermore we define weight matrix W of graph $\mathcal{G}$ using the method proposed by [5]. Finally prior covariance matrix is defined by the inverse graph Laplacian $\widetilde{\Delta}^{-1}$.

Because of the sigmoid noise label output model, the $P(l_A, l_U | \mathcal{X}_U, \mathcal{D}_A, \mathcal{D}_T)$ is no longer Gaussian and has no closed form solution. Assuming $P(l_A, l_U | \mathcal{X}_U, \mathcal{D}_A, \mathcal{D}_T)$ is a uni-modal function, we use its Laplace approximation to get the optimal estimation of l_A and l_U.

Because we construct prior covariance matrix depending on all samples, the correlated structure of the labelled samples and unlabelled samples has a significant effect on the latent real-valued output. The latent variable l_A is the re-weighting knowledge extracted from the Regression can be a soft replacement of indicator label y_A, and is better for ameliorating sample misalignment problem, less sensitive to noisy compare with the indicator variable.

2.3 Tracker's birth and death

For the purpose of maintaining the tracker, we divided trackers into two groups based on the template it owes. Once a tracker is born we call it Novice, it will accumulate templates throughout the tracking process, after **K** template accumulated over a period of robust tacking. Novice would be promoted to Expert, conversely an Expert demoted to a Novice when it loses template less than **K**, we set K to 5. Each tracker keep at most N_{max} reliable templates by discarding the lower score template, we set N_{max} to 10.

A tracker candidate is activated when its detection rate is above the ξ_{init}. On the contrary, a tracker would be killed when its detection rate is less than ξ_{term}. Both of them is given by in equation (5) and (6).

$$\xi_{term} = \Omega - \gamma_2 \sqrt{\Omega} \quad \xi_{init} = \Omega - \gamma_1 \sqrt{\Omega} \tag{5, 6}$$

where γ_1 and γ_2 is the scale factor, we set to 1 and 2 respectively, for each tracker's detection rate is defined as equation (7).

$$\Omega_i = \frac{\Delta N_i^{matched}}{\Delta t} \tag{7}$$

where $\Delta N_i^{matched}$ is the number of detections matched with T_i in a sliding window of length Δt

Fig. 1. Representative tracking results of PET09 sequence.

3 Experiments

Through several experiments, compared with the state-of-the-arts, our approach shows his unique advantages.

3.1 Datasets and Ground Truth

We currently test our algorithm on two sequences, one is the sequence S2L1, which is taken from the VS-PETS benchmark 2009 [6], representative tracking results can be seen in Fig 1. This sequence is filmed by 7 cameras and show up to 8 people the resolution is (768×576), we only use the first viewpoint. Most people wear similar dark clothes, which make colour-based observation model for tracking difficult. We also test the new algorithm under a crowed environment (S2L2), which has a lot of pedestrians within a confined space make tracking difficulty even for the individual detect. A brief description for these two sequences is as shown in Table 1, the ground-truth used for evaluation is public available[1].

Table 1. Datasets description.

	Frame Rate	Number of Frames	Number of Id	HOG Detector	
				Precision	Recall
PETS2009 S2L1	7	795	19	0.87	0.81
PETS2009 S2L2	7	436	43	0.90	0.59

3.2 Experimental Environments

All experiments were tested on a computer with 2.8GHz Octa-core CPU, 16GB memory. We use C++ implementation and rely on the OpenCV and Eigen library. Runtime performance about 2 fps per second with the new observation model employed for tracking. We believe that with GPU implementation or more optimized code could achieve real-time performance.

3.3 Evaluation Metrics

There is no standard established protocol to measure multi-object tracking performance, we use the current best practice which calculates the CLEAR-MOT metrics proposed in [7], FP means false positive, MS considers number of missed detections and ID.S takes account for

[1] http://www.gris.informatik.tu-darmstadt.de/%CB%9Caandriye/data.html

the switches of identities. The multiple objects tracking accuracy (MOTA) was defined as equation (8).

$$MOTA = 1 - \frac{\sum_{t=1}^{N_{frames}} \left(c_m(m_t) + c_f(fp_t) + c_s(ID-SWITCHES_t)\right)}{\sum_{t=1}^{N_{frames}} N_G^{(t)}} \tag{8}$$

The multiple objects tracking precision (MOTP) was given by equation (9).

$$MOTP = \frac{\sum_{i=1}^{N_{mapped}} \sum_{t=1}^{N_{frames}^{(t)}} \left[\left| \frac{G_i^{(t)} \cap D_i^{(t)}}{G_i^{(t)} \cup D_i^{(t)}} \right| \right]}{\sum_{t=1}^{N_{frames}} N_{mapped}^{(t)}} \tag{9}$$

Note that both MOTA and MOTP, higher values of the output indicate better performance (see [7] for detail).

Table 2. Results Comparison.

Sequence	Method	MOTP(%)	MOTA(%)	FP	MS	ID.S
PETS2009-S2L1	Ours	62.53	67.69	537	839	126
	[8]	68.22	60.27	28	1804	15
	[9]	-	67	-	-	-
PETS2009-S2L2	Ours	52.97	49.45	664	4118	420
	[8]	61.34	24.46	31	7696	47

As shown in table 2, we compare our method with [8] on the PETS2009 sequence the results of [8] is tested by ourselves, for fair comparison we use the same detector as our method and same protocol to evaluate the outputs. Note that the results of [8] are slightly different from the original paper. It may be influenced by the parameter tuning, pre-treatment optimization and other factors. We also show the results in [9] when available. As shown method [9] made the calibration action and perform batch processing of the data, our method achieve higher MOTA score success surpass both [8] and [9] on the S2L1 and S2L2 sequence. Compare with [8] our method have less missing detection and potentially increases the number of false positive. We believe the slightly lower MOTP score was caused by the update of the sample set not perfectly adapt the scale change over time. We also noticed that with the increase of density of people in the sense there are few veterans the ratio of veterans is much higher in sequence S2L1 than S2L2, it can be explained by there is more occlusion issues in S2L2 than S2L1.

Table 3. Methods Comparison.

Sequence	Tracker	Rcll(%)	Prcn(%)
PETS2009-S2L1	[10]	83.9783	80.9783
	Ours	81.957	87.6495
	[11]	81.6559	83.4139
	[12]	81.2903	84.2434
	[13]	75.1828	91.5663
PETS2009-S2L2	[14]	54.8387	90.4224
	Ours	59.9883	90.2896
	[15]	49.611	82.4797
	[16]	38.2014	82.045
	[17]	50.8039	81.3081
	[18]	35.5461	78.3673

Table 4. Results Comparison with the most recent state-of-the-arts.

Sequence	Tracker	FP	IDSW	MOTA(%)	MOTP(%)
PETS2009 -S2L1	[10]	910	348	56.2581	71.119
	Ours	537	126	67.6989	62.5369
	[11]	755	31	64.7527	70.1903
	[12]	707	239	60.9462	71.1903
	[13]	322	99	66.129	71.5962
PETS2009 -S2L2	[14]	560	238	46.5616	67.6273
	Ours	664	420	49.4559	52.9748
	[15]	1016	139	37.631	65.9038
	[16]	806	240	27.3519	67.361
	[17]	1126	190	37.1538	67.6956
	[18]	946	162	24.0535	67.5983

We further compare our method with most recently state-of-the-arts [10] [11] [12] [13] [14] [15] [16] [17] [18], the comparison results are as shown in Table 3 and Table 4. Compare with other methods, the Recall rate and Precision rate of our method success surpass most methods in both S2L1 and S2L2 sequence. We also got the best MOTA score compare with other methods. Because the tracking decision mainly depend on the observation model, the observation model update process has a significant impact on MOTP score, we will do further optimization to improve the MOTP performance. The speed of this algorithm is proportional to the sample sampling size. While reducing the number of samples, we also consider the balance of accuracy of tracking and algorithm speed. In order to test the sensitivity of our algorithm to parameters, we conduct experiments which parameter setting from the baseline each parameter floating up and down 40% while keeping other parameters fixed. As shown in figure 2, the performance of our algorithm changes within a reasonable range for all the sequence. This indicates that our algorithm is relatively robust to the setting of parameters.

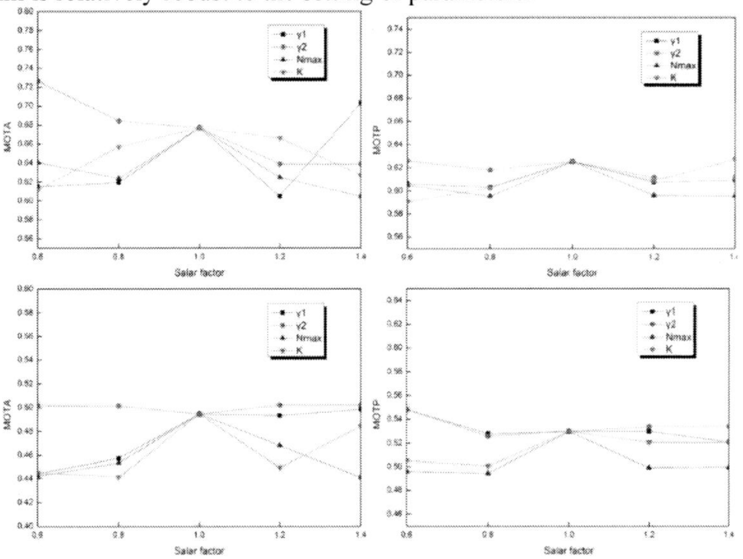

Fig. 2. Algorithm Sensitivity test on several major parameters. First row is the effect of parameters on MOTA and MOTP of S2L1 sequence, Second row is the effect of parameters on MOTA and MOTP of S2L2 sequence.

4 Conclusion

In this paper, we have presented a novel multi-person tracking algorithm. The new observation model adopts graph Laplacian, meanwhile prior gram matrix is constructed based on all samples. In this way unlabelled samples have strong influence on the prior can be viewed as a transfer learning strategy. We divided trackers into two categories base on the number of templates it holds, experimental results show that our algorithm hold obvious advantage compared with other methods.

A future work will research on re-identification scheme in our algorithm to help account for people re-identification. We will extend this framework to Multi-Target, Multi-Camera Tracking.

Acknowledgment

This work was supported by the Principal Foundation of Xiamen University under grant 20720180075.

References

1. Breitenstein, M. D., Reichlin, F., Leibe, B., Koller-Meier, E., & Van Gool, L. (2011). Online multiperson tracking-by-detection from a single, uncalibrated camera. *IEEE transactions on pattern analysis and machine intelligence*, **33**(9), 1820-1833.
2. Avidan, S. (2007). Ensemble tracking. *IEEE transactions on pattern analysis and machine intelligence*, **29**(2).
3. Andriyenko, A., Roth, S., & Schindler, K. (2011, November). An analytical formulation of global occlusion reasoning for multi-target tracking. *In Computer Vision Workshops (ICCV Workshops)*, 2011 IEEE International Conference on (pp. 1839-1846). IEEE.
4. Gao, J., Ling, H., Hu, W., & Xing, J. (2014, September). Transfer learning based visual tracking with gaussian processes regression. *In European Conference on Computer Vision* (pp. 188-203). Springer, Cham.
5. Hu, W., Li, X., Luo, W., Zhang, X., Maybank, S., & Zhang, Z. (2012). Single and Multiple object tracking using log-Euclidean Riemannian subspace and block-division appearance model. *IEEE Transactions on Pattern Analysis and Machine Intelligence*, **34**(12), 2420-2440.
6. Ferryman, J., & Shahrokni, A. (2009, December). Pets2009: Dataset and challenge. In Performance Evaluation of Tracking and Surveillance (PETS-Winter), *2009 Twelfth IEEE International Workshop* on (pp. 1-6). IEEE.
7. Kasturi, R., Goldgof, D., Soundararajan, P., Manohar, V., Garofolo, J., Bowers, R., ... & Zhang, J. (2009). Framework for performance evaluation of face, text, and vehicle detection and tracking in video: Data, metrics, and protocol. *IEEE Transactions on Pattern Analysis and Machine Intelligence*, **31**(2), 319-336.
8. Zhang, J., Presti, L. L., & Sclaroff, S. (2012, September). Online multi-person tracking by tracker hierarchy. *In Advanced Video and Signal-Based Surveillance (AVSS)*, 2012 IEEE Ninth International Conference on (pp. 379-385).
9. Leal-Taixé, L., Pons-Moll, G., & Rosenhahn, B. (2011, November). Everybody needs somebody: Modeling social and grouping behavior on a linear programming multiple people tracker. *In Computer Vision Workshops (ICCV Workshops)*, 2011 IEEE International Conference on (pp. 120-127). IEEE.

10. Pirsiavash, H., Ramanan, D., & Fowlkes, C. C. (2011, June). Globally-optimal greedy algorithms for tracking a variable number of objects. In Computer Vision and Pattern Recognition (CVPR), 2011 IEEE Conference on (pp. 1201-1208). IEEE.
11. Bae, S. H., & Yoon, K. J. (2014). Robust online multi-object tracking based on tracklet confidence and online discriminative appearance learning. *In Proceedings of the IEEE conference on computer vision and pattern recognition*(pp. 1218-1225).
12. Geiger, A., Lauer, M., Wojek, C., Stiller, C., & Urtasun, R. (2014). 3d traffic scene understanding from movable platforms. *IEEE transactions on pattern analysis and machine intelligence*, **36**(5), 1012-1025.
13. Dicle, C., Camps, O. I., & Sznaier, M. (2013, December). The way they move: Tracking multiple targets with similar appearance. In Computer Vision (ICCV), 2013 IEEE International Conference on (pp. 2304-2311). IEEE.
14. Leal-Taixé, L., Fenzi, M., Kuznetsova, A., Rosenhahn, B., & Savarese, S. (2014). Learning an image-based motion context for multiple people tracking. *In Proceedings of the IEEE Conference on Computer Vision and Pattern Recognition* (pp. 3542-3549).
15. Rezatofighi, S. H., Milan, A., Zhang, Z., Shi, Q., Dick, A. R., & Reid, I. D. (2015, December). Joint Probabilistic Data Association Revisited. *In ICCV* (pp. 3047-3055).
16. Bewley, A., Ge, Z., Ott, L., Ramos, F., & Upcroft, B. (2016, September). Simple online and realtime tracking. In Image Processing (ICIP), 2016 IEEE International Conference on(pp. 3464-3468). IEEE.
17. Yoon, J. H., Yang, M. H., Lim, J., & Yoon, K. J. (2015, January). Bayesian multi-object tracking using motion context from multiple objects. *In Applications of Computer Vision (WACV)*, 2015 IEEE Winter Conference on (pp. 33-40). IEEE.
18. Fagot-Bouquet, L., Audigier, R., Dhome, Y., & Lerasle, F. (2015, September). Online multi person tracking based on global sparse collaborative representations. In Image Processing (ICIP), 2015 IEEE International Conference on(pp. 2414-2418). IEEE.

MATEC Web of Conferences **277**, 01004 (2019) https://doi.org/10.1051/matecconf/201927701004
JCMME 2018

Payload dropping control of an unmanned quadrotor helicopter based on backstepping controller

Jing Qiao*, *Zhixiang* Liu, and *Youmin* Zhang

Concordia University, Montreal, H3G 1M8, Quebec, Canada

Abstract. Quadrotors have generated considerable interest in both the control community due to their simple dynamics and widely applications because of their advantages over regular air vehicles. Unlike other control algorithms that tend to linearize nonlinear systems, backstepping control does not work to cancel the nonlinearities in the system. This leads to more flexible designs since some of the nonlinear terms can contribute to the stability of the system. The backstepping control is implemented in this paper for the robustness of the quadrotor helicopter in the payload dropping process. This control algorithm is implemented on the parameters of an unmanned quadrotor helicopter platform known as Qball-X4 available at the Networked Autonomous Vehicles Lab of Concordia University. Simulation results carried out using a nonlinear model, and wherein the performance achieved with this control strategy is shown.

1 Introduction

Unmanned helicopters have become increasingly popular for various purposes concerning future commercial and military applications [1,2], either integrated with current manned aircraft or playing the role individually. The investigated and potential application areas include aerial surveillance [3,4], border patrol [5], environmental monitoring, search and exploration for unsecured regions [6], and search and rescue in the events of accidents and natural disasters in flood, earthquake, nuclear and radiation accidents [7,8]. Furthermore, monitoring and preventing forest fires and high building fires in urban areas [9,10], surveillance of pipelines leakage [11], and other circumstances under which are highly risky for human pilots have also raised the need for the employment of unmanned systems for such applications.

Trajectory tracking with payload dropping is typical flight maneuver in using unmanned helicopters. Successful fulfillment of this mission is closely tied to the intervention of autopilot or autonomous control systems. This fact has motivated the research and development of new control algorithms for such applications with UAVs. Initially started with simple control algorithms such as PID control [12,13] or LQR control [14], however, simple linear methods were proved not to be promising for the nonlinear quadrotor system. The problem of nonlinear control design has been addressed using several methods such as

* Corresponding author: josephchiao250@gmail.com

feedback linearization [15], sliding mode control [16], model predictive control (MPC) [17], and back-stepping control [18].

Nevertheless, among most of the studies concerning these control approaches, maximum take-off weight has always been assumed to be constant during the entire flight and the effects of either gradual or abrupt mass variation over the period of flight have not been widely investigated in the existing literature. The issue of mass variation caused by payload dropping is of much concern since some specific applications like search and rescue or fire-fighting, abrupt or gradual mass variation is inevitable. The study in this aspect for the benefits of such a practical application of the quadrotor was the main motivations for this work. In view of popularity and advantages of backstepping control strategy, in this paper, a backstepping controller is designed and implemented on a quadrotor helicopter UAV.

2 System modeling of quadrotor

2.1 Qball-X4 quadrotor UAV

The quadrotor UAV which is as the object of this study is the Qball-X4 [13] as shown in Fig. 1 which is at the Network Autonomous Vehicle (NAV) Lab in the Department of Mechanical and Industrial Engineering of Concordia University and was developed by Quanser Inc. partially under the financial support of NSERC (Natural Sciences and Engineering Research Council of Canada) in association with an NSERC Strategic Project Grant led by Concordia University since 2007. The quadrotor UAV is enclosed within a protective carbon fiber ball-shape cage (therefore a name of Qball-X4) to ensure safe operation. It uses four 10×4.7 inch propellers and standard RC motors and speed controllers.

The motors of the Qball-X4 are out-runner brushless motors. The generated thrust T_i of the *No.i* motor is related to the *No.i* PWM input u_i by a first-order linear transfer function as follows:

$$T_i = k \frac{\omega}{s + \omega} u_i \tag{1}$$

where $i = 1, 2, 3, 4$ and k is a positive gain and ω is the motor bandwidth. k and ω are theoretically the same for the four motors but this may not be the case in practice. It should be noted that $u_i = 0$ corresponds to zero thrust and $u_i = 0.05ms$ corresponds to the maximal thrust that can be generated by the *No.i* motor.

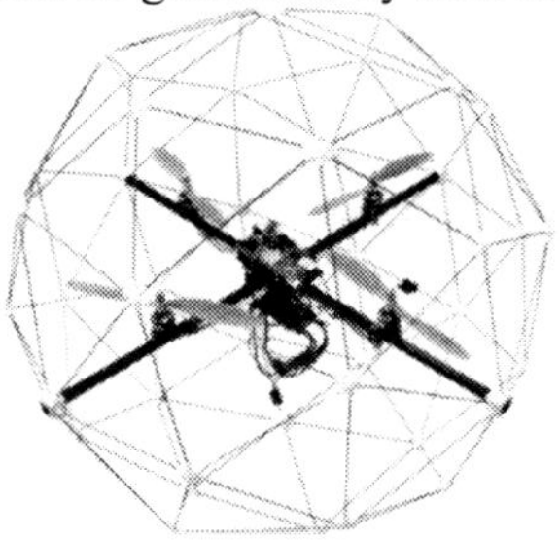

Fig. 1. The Qball-X4 quadrotor UAV.

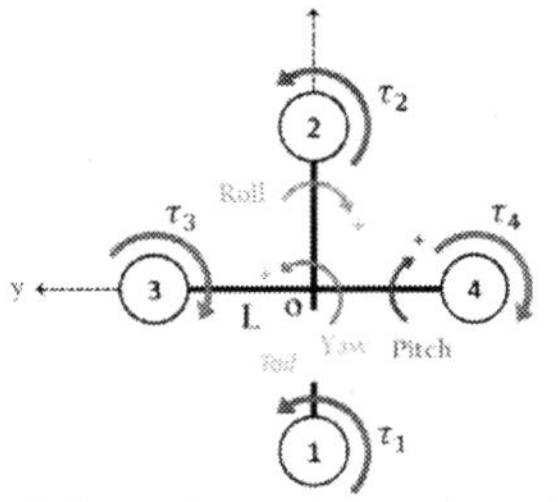

Fig. 2. Schematic representation of the Qball-X4

2.2 System modeling of quadrotor helicopter

A quadrotor is an under actuated aircraft with fixed pitch angle as shown in Figure 2. Having Quadrotors with fixed angles makes quadrotor has four input forces which are basically the thrust provided by each propellers. The motors and propellers are configured in such a way that the back and front (1 and 2) motors spin clockwise (thus inducing two counterclockwise torques on the body) and the left and right (3 and 4) spin counterclockwise (thus inducing two clockwise torques on the body). Each motor is located at a distance L from the center of mass o (0.2 m) and when spinning, a motor produces a torque τ_i which is in the opposite direction of motion of the motor as shown in Fig. 2. Forward (backward) motion is maintained by increasing (decreasing) speed of front (rear) rotor speed while decreasing (increasing) rear (front) rotor speed simultaneously which means changing the pitch angle. Left and right motion is accomplished by changing roll angle by the same way. The front and rear motors rotate counter-clockwise while other motors rotate clockwise so yaw command is derived by increasing (decreasing) counter-clockwise motors speed while decreasing (increasing) clockwise motor speeds.

Suppose the lift force generated by the *i-th* propeller in the quadrotors is $F_i(i=1,2,3,4)$, and $\tau_1, \tau_2, \tau_3, \tau_4$ are torque generated by each motor. The force acting on the UAV in the airframe coordinate system is:

$$F_B = [0,0, U_1]^T \tag{2}$$

where

$$U_1 = F_1 + F_2 + F_3 + F_4 \tag{3}$$

Based on Newton's second law, the rectilinear motion model and angular motion model of the quadrotors were obtained as,

$$\begin{bmatrix} \ddot{X} \\ \ddot{Y} \\ \ddot{Z} \end{bmatrix} = \frac{1}{m} \begin{bmatrix} U_1(\sin\phi\sin\psi + \cos\phi\cos\psi\sin\theta) - K_1\dot{X} \\ U_1(\cos\phi\sin\theta\sin\psi - \cos\psi\sin\phi) - K_1\dot{Y} \\ U_1(\cos\phi\cos\theta) - K_2\dot{Z} - mg \end{bmatrix} \tag{4}$$

$$\begin{cases} \ddot{\varphi} = \left[L(F_4 - F_2) + \dot{\theta}\dot{\psi}(I_y - I_z)\right] / I_y \\ \ddot{\theta} = \left[L(F_3 - F_1) + \dot{\varphi}\dot{\psi}(I_z - I_x)\right] / I_y \\ \ddot{\psi} = \left[(\tau_2 + \tau_4 - \tau_1 - \tau_3) + \dot{\varphi}\dot{\theta}(I_x - I_y)\right] / I_z \end{cases} \tag{5}$$

The quadrotor is a 4 input, 6 output system and the coupling of 4 inputs to 6 outputs is too strong, so, we set intermediate inputs as:

$$\begin{cases} U_1 = F_1 + F_2 + F_3 + F_4 \\ U_2 = F_4 - F_2 \\ U_3 = F_3 - F_1 \\ U_4 = \tau_2 + \tau_4 - \tau_1 - \tau_3 \end{cases} \tag{6}$$

Using above formulas, the nonlinear dynamic model of the quadrotor can be obtained as,

$$\begin{aligned} &\ddot{X} = \left[U_1(\cos\varphi\sin\theta\cos\psi + \sin\varphi\sin\psi) - K_1\dot{X}\right] / m \\ &\ddot{Y} = \left[U_1(\sin\psi\sin\theta\cos\varphi - \cos\psi\sin\varphi) - K_2\dot{Y}\right] / m \\ &\ddot{Z} = \left[U_1(\cos\varphi\cos\theta) - mg - K_3\dot{Z}\right] / m \\ &\ddot{\varphi} = \left[U_2 L + \dot{\theta}\dot{\psi}(I_y - I_z)\right] / I_x \\ &\ddot{\theta} = \left[U_3 L + \dot{\varphi}\dot{\psi}(I_z - I_x)\right] / I_y \\ &\ddot{\psi} = \left[U_4 + \dot{\varphi}\dot{\theta}(I_y - I_z)\right] / I_z \end{aligned} \tag{7}$$

The dynamic equation (7) shows that the quadrotor UAV flight dynamics model has the characteristics of underactuated, nonlinear, uncertainty and strong coupling.

3 Backstepping controller design

Unlike other control algorithms that tend to linearize nonlinear, backstepping does not work to cancel the nonlinearities in the system. This leads to more flexible designs since some of the nonlinear terms can contribute to the stability of the system. The backstepping control requires that the dynamic model of the system satisfies the strict feedback form, that is, the form of the inverted triangle. The dynamic model of the horizontal position subsystem of the quadrotor unmanned aerial vehicle shows that the control input of the system directly affects the roll angle and pitch angle of the aircraft, and the position and line velocity signals are not included in the dynamic model of the attitude angle. The dynamic model of the aircraft's horizontal position is directly related to the attitude angle of the aircraft, so it can be shown that the horizontal position subsystem of the quadrotor unmanned aerial vehicle satisfies the strict feedback form.

3.1 Backstepping controller design for quadrotor UAV

Due to the strict symmetry of the quadrotor helicopter structure, the Quadrotor height control channel and yaw control channel are relatively independent, and its horizontal control is controlled by the underdrive of rolling angle and pitch angle. When virtual input vector $U=(U_2,U_3,U_4)$ is known, three attitude angles and their angular velocities can be obtained. If the U_1 is also known, then the position and speed of the aircraft will be obtained. Obviously, the attitude loop and the position loop are semi-coupled. The position loop is the inner loop and the position is the outer loop.

Take the state variables as $x=(\phi,\dot{\phi},\theta,\dot{\theta},\psi,\dot{\psi},z,\dot{z},x,\dot{x},y,\dot{y})$,that is $x_1=\phi, x_2=\dot{x}_1=\dot{\phi}; x_3=\theta, x_4=\dot{x}_3=\dot{\theta}; x_5=\psi, x_6=\dot{x}_5=\dot{\psi};$
$x_7=z, x_8=\dot{x}_7=\dot{z}; x_9=x, x_{10}=\dot{x}_9=\dot{x}; x_{11}=y, x_{12}=\dot{x}_{11}=\dot{y}$

The mathematical model of the Quadrotor is written as a strict feedback format,

$$f(x,U)=\begin{cases} x_2=\dot{x}_1 \\ \dot{x}_2=a_1\cdot x_4\cdot x_6+b_1\cdot U_2 \\ \dot{x}_3=x_4 \\ \dot{x}_4=a_2\cdot x_2\cdot x_6+b_2\cdot U_3 \\ \dot{x}_5=x_6 \\ \dot{x}_6=a_3\cdot x_2\cdot x_4+b_3\cdot U_4 \\ \dot{x}_7=x_8 \\ \dot{x}_8=\cos x_1\cdot\cos x_3\cdot U_1/m-g \\ \dot{x}_9=x_{10} \\ \dot{x}_{10}=U_x\cdot U_1/m \\ \dot{x}_{11}=x_{12} \\ \dot{x}_{12}=U_y\cdot U_1/m \end{cases} \tag{8}$$

where $a_1=(I_y-I_z)/I_x$, $b_1=L/I_x$; $a_2=(I_z-I_x)/I_y$, $b_2=L/I_y$;
$a_3=(I_x-I_y)/I_z$, $b_3=L/I_z$; $U_x=(\cos\phi\sin\theta\cos\psi+\sin\phi\sin\psi)$;
$U_y=(\cos\phi\sin\theta\sin\psi-\sin\phi\cos\psi)$.

In order to reduce the difficulty of the design of the system controller, the whole system is divided into 6 relatively independent control channels. Where, position loop contains a height controlled channel, horizontal *x, y* control channel. And the attitude loop contains the roll control channel, the pitch control channel and the yaw control channel. The backstepping method is applied in each relatively independent channel.

3.2 System parameters

The MATLAB/Simulink simulation tool is used to simulate the control algorithm for quadrotor helicopter. Some of the basic parameters for QballX4 quadrotor UAV model simulation are shown in Table 1.

Table 1. Basic parameters of the quadrotor UAV Qball-X4.

Parameter	value
K	120N
ω	15 (rad/sec)
Ix	0.03 (kg.m^2)
Iy	0.03 (kg.m^2)
Iz	0.04 (kg.m^2)
m	1.4 (Kg)
K'	4 (N.m)
l	0.2 (m)
g	9.8 (N/kg)

Normally, if the desired input signal is small, select the larger R matrix, which forces the input signal to be smaller. Otherwise, the objective function value will increase and will not achieve the optimization requirements. Compared with the results of simulation tests, a larger c_i is likely to lead to a smaller overshoot. At the same time, the overshoot of position will also increase. Considering the safety of actual flight, take

$$\begin{aligned} c_i &= 1.5, (i = 1, 2, 3, 4, 5, 6) \\ c_j &= 1.25, (j = 7, 8, 9, 10, 11, 12) \end{aligned} \tag{9}$$

4 Simulation results

In this paper, the system simulation based on the Matlab/Simulink is carried out to verify the design result of backstepping controller. In the simulation experiment, the initial state of the Qball-X4 is set as, altitude z is 0 m, roll angle ϕ, pitch angle θ, and yaw angle ψ are all 0 degree. And while the Qball-X4 finishes climbing and is in a rotating trajectory, the desired altitude z_d is 1m and the rotating diameter is 1m. The payload of 300g in the Qball-X4 is dropping at the 20 second. Simulation time is set to 40 seconds. The command for the backstepping controller of the Qball-X4 helicopter is shown in figure 3. The focus of this study is mainly on the flight stability when the payload is dropped from the Qball-X4.

Figure 4 to figure 7 show the flying path, flying attitude and the total thrust of Qball-X4 when payload is dropping with the backstepping controller. It could be seen from simulation results that the backstepping controller was capable of keeping the desired height and trajectory.

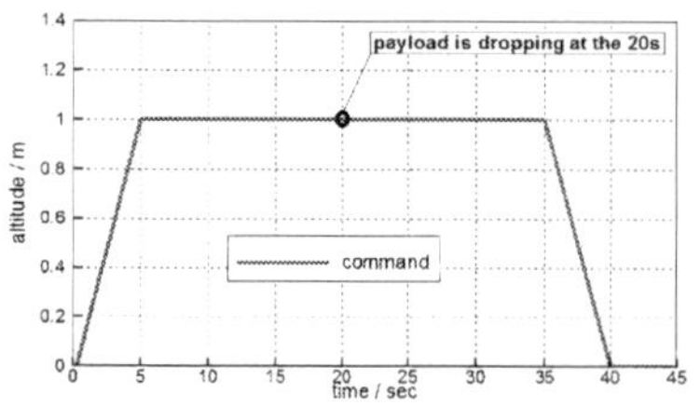

Fig. 3. Command for the Qball-X4 controller.

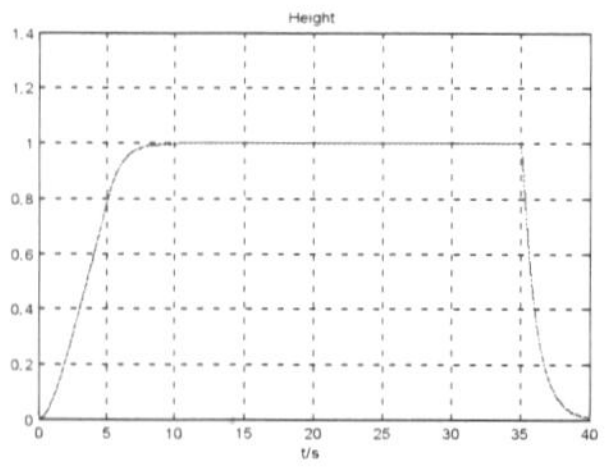

Fig. 4. The flying path of Qball-X4 for payload dropping at the 20s with backstepping controller.

The flying height in figure 4 indicates that the Qball-X4 eventually reach the desired altitude at the10s. It could also been found from the flying attitude of Qball-X4 that, at the moment when Qball-X4 arriving at desired height and the moment of Qball-X4 beginning landing, the roll angle and pitch angle of Qball-X4 are fluctuating. However these flying attitude fluctuating is restricted to a small extent, the fluctuating of the roll angle is smaller than 0.6 degrees, and the fluctuating of the pitch angle is in the range ±2 degrees. And the roll angle also fluctuates at the moment of payload drop.

It could also be obviously seen from figure 8 that total thrust abrupt change of Qball-X4 when payload drops. Figure 8 is a three dimensional trajectory map of a Quadrotor helicopter trajectory tracking. As can be seen from this figure, the aircraft starts at points (0, 0, 0), then climbs up to the plane of z= 1, and this plane travels along a circle of 1 meters in diameter. And, after the payload drop, the quadrotor helicopter gradually descends to starts-points (0, 0, 0). Therefore, the control design makes the Quadrotor UAV achieve the trajectory tracking effect.

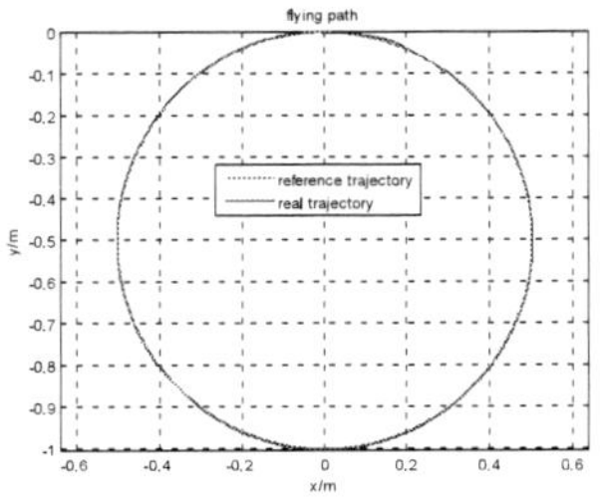

Fig. 5. The x-y position change of Qball-X4 for payload dropping at the 20s with backstepping controller.

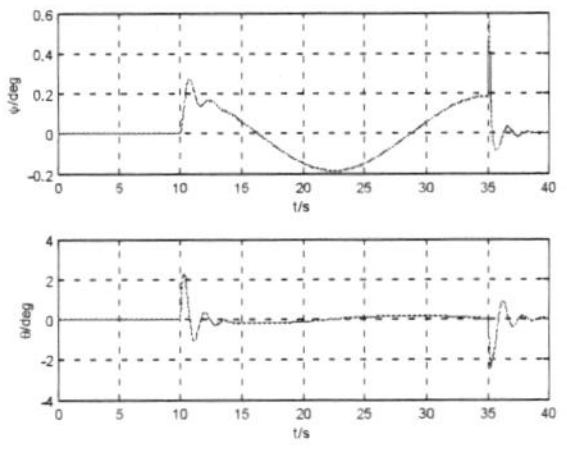

Fig. 6. The flying attitude of Qball-X4 for payload dropping at the 20s with backstepping controller.

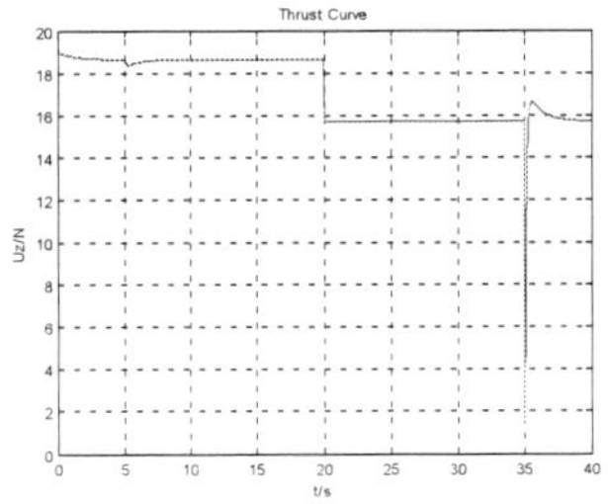

Fig. 7. The total thrust of Qball-X4 for payload dropping at the 20s with backstepping controller.

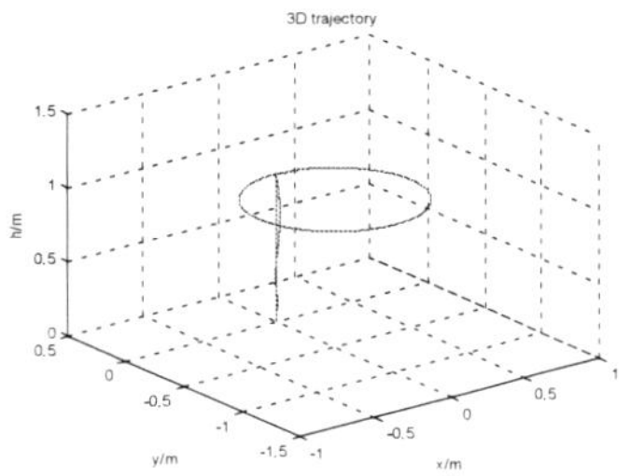

Fig. 8. 3D trajectory of Qball-X4.

5 Conclusion

In this paper, a robust backstepping control strategy has been presented for the quadrotor to obtain trajectory tracking and payload dropping by controlling its attitude and position simultaneously. The flight simulation experiments of the quadrotor UAV was carried out in the paper. The simulation results verify the effectiveness of the proposed backstepping controller. It is found that the backstepping controller is able to control the quadrotor in the presence of payload dropping and to make the quadrotor UAV achieve the trajectory tracking effect. With the backstepping control, the position {x y, z} and attitude tracking was successfully achieved in the climbing, hovering, payload dropping and descending of a quadrotor helicopter.

It should be indicated that it is difficult to model the quadrotor UAV accurately when there is uncertainty in the model such as the abrupt mass variation over the period of flight for payload dropping. The control design of this paper only investigated the problem of aircraft control when there is a payload dropping in certain smooth flight trajectory using backstepping controller. The simulation results indicate that the roll angle will fluctuates at the moment of payload drop.

It is well known that the system should maintain acceptable performance with partial actuator faults using fault tolerant control strategies for the control of UAV. Therefore, in the future, with adopting the backstepping control approach and utilizing the quadrotor mathematical model, a fault tolerant control law could be further designed and evaluated by single, triple or even quadruple partial actuator faults.

References

1. Y. M. Zhang, A. Chamseddine, C. A. Rabbath, B. W. Gordon, C.-Y. Su, S. Rakheja, C. Fulford, J. Apkarian and P. Gosselin, Development of advanced FDD and FTC techniques with application to an unmanned Quadrotor helicopter testbed, J. Franklin Inst. 350(9) (2013) 2396– 2422.
2. F. Lin, K. Z. Y. Ang, F. Wang, B. M. Chen, T. H. Lee, B. Yang, M. Dong, X. Dong, J. Cui, S. K. Phang, B. Wang, D. Luo, K. Peng, G. Cai, S. Zhao, M. Yin and K. Li, Development of an unmanned coaxial rotorcraft for the DARPA UAVForge challenge, Unmanned Syst. 1(2) (2013) 211–245.
3. J. Keller, D. Thakur, V. Dobrokhodov, K. Jones, M. Pivtoraiko, J. Gallier, I. Kaminer and V. Kumar, A computationally efficient approach to trajectory management for coordinated aerial surveillance, Unmanned Syst. 1(1) (2013) 59–74.
4. P. C. Niedfeldt, B. T. Carroll, J. A. Howard, R. W. Beard, B. S. Morse and S. Pledgie, Enhanced UAS surveillance using a video utility metric, Unmanned Syst. 1(2) (2013) 277–296.
5. C. C. Haddal and J. Gertler, Homeland security: Unmanned aerial vehicles and border surveillance, CRS Report for Congress, Congressional Research Service (2010).
6. J. Hu, J. Xu and L. Xie, Cooperative search and exploration in robotic networks, Unmanned Syst. 1(1) (2013) 121–142.
7. M. K. Habib and Y. Baudoin, Robot-assisted risky intervention, search, rescue and environmental surveillance, Int. J. Adv. Robot. Syst. 7(1) (2010) 1–8.
8. T. Skrzypietz, Unmanned aircraft systems for civilian missions, BIGS Policy Paper No. 1 (2012).
9. R. W. Beard, T. W. McLain, D. B. Nelson, D. Kingston and D. Johanson, Decentralized cooperative aerial surveillance using fixed-wing miniature UAVs, Proc. IEEE 94(7) (2006) 1306–1324.
10. D. W. Casbeer, D. B. Kingston, R. W. Beard and T. W. McLain, Cooperative forest fire surveillance using a team of small unmanned air vehicles, Int. J. Syst. Sci. 37(6) (2006) 351–360.
11. H. Dieter, Z. Werner, S. Gunter and S. Peter, Monitoring of gas pipelines: A civil UAV application, Aircraft Eng. Aerosp. Technol. 77(5) (2005) 352–360.
12. B. Erginer and E. Altug, Modeling and PD control of a Quadrotor VTOL vehicle, IEEE Intelligent Vehicles Symp. (2007), pp. 894–899.
13. A. Wahyudie, T. B. Susilo and H. Noura, Robust PID controller for quadrotors, J.

Unmanned Syst. Technol. 1(1) (2013) 14–19.

14. K. Oner, E. Cetinsoy, M. Unel, M. Aksit, I. Kandemir and K. Gulez, Dynamic model and control of a new Quadrotor UAV with tilt-wing mechanism, World Academy of Science, Engineering and Technology (2008).
15. M. O. Efe, Robust low altitude behavior control of a Quadrotor rotorcraft through sliding modes, Mediterranean Conf. on Control and Automation (2007).
16. H. Bouadi, M. Bouchoucha and M. Tadjine, Sliding mode control based on backstepping approach for an UAV type-Quadrotor, World Academy of Science, Engineering and Technology (2007).
17. M. Abdolhosseini, Y. M. Zhang and C. A. Rabbath, An efficient model predictive control scheme for an unmanned Quadrotor helicopter, J. Int. Robot. Syst. 70(1) (2013) 27–38.
18. Madani, T. and Benallegue, A. (2006) Backstepping Control for a Quadrotor Helicopter. Proceedings of the 2006 IEEE/RSJ International Conference on Intelligent Robots and Systems, Beijing, 9-15 October 2006, 3255-3260.

MATEC Web of Conferences **277**, 01005 (2019) https://doi.org/10.1051/matecconf/201927701005
JCMME 2018

Modelling of cross organizational manufacturing resource service chain based on service supply-demand dynamic matching network

Qingqing Yang*, *Jia* Liu, and *Kewei* Yang

College of systems engineering, National University of Defense Technology, Changsha, China, 410073

Abstract. In the cloud manufacturing systems, both manufacturing tasks and manufacturing services are in a dynamic environment. How could cloud manufacturing platform optimizes manufacturing cloud services based on QoS, matching an optimal service composition for manufacturing tasks has become an urgent problem at present. In view of this problem, we study the matching of manufacturing tasks and manufacturing services from the perspective of complex network theory. On the basis of manufacturing task network and manufacturing service network, a dynamic matching network theory model of manufacturing task-service is constructed. And then, we take a dynamic assessment of QoS. Finally, we use load and dynamic QoS as the optimization objectivities, transform the optimal manufacturing service composition problem into the shortest path problem, and the dynamic scheduling of manufacturing services is realized.

1 Introduction

Because of the diversification of customer needs, the manufacturing industry is increasingly trying to integrate products and service components together to quickly respond to customer needs, shorten product development cycles and maintain agility by creating more complete solutions. However, shifting from a product-centric perspective to a solution-oriented perspective is challenging. [1] manufacturing tasks and manufacturing services are in a dynamic and changing environment. Because of the dynamic nature of manufacturing resources, manufacturing services are in a dynamic environment, including the addition of new services, the evolution or extinction of old services. Exceptions may also occur to the services being invoked. [2] In view of dynamic and changeable operating environment, there are many different researchers study the issue of supply demand matching from different angles. Moon etc. building an integer programming model to match the supply and demand of manufacturing resource services to ensure the optimality of QoS. [3] Zeng ets put forward the idea of Global re-planning to cope with the anomalies in service composition. [4] However, re-planning will bring additional costs. For this reason, Dai et al.

* Corresponding author: qingqingyang1015@163.com

proposed a self-healing method for service composition, which reduces the number of re selection through performance prediction. [5] Dotoli etc. design an integrated E-supply chain for agile and environmentally conscious manufacturing. [6] Most of these methods cannot be well applied to manufacturing services. Firstly, there is no systematic approach to considering changes in demand and services. Secondly, compared with the traditional Web service environment, the complexity and uncertainty of cloud manufacturing environment are more prominent. Thirdly, the variety of manufacturing resources is more complex and diverse. As the result, the existing researches on supply and demand matching and optimal allocation of manufacturing resources (for example, the problem of matching for simple demand or primitive task[7] and matching for complex demand or compound task[8]) are still far from the service-oriented practice.

The dynamic changes of manufacturing tasks and manufacturing services pose challenges to the management of cloud manufacturing systems. At present, there is a lack of in-depth research on the supply-demand dynamic matching method in the cloud manufacturing environment, and no mature scheme has been formed. We will use the theory of complex network to construct the service-demand dynamic matching network model. The model will provide a theoretical basis for the dynamic management of manufacturing tasks and manufacturing services.

2 Problem description

In service-oriented manufacturing systems, the supply is the set of various social available manufacturing resources and capabilities in the form of services.Nodes load of service manufacturing, which are denoted as L, referring to the number of manufacturing tasks currently processed by manufacturing services. Quality of Service (QoS), which refers to the non-functional attributes of manufacturing services. QoS is an important basis for the service composition.

In the traditional manufacturing service composition, we need to pick up one suitable service from the service classes of C_1 , C_2 and C_3 to build the service composition for a manufacturing task sequence $T_1 \to T_2 \to T_3$,. Each service class contains several services which having same or similar function but different in QoS, as shown in fig.1.

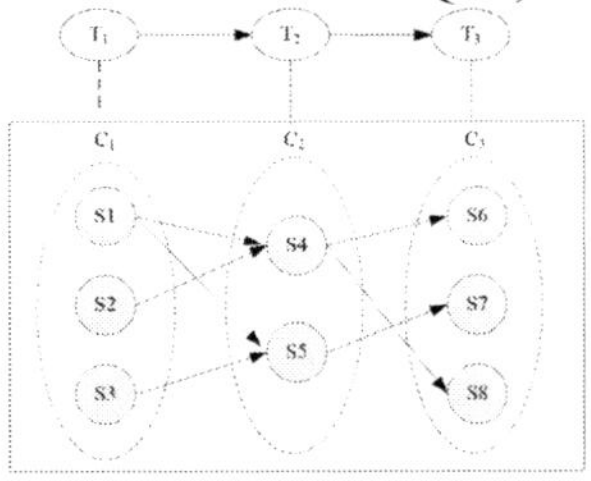

Fig.1. Problem scenario of manufacturing service composition.

Based on the potential call relationship between services, we can choose different service composition to carry out the task, e.g. $S_1 \to S_4 \to S_8$ is a set of solution. Assuming that the user's requirement changes in t , so we need to adjust the manufacturing task T_2 that is in the state of execution, which causing the service S_4 unable to meet the requirement and the service needs to be reselected. It can be seen that traditional service composition methods need to replace multiple services simultaneously, and it is difficult to adapt to changes in dynamic environment, resulting in Qos degradation.

3 Models of manufacturing service-demand dynamic matching network

3.1 Models of manufacturing service network (MSN)

Manufacturing service network (MSN) is the virtual network topology structure composed of a large number of service nodes and a variety of edges. Atomic manufacturing services (AMS) refer to fine-grained manufacturing services that can be invoked independently and are functionally indivisible, and is recorded as S . Manufacturing services can be either mapped to manufacturing resources with a single function, but also can be mapped to manufacturing resources with complex functions. The former is called fine- grained manufacturing service, that is, atomic manufacturing service. The latter, which we call coarse-grained manufacturing services, is denoted as S^C . We divided coarse-grained services S^C into composite services made up of multiple atomic manufacturing services. Thus, there are only atomic manufacturing services in MSN. The manufacturing service dependency edge (MSDE) refers to the weighted directed edges based on the potential call relationship between atomic manufacturing services, which is denoted by $E^S = \left\{ w_{ij}^S \right\}$, in which $w_{ij}^S = 1$ indicates that there is an execution sequence relationship between manufacturing service S_i and S_j .

Manufacturing service network (MSN) can be represented as directed acyclic network composed of AMS and MSDE. The basic service description is expanded as shown in formula.

$$Service = \langle S - ID, B - attri, F - attri, U - attri, N - attri, MRP \rangle \quad (1)$$

Which, $B - attri = \langle name, type, provider, address, \cdots \rangle$,

$F - attri = \langle Ftype, input, output, constraints, results \cdots \rangle$,

$U - attri = \langle QoS, State, S - time, E - time, executionrecords \cdots \rangle$

$N - attri = \langle S - ID, E^S - type, E^S, W^S \cdots \rangle$, $MRP = \langle R - ID, description, capability \cdots \rangle$.

The parameters and notations in formula referred to the expanded manufacturing services description model are shown in tab.1

Table1. Parameters and notations in manufacturing service description.

Notations	Explanation	Notations	Explanation
$S - ID$	Identification of manufacturing service	$B - attri$	Basic attributes of manufacturing service
$F - Attri$	Functional attributes of manufacturing service	$U - attri$	Using attributes of manufacturing service
$N - attri$	Network-related attributes of manufacturing service	MRP	Manufacturing resource mapped to the service
$Ftype$	Type of functions of manufacturing service	$S - time$	Start time of manufacturing service execution
$E - time$	End time of manufacturing service execution	$R - ID$	Identification of manufacturing resource mapped to the service
$E^S - type$	Type of correlation edges in MSN	E^S	Judgment matrix of correlation edges in MSN

W^S	Weight of correlation edges in MSN		

The state of manufacturing services S_{status} can be divided into two categories: not being invoked ($S_{status}=0$) and being invoked ($S_{status}=1$).

3.2 Models of manufacturing task network (MTN)

Similar to MSN, manufacturing task network (MTN) is also a virtual network topology structure composed of a large number of manufacturing task nodes and a variety of correlation edges among those task nodes. The cloud manufacturing platform can be decomposed into a series of atomic manufacturing tasks (AMT) with sequential relations through a certain strategy. AMT is an indivisible fine-grained manufacturing task, which can be executed by calling an atomic manufacturing service, and is recorded as T . The manufacturing task Executive Edge (MTEE) indicates the execution sequence relationship between atomic manufacturing tasks. It is obtained by the decomposition of coarse-grained manufacturing task T^c . MTEE is a weighted directed edge, which is recorded as $E^T=\{w_{ij}^T\}$, in which $w_{ij}^T=1$ indicates that there is an execution sequence relationship between manufacturing tasks T_i and T_j .The basic executive relations of manufacturing tasks can be divided into 4 structures: order, concurrency, selection and circulation. Similarly, the description of manufacturing tasks oriented to the modelling of MTN is expanded as shown in formula.

$$Task=\langle T-ID, B\inf, F\inf, E\inf, N\inf,\cdots\rangle \tag{2}$$

Which, $B\inf=\langle object, feature, material, comsumer, address,\cdots\rangle$

$F\inf=\langle Ftype, precondition, effect, constraints,\cdots\rangle$, $N\inf=\langle T-ID, E^T-type, E^T, W^T\rangle$

The parameters and notations in formula referred to the expanded manufacturing task description model are shown in tab.2

Table 2. Parameters and notations in manufacturing task description.

Notations	Explanation	Notations	Explanation
$T-ID$	Identification of manufacturing task	$B\inf$	Basic information of manufacturing task
$F\inf$	Functional information of manufacturing task	$E\inf$	Evaluation information of manufacturing task
$N\inf$	Network-related information of manufacturing task	E^T-type	Type of correlation edges in MTN
E^T	Judgment matrix of correlation edges in MTN	W^T	Weight of correlation edges in MTN

At different times, atomic manufacturing tasks are in different execution states T_{status} , and they are divided into four states: already execution ($T_{status}=0$), prepare execution ($T_{status}=1$), being execution ($T_{status}=3$), and not executed ($T_{status}=4$).

3.3 Modeling of dynamic matching network

3.3.1 Model description

Based on the models of MSN and MTN described in sections 3.1 and 3.2, the key point to construct the model of dynamic matching network (DMM), is to work out the existence of inter-network dynamic matching edge (INDME) which connect service nodes and task nodes. Inter-network dynamic matching edge indicates the dynamic matching relationship between atomic manufacturing task T and atomic manufacturing service, which are defined as formula $E^{TS}=\left\{w_{ij}^{TS}\right\}$. When $w_{ij}^{TS}=1$, it is indicates that the manufacturing task T_i and the manufacturing service S_i are in a dynamic binding state. There is a dynamic matching edge between T_i and S_i. When the manufacturing task T_i is completed, the matching edge disappears. The time varying characteristics of the network are realized.

Dynamic matching network is a double-layer network formed by the dynamic matching relationship between manufacturing task network and manufacturing service network. It is recorded as $Net^{TS}=\left\{Net^{T},Net^{S},E^{TS}\right\}$.

3.3.2 Dynamic scheduling of manufacturing services

QOS is an important evaluation standard for dynamic scheduling of manufacturing services. Due to the diversity of QoS indicators of manufacturing services, such as response time, cost, reliability, and so on, there are great differences between units and ranges. Therefore, it is usually necessary to normalize the QoS value, as shown in formula (3).

$$U\left(S_{ji}\right)=\sum_{k=1}^{r}\frac{q_k\left(S_{ji}\right)-Q_{j,k}^{\min}}{Q_k^{\max}-Q_k^{\min}}\cdot w_k \tag{3}$$

in which, w_k is the weight of each attribute value of the QoS, which indicate the degree of preference for each attribute of a user. $Q_{j,k}^{\max}$ represents the maximum k dimension attribute of QoS in all manufacturing service communities C_j. Manufacturing service community refers to a group of service classes that have the same or similar function but different QoS, which can be clustered according to the functional similarity between the services, $C^{S}=\left\{S_1^{C},S_2^{C},\cdots,S_n^{C}\right\}$, in which n represents different atomic manufacturing services. $Q_{j,k}^{\min}$ represents the minimum k dimension attribute of QoS in all manufacturing service communities C_j. $q_k\left(S_{ji}\right)$ represents the K dimension attribute value of Qos in the candidate service S_{ji}.

Dynamic scheduling of manufacturing services has two optimization objectives. Firstly, from the perspective of users, cloud manufacturing platform needs to provide users with QoS optimal manufacturing services or service composition. Secondly, from the perspective of cloud manufacturing platform management, in a multi-user environment, how to realized the load balancing of various manufacturing service nodes through reasonable scheduling of resources and optimized the load queue is the other objectives. Suppose that the number of manufacturing service community that need to be invoked in the current manufacturing process is N, each manufacturing service community contains services number are n. We can model the dynamic scheduling of manufacturing services as formula (4).

$$\begin{cases} ObjectMin\left(W_1 \cdot \sum_{j-1}^{N}\sum_{i=1}^{n} U\left(S_{ji}\right)\right) + W_2 \cdot \sum_{j=1}^{N}\sum_{i=1}^{n} L\left(S_{ji}\right) \cdot x_{ji} \\ Subjectto \sum_{j=1}^{N}\sum_{i=1}^{n} q_k\left(S_{ji}\right) \cdot x_{ji} \le Cons_k, 1 \le k \le r \\ \sum_{i=1}^{n} x_{ji} = 1, x_{ji} = \{0,1\} \end{cases} \tag{4}$$

in which, r represents the number of QoS properties of the service, k represents the k dimension property, $U\left(S_{ji}\right)$ represents the utility function value of the manufacturing service S_{ji}, which can be concluded by formula (5). $L\left(S_{ji}\right)$ represents the current load of the manufacturing service S_{ji}, $Cons_k$ represents the constraint value of the k dimension property. x_{ji} is a decision variable, $x_{ji} = 1$ indicates that the manufacturing service S_{ji} is selected, in contrast $x_{ji} = 0$ indicates that the manufacturing service S_{ji} is not selected. Both W_1 and W_2 are both weights, and $W_1 + W_2 = 1$.

Next, we will use shortest path routing strategy of complex network to solve the above models. The globally optimal service composition can be transformed to search for a least cost service composition path from the manufacturing service network, as shown in algorithm 1.

Algorithm 1 Dynamic scheduling algorithm for manufacturing services

Input: T_{status}, $Cons_{n-m}$, Service composition cost matrix $MatrixC$.

Output: S_{ji}.

1. $T = getExetask\left(T_{status}\right)$
2. $C^S = getComunity\left(T\right)$
3. FOR $j = m$ to n do
4. $\quad len = getLength\left(C_j\right)$
5. $\quad$ FOR $i = 1$ to len do
6. $\quad\quad L_{ji} = getLoad\left(S_{ji}\right)$
7. $\quad\quad QoS_{ji} = updateQoS\left(L_{ji}\right)$
8. $\quad$ END FOR
9. END FOR
10. $MatrixC = updateMatrix\left(L_{ji}, QoS_{ji}\right)$
11. $MinPath = \min Path\left(MatrixC, Cons_{n-m}\right)$
12. $SC \leftarrow SC[1]$
13. Return S_{ji}

4 Conclusions

As the common issue to be addressed in advanced manufacturing systems, the supply-demand matching of manufacturing service in service-oriented manufacturing system is becoming to the decisive factor for its application and operation. Especially under the current socialization and cloud environment, there are different kinds of supply-demand matching problems considering the uncertainty and dynamics in service-oriented manufacturing systems. The theory of complex network provides an alternative method to analyse and address these problems.

From the perspective of complex network, this paper studies the dynamic management of manufacturing tasks and manufacturing services in cloud manufacturing systems. A dynamic matching network theory model has been built. The model takes the load and dynamic QoS as the optimization objectives, and transforms the optimal manufacturing service composition problem into the shortest path search in the manufacturing service network, so as to realize the dynamic scheduling of the manufacturing service.

Further research will improve the reliability of QoS's other indicators, such as the relationship between reliability and load, to improve the generality of the dynamic QoS model. Furthermore, we will study the impact of statistical properties of the service network on the service composition, such as the degree of nodes, and so on.

Acknowledgments

This work was supported by the National Natural Science Foundation under Grant number 71690233 and 71571185.

References

1. Baines, T., Bigdeli, A. Z. and Bustinza, O. Servitization: Revisiting the state-of-the-art and research priorities. *International Journal of Operations& Production Management*, 2017, 37(2), pp 54-567.
2. F Tao., L. Zhang, Y. K. Liu, Y. Cheng etc. Manufactruing service management in cloud manufacturing: Overview and future research directions, *Journal of manufacture science engineering-transaction*, 2015,(4), pp:09-12.
3. Moon C, Lee Y, Jeong C, etc. Integrated process planning and scheduling in a supply chain. *Computers&Industrial Engineering*, 2008, 54(4), pp. 1048-1061.
4. Zeng L, Benatallah B, Ngu AH, etc. QoS-aware middleware for Web services compositon. IEEE Transactions on Software Engineering, 2004,30(5), pp. 311-327.
5. Dai Y, Yang L, Zhang B. QoS-driven self-healing web service composition based on performance prediction. Journal of Computer Science and Technology, 2009, 24(2), pp. 250-261.
6. M. Dotoli, M.P. Fanti. C.Meloni, M.C. Zhou. Design and optimization of integrated E-supply chain for agile and environmentally conscious manufacturing. *IEEE Transaction systems manufacturing*, 2006, 36(1), pp. 62-75.
7. S.X. Huang, S. Zeng, Y.S. Fan. Optimal service selection and composition for service-oriented manufacturing network. *International Journal Computer integrate manufacture*, 2011, 24(5), pp. 416-430.
8. F.Tao, D. Zhao, Y.F. Hu etc. Resource service composition and its optimal-selection based on particle swarm optimization in manufacturing grid system, *IEEE Transactions and information*, 2008, 4(4), pp. 315-327.

Similarity attraction-based mobile advertising: designing and testing

M Shin[1], and *KW* Lee[2,*]

[1]Department of Herb Science, Dong-Eui Institute of Technology, Busan 47230, Korea
[2]Hallym University Sacred Heart Hospital, Gwanpyeong-ro, Dongan-gu, An-yang, South Korea

Abstract. This study explores the possibility of applying the socio-psychological phenomenon known as similarity attraction to the creation of more customized and persuasive mobile advertising without privacy concerns, and for this purpose, we develop two applications, SensPlus and PerAds. While the SensPlus app accumulates the smartphone usage profiles of participants and constructs a regression model to predict their personality, the PerAds app presents user-created content according to the user's personality. The participants' response to a smartphone ad is measured in terms of invitation acceptance of the cherry blossom festival and classical music concert. The χ^2 statistics imply participants respond differentially to the various types of smartphone ads according to the similarity between the personalities of the user and ad types. Our results indicate that the socio-psychological phenomenon can be introduced into practical application by replacing the personality trait factors with the smartphone usage pattern.

1 Introduction

The advertisement landscape is undergoing rapid changes due to the decline in non-individualistic mass communication and rise in customized and personalized communication. Smartphones are at the core of this change. However, most of current customized advertising strategies like ads based on user's location [1], internet search [2] and personal profiles [3], [4] paid little attention to socio-psychological aspects.

We investigate the possibility of a customized advertising strategy based on the interaction between smartphone usage profiles and ad types, known as the "similarity-attraction effect," that is, the tendency of people to be attracted by those similar to themselves [5], [6]. Most of the similarity-attraction studies based on personality employ personality questionnaire, but in real-world we cannot ask users to complete a personality questionnaire [7]. Smartphones can provide us with sufficient information to extract or infer the users' psychological states or personality traits [8], [9].

In this study, we test the opportunity by designing a smartphone ads system in which (1) the smartphone user's personality is based on smartphone usage pattern and not a questionnaire response; (2) the presentation of smartphone ads is customized to suit the

* Corresponding author: kangwooster@gmail.com

smartphone user's personality; and (3) the smartphone user's personality information should not be collected, implying that the privacy-concern should be reduced.

2 System model and methods

2.1 Personality-based smartphone ads system

Some text. The general scheme of the personality-based smartphone ads system is presented in figure 1. The scheme consists of two parts, regression modeling and customized advertisement.

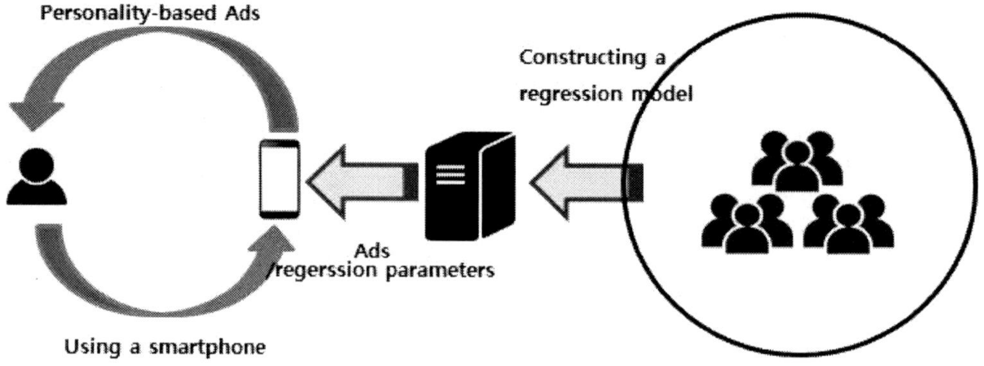

Fig. 1. General scheme of the personality-based smartphone ads system.

Our previous study [10] constructed a regression model using the SensPlus app and support vector machine (SVM) regression. The SensPlus app collects smartphone usage profiles such as number and time of incoming and outgoing calls and number of SMSs (input), and then transfers them to a server. A regression model can be established by mapping the smartphone usage profile to the extroversion trait (output), a modified version of the Wiggins personality test.

For customized advertisement, the PerAds app which selectively presents personality-based ads was developed on the Android platform. Two different types of ads are used – one to promote extroversion, and the other to promote introversion. The decision of which type of ad is presented to the user depends on the threshold values set by the experimenter. To avoid misclassification, we used dual thresholds T_1 and T_2. The ad is short and can be interrupted by the user anytime. The detailed procedure of PerAds is given below:

- Accumulate the user's smartphone usage profiles and average the profiles.
- Calculate the user's extroversion value by setting the averaged smartphone usage profile in the regression model.
- Compare the estimated extroversion value with the threshold value.

  value $> T_1$: extrovert
  value $< T_2$: introvert
  $T_2 \leq$ value $\leq T_1$: ambivert

2.2 Methods

2.2.1 Participants

A total of 176 undergraduate and graduate students with an Android smartphone enrolled in the experiment and paid 10,000 Korean won (approximately USD 9). They were aged from

20 to 35 years, with an average age of 28.7 (SD = 6.7 years). They signed a consent form and were informed of the experiment procedure.

2.2.2 PerAds apps

Two animation video clips (figure 2) were used for the experiment, one to advertise a cherry blossom festival for extroverted participants and the other to advertise a classical music concert for introvert participants. An animated avatar introduces each event and invites a smartphone user. In order to create the extrovert and introvert ads, three visual cues of the avatars were manipulated, that is, clothe types, movement, and invitation phrases. The extrovert avatar wore a casual suit, actively moved her body, and used informal language, whereas the introvert avatar wore a formal suit, rarely moved her body, and used formal language. Each video clip had a 320 x 240 pixel resolution and lasted five seconds. Text-only ads were also used that contained information such as "invitation for cherry blossom festival (for extroverts)," "invitation for classical music concert (for introverts)," date, and location. The date and location are the same for both the extrovert and introvert clips.

Fig. 2. The animation video clips used for advertising stimulus. (a) Invitation for cherry blossom festival (b) Invitation for classical music concert.

2.2.3 Procedure

The participants were randomly assigned to two groups, the text-only group (80 participants) and the animation group (96 participants). Text-only ads were given to the text-only group, and animation ads were given to the animation group. The participants were asked to install the mobile app PerAds in their phone, with instruction not to delete the application until 15 days later, that is, until the end of the experiment period. One day after the usage data accumulation period, an invitation ad is presented differentially to each participant according to his/her estimated personality. After the participants watched the ads, they are asked whether they would accept the invitation or not. The responses are sorted by the invitation and media type.

3 Results

We carried out chi-square analysis for each group using the number of participants who responded "accept" or "reject" to the invitation of events. The participants who did not indicate either extroverted or introverted traits (between T_1 and T_2) were not included in the statistical analysis. The results are summarized in tables 1, 2, and 3.

For the text-only group, we expected number of accept (or reject) as 50% of participants, because we did not expect any difference between accept and reject responses to the invitations. The result is given in table 1. For the festival condition, the chi-square (χ^2)

statistic is 1.324 and the p-value is 0.2498. For the concert condition, the chi-square is 0.533 and the p-value is 0.4652. Both results are not statically significant at $p < .05$.

Table 1. Chi-square results for the participants' responses to the cherry festival invitation (top) and the classical music concert invitation (bottom) of text-only stimulus (1).

Festival	Observed	Expected #	Expected
Accept	22	18.5	50%
Reject	15	18.5	50%
Column Totals	37	37	100%
Concert	Observed	Expected #	Expected
Accept	17	15	50%
Reject	13	15	50%
Column Totals	30	30	100%

For the animation group, we again set the expected number of accept (or reject) to half the number of each personality trait, and then carried out the chi-square analysis. The result is given in Table 2. For the festival condition, the chi-square statistic is 6.422 and the p-value is 0.0113. This result is statistically significant at $p < .05$. However, for the concert condition, the chi-square is 3.457 and the p-value is 0.0630.

Table 2. Chi-square results for the participants' responses to the cherry festival invitation (top) and the classical music concert invitation (bottom) of animated stimulus (2).

Festival	Observed	Expected #	Expected
Accept	31	22.5	50%
Reject	14	22.5	50%
Column Totals	45	45	100%
Concert	Observed	Expected #	Expected
Accept	23	17.5	50%
Reject	12	17.5	50%
Column Totals	35	35	100%

The assumption of no difference between 'accept' and 'reject' responses to the invitations can be problematic. For the text-only condition, though the ad stimulus is relatively neutral, the participants are more likely to accept the invitation, but the tendency is not statistically significant. This reflects the participants' attitude or response tendency to the events and can be considered as the baseline for the effect of animated advertising. Thus, we re-calculate the expected numbers by multiplying the total number of participants in each condition with the proportion of each response in the text-only group: 45x22/37(26.775), 45x15/37(18.225) for the festival invitation, 35x17/30(19.845), 35x13/30(15.155) for the concert invitation. The result is given in Table 3. For the festival condition, the chi-square statistic is 1.646 and the p-value is 0.1995. For the concert condition, the chi-square is 1.158 and the p-value is 0.2818. Both results are not statistically significant at $p < .05$.

Table 3. Chi-square results for the participants' responses to the cherry festival invitation (top) and the classical music concert invitation (bottom) of text-only stimulus (3).

Festival	Observed	Expected #	Expected
Accept	31	26.775	59.5%
Reject	14	18.225	40.5%
Column Totals	45	45	100%
Concert	Observed	Expected #	Expected
Accept	23	19.845	56.7%
Reject	12	15.155	43.3%
Column Totals	35	35	100%

4 Discussion and conclusion

The scheme of personality-based ads that use smartphone usage profiles to estimate the user's personality not requiring personal information by separating regression modeling from personality estimation was proposed. Furthermore, we suggest the non-privacy invasive evaluation procedure to test the effect of the scheme. Usual 2x2 factorial design allows for investigating the interaction between independent factors, but needs to know the properties of individuals that are privacy-sensitive. Thus, the factorial design is impractical for a real world application. In contrast, our experiment design does not require privacy-sensitive information of participants and can thus be applied more freely to smartphone application.

Throughout the experiment, we tried to demonstrate the effect of the personality-based ad scheme. Our results indicate that the participants are more likely to accept than reject the invitation. This study introduces the socio-psychological phenomenon of similarity attraction to the customized advertisement application domain by establishing the relationship between extroversion and smartphone usage pattern. However, we could not get significant difference of acceptance rate using animated ads from acceptance rate using text-only ads. That might be due to the quality of ads. Advertisement is field of art, so acceptance rate might be more influenced by sense, concept, graphic etc. Further try with more expressive animated ads is expected.

Acknowledgments

This research was supported by Basic Science Research Program through the National Research Foundation of Korea (NRF) funded by the Ministry of Education (2016R1A6A3A11933734).

References

1. U. Varshney, "Location management for mobile commerce applications in wireless internet environment", *ACM Transactions on Internet Technology*, vol. 3, no. 3, pp.236-255, 2003.
2. K. Partridge, B. Begole, P. Alto and C. H. Road, "Activity-based Advertising: Techniques and Challenges", in *Proc of the First Workshop on Pervasive Advertising*, May 11, pp. 2-5, 2009.

3. D. J. Xu, S. S. Liao and Q. Li, “Combining empirical experimentation and modeling techniques: A design research approach for personalized mobile advertising applications”, *Decision Support Systems*, vol. 44, no. 3, pp. 710–724, 2008.
4. J. B. Todd, Y. Christine and S. Esther, “From firm-controlled to consumer-contributed: consumer co-production of personal media marketing communication”, *Journal of Interactive Marketing*, vol. 28, no. 2, pp.117-133, 2014.
5. E. C. Klohnen and S. Luo, “Interpersonal attraction and personality: What is attractive – self similarity, ideal similarity, complementarity, or attachment security?”, *Journal of Personality and Social Psychology*, vol. 85, no. 4, pp. 709–722, 2003.
6. M. Matthew, S. H. Robert and K. Jeffrey, "Is actual similarity necessary for attraction? A meta-analysis of actual and perceived similarity", *Journal of Social and Personal Relationships*, vol. 25, no. 6, pp. 889–922, 2008
7. F. Evanthia, T. Athanasios and T. Giannis, “An integrated e-recruitment system for automated personality mining and applicant ranking", *Internet Research*, vol. 22, no. 5, pp.551–568, 2012.
8. W. Lane and C. Manner, “The Impact of Personality Traits on Smartphone Ownership and Use”, *International Journal of Business and Social Science*, vol. 2, no. 17, pp. 22–28, 2003.
9. S. Butt and J. G. Phillips “Personality and self reported mobile phone use,” *Computers in Human Behavior*, vol. 7, no. 3, pp. 346-360, 2008.
10. K. Lee. and H. Choo, “Predicting user attitudes toward smartphone ads using support vector machine”, *International Journal of Mobile Communications*, vol. 14, no. 3, pp. 226–243, 2016.

Adaptive fractional PID control of biped robots with time-delayed feedback

IEEE MEMBER *Joel Perez* P*, *Jose P.* Perez, *Mayra Flores* Guerrero, *Ruben Perez* P. , *Oscar Rangel* Aguilar, *Daniel* Ramirez-V, *Joey Perez* F., and *J. Javier Perez* D.

Department of Physical and Mathematical Sciences, Autonomus University of Nuevo Leon (FCFM,UANL), Cd. Universitaria, San Nicolas de los Garza, Nuevo Leon, 66450, Mexico

Abstract. This paper presents the application of Fractional Order Time-Delay adaptive neural networks to the trajectory tracking for chaos synchronization between Fractional Order delayed plant, reference and Fractional Order Time-Delay adaptive neural networks. The proposed new control scheme is applied via simulations to control of a 4-DOF Biped Robot [1]. The main methodologies, on which the approach is based, are Fractional Order PID the Fractional Order Lyapunov-Krasovskii functions methodology. The structure of the biped robot is designed with two degrees of freedom per leg, corresponding to the knee and hip joints. Since torso and ankle are not considered, it is obtained a 4-DOF system, and each leg, we try to force this biped robot to track a reference signal given by undamped Duffing equation. The tracking error is globally asymptotically stabilized by two control laws derived based on a Lyapunov-Krasovski functional.

1 Introduction

Fractional calculus is a generalization of differential and integral calculus which involves generalized functions. The first to work this new branch of mathematics was Leibniz. Due to the growing interest in the applications of fractional calculation, in this work we obtain conditions that guarantee the tracking of trajectories of nonlinear systems generated by differential equations of fractional order which we will call plants (This term is widely used in engineering), which in our case will be a mechanical arm, a helicopter, a plane or limbs of a humanoid, all of fractional order.

The problem of tracking control of trajectories is very important, since the control function allows the non-linear system to carry out a previously assigned task, work or trajectory, for example, a mechanical arm and its objective is to cut a piece with a previously generated form, or the coupling of two aircraft in space. We include mathematical models with time delay, since the processing and transmission of information is important in this type of systems, which depending on the delay, these systems can generate undesirable oscillatory or chaotic dynamics, and cause instability in the mathematical model that describes the trajectory tracking error.

* Corresponding author: joel.perezpd@uanl.edu.mx

The chapter is organized as follows: first, the general mathematical model of non-linear systems is proposed, as a second part, the Neural Network is proposed that will adapt to the non-linear system and the reference signal that both must follow, as a third part obtains the dynamics of the tracking error between the non-linear system and the reference, after obtaining conditions in the laws of adaptation of weights in the Neural Network and obtaining the control law that guarantees that the tracking error converges to zero , so that the non-linear system will follow the indicated reference signal, which is what was wanted to be demonstrated. Finally simulations are presented, which illustrate the theoretical results previously demonstrated.

There are several ways to define the fractional calculation, in this research we will use the well-known derivative of Caputo, which has equation (1):

$$aD_t^{\alpha} f(t) = \frac{1}{\Gamma(\alpha-n)} \int_a^t \frac{f^{(n)}(\tau)}{(t-\tau)^{\alpha-n+1}} d\tau ,(n-1<\alpha<n) \quad (1)$$

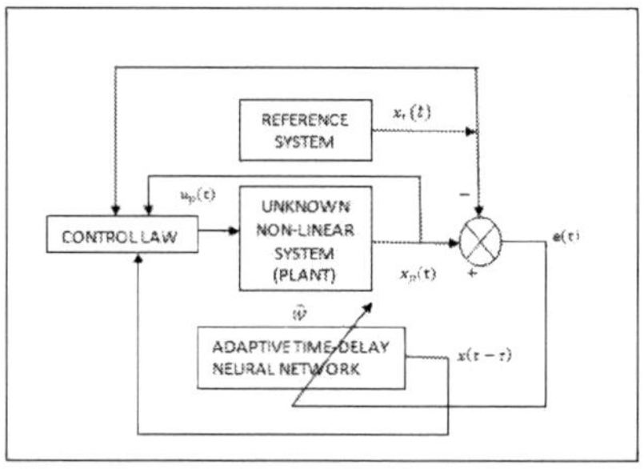

Fig. 1. Adaptive recurrent control diagram.

We use the scheme of figure 1 to indicate the procedure used in the obtaining of the laws of adaptation of weights and the laws of control that guarantee that the tracking error between the non-linear system, the neural network and the reference signal converges to zero.

2 Time-delay adaptive neural network and the reference

The nonlinear system, equation (5), which is forced to follow a reference signal:

$$\boldsymbol{aD_t^{\alpha} x_p = f_p(t, x_p(t) + x_p(t-\tau)),\ t \in [0,T],\ 0 < \alpha \leq 1,} \quad (2)$$

$$x_p(\mathrm{t}) = \mathrm{g}(\mathrm{t}), \text{ where } x_p, f_p \in \Re^n,\ u \in \Re^m,\ g_p \in \Re^{nxn}.$$

The differential equation will be modeled by the neural network [2]: $aD_t^{\alpha} x_p = \mathrm{A}(x) + \mathrm{W}^{*}\Gamma_{\mathrm{z}}(x(t-\tau) + \Omega_{\mathrm{u}}$

The tracking error between these two systems:

$$w_{per} = x - x_p \text{ and } aD_t^{\alpha} w_{per} =- k w_{per} \quad (3)$$

We use the next hypotheses. The nonlinear system is [3]:

$$aD_t^{\alpha} x_p = aD_t^{\alpha} x + w_{per} = \mathrm{A}(x) + \mathrm{W}^{*}\Gamma_{\mathrm{z}}[x(t-\tau)] + w_{per} + \Omega_{\mathrm{u}} \quad (4)$$

where the W^{*} is the matrix weights.

3 Tracking error problem

In this part, we will analyze the trajectory tracking problem generated by

$$aD_t^{\alpha} x_r = f_r(x_r, u_r),\ w_r, x_r \in \Re^n \quad (5)$$

The time derivative of the error is:

$$aD_t^{\alpha} e_n = aD_t^{\alpha} x - aD_t^{\alpha} x_r = \mathrm{A}(x) + \mathrm{W}^{*}\Gamma_{\mathrm{z}}[x(t-\tau)] + w_{per} + \Omega_{\mathrm{u}} - f_r(x_r, u_r) \quad (6)$$

The equation (6), can be rewritten as follows, and the unknown plant will follow the fractional order reference signal, if:

$$aD_t^{\alpha}e = Ae + W^{*}\Gamma_z(x(t-\tau)) - \widehat{W}\Gamma_z(x_r(t-\tau)) - Ae + (A+I)(x-x_r) + \Omega(u - \alpha_r(t,\widehat{W})) \quad (7)$$

Now, $\widehat{W}$ is part of the approach, given by W^{*}. The Eq. (7), can be expressed as Eq. (8)

$$aD_t^{\alpha}e = Ae + (W^{*} - \widehat{W})\Gamma_z(x(t-\tau)) + \widehat{W}\Gamma(z(x(t-\tau)) - z(x_r(t-\tau))) + (A+I)(x-x_r) - Ae + \Omega(u - \alpha_r(t,\widehat{W})) \quad (8)$$

$$\widetilde{W} = W^{*} - \widehat{W} \text{ and } \widetilde{u} = u - \alpha_r(t,\widehat{W}) \quad (9)$$

And by replacing Eq. (9) in Eq. (8), we have:

$$aD_t^{\alpha}e = (A+I)e + \widetilde{W}\sigma(x(t-\tau)) + \widehat{W}\emptyset_\sigma(t-\tau) + \Omega u_2 \quad (10)$$

The control law, we will obtain using the fractional order Lyapunov-Krasovskii methodology.

4 Study of trajectory tracking error

Our mathematical model of the dynamics in the tracking error is described in (10). In this equation we can see that an equilibrium state of this system is $(e,\widetilde{W}) = 0$.

Without loss of generality we can assume that the matrix A is given $A = -\lambda I, \lambda > 0$, where I is the identity matrix of order nxn.

For the study of the stability of the tracking error we propose the following PID control law [4], widely used in science and engineering.

We will determine conditions in the parameters that guarantee that the tracking error converges to zero, and we will also use the following control law [5].

$$\Omega u_2 = K_p e + K_i aD_t^{-\alpha}e + K_v aD_t^{\alpha}e - \gamma\left(\frac{1}{2} + \frac{1}{2}\|\widehat{W}\|^2 L_\phi^2\right)e \quad (11)$$

We also include the following control law, $PI^{\lambda}D^{\alpha}$ [6]: $u(t) = K_p e(t) + K_i aD_t^{-\lambda}e(t) + K_d aD_t^{\alpha}e(t)$,

Substituting Eq. (11) in Eq. (10):

$$aD_t^{\alpha}e = \frac{-1}{a}(\lambda - 1 + K_p)e + \frac{1}{a}\widetilde{W}\sigma(x(t-\tau)) + \frac{1}{a}\widehat{W}\emptyset_\sigma(t-\tau) + \frac{1}{a}K_i aD_t^{-\alpha}e - \frac{\gamma}{a}\left(\frac{1}{2} + \frac{1}{2}\|\widehat{W}\|^2 L_\phi^2\right)e \quad (12)$$

And if $w = \frac{1}{a}K_i aD_t^{-\alpha}e$, then $aD_t^{\alpha}w = \frac{1}{a}K_i e(t)$, [7], then Eq. (12) we rewrite as:

$$aD_t^{\alpha}e_n = \frac{-1}{a}(\lambda - 1 + K_p)e + \frac{1}{a}\widetilde{W}\sigma(x(t-\tau)) + \frac{1}{a}\widehat{W}\emptyset_\sigma(t-\tau) + w - \frac{\gamma}{a}\left(\frac{1}{2} + \frac{1}{2}\|\widehat{W}\|^2 L_\phi^2\right)e \quad (13)$$

Let V be, the next candidate Lyapunov function as [8] and [9] :

$$V = \frac{1}{2}(e_n{}^T, w^T)(e_n, w)^T + \frac{1}{2a}tr\{\widetilde{W}^T\widetilde{W}\} + \frac{1}{a}\int_{t-\tau}^{t}[\emptyset_\sigma^T(s)\,\widehat{W}^T\widehat{W}\emptyset_\sigma(s)]ds \quad (14)$$

The fractional order time derivative of (14) along the trajectories of Eq. (13), and we select the next learning law from the neural network weights as in [10] and [11]:

$$tr\{aD_t^{\alpha}\widetilde{W}^T\widetilde{W}\} = -e^T\widetilde{W}\sigma(x(t-\tau)) \quad (15)$$

Then Eq. (13) is reduced to

$$aD_t^{\alpha}V = \frac{-1}{a}(\lambda - 1 + K_p)e^Te + \frac{e^T}{a}\widehat{W}\emptyset_\sigma(t-\tau) + \left(1 + \frac{K_i}{a}\right)e^Tw - \frac{\gamma}{a}\left(\frac{1}{2} + \frac{1}{2}\|\widehat{W}\|^2 L_\phi^2\right)e^Te + \frac{1}{a}[\emptyset_\sigma^T(t)\widehat{W}^T\widehat{W}\emptyset_\sigma(t) - \emptyset_\sigma^T(t-\tau)\widehat{W}^T\widehat{W}\emptyset_\sigma(t-\tau)] \quad (16)$$

Next, let's consider the following inequality proved in [12]

$$X^TY + Y^TX \leq X^T\Lambda X + Y^T\Lambda^{-1}Y \quad (17)$$

Which holds for all matrices $X,Y \in \Re^{nxk}$ and $\Lambda \in \Re^{nxn}$ with $\Lambda = \Lambda^T > 0$. Applying (22) with $\Lambda = I$ to the term $\frac{e^T}{a}\widehat{W}\emptyset_\sigma(t-\tau)$ from Eq. (16), where, we get

Here, we select $\left(1+\frac{K_i}{a}\right)=0$ and $K_v = K_i + 1$, with $K_v \geq 0$ then $K_i \geq -1$, with this selection of the parameters from Eq. (16) is reduced to:

$$aD_t^\alpha V \leq \frac{-1}{a}(\lambda - 1 + K_p)e^T e - \frac{(\gamma-1)}{a}\left(\frac{1}{2}+\frac{1}{2}\|\widehat{W}\|^2 L_\phi^2\right)e^T \quad (18)$$

If $(\lambda - 1 + K_p) > 0$, $a > 0$, $(\gamma - 1) > 0$, so that: $aD_t^\alpha V \leq 0, \forall\, e, w, \widehat{W} \neq 0,\, e \neq 0$, is wanted to be demonstrate.

The control law is given by Eq. (19)

$$u_n = \Omega^\dagger[-\widehat{W}\Gamma\big(z(x_n(t-\tau)) - z(x_p(t-\tau))\big) - (A+I)(x - x_p) + K_p e + K_i aD_t^{-\alpha} e + K_v aD_t^\alpha e - \Gamma\left(\frac{1}{2}+\frac{1}{2}\|\widehat{W}\|^2 L_\phi^2\right)e_n + f_r(x_r,u_r) - Ax_r - \widehat{W}\Gamma_z(x_r(t-\tau)) - x_r + x_p] \quad (19)$$

Theorem: The control law Eq. (19) and the neuronal adaptation law given by Eq. (15) guarantee that the fractional tracking error converges to zero, by which the tracking of trajectories of the non-linear system is guaranteed Eq. (4).

5 Identification of the unknown non-linear system by the neural network

Following the procedure above, we obtain the following control law. Let V be, the next candidate Lyapunov function as

$$V = \frac{1}{2}(e_p{}^T, w^T)(e_p, w)^T + \frac{1}{2a}tr\{\widetilde{W}^T\widetilde{W}\} + \frac{1}{a}\int_{t-\tau}^{t}[\emptyset_\sigma^T(s)\,\widetilde{W}^T\widetilde{W}\emptyset_\sigma(s)]ds \quad (20)$$

Then, (21) is reduced to

$$aD_t^\alpha V \leq \frac{-1}{a}(\lambda - 1 + K_p)(e_p{}^T)(e_p) - \frac{(\gamma-1)}{a}\left(\frac{1}{2}+\frac{1}{2}\|\widehat{W}\|^2 L_\phi^2\right)(e_p{}^T)(e_p) < 0 \quad (21)$$

The previous inequality guarantees that the identification of the non-linear system is satisfied, that is, the approach error converges to zero asymptotically

$$u_p = \Omega^\dagger[\widehat{W}\Gamma z(x_r(t-\tau)) - \widehat{W}\Gamma\big(z(x_n(t-\tau)) - z(x_p(t-\tau))\big) - (A+I)(x - x_p) + K_p e + K_i aD_t^{-\alpha} e + K_v aD_t^\alpha e - \Gamma\left(\frac{1}{2}+\frac{1}{2}\|\widehat{W}\|^2 L_\phi^2\right)e_n - \Gamma\left(\frac{1}{2}+\frac{1}{2}\|\widehat{W}\|^2 L_\phi^2\right)e_p + f_r(x_r,u_r) - f_p(x_p) + Ax_p - Ax_r + \widehat{W}\Gamma_z(x_p) - x_r + x_p] \quad (22)$$

6 Simulation

The mathematical model, which describes the movement dynamics of the bipedal robot (figure 2), is obtained using the Euler-Lagrange equations [1], [13].

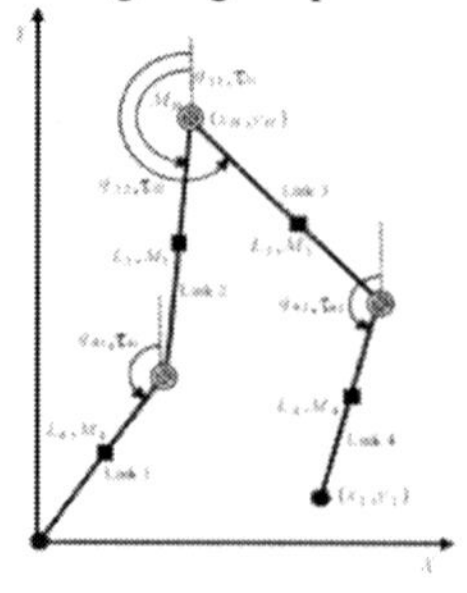

Fig. 2. Dynamic model of Biped Robot.

$$D(q(t)\ddot{q}(t) + C(q(t),\dot{q}(t))\dot{q}(t) + G(q(t)) = B\tau(t)$$

where $q(t) = [q_{31}(t)q_{32}(t)q_{41}(t)q_{42}(t)]^T$, is the generalized coordinates vector. As usual, $D(q(t))$ is the inertia matrix, bounded and positive definite, and $C(q(t),\dot{q}(t))$ is the matrix of Coriolis and centripetal forces. $G(q(t))$ represents a matrix of gravitational effects and B defines the input matrix. The vector $\tau(t) = [\tau_{31}(t)\tau_{32}(t)\tau_{41}(t)\tau_{42}(t)]^T$, defines the applied joint torques of the robot.

To illustrate the theoretical results obtained, we propose an example, which, as can be seen in the simulations, trajectory tracking is guaranteed and the neural network is described by:

$aD_t^\alpha x_p = \mathrm{A}(x) + \mathrm{W}^*\Gamma_z(x(t-\tau)) + \Omega_u$, with τ=25 sec, $A =- 20I, I \in \Re^{4x4}$, and, W^* is estimated using the learning law given in (15).

$\Gamma_z(x(t-\tau)) = (\tanh(x_1(t-\tau)), \tanh(x_2(t-\tau)),..., \tanh(x_n(t-\tau)))^{\mathrm{T}}$, $\Omega = \begin{pmatrix} 0 & 0 & 1 & 0 \\ 0 & 0 0 & 1 \end{pmatrix}^{\mathrm{T}}$ and the u is obtained using (22), and the reference signal that they have to follow, both the non-linear system and the neural network is given by the Duffing equation [14].

$$\ddot{x} - x + x^3 = 0.114\cos(1.1t) : x(0) = 1, \dot{x}(0) = 0.114$$

$$\frac{x(t)}{dt} = y(t)$$

$$\frac{y(t)}{dt} = x(t) - x^3(t) - \alpha y(t) + \delta\cos(\omega t)$$

Here, the conventional derivatives are replaced by the fractional derivatives as follows:

$$aD_t^\alpha x(t) = y(t)$$

$$aD_t^\alpha x(t) = x(t) - x^3(t) - \alpha y(t) + \delta\cos(\omega t)$$

where α, ω, δ, are the parameters of the Duffing differential equation, which we will use as a reference trajectory, that the non-linear system and the neural network have to follow.

As can be seen in figures 4 and 5, the tracking of trajectories in the states of the system are performed with satisfaction, while figure 3 shows the plane phase of the same fractional order differential equation. Figures 6, 7 show the torques applied to the ends of the bipedal robot. Parameter values of the fractional order, alpha (0.0001) and beta (0.0001) are included.

$$\alpha = 0.001, \ \beta = 0.001$$

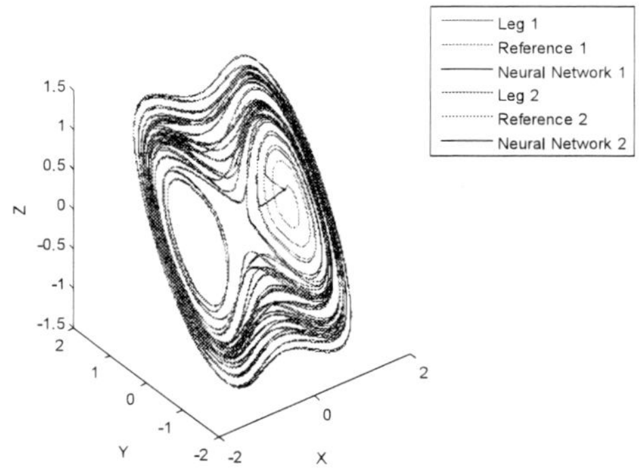

Fig. 3. A phase space trajectory of Duffing equation.

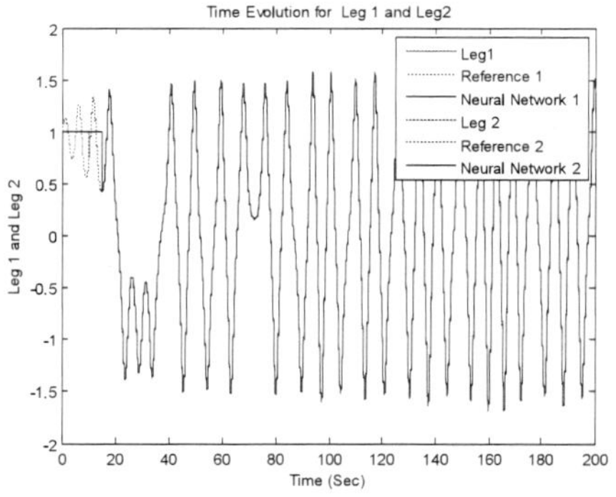

Fig. 4. Time evolution for the angular Position Leg 1 and Leg 2 (rad) of Link 1.

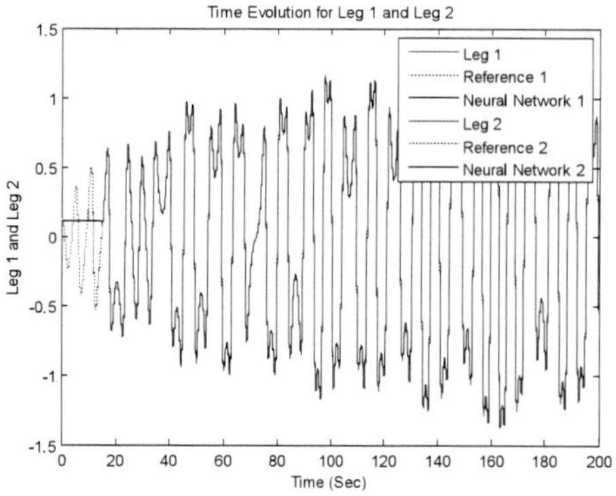

Fig. 5. Time evolution for the angular Position Leg 1 and and Leg 2(rad) of Link 2.

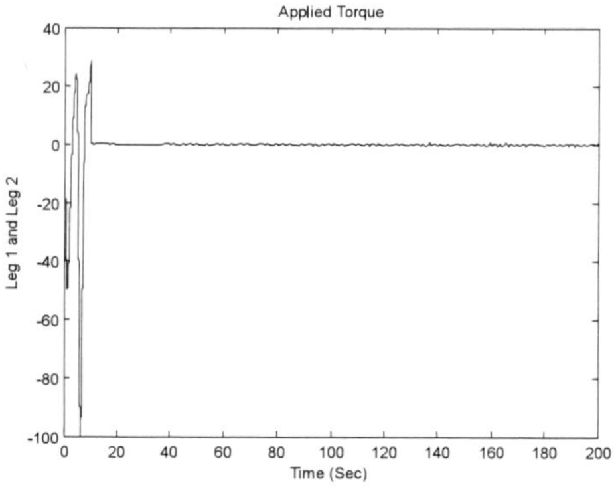

Fig. 6. Torque (Nm) applied to Leg 1 and Leg 2 of Link 1.

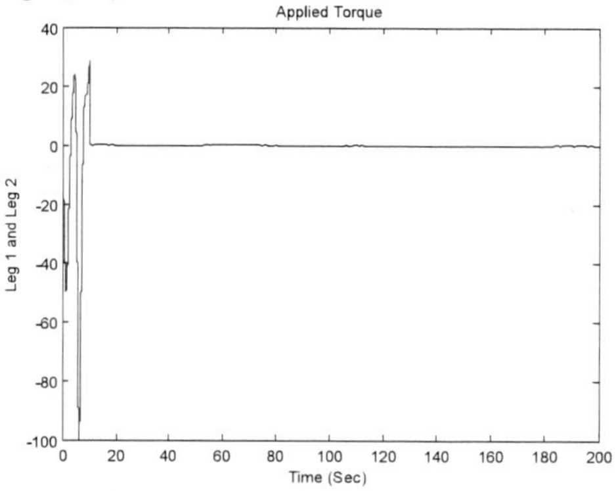

Fig. 7. Torque applied to Leg 1 of Link 2.

7 Conclusions

In this research work, conditions have been obtained in the parameters of the adaptive recurrent neural network, as well as laws of control and laws of neuronal adaptation, which, together, guarantee that the tracking error of trajectories between the non-linear system and the reference signal converges asymptotically to zero, so that trajectory tracking is develops with satisfaction.

Acknowledgements

Authors thank, Mexican National Science and Technology Council, (CONACyT), Mexico, to theAutonomous University of Nuevo Leon, Dynamical Systems Group of the Department of Physical and Mathematical Sciences, (FCFM-UANL), Mexico.

References

1. Design and Real-Time Control of a 4-DOF Biped Robot, Jose Alejandro Vazquez and Martin Velasco-Villa, International Journal of Advanced Robotic Systems, Intech, 2013.
2. Nonlinear Dynamics and Chaos in Fractional-Order Hopfield Neural Networks with Delay, Xia Huang, Zhen Wang, and Yuxia Li, Hindawi Publishing Corporation,

Advances in Mathematical Physics, Volume 2013, Article ID 657245, 9 pages, http://dx.doi.org/10.1155/2013/657245

3. Stability analysis of fractional-order Hopfield neural networks with discontinuous activation functions, Shuo Zhang, Yongguang Yu, Qing Wang, Neurocomputing 171 (2016) 1075–1084, www.elsevier.com/locate/neucom.
4. Integer & Fractional Order PID Controller for Fractional Order Subsystems of AUV, Mrs. Sneha D. Joshi, Dr. D. B. Talange, 2013 IEEE Symposium on Industrial Electronics & Applications (ISIEA2013), September 22-25, 2013, Kuching, Malaysia.
5. R. Kelly, R. E. Haber, R.E. Haber Guerra and Fernando Reyes, "Lyapunov Stable Control of Robot Manipulators: a Fuzzy Self-TuningProcedure", Intelligent Automation and Soft Computing, Vol. 5, No. 4, pp. 313-326, 1999.
6. Arturo Rojas.Moreno, Victor Jara.Sandoval , Fractional Order PD and PID Position Control of an Angular Manipulator of 3DOF, arojas@utec.edu.pe, jaravictor2000@yahoo.com.
7. Fractional Versions of the Fundamental Theorem of Calculus, Eliana Contharteze Grigoletto, Edmundo Capelas de Oliveira, Applied Mathematics, 2013, 4, 23-33, Scientific Research, http://dx.doi.org/10.4236/am.2013.47A006, (http://www.scirp.org/journal/am).
8. Fractional order Lyapunov stability theorem and its applications in synchronization of complex dynamical networks, Diyi Chen, Runfan Zhang, Xinzhi Liu, Xiaoyi Ma, Commun Nonlinear Sci Numer Simulat 19 (2014) 4105–4121, www.elsevier.com/locate/cnsns
9. Disturbance Rejection for Fractional-Order Time-Delay Systems, Hai-Peng Jiang and Yong-Qiang Liu, Hindawi Publishing Corporation, Mathematical Problems in Engineering, Volume 2016, Article ID 1316046, 8 pages, http://dx.doi.org/10.1155/2016/1316046
10. Rovitahkis G. A. and M. A. Christodoulou, Adaptive Control with Recurrent High-Order Neural Networks, Springer Verlang, New York,USA, 2000.
11. Petros A. Ioannou, Jing Sun, "Robust Adaptive Control", PTR Prentice-Hall, Upper Saddle River, NJ 07458. ISBN 0-13-439100-4.
12. Adaptive Synchronization of Fractional Neural Networks with Unknown Parameters and Time Delays, Weiyuan Ma, Changpin Li, Yujiang Wu and Yongqing Wu, Entropy 2014, 16, 6286-6299; doi:10.3390/e16126286, ISSN 1099-4300.
13. Mark W. Spong and M. Vidyasagar, Robot Dynamics and Control. John Wiler and Sons, USA, 1989.
14. Ivo Petras, Nonlinear Physical Science, Fractional-Order Nonlinear Systems, Modeling, Analysis and Simulation, Springer,Springer Heidelberg Dordrecht London New York, ISSN 1867-8440, ISBN 978-3-642-18100-9.

Auto-tuning for cascade PID height position controller of rotorcraft

Guiqiu Liao[1,2,*], and *Jiankang* Zhao[1,2]

[1]Department of Instrument Science and Engineering, Shanghai Jiao Tong University, China
[2]Key Laboratory of Aerospace Intelligent Control Technology, Shanghai, China

Abstract. In this article, we present a method for tuning controller parameters for cascade PID based on the step response performance characteristics of a closed loop system with application to an unmanned aerial vehicle(UAV). With specifically designed system identification procedures, a neural network mapping is obtained automatically when the UAV is flying around the target height. With this network system model, a gradual regression and optimization algorithm is proposed to tune the controller. The regression model primly illustrated the relation between PID parameters with controller performance, and the construction for optimization cost function takes the physical significance of step response performance of flying machine into account. Experimental data collected from fight experiment when auto-tuner is implemented on a quadrotor-craft demonstrate the efficiency of the proposed method.

1 Introduction

In the past decades, the automatic parameter tuning for the controller has been a pursuit for researchers and engineers [1][2][3][4]. Amount of these studies are based on adequate prior knowledge of the control plant, for example the system model has always been used as a key information to solve the problem of auto-tuning [5]. The system model with explicit expression such as transfer function or state transition matrix has several advantages. In some circumstance, the parameters of controller can be obtained directly by certain calculation (usually there is no need to do any iteration). Some researchers use the identified equation of open-loop system directly to calculate the feedback coefficients of Linear Quadratic Regulator (LQR) controllers [6][7], which is a representative approach to obtain the optimal control parameters without iteration computation. Another approach is to use the identified transfer function of plant to analyze the amplitude and phase figures of open-loop and closed-loop systems, so that PID controller with desired amplitude and phase characteristics can be designed [8]. In many situations, however, the system model (explicit functions) can not be acquired, even a linearized approach will make enormous effort to implement open loop response experiment to identify an approximate system transfer function. A simplified approximate model may be useful to determine rough controller parameters, but the desired result usually can not be realized the optimized parameters are applied to a practical system.

* Corresponding author: guiqiu@sjtu.edu.cn

Some researchers considered about tuning the controller without any system model, this form of method can be conclude as obtaining the optimal control parameters by valuating a series of practical experiment results corresponding to different system settings [9]. Relay feedback tuning is one good application of this kind of idea [10]. By temporarily replacing feedback controller by a relay, most systems will oscillate and a point on Nyquist curve can be obtained. Based on the positions of these points, the designer can choose the ideal PID parameters according to the phase and amplitude margin. This method has also been implemented in the system with cascade controller [11]. This method is a considerable choice for systems with difficulty in system identification, it requires the system to oscillate under various magnitudes in the open-loop control process, which is unrealistic for some systems and easy to cause safety problems in the process of tuning.

The key motivation of this paper is to design a highly matching auto-tuning algorithm for cascade controller which is applied on integrating system like UAV. Step response performance characteristics provide available criteria for controller design [12]. A nonlinear black-box system modeling [13] is a suitable bridge effectively for obtaining the mapping from controller parameters to performance characteristics. The black-box identification for UAV height input-output response is well designed to ensure the safety of all the procedures. The procedure of optimal parameter seeking depends on varying controller parameters and collecting the step response performance data in a virtual environment, so there is no risk of controlling the UAV under unstable parameter setting. This new optimal seeking method combines the physical features of different performance characteristics to achieve visualized ideal results, and a gradual regression and optimization method contributes to the accuracy and efficiency simultaneously.

2 Controller and system structure

In the auto-tuner design, a specific cascade closed-loop controller for the rotorcraft height control is built. A black-box identification algorithm using Gaussian Random Signal is embedded in to the control system. The complete system is presented in figure 1, the outer loop is controlled by a PD controller with limited output, and the inner loop is controlled by a PID controller. This choice is made according to the different physical features between outer loop and inner loop. Compared with the position stabilization, the inner loop control determines the value thrust, which compensates the varying of system bias error due to electrical energy consumption. Based on these, the integral item is essential while it is not required in the outer loop.

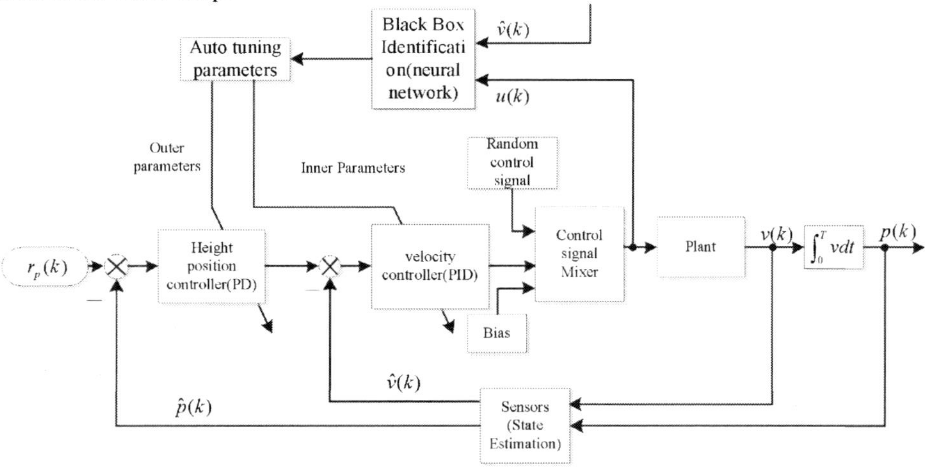

Fig. 1. UAV auto-tuning control system block diagram.

2.1 Initial setting for controller

By setting a group of rough height controller parameters, the aircraft can fly stably around a target altitude (with no consideration about response speed or other performance characteristics). These initial parameters can be calculated regarding the take-off thrust which is equal to the weight of UAV. In the calculation of rough initial inner loop PID parameters, the differential coefficient is set to 0, the relation between proportional coefficient, integral coefficient and the basic take-off thrust is:

$$\begin{cases} Kp_{in} \cdot e_{v\max} = 0.2Th_b \\ \int_0^{t_0} Ki_{in} \cdot 0.05 e_{v\max} dt = 0.02Th_b \\ Kd_{in} = 0 \end{cases} \tag{1}$$

where Kp_{in} , Ki_{in} , Kd_{in} are the rough proportional, integral and differential coefficient of inner controller; $e_{v\max}$ is the output limit of outer loop(target velocity limit of inner loop) ; Th_b is the basic bias control value which can generate a thrust to meet the weight of rotorcraft. t_0 is the integral time, the value of it determine the speed of removing the constant control system error, here it is set to 30 seconds, because in the procedure of system identification, converging time will not affect the result.

As for the outer loop, the idea for rough setting is quite similar, where the reference value is the target speed allowed for system self-stabilization. The relation can be illustrated as follows

$$\begin{cases} Kp_{out} \cdot e_{p\max} = 0.2Aim_{v\max} \\ Kd_{out} \cdot de_{p\max} = 0.02Aim_{v\max} \end{cases} \tag{2}$$

where Kp_{out} , Ki_{out} are the rough proportional, integral of outer controller; $e_{p\max}$ is the input limitation (position error limitation) for the outer loop controller, $de_{p\max}$ is the error differential limitation; $Aim_{v\max}$ is the allowed maximum movement velocity for position stabilization.

The formula(1) and (2) provide an estimation for rough control parameters to make sure the UAV can keep stable before tuning algorithm. After the calculation, an insurance procedure should be implemented by disabling the outer loop control and set target velocity to zero. Then the initial parameter will be validated when the UAV can be controlled stably with no vertical speed.

2.2 Neural network black-box system identification

Now, consider about the black box system identification block in figure 1, the control signal $u(k)$ and estimation velocity value $\hat{v}(k)$ will be used as the identification data. Considering the control plant as a nonlinear dynamic system, it can be represent as following discrete equation [14],

$$\begin{cases} x(k+1) = f[x(k), u(k), k] \\ v(k) = g[x(k), k] \end{cases} \tag{3}$$

where $x(k) \in R^n$, $u(k), y(k) \in R$, $f, g \in C$ are unknown function. If the linearized system around the equilibrium point is observable, the input-output equation can be represented by[15]:

$$v(k+1) = \beta[\boldsymbol{V}(k), \boldsymbol{U}(k)] \tag{4}$$

where

$$\begin{cases} V(k) = [v(k), v(k-1), \cdots, v(k-n+1)] \\ U(k) = [u(k), u(k-1), \cdots, u(k-n+1)] \end{cases} \quad (5)$$

Equation(4) means that there is a black box β that can map from the historical system outputs (historical speed values), current input and historical input to the current output value. Back Propagation Neural Network (BPNN) is a model that is used to illustrate the mapping. The training of model is shown in figure 3. The BPNN model mainly contains input layer, hidden layer and output layer. The length of buffers for control signal and UAV velocity determine the size of input layer.

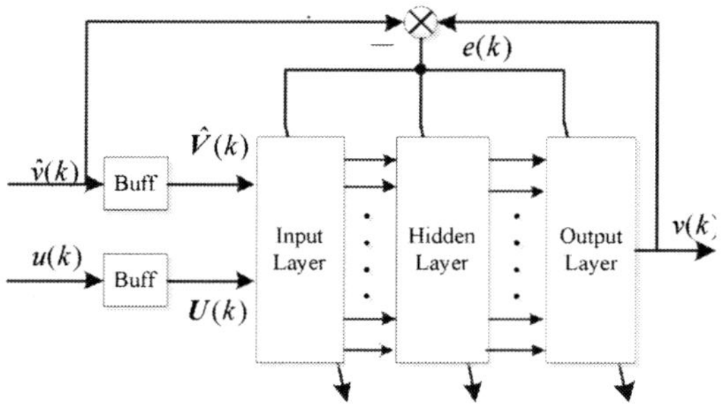

Fig. 2. Neural network system identification model.

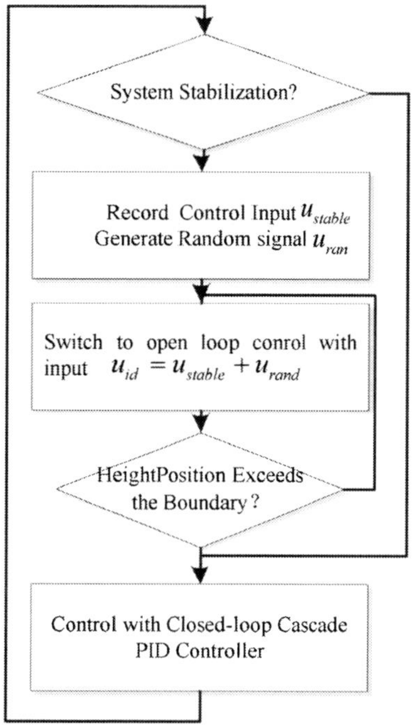

Fig. 3. Signal mixer logic block diagram.

The Data Set (DS) collecting is a crucial part for the neural network training. In order to cover a variety of input and motion situations[16], the Gauss random signal is added to a constant value as identification input signal. The control signal mixer can judge whether the aircraft has reached the steady state, and then switch to the identification signal to disturb the system. The logic for the signal mixer is presented in figure 3. There is a switch between the open-loop control and closed-loop control to ensure system safety. When the controller is switched to open-loop, the constant value used for mixer is the control input value which stabilizes the system around the equilibrium point. In this way, the constant value always compensates the drifting of system bias (it is caused by the consuming of electrical energy).

3 Auto tuning algorithm

The predominant idea of tuning algorithm is to obtain the relation (a function with explicit equation) between PID parameters and the control performance characteristics. For the selection of performance criteria, some previous existed controllers had concerned about the physical meanings of performance characteristics, no matter in auto-tuning [17] or experimental tuning [18] . The value of overshoot reflects the stability and security of a system [19]. The settling time is another important performance characteristic of the control system, which reflects the response speed of the controller, it is often referred in systems requiring dynamic tracking [20]. Integral square error(ISE) can also reflect the stability of a system, which is a more comprehensive value. ISE considers both the time and tracking error. Moreover it reflects the total energy consuming during the control process [21], so it is a frequently-used criteria in control design and tuning.

The proposed tuner synthetically utilizes the above three characteristics. Since the importance of these performance characteristics are different to the inner loop and outer loop of the proposed controller, the cost function will be designed depending on all of them. The physical meaning of performance characteristics is not the only one that needs to be considered about in cost function choosing, but the precision of fitting and extreme value problem should also be taken into account.

3.1 Objective functions for optimization

As for the inner loop, the speed of response is the main concern, so the settling time state is the most important criteria to the PID controller. It is not necessary to overemphasis the stability of the inner loop, and the inner loop has one more parameter (integral parameter), which means it needs to reduce the computation. Based on that, the overshoot is used as the restriction So the optimization problem of inner loop controller is:

$$\begin{cases} min & t_s = w(P_{in}, I_{in}, D_{in}) \\ subject\,to & e_{over} = z(P_{in}, I_{in}, D_{in}) < 0.1r_v \end{cases} \tag{6}$$

where w and z are the functions of settling time and overshoot with respect to PID controller parameters. This objective function will lead to a minimized settling time t_s with the overshoot e_{over} no more than 10% of the target velocity.

As for the outer loop, it determines the final overall performance of the UAV height control. Even though the inner loop control still has some defects, such as the settling time and overshoot can not be satisfied at the same time, the optimal outer loop control parameters can make sure the control system performs as desired(it can set reasonable target speed for inner loop). The objective function of outer loop is:

$$\begin{cases} min & t_s = y(P_{out}, D_{out}) \\ subject\,to & e_c = \varphi(P_{out}, D_{out}) < 0.05r_p^2 \end{cases} \tag{7}$$

where the settling time t_s is still the key criteria, y and φ are the functions of settling time and ISE with respect to PD parameters. The aim of this objective function is to ensure the response speed, smoothness and stability simultaneously.

3.2 Performance characteristic functions regression

Based on the black-box system model, two groups of Data Sets are created by controlling the plant model in a virtual environment. The first group is generated with a step response on input-output speed model, which is used for inner loop performance functions regression and tuning.

Choosing an appropriate regression hypothesis can ensure the fitting accuracy. We have tried several hypothesis including linear regression model and nonlinear regression model. Commonly, a linear model should be tested prior to a nonlinear one because it is time saving in computation(no iteration is needed). The model for inner loop overshoot is a linear regression model, the fitting results show that the precision is sufficient for optimization limitation. This linear regression model is:

$$e_{over} = z(P_{in}, I_{in}, D_{in}) = \sum_{i=1}^{M} \alpha_i f_i (P_{in}, I_{in}, D_{in}) \tag{8}$$

where the functions f_i are square polynomials, logarithms and exponential functions. The α_i coefficient of these functions can be calculated by least square method. According to the data set of inner loop:

$$\boldsymbol{E} = \begin{bmatrix} e^1_{over} \\ e^2_{over} \\ \vdots \\ e^n_{over} \end{bmatrix} \quad \boldsymbol{F} = \begin{bmatrix} f_1^1 & f_2^1 & \cdots & f_M^1 \\ f_1^2 & f_2^2 & \cdots & f_M^2 \\ \vdots & \vdots & \vdots & \vdots \\ f_1^n & f_2^n & \cdots & f_M^n \end{bmatrix} \quad \boldsymbol{\alpha} = \begin{bmatrix} \alpha_1 \\ \alpha_2 \\ \vdots \\ \alpha_M \end{bmatrix} \tag{9}$$

$$\boldsymbol{E} = \boldsymbol{F\alpha} \tag{10}$$

Solve the equation with least square method [22], the hypothesis coefficients is obtained:

$$\boldsymbol{\alpha} = \left(\boldsymbol{F}^T \boldsymbol{F}\right)^{-1} \boldsymbol{F}^T \boldsymbol{E} \tag{11}$$

However, for the regression for settling time function, linear regression does not have an ideal result in practical application. In order to realize a highly fitted regression, a non-linear regression model is used to improve the accuracy:

$$t_{stable} = w(P_{in}, I_{in}, D_{in}) = \frac{\sum_{i=1}^{N} \beta_i z_i (P_{in}, I_{in}, D_{in})}{\sum_{i=1}^{N} \eta_i z_i (P_{in}, I_{in}, D_{in})} \tag{12}$$

This is a ratio model which can better reflect complex relationships. The drawback is that it need iteration computation to solve the problem of regression, Gauss-Newton method is applied in the solution [23], the algorithm is illustrated in table 1.

Table 1. Gauss-Newton integration algorithm.

1. Set initial parameters $\boldsymbol{\gamma}^1 = [\boldsymbol{\beta}^1 \; ; \; \boldsymbol{\eta}^1]$ and converging threshold $\boldsymbol{\varepsilon}$.
2. Calculating residual error vectors by all data $\boldsymbol{r}^k = [t^1 \quad t^2 \quad \cdots \quad t^n] - [w_k^{\ 1} \quad w_k^{\ 2} \quad \cdots \quad w_k^{\ n}]$.
3. Compute Jacobi matrix $\boldsymbol{J}^k$ of model w^k with respect to $\boldsymbol{\gamma}$.
4. Solve $\boldsymbol{\gamma} - \boldsymbol{\gamma}^k = -(\boldsymbol{J}^{kT} \boldsymbol{J}^k)^{-1} \boldsymbol{J}^{kT} \boldsymbol{J}^k \boldsymbol{r}^k$,obtain $\boldsymbol{\gamma}^{k+1}$.
5. if $\left\Vert \boldsymbol{\gamma}^{k+1} - \boldsymbol{\gamma}^k \right\Vert < \boldsymbol{\varepsilon}$, break; else, repeat 2,3,4and5.

As for the outer loop, the structure is similar to that of the inner loop. The outer loop only has PD parameters, and the settling time is used for the cost function, the regression model is:

$$t_{stable} = y(P_{out}, D_{out}) = \frac{\sum_{i=1}^{N} \beta_i z_i (P_{out}, D_{out})}{\sum_{i=1}^{N} \eta_i z_i (P_{out}, D_{out})} \tag{13}$$

The limitation function for outer loop refers to the ISE value, data set of ISE in control process is calculated by:

$$e_c = \frac{1}{k_s - k_0} \sum_{k=k_0}^{k_s} e^2 (k, P_{out}, D_{out}) \tag{14}$$

where $e(k, P_{out}, D_{out}) = r_p(k) - p(k)$ is the control error of step response. k_0 is the discrete time point when system reaches its peak, k_s is the point of settling time. Similarly, since the requirement for limitation function is not demanding, a linear regression model is enough:

$$e_c = \varphi(P_{out}, D_{out}) = \sum_{i=1}^{M} \alpha_i f_i(P_{out}, D_{out}) \tag{15}$$

Regarding the importance of the outer loop control parameters, the optimization algorithm here will be improved to enhance the accuracy. An iterative gradual regression and optimization method is a good one which can maintain the efficiency of the algorithm.

Table 2. Iterative regression and optimization algorithm.

1. Initialize parameter varying step $\boldsymbol{l}_1$, controller parameters $\boldsymbol{\theta}_1$, converging threshold $\boldsymbol{\delta}$.
2. In the boundary $[\boldsymbol{\theta}_k - 100\boldsymbol{l}_k, \boldsymbol{\theta}_k + 100\boldsymbol{l}_k]$, changing control parameter by step $\boldsymbol{l}_k$, generate data set of control performance characteristics.
3. Regression algorithm, get $y(P_{out}, D_{out})$ and $\varphi(P_{out}, D_{out})$; Optimization algorithm, get $\boldsymbol{\theta}_{k+1}$.
4. if $\lVert\boldsymbol{\theta}_{k+1} - \boldsymbol{\theta}_k\rVert < \boldsymbol{\delta}$, break; else, $\boldsymbol{l}_{k+1} = 0.1\boldsymbol{l}_k$, repeat 2,3 and 4.

As shown in table 2, firstly, the rough control parameters $\boldsymbol{\theta} = [Kp\ ,Kd\ ]^T$ calculated by equation(2) are used as the initial parameters of the optimization algorithm, and a relatively larger vector $\boldsymbol{l}_1$ is set as initial varying step , and a certain threshold (which is small enough to ensure the accuracy) is given as the break condition. By gradual decreasing the length of boundary and increase the sample density, less point of data set is required, that means there is less step responses is generated and evaluated in the tuning algorithm.

4 Experimental validation

The quadrotor-craft consists of four main components: RC transmitter/receiver, IMU/GPS/barometer sensor navigation system, micro-processor and actuators. The RC transmitter is to send control-position information and receive controller parameters. The navigation system provides the sensor information for aircraft position and attitude estimation. The micro-processor is to process data and generate control signals. Actuators are to generate thrust and torques, which consist of motor drives, DC motors, gearboxes and blades. The experimental platform is shown in figure 4.

Fig. 4. Rotorcraft for flight experiment.

4.1 Fitting and optimization

The performance characteristics functions of the inner loop (settling time and overshoot) have three variables (Kp, Ki and Kd), so the 3D thermal chart is used to show the relation between controller parameters and performance characteristics, which is shown in figure 5. The brightness of a point repents the magnitude of a dependent variable value, it can be seen that the figure of inner settling time function has a brighter outside, it's a convex function which can be used for extreme point seeking; The overshoot function is not a convex one, but it's used for limitation. The regression Root Mean Squared Error (RMSE), optimal controller parameters, step response settling time and overshoot of inner loop are presented in table 3.

Table 3. Inner loop regression and optimization result.

Regression RMES		Controller Parameters			Performance result	
Settling(s)	Overshoot(%)	*Kp*	*Ki*	*Kd*	Settling(s)	Overshoot (%)
0.17	0.2	0.71	0.033	-0.73	2.08	7

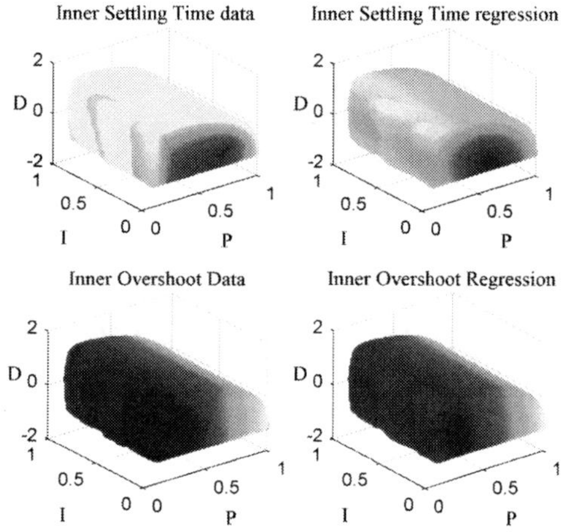

Fig. 5. Inner loop control performance regression.

The optimized inner loop parameters can perform well in the test on a velocity model. Then it is fixed when outer parameters is optimized on a position model(integral of velocity model). For this iteration method, the dataset generation, regression and optimization will be repeated in the extremum seeking process. A rational initial setting will ensure the convergence of the algorithm. The regression result of PD controller performance characteristics (settling time and Integral Square Error) are shown in figure 6. Because of limited regression accuracy, the result of first step may be not the most ideal one. Figure 7 shows the difference value of parameter optimization results of each step, the results converged after about 200 steps (about half a minutes). The ideal PD parameters lead to a settling time of 4.3 second and ISE 0.009(normalized) in simulation on the black box neural network system model.

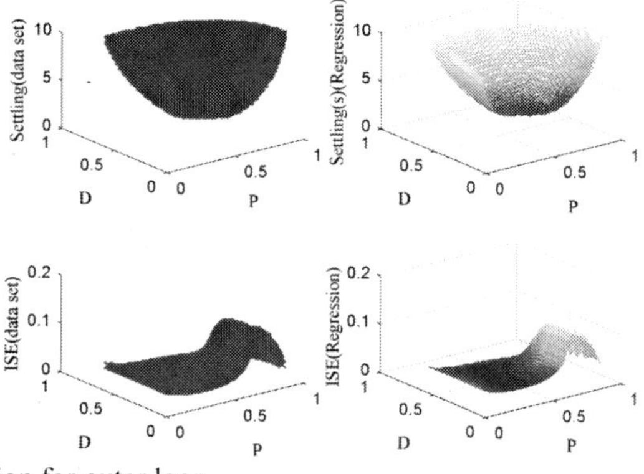

Fig. 6. First step regression for outer loop.

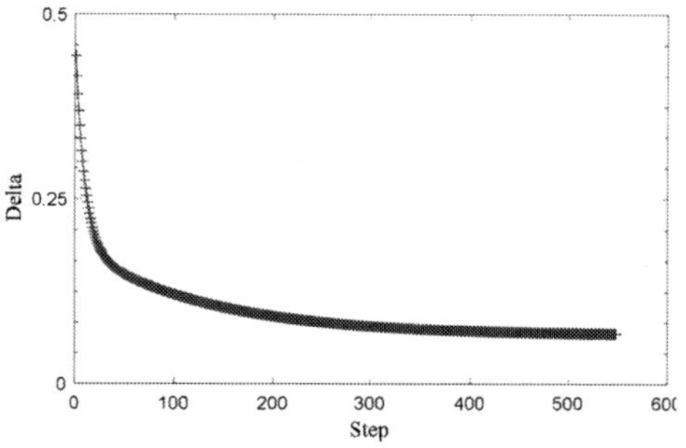

Fig. 7. Iterations result.

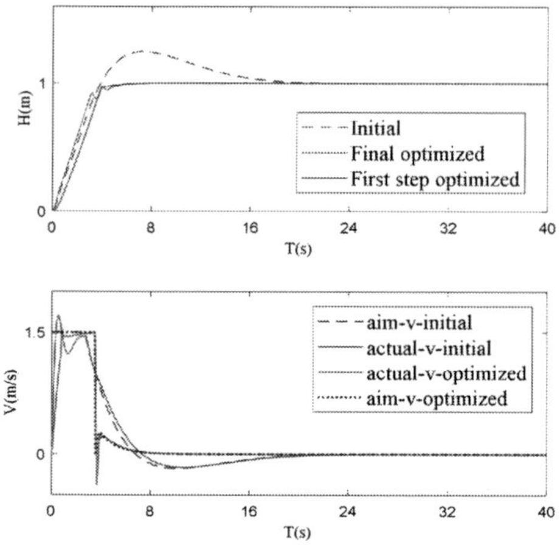

Fig. 8. Outer loop step response.

The optimized parameters are verified with step response, which is shown in figure 8. The first chart compares the control results by initial parameters, parameters optimized by first regression step, and parameters of final tuning result. The second chart illustrates the variation of target speed and actual speed tracking in the control process. It is worth mentioning that the optimized result of first step has reduced the settling time and ISE value significantly, however, after the iterated optimization procedure, the final group of parameters given by tuner are more ideal than that of the first step.

4.2 Parameters application on practical UAV system

The rotorcraft is controlled with designed cascade PID controller and flying in an outdoor open space. Before the tuner works, the UAV is given a varying target height, the height and velocity tracking result is shown in figure 9. The flight tracking performance is improved after the auto-tuning algorithm, which is shown in figure 10. It is apparent that both in the outer loop (height tracking) and inner loop(velocity tracking), the response speed is increased. The tracking error is also minished, the height tacking RMSE has been declined from 0.53 to 0.25, which is half less than before, and the velocity tacking RMSE has been declined from 0.24 to 0.085.

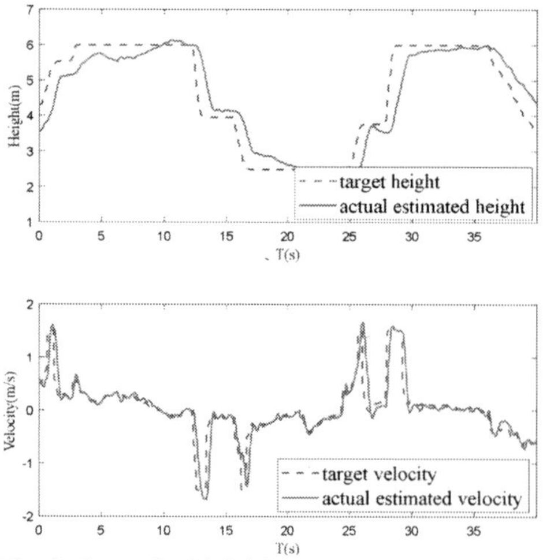

Fig. 9. Control with initial parameters.

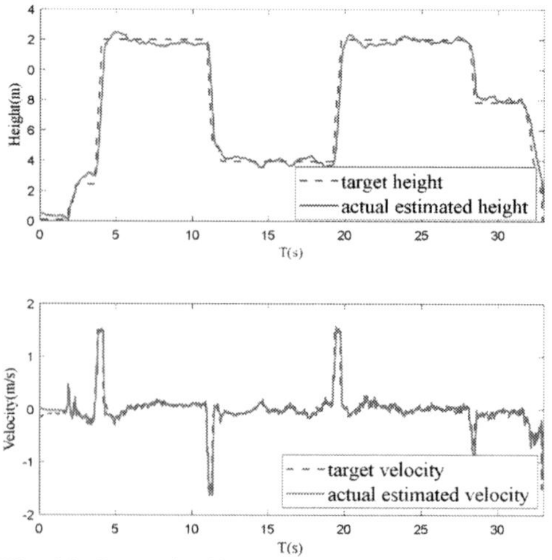

Fig. 10. Control with optimized parameters.

Table 4. Control results comparison.

Initial tracking RMSE			Optimized tacking RMSE	
$H(\sqrt{m^2})$	$V(\sqrt{m/s}^2)$		$H(\sqrt{m^2})$	$V(\sqrt{m/s}^2)$
0.53	0.24		0.25	0.085

5 Conclusion

This paper proposed a safe auto tuner with high computation efficiency. The foundation of this tuner is a reasonable black box system identification implemented on the flying machine. The ultimate aim of the tuner is to realize desired physical performance, by generating control Data Set based on the system model, the mapping from controller parameters to performance characteristics is obtained. Specifically, in order to improve the optimization accuracy and efficiency, a gradual regression and extremum seeking method is applied as the key process for the tuning algorithm. A direct application of the auto-tuning algorithm is shown in the cascade control of a self-developed laboratory quadrotor platform, followed by fight experimental evaluation in the real environment.

References

1. Astrom, K. J., Hagglund, T. Automatic tuning of simple regulators with specifications on phase and amplitude margins[J]. Automatic, 1984, 20(5):645-651.
2. Rivera D E . PID Control: New Identification and Design Methods - [Book Review][J]. Control Systems IEEE, 2006, 26(1):95-97.
3. Yu C C . Autotuning of PID Controllers[M]. Springer London, 2006.
4. Hang C C , Astrom K J , Ho W K . Refinements of the Ziegler-Nichols Tuning Formula[J]. IEEE Proceedings-D, 1991, 38(2):111-118.
5. Astrom K J , Hagglund T , Hang C C , et al. Automatic tuning and adaptation for PID controllers - a survey[J]. Control Engineering Practice, 1993, 1(4):699-714..
6. Amir Torabi, 2 Amin Adine Ahari, 3 Ali Karsaz, 4 Seyyed Hossin Kazemi Intelligent pitch controller identification and design
7. Lian F L, Moyne W, Tilbury D. Analysis and modeling of networked control systems: MIMO case with multiple time delays[C]. American Control Conference, 2001. Proceedings of the. IEEE, 2001:4306-4312 vol.6.
8. Ho W K , Hang C C , Cao L S . Tuning of PID controllers based on gain and phase margin specifications[J]. Automatica, 1995, 31(3):497-502.
9. Koivo H N , Tanttu J T . Tuning of PID Controllers: Survey of SiSo and MiMo Techniques[J]. 1991.
10. Leva A . PID autotuning algorithm based on relay feedback[J]. IEEE Proceedings-D, 1991, 38..
11. Chen X , Wang L . Quadrotor cascade PID controller automatic tuning[C]. Control Conference. IEEE, 2017.
12. Zhuang M, Atherton D P. Automatic tuning of optimum PID controllers[J]. 1993, 140(3):216-224.
13. Truong N V . Nonlinear System Identification[J]. Journal of Northwestern Polytechnical University, 1986, 21(1):109-122.

14. Levin A U, Narendra K S. Control of nonlinear dynamical systems using neural networks: controllability and stabilization.[J]. IEEE Transactions on Neural Networks, 1993, 4(2):192-206.
15. Levin A U . Control of Nonlinear Dynamical Systems Using Neural Networks-Part II : Observability, Identification, and Control[J]. IEEE Transactions on Neural Networks, 1996, 7(1):30-42.
16. Haber R , Unbehauen H . Structure identification of nonlinear dynamic systems—A survey on input/output approaches[J]. Automatica, 1990, 26(4):651-677.
17. Concepcion A. Monje, Vinagre B M , Feliu V , et al. Tuning and auto-tuning of fractional order controllers for industry applications[J]. Control Engineering Practice, 2008, 16(7):798-812.
18. Wang Q G . PID Tuning for Improved Performance[J]. IEEE Trans.contr.syst.technol, 1999, 7(4):457-465.
19. Shamsuzzoha M , Skogestad S . The setpoint overshoot method: A simple and fast closed-loop approach for PID tuning[J]. Journal of Process Control, 2010, 20(10):1220-1234.
20. Killingsworth N J , Krstic M . PID tuning using extremum seeking - Online, model-free performance optimization[J]. Control Systems IEEE, 2006, 26(1):70-79.
21. Killingsworth, Krstic. PID Tuning Using Extremum Seeking[J]. IEEE Control Systems Magazine, vol. 26, no. 1, February 1, 2006, pp. 70-79, 2005.
22. Gong Z W. Least-square method to priority of the fuzzy preference relations with incomplete information[M]. Elsevier Science Inc. 2008.
23. Cornell J . Classical and Modern Regression With Applications[J]. Technometrics, 1987, 29(3):2.

Research on IFLP optimization model for carbon emission reduction

Qiannan GUO*

Northeast Forestry University, China

Abstract. This study proposed an optimization model combined with Binding Interval Linear Programming (ILP) and Fuzzy linear programming (FLP) Methods and further analyzed from the views of economic output, energy consumption, carbon dioxide emission and emission cost. The optimization model results demonstrated that the heavy energy consumption industries will significantly reduce, and the target of the carbon dioxide emissions per unit of GDP reduction decrease by 40%-45% from 2005 to 2020. In addition, the economic development model will trend to optimize the allocation of resources and green economy. However, a single low-carbon economic policy will always have shortcomings and low efficiency in emission reduction. Hence, the government should combine all emission reduction policies together and establish a low-carbon economy system that practice production activities need.

1 Research background

China puts a positive spin on reducing CO_2 emissions by introducing reduction, capture and storage technologies and issuing related policies on these technical means [1]. However, the domestic resource structure and production mode are hindering CO_2 emissions reduction, causing contradiction between rapid economic growth and environmental protection [2].

To balance the economy-environment-energy relationship, the system feedback is simulated as an adjustable mechanism in the model or external adjustment [3-5], and the interval linear programming (ILP) is used to handle the uncertain information. To address the subjective uncertainty, the fuzzy linear programming (FLP) is introduced for economic restructuring and energy planning as the probability distribution or membership extent of variables (or parameters) are hard to obtain in the real world [6-9]. ILP and FLP were coupled to cope with a considerable amount of uncertain information among economy, energy and environment, leading to more stable decision-making results of the optimization model [10, 11].

2 Determination of objective function and constraints

Aiming to minimize the cost of CO_2 emission reduction in the target year, the target total CO_2 emissions, energy consumption, domestic economic growth and industrial economic

* Corresponding author: 574943180@qq.com

growth are selected as constraints in this study[12]. The satisfaction should be precisely limited by IFLP method with certain confidence level.

2.1 Setting of objective function

$$Min\ f = \sum_{i=1}^{I}\sum_{j=1}^{J} \alpha \cdot XGO_i^{\pm} \cdot UEC_{ij}^{\pm} \cdot \widetilde{EC}_{ij}^{\pm} \cdot \theta_j^{\pm} \tag{1}$$

In the formula, f is total cost of CO_2 emission reduction in 2020 (in one hundred million Yuan); i is industry, with 27 in total; j is fossil energy, with j=1 as coal, j=2 as natural gas, j=3 as oil and j=4 as electricity; XGO_i is target output value of each industry in 2020 (in one hundred million Yuan), which is the decision-making variable; UEC_{ij} is consumption of energy j per unit output value of industry i (in one million tons of standard coal); α-1 is CO_2 emissions per one million tons of standard coal (in one million tons of CO_2/one million tons of standard coal), with the value of 2.4567 tons of CO_2 per one ton of standard coal recommended by Energy Research Institute, National Development and Reform Commission; EC_{ij} is cost of CO_2 emissions reduction for energy j in industry i (in one hundred million Yuan/one million tons); θ_j is ratio of target CO_2 emission reduction of industry in 2020.

2.2 Determination of constraints

- Constraint on target total CO2 emissions:

China unveiled at the Copenhagen conference its target to reduce greenhouse gases emitted per unit GDP by 40-45% from that in the benchmark year, reaching the 2020 control requirement on total CO2 emissions after industrial restructuring.

$$\sum_{i=1}^{I}\sum_{j=1}^{J} \alpha^{\pm} \cdot UEC_{ij}^{\pm} \cdot XGO_i^{\pm} \cdot 10^{-4} \le \left(1-\beta^{\pm}\right) \cdot BCI^{\pm} \cdot TGDP^{\pm} \tag{2}$$

In the formula, β is target CO_2 emissions reduced per unit GDP in 2020, which is 40%-45%; *BCI* is CO_2 emission intensity per unit GDP (in one million tons of CO_2/one hundred million Yuan per unit GDP) in 2005; TGDP is target GDP in 2020 (in one hundred million Yuan).

- Constraint on target total energy consumption:

According to the *Notice of Strategic Action Plan for Energy Development (2014-2020)* released by the General Office of the State Council ,the primary energy consumption will be capped at 4.8 billion tons of standard coal by 2020, of which the fossil energy should be less than 85%, about 4.08 billion tons.

$$\sum_{i=1}^{I}\sum_{j=1}^{J} UEC_{ij}^{\pm} \cdot XGO_i^{\pm} \cdot 10^{-4} \le TEC^{\pm} \tag{3}$$

In the formula, TEC is target total fossil energy consumption in 2020 (in one million tons of standard coal).

- Constraint on domestic economic growth:

The sum of output values of all industries is expected to meet the economic growth demand in China. According to the planning requirements for overall economic development in the future, this constraint is to avoid the decline in economic growth through model optimization.

$$\sum_{i=1}^{I} XGO_i^{\pm} \geq \left(1+\delta^{\pm}\right)^5 \cdot IGDP \tag{4}$$

In the formula, δ is target domestic annual economic growth (%); IGDP is domestic GDP in 2015 (in one hundred million Yuan).

- Constraint on gross industrial economic growth:

The constraint on gross industrial economic growth is that industries should appropriately adjust its economic growth within a certain range.

$$XGO_i^{\pm} \geq LGO_i^{\pm} \text{ and } XGO_i^{\pm} \leq UGO_i^{\pm}, \forall i \tag{5}$$

In the formula, LGO is lowest gross output (in one hundred million Yuan); UGO is uppermost gross output (in one hundred million Yuan).

- Constraint on economic growth of each industry:

Constraints on energy consumption in various industries means that optimizing the composition of energy consumption needs to meet the energy needs of the industry without affecting economic development.

$$XGO_i^{\pm} \geq DIE_i^{\pm} \text{ and } XGO_i^{\pm} \leq UIE_i^{\pm}, \forall i \tag{6}$$

In the formula, DIE is the lowest industry output (in one hundred million Yuan); UIE is uppermost industry output (in one hundred million Yuan).

- Non-negative constraint:

$$XGO_i^{\pm} \geq 0, \forall i \tag{7}$$

3 Analysis of model calculations

3.1 Analysis of economic output structure

As shown in Figure 1, 27 industries are classified into energy intensive dependence, equipment manufacturing and other textile ones respectively. Optimization model results reveal that the target year (2020) will see economic consumption ratio of energy intensive dependence industries significantly decreasing [12.59, 14.61] %, and of equipment manufacturing and other textile industries rising [5.6, 7.25] % and [5.31, 9.01] %, respectively, from the 2010 levels. Therefore, the IFLP optimization model can reasonably adjust the economic structure and provide more stable decision-making and guidance for improving the modern energy market system and deepening the key emission reduction industries.

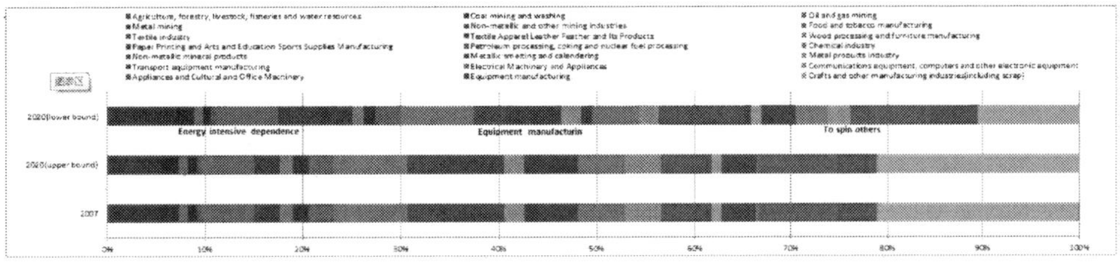

Fig. 1. Uncertainty ratio analysis of economic output in main industries in 2010 and the target year (2020)

3.2 Analysis of energy consumption structure

China's energy consumption ranking is coal > oil > electricity > natural gas. According to the optimization model, the total energy consumption in 2020 is [47.57, 47.78] million tons, and the optimization results meet the national planning requirements within a moderate range. As shown in Figure 2, model optimization indicates that energy consumption ranking will be constant in 2020. However, as top three total energy-consuming industries ,oil processing, coking and nuclear fuel processing industry, metal smelting and rolling processing industry, and electricity and heat production and supply industry accounted for 68.82% coal consumption in 2010.Their energy consumption decreased [29.38, 32.45] %. Meanwhile, after adjusting the deterministic model using IFLP optimization model, the energy consumption of four industries including transportation equipment manufacturing; communications equipment; computers and other electronic equipment; transportation, storage and post, will rise significantly [14.33, 15.81] % from the 2010 level until 2020.

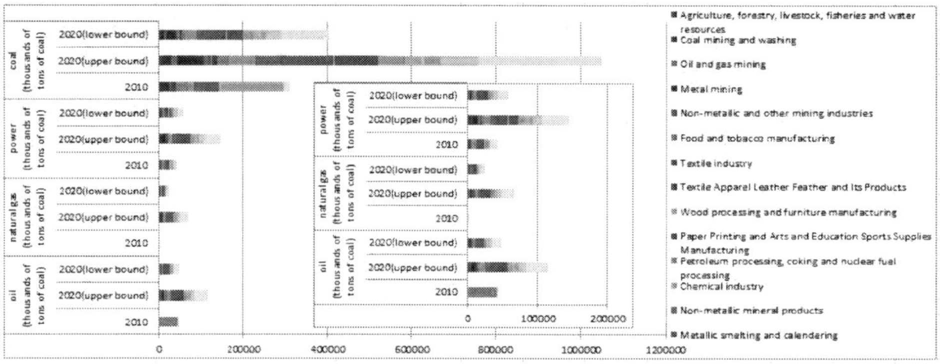

Fig. 2. Uncertainty ratio analysis of energy consumption in 2010 and the target year (2020)

Economic restructuring puts a positive spin on energy restructuring and blindly pursuing carbon emission reduction in the energy structure will lead to economic disorder. Thus, adjusting of economic output structure and energy consumption structure should be limited within a certain range. In this paper, the IFLP optimization model for carbon emission reduction has offered a more stable decision basis for the relationship among economic structure, energy consumption and carbon emission reduction, making results more reliable.

3.3 Analysis of CO_2 emissions

Considering data of 2010 as an example, China's CO_2 emissions level is consistent with its energy consumption ranking. About 90% of emission came from burning fossil fuels, which indicate a strong positive correlation between them. This positive correlation is related to China's current energy structure dominated by fossil energy. China's policy for CO_2 emission reduction aims to promote clean and efficient exploitation and utilization of coal. China will emit [341, 372] billion tons of CO_2 in 2020 resulted by optimization model (shown in Figure 3). This target can be achieved adequately if the guidance of this model is strictly implemented.

Relevant policies point out that promotion of the natural gas industry and alternative energy should be made continuously. In the IFLP optimization model for carbon emission reduction, it can be projected that by 2020, the energy use ratios of natural gas and electricity will rise by [1.85, 2.03] % and [2.40, 2.68] %, respectively, from 2010 to 2020. According to the *Reply of the State Council on China's National Plan for Climate Change (2014-2020)* (G.H. [2014] No. 126) by 2020, the CO_2 emissions per unit GDP in 2020 will fall almost 40-45% compared with 2005 : the ratios of using non-fossil fuel resources in primary energy, forest area and storage volume can reach the standards expected in the optimized deterministic model. This model is devoted to transforming energy intensive dependence industries which contribute higher emissions to equipment manufacturing industries and other textile industries.

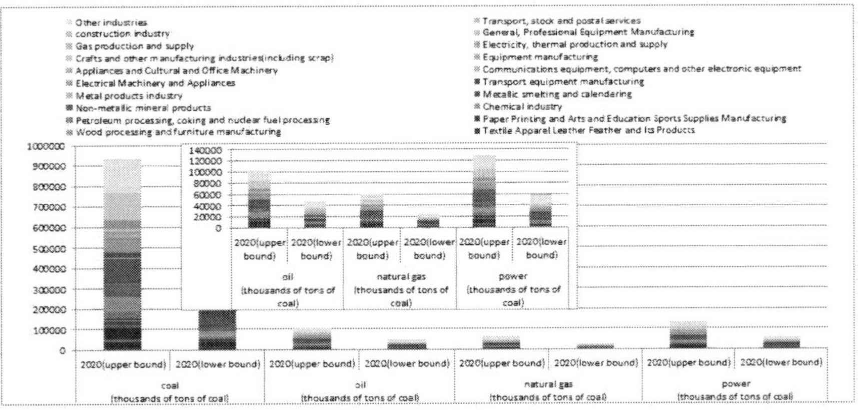

Fig. 3. Uncertainty analysis of CO_2 emissions in the target year (2020)

3.4 Analysis of cost for CO_2 emission reduction

Reducing the cost of carbon dioxide emission reduction can better inhibit the non-benign development of the company. Therefore, it is necessary to increase the technological investment in petrochemical and other high-energy-consuming industries, accelerate the transformation and upgrading of technology equipment, reduce energy consumption, improve the utilization efficiency, and reduce the carbon dioxide emissions. It has made significant contributions to the environmental protection and control of the malignant development of high-emission companies. As shown in Figure 4, according to the IFLP optimization model results, the chemical industry, non-metallic mineral product industry, metal smelting and rolling processing industry, construction industry, and transportation, storage and post industry are high-cost industries for carbon emission reduction, accounting for [57.20, 58.00] %. Thus, they are all industries with potential for emission reduction.

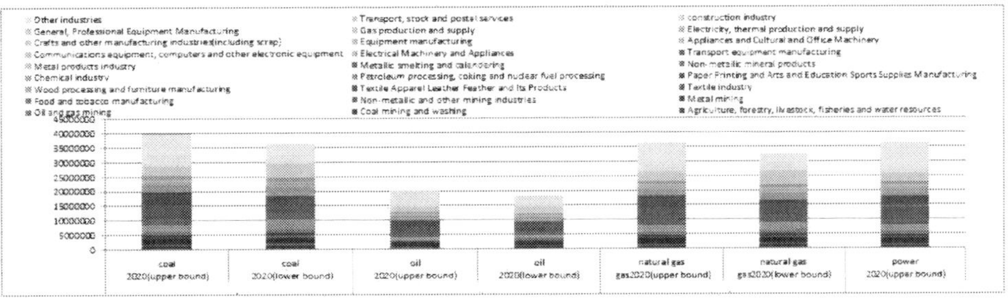

Fig. 4. Cost uncertainty measurement analysis of CO_2 emissions in the target year (2020)

4 Conclusion

In this paper, based on low-carbon economic development and IFLP theory, the following main options can be concluded:

- In the target year (2020), the energy consumption in intensive dependence industries should have a significant adjustment, from extensive economic growth to relying on the optimal allocation of resources and clean energy. The optimization model enables the market economy to play a guiding role in allocating resources. Precisely, adjusting to improve the equipment manufacturing and other textile industries fully verifies the optimization of resources.

- China's energy consumption ranking in 2010 was coal > oil > electricity > natural gas, which will be unchanged in 2020 after optimization. In the national energy strategy for development, it is emphatically pointed out that in order to promote the basic principles of clean and efficient development and utilization of coal, domestic production of petroleum should be steadily increased, and natural gas should be vigorously developed to replace energy.
- China's energy consumption ranking in 2020 is expected to be fixed, although the consumption of natural gas and electricity will rise. The optimization model is devoted to transforming energy intensive chemical industries which contribute higher emissions to equipment manufacturing industries and other textile industries.
- The cost measurement analysis of CO_2 emission reduction is the scientific basis for formulating and implementing carbon emission reduction policy. Nowadays, China relies heavily on fossil energy sector and the forced emission reduction will bring a serious impact on macro economy, i.e. CO_2 emission reduction has a high macroeconomic price or cost.

References

1. Zou Wei, Hu Li, Lin Shumin, et al. Intrinsic driving force of rural land integrated consolidation [J]. Chinese Society of Agricultural Engineering, 2013, 29(8): 224-231.
2. Li Yu'e, Ma Xin and He Xiaojia. The Progress on adaptation negotiations since adoption of bali action plan and future needs analysis [J]. Advances in Climate Change Research, 2014, 10(2): 135-141.
3. The State Council. Circular of the State Council concerning issuance of the 12th five-year plan for energy conservation and emission reduction (G.F. [2012] No. 40) [J]. http.*//www. gov. cn/zwgk2012-08/21/*content_2207867. htm, 2012.
4. Cai Bofeng. Research on greenhouse gas emissions inventory in the cities of china [j]. china population, Resources and Environment, 2012, 22(1): 21-27.
5. Lin Boqiang and Liu Xiying. Carbon emissions at the urbanization stage: influencing factors and emission reduction strategies [J]. Economic Research Journal, 2010, 8(1): 22.
6. Du Tingting, Mao Feng and Luo Rui. Study on China's economic growth and co2 emissions [j]. china population, resources and environment, 2007, 17(2): 94-99.
7. Birge J R, Louveaux F. Introduction to stochastic programming[M]. Springer Science & Business Media, 2011.
8. Zeng Guangming and Huang Guohe. Uncertainty theory and methods in environmental science and engineering [M]. Changsha: Hunan University Press, 2002.
9. Chakraborty.D, Redefining chance-constrained programming in fuzzy environment, Fuzzy Sets and Systems 2002, (125), 327-333.
10. Yang N, Wen F. A chance constrained programming approach to transmission system expansion planning[J]. Electric Power Systems Research, 2005, 75(2): 171-177.
11. Nam K M, Selin N E, Reilly J M, et al. Measuring welfare loss caused by air pollution in Europe: A CGE analysis[J]. Energy Policy, 2010, 38(9): 5059-5071.
12. Li Zengfu and Zheng Youhuan. Research in the implementation mechanism of the "low -carbon city" [J]. Economic Geography, 2010 (6): 949-954.

Forecasting N_2O emission and nitrogen loss from swine manure composting based on BP neural network

Haotian Chen, *Shaoze* Sun, and *Baoli* Zhang*

College of Resources and Environmental Sciences, China Agricultural University, Beijing 100193, China

Abstract. Nitrogen loss and greenhouse gas emission during compost will cause secondary pollution and waste nutrients. To address this issue, a predictive model was set up to obtain a clear knowledge of the N_2O emission and nitrogen loss from swine manure composting. This paper collected 68 group data from 11 published papers about pig manure composting N_2O emission and total nitrogen loss. Select 4 indexes were taken as predicted indexes include aeration rate, moisture content, C/N, and the amount of superphosphate to establish a BP neural network for forecasting the N_2O emission and total nitrogen loss from composting. The analyses show that the mean error of N_2O emission forecasting model is 1.17; the value of MAPE is 138.85%. As for nitrogen loss, the mean error is 24.72 and the mean absolute percentage error is 11.06%. Compare to the traditional linear regression, the BP neural network model has good accuracy on forecasting N_2O emission and TN loss from manure composting. BP neural network has considerable application prospect in forecast nitrogen loss and greenhouse gas emission from composting.

1 Introduction

Composting is an effective way to reuse nutrients in livestock manure. However, N_2O emission from composting will cause secondary environmental pollution and also waste nitrogen in the manure. It was reported that 0.2–9.9% of initial total N is lost as N_2O from nitrification or denitrification [1-3]. Several studies have concluded that add superphosphate will fix nitrogen and reduce N_2O emission from manure composting [4-6]. Besides, other parameters during composting like C/N ratio, aeration rate and moisture content also have effect on N_2O emission and nitrogen loss [7, 8]. It is rather difficult to build a valid model for forecasting TN loss and N_2O emission of the composting system based on traditional model. Neural network is a mathematical model that is inspired by the structure and functional aspects of biological neural networks, it provides a new approach for the nonlinear problems. Artificial neural networks (ANNs) have received great interest in various research fields include agriculture [9] and environment science [10]. BP neural network, which have a neural propensity for storing experimental knowledge, is a neural

* Corresponding author: baoli@cau.edu.cn

network model that is most widely used in nonlinear forecast [11]. This paper presents a new method applying the technology of neural network to forecast the N_2O emission and nitrogen loss during the swine manure composting. With this model, scientists and policy makers can use this model to set a reasonable parameter for composting program.

2 The establishment of BP neural network model

2.1 Data extraction

Data were extracted from peer-reviewed journal papers published until July 2018. Totally 68 valid data of N_2O emission and nitrogen loss from 11 paper were included in this study. To minimize bias, all studies use swine manure and straw in composting. Detailed data are listed in table 1, giving the following information: proportion of N_2O-N emission on total nitrogen (%), TN loss rate (%), C/N ratio, Moisture content (%), aeration rate (L/kg min^{-1}, dry matter), and superphosphate content (%, dry matter).

Table 1. Data of nitrogen loss and N_2O emission.

N_2O-N /TN	TN loss	Moisture content	C/N	Aeration rate	SSP	Reference
0.68~1	14.3~39.2	67.6	15.4	0	0~13.2	[12]
0.04~0.65	24~50.8	55~65	15~21	0.12~0.36	10	[13]
1.5~7.3	18.4~45.9	65~75	15~21	0.24~0.72	0	[14]
0.73	33.01	70	20	0.38	0	[15]
1.37	13.6	65	22.5	0	0	[16]
1.5	43.9	52.2	25	0.25	0	[17]
0.24~0.55	25.44~42.34	60	15~30	0.24	0	[18]
0.15~0.35	30.50~42.34	70	15~30	0.5	0	[19]
0.03~0.24	29.5~45.2	55~65	15~21	0.12~0.36	0~24	[20]
1.58~6.69	18.45~54.54	63.4~75	15~21	0~0.72	0	[21]
0.5~4.36	19.9~44.13	63.7~70	17.3~18.4	0~0.54	0	[22]

2.2 Set the parameter of neural network model

In this study, we use three-layered neural network to establish forecasting model. Existing theory confirmed that arbitrary continuous functions could be approximated by three-layered neural network with feed forward structure [23]. The input variables of neural network model of nitrogen loss and N_2O emissions are C/N, moisture content, aeration rate and superphosphate content, so the number of input neurons is 3; the output parameters are the proportion of N_2O-N emission in total nitrogen (N_2O-N) and the proportion total nitrogen loss (TN loss), so the number of output neurons in the output layer is 2.The principle of forecast of neural network is described by a 4-N-2 ANN. That topology is shown in figure 1.

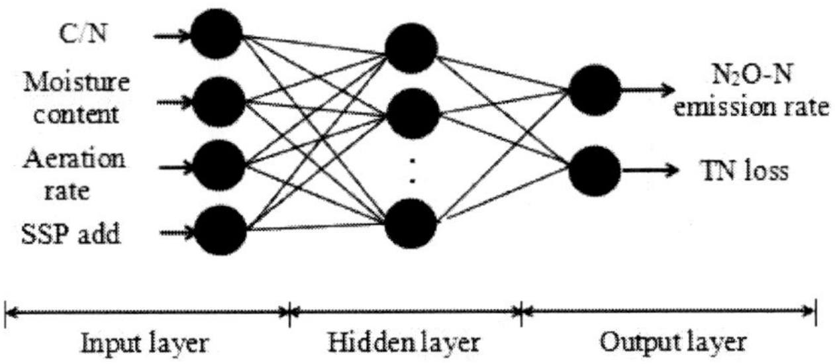

Fig. 1. The sketch map of BP network.

The neural network training program is processed under Matlab2016a. This paper used a fast training function Levenberg-Marquardt algorithm. This algorithm typically requires more memory but less time. Select 68 valid data, 70% of them (48 group) is used to training Neural Network, 15% is the share of testing data (10 group), the rest of them (10 group) is used to valid Neural Network. Other model parameters used in the BP neural network prediction model is shown in table 2.

Table 2. The BP neural network prediction model parameters.

The parameter name	Value / Option
Hidden layer nodes transfer function	tansig
Output layer neuron excitation function	purelin
Training function type	trainlm
Max failure	6
Max iteration number	1000
The network learning rate	0.001
Network training goal error	0.00001

2.3 The number of neurons in the hidden layer

The number of neurons in the hidden layer will affect the prediction precision of the model. Until now, it's still lack of theoretical basis to figure out the number of neurons in the hidden layer, researchers usually determined it empirically. In this paper, the numbojer of hidden layer nodes is initially set according to the below empirical formula:

$$P = \sqrt{m+n} + \alpha \tag{1}$$

P is the number of the hidden layer nodes; m represents the number of input layer nodes; n means the number of output nodes; α is a natural number between 1-10.

Under the MATLAB software platform and according to the training process above, the BP model was simulated and trained several times to find the number of hidden layers. Table 3 shows the prediction performance of BP neural network with changing of the hidden layer unit number. According to this Table, we found that when the hidden layer N=8, the function is more feasible, so in this paper the number of ANN hidden layer is selected to be 8.

Table 3. Prediction error under different hidden nodes of neural network.

	N=4	N=6	N=8	N=10	N=12
MSE	20.3570	14.6735	12.9411	13.2661	15.5303
R	0.6955	0.7849	0.8336	0.8037	0.7406

2.3.1 Error evaluation

In order to evaluate the obtained results, the mean absolute percentage error (MAPE) and mean squared error (MSE) are used, which is given in the below formula (2), (3). The error index provides information on the short-term performance and represents a measure of the variation of predicted values around the measured data.

$$\text{MAPE} = \frac{1}{n}\sum_{i=1}^{n}\left|\frac{y_i-\hat{y}_i}{y_i}\right| \times 100\% \tag{2}$$

$$\text{MSE} = \frac{1}{n}\sum_{i=1}^{n}(y_i-\hat{y}_i)^2 \tag{3}$$

where y_i is the measured value of power out of the system; $\hat{y}_i$ is the forecasting value of the power output; n represents the number of model forecasting samples. Different simulations have been performed in order to compare errors. And the error analysis is used to select the optimal number of hidden nodes.

3 Simulation and analysis

3.1 Regression value

The BP model synchronously predicts the training sample, validation sample, testing sample, and all samples while training by the regression equation. Figure 2 shows the regression results. The abscissa represents the target value; the ordinate represents the output value of the BP model. Regression R value is the regression coefficient, it measure the correlation between outputs and targets. The closer Regression value to 1, the better is the BP model and the higher is the prediction accuracy. In this paper training sample R=0.98588; validation sample R=0.97489; testing sample R=0.94657, the total R=0.97728. The regression results show prediction is feasible and reliable.

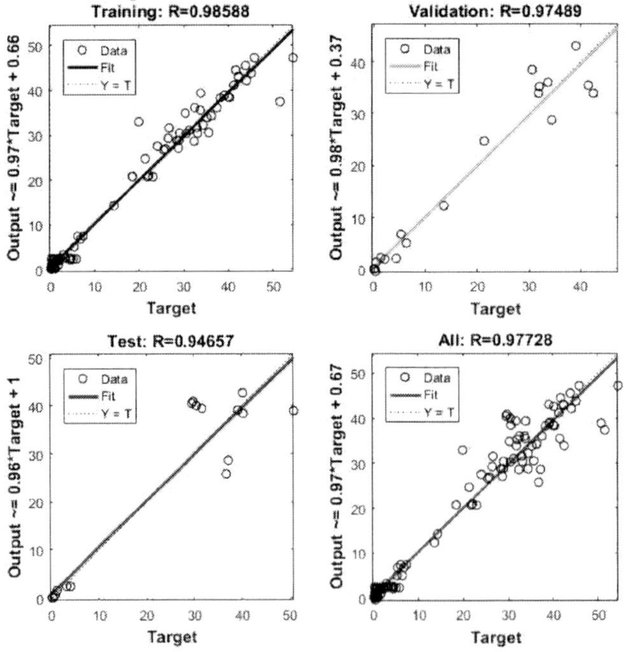

Fig. 2. Regression curve of model error.

3.2 The mean squared error

Mean Squared Error is the average squared difference between outputs and targets. Figure 3 shows the change of MSE during the process of net training, abscissa represents the number of iterations and the ordinate represents the mean squared error (MSE) between the model output value and the actual value. Mean square error (MSE) of the training sample becomes small gradually as iterative steps increase. Due to BP network's own restrictions, at the initial stage, the curve comes to faster convergence, at the latter it becomes more gently. When the forecasting model output value is close to actual value, the value of MSE is close to 0. The MSE performance of train, validation and test is respectively shown in blue, green and red line. The dotted line shows that the validation results are the most ideal after the 15th iteration. This dotted line represents one of the set targets for the BP model to stop training.

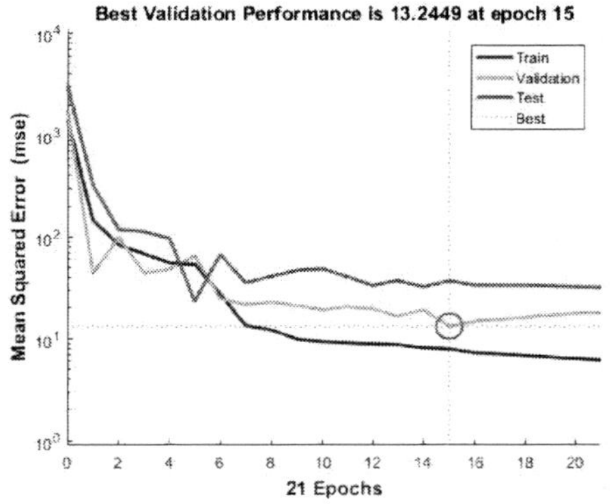

Fig. 3. Mean Squared Error of data in training process.

3.3 BP Fitting result

The predicted value and actual value of N_2O-N emissions and TN loss are shown in figure 4 and 5, respectively. Besides, we use traditional linear regression to forecast N_2O-N emissions and TN loss, the performance comparison is show in Table 4. It can be observed that in BP model, the predicted values are quite close to the actual. As for TN loss rate, the mean squared error (MSE) of Neural Network is 24.72, the mean absolute percentage error (MAPE) between the predicted and actual values arrived at 11.06%. Better than linear regression, which the mean squared error (MSE) is 19.28 and the mean absolute percentage error (MAPE) 50.49%. For N_2O-N emission predicts, the mean squared error (MSE) of Neural Network is 1.17, the mean absolute percentage error (MAPE) between the predicted and actual values arrived at 138.85%. Better than linear regression, which the mean squared error (MSE) is 1.43 and the mean absolute percentage error (MAPE) 250.82%.

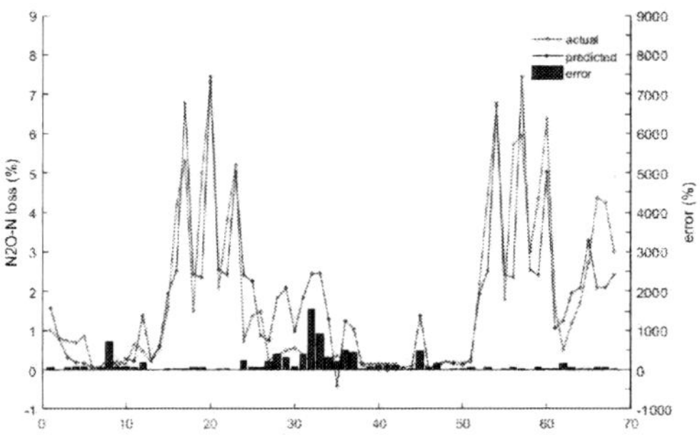

Fig. 4. Compensation results of N_2O-N emission.

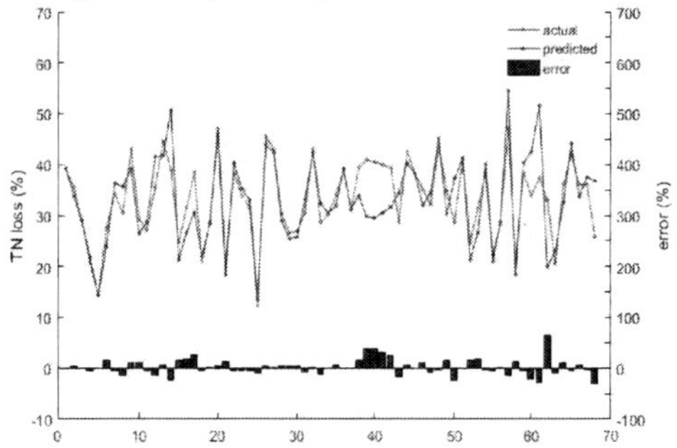

Fig. 5. Compensation results of TN loss.

Table 4. The performance comparison between two prediction models.

	Model	N_2O-N emission	TN loss
MAPE (%)	BP network	138.85	11.06
	Linear regression	250.82	19.28
MSE	BP network	1.17	24.72
	Linear regression	1.43	50.49

4 Conclusion

Compare to traditional linear regression, the BP neural network prediction values have a good agreement with the experimental test values. The mean squared error (MSE) of nitrogen in BP model is 24.72; the mean absolute percentage error (MAPE) is 11.06%, which is presented to explain the prediction is feasible and reliable. However, the mean absolute percentage errors of N_2O emission arrive at 138.85%, which is impermissible in the engineering application. In this study, the experimental data only involves 4 indexes, there is a further need more indexes such as initial pH and the volume of vessel, which also influence swine manure composting program. This will supplement and improve the accuracy of the neural network predicting N_2O emission and nitrogen loss. Besides, more research is needed for other place or situation to enlarge the neural network.

Acknowledgments

This work was funded by Research on Key Technologies of Quantitative Evaluation of Non-CO_2 Greenhouse Gas Emission Reduction in aquaculture Industry (2017YFF0211701).

References

1. Fukumoto, Y., Osada, T., Hanajima, D., & Haga, K. (2003). Patterns and quantities of NH_3, N_2O and CH_4 emissions during swine manure composting without forced aeration--effect of compost pile scale. *Bioresource Technology*, 89(2), 109-114.
2. Szanto, G. L., Hamelers, H. V. M., Rulkens, W. H., & Veeken, A. H. M. (2007). NH_3, N_2O and CH_4, emissions during passively aerated composting of straw-rich pig manure. *Bioresource Technology*, 98(14), 2659-2670.
3. Morand, P., Peres, G., Robin, P., Yulipriyanto, H., & Baron, S. (2005). Gaseous emissions from composting bark/manure mixtures. *Compost Science & Utilization*, 13(1), 14-26.
4. Luo, Y., Li, G., Luo, W., Schuchardt, F., Jiang, T., & Xu, D. (2013). Effect of phosphogypsum and dicyandiamide as additives on NH_3, N_2O and CH_4 emissions during composting. *Acta Scientiae Circumstantiae*, 25(7), 1338-1345.
5. Hao, X., Larney, F. J., Chang, C., Travis, G. R., Nichol, C. K., & Bremer, E. (2005). The effect of phosphogypsum on greenhouse gas emissions during cattle manure composting. *Journal of Environmental Quality*, 34(3), 774-781.
6. Yang, F., Li, G., Shi, H., & Wang, Y. (2015). Effects of phosphogypsum and superphosphate on compost maturity and gaseous emissions during kitchen waste composting. *Waste Management*, 36, 70-76.
7. Schuchardt, F. (2011). Effect of C/N ratio, aeration rate and moisture content on ammonia and greenhouse gas emission during the composting. *Journal of Environmental Sciences*, 23(10), 1754-1760.

8. Zhu, N. (2007). Effect of low initial C/N ratio on aerobic composting of swine manure with rice straw. *Bioresource Technology*, 98(1), 9-13.
9. Topuz, A. (2010). Predicting moisture content of agricultural products using artificial neural networks. *Advances in Engineering Software*, 41(3), 464-470.
10. Nabavi-Pelesaraei, A., Rafiee, S., Hosseinzadeh-Bandbafha, H., & Shamshirband, S. (2016). Modeling energy consumption and greenhouse gas emissions for kiwifruit production using artificial neural networks. *Journal of Cleaner Production*, 133, 924-931.
11. Tan, W. H., Ooi, K. B., Leong, L. Y., & Lin, B. (2014). Predicting the drivers of behavioral intention to use mobile learning: a hybrid sem-neural networks approach. *Computers in Human Behavior*, 36(C), 198-213.
12. Luo, Y., Li, G., Frank, S., Wang, K., Jiang, T., & Luo, W. (2012). Effects of additive superphosphate on nh3, n2o and ch4 emissions during pig manure composting. *Transactions of the Chinese Society of Agricultural Engineering*, 28(22), 235-242.
13. Wu, J., He, S., Li, G., Li, Z., Bao, Y., & Liang, Y. (2017). Process optimization of pollutant gases emission reduction with superphosphate addition during pig manure composting. *Transactions of the Chinese Society for Agricultural Machinery*.48(5), 304-311
14. Jiang, Frank, Schuchardt, Guoxue, Yuanqiu, & Zhao. (2011). Effect of c/n ratio, aeration rate and moisture content on ammonia and greenhouse gas emission during the composting. *Journal of Environmental Sciences*, 23(10), 1754-1760.
15. Li, D., Li, S., Li, G., & Wang, K. (2016). Effects of additive on nitrogen loss during composting of pig manure and corn straw. *Transactions of the Chinese Society of Agricultural Engineering*. 32 (s2), 260-267.
16. Fu, X., Liu, Q., Li, L., Pan, et al. (2017). Effects of biochar on nitrogen transformation and greenhouse gas emissions during swine manure composting. *Journal of Agro-Environment Science*. 36(9), 1893-1900.
17. Jiang, T., Ma, X., Tang, Q., Yang, J., Li, G., & Schuchardt, F. (2016). Combined use of nitrification inhibitor and struvite crystallization to reduce the NH_3 and N_2O emissions during composting. *Bioresource Technology*, 217, 210-218.
18. Shen, Y. J. (2009). The Effect of Different Parameters on Gas Emission during Aerobic Composting. (Doctoral dissertation, *China Agricultural University*).
19. Ren, L. M. (2009). Study on Mechanism of Carbon and Nitrogen Loss and Control during Composting (Doctoral dissertation, *China Agricultural University*).
20. He, S. Z. (2016). The investigation of adding proportion of Superphosphate and control condition during composting based on the reduction of CO_2 (Doctoral dissertation, *China Agricultural University*).
21. Zhao, Y.Q. (2010). The study on gases nitrogen emission rules during aerobic composting (Doctoral dissertation, *China Agricultural University*).
22. Jiang, T. (2011). The Greenhouse Gas Formation Mechanism During Composting and Mitigation Technologies Research (Doctoral dissertation, *China Agricultural University*).
23. Derks, E. P. P. A., Pastor, M. S. S., & Buydens, L. M. C. (1995). Robustness analysis of radial base function and multi-layered feed-forward neural network models. *Chemometrics & Intelligent Laboratory Systems*, 28(1), 49-60.

MATEC Web of Conferences **277**, 01011 (2019) https://doi.org/10.1051/matecconf/201927701011
JCMME 2018

Study on the ecological wisdom of the grape drying room in Mazar village of Turpan from the perspective under the constraint of regional resources, China

Fuli Meng[1,2], and *Yi* He[1,*]

[1]Huazhong University of Science & Technology, Wuhan Hubei 430074, China
[2]School of Literature and Art, Shihe-zi University, Shihe-zi832003, Xinjiang, China

Abstract. Productive grape-dry room in Turpan of Xinjiang is the core element of the spirit of the grape culture craftsman. Grape-drying house has 2 aspects of internal and external properties:(a)As productive building: which the process of using in regional resources about the practice of ecological agriculture；(b)As an oasis agricultural cultural heritage in Arid Areas:witnessing the history, culture and technology transmission and development of the ancient Silk Road in the Western is a living building sample, which has universal cultural and practical value in the current production and life, culture, etc.This paper is select studying of the traditional grape growing area of Mazar Village, which using the approach of interdisciplinary to cultural geography and ecology,and it is combed the characteristics of 4 aspects of ecological economic artistic and aesthetic of the Grape-drying House from the perspective of the Craftsman's Spirit. The artcle is analysised of 3 aspects from coping with climate resources, coordination of human land relations and building schem construction . The results show that: Concising 7 types of special regional environment to create unique ecological practice model (The upper layer is the Grape-drying room and the lower is the Living room;Building around the dwellings; Building in high altitude ventilation and shelter;Low technology and lower consumption ;The principle of appropriate balance of land using, etc. This study provides some reference for the later research.

1 Introduction

Under the background of Xinjiang's agricultural civilization, the grape drying house is the product of the joint role of the craftsman's wisdom and the nature and the humanities. It is a great significance to study the current green building in the arid area. The research results are mainly focused on the Micro-leve of the form, visual art and historical development of the architecture[1-4]. It is still insufficient to study the multi-layer property and multi-scale ecological wisdom of production and architectural heritage in the context of human settlements in ethnic minority areas.

* Corresponding author: 2188728992@qq.com

The study area Shanshan oasis has the independence and typicality of the geographical unit and cultural unit. It has the advantageous geographical and resource conditions to cultivate the grape culture tourism town with national characteristics. The multi-dimensional study of the grape drying house, which is an important cultural landscape type and tourism resource, is particularly important. This paper selects the universal village of Mazar as the research object, through the case study of the typical grape growing area of Mazar village.

2 Methodology

2.1 Study areas

2.1.1 Natural elements——The formation of material base in the grape drying room

The Turpan basin is located in the eastern part of Xinjiang. The mountain basins in the eastern Tianshan Mountains are surrounded by three hills. By aerial photographs, Turpan basin has a large number of alluvial fans, oasis plains and valley oasis basins, which provide good land resources for grape cultivation. The Shanshan oasis basin is located in the ashinterland of the Eurasian continent, which belongs to the continental warm temperate desert climate, with the characteristics of climate, dry rain, large evaporation, and large day and night temperature difference. Winter is cold, summer heat, the surface temperature is up to 70~80 degrees, the annual sunshine time is about 3000 - 3200h[5]. The above climate features provide a natural advantage and a comprehensive basis for the planting of grapes in Turpan oasis, quality and processing of grapes.

2.1.2 Human elements——The cultural foundation of the grape drying room

Shanshan oasis is located in the ancient Silk Road cultural hub, at this intersection, cultural space advantages, provide social conditions richly endowed by nature and its frequent regional prosperity and sustainable interaction, but also for the processing and fruit grape planting in Turpan area to provide a communication and learning platform provides a humanistic environment in the Shanshan region prominent. The mazar village is one of the ancient village under the Shanshan County tuyugou township (See Figure 1), the unique humanity environment is one of the most important factors in the development of the typical grape culture landscape: Since ancient times, the region has a tradition of mulberry spinning, and multiple religions converge and change here, and the economy prosperity. There are historical cultural landscapes distributing on both sides of the valley of the Tuyugou gully: "Ding Gu Temple" is also called thousand -Buddha cave and the ancient Mazha group, to witness the cultural changes of the village. At the bottom of the valley, there is formationing an ancient road to form a flat natural oasis, which provides good land, water and human foundation for the planting of traditional grapes in the region. The village of Mazar preserves a large number of planting and production space for grapes, which protects the ecological environment of the oasis in this area. improves rural living environment, and creates a natural ecological law, to create the traditional settlements environment technology etc[6], has become a symbol of high identification of Grape Agricultural Cultural Landscape in Turpan.

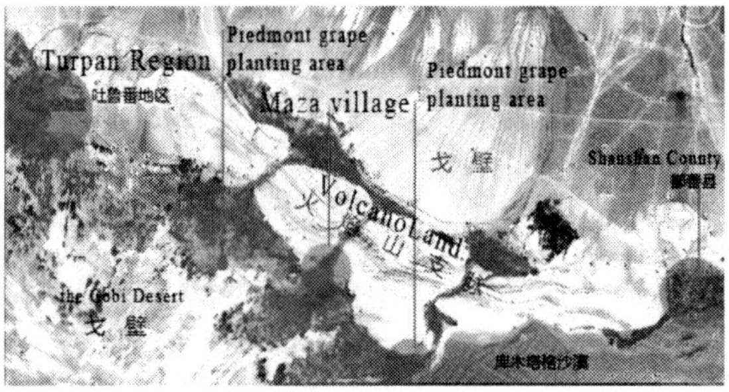

Fig. 1. Regional relation diagram of research objects.

2.2 Data descriptions and pre-processing

In this study, the traditional settlement of Turpan typical grape planting area was selected as the research sample, and the ArcGIS10.2 geographic information analysis platform was used to make a comprehensive analysis of the main driving factors affecting the construction of the grape drying room, such as the humanistic background, the natural conditions, the economic drive and so on, and summarized the ecological construction mode of the grape drying room.

3 Results

The cultural landscape characteristics of the grape drying room from the perspective of the ecological wisdom.

As a typical sample of the study of oasis agricultural cultural landscape, mazar village reflects the mappings of the craftsman's wisdom in the regional production architecture. The common characteristics of Turpan grape planting cultural landscape. As a core element of grape culture landscape with regional characteristics, grape room is discussed in terms of ecological dimension, economic dimension, art and aesthetics dimension etc.

3.1 Ecological characteristics of the grape drying room

The grape drying room is to cope with the dry and hot weather conditions in Xinjiang, and to improve the microclimate of the Human living environment in Oasis. Adapting to the shortage of water resources and arable land resources in Turpan oasis, selecting grape crops and building and producing airhouses as the important way of survival and ecology in this area, we summarize 2 aspects of ecological characteristics.

3.1.1 Improving the ecological nature of the oasis environment

The organic combination of grape drying room and houses, not only meet the grape drying function, but also improve the living environment, create a suitable living environment, grape dry room, Grape Garden, semi basement and other factors to help local residents to solve the Heat insulation, heat insulation, and sunscreen etc. Constructioning and using the grape drying room to follow the natural conservatism principle in a certain extent, to protect the fragile ecological environment of oasis, embodies the materialism follow the natural law of survival of the simple ecological view[7] . The climate conditions of drought and less rain provide the prerequisites for the construction of the grape drying room, which also provides an effective ecological model to adjust the dry and hot climate.

3.1.2 The ecology of construction materials

Building materials in the use of recycled materials, the life cycle of the application in barn building material, the material itself ecological recycling characteristics, the local people respect the natural resources recycling principle, realize real resource sustainable.

3.2 The economic characteristics of the grape drying room

3.2.1 The economic characteristics of the grape drying room are in the construction and Material

The materialized form transformed by production technology —the economic characteristics of the grape drying room. ① The economical of the grape drying room are in the construction and Material of obtain raw material locall. The local building materials taken from the foot of the dry earth by the people. As the Productive of earth buildings, yellow clay and ryegrass is the main material of adobe bricks, mix together, made of different sizes for different parts of the building brick, respectively. Simple pylons is arranged in the rafters inside the Drying room, soliding and fastening with branches, hooks, to hang the grapes. ② The simple principles of the construction process. The grape drying room in mazar village follows this rule. The selection of the material is simple and the plasticity is high. Low cost, excellent heat absorption and heat dissipation, simple construction. Based on the construction material is simple, low cost, suitable for the development of local economy, to promote the use of large-scale.

3.2.2 The economic characteristics of the development of the courtyard economy

① As a traditional Grape courtyard Economy. The economy of the Grape Garden in mazar village has become an important part of the economic structure of grape planting in this area. In the field research of Mazar village, it is found that there are custom of planting mulberry grapes in traditional courtyard. Grape yard planting and processing meet the demand of self food, and the sale of external market is also one of the main sources of family income to solve livelihood problems. Grape trellis and grape drying room have become a necessary building for local families, as an important part of the function of the courtyard.

② The value of tourism resources as an oasis architectural cultural landscape. With the nineteen reports, the implementation of the revitalization of the rural strategy is proposed, and the regional characteristics of the oasis are excavated. It is one of the most important measures to solve the problem of rural poverty, economic development and social stability to cultivate and build a project of "one village, one product" and support the economic characteristics of Grape Garden[8]. In the new era, it gradually evolved into the grape cultural landscape resources with multiple values. In the face of the historical opportunity of the strategy of rural revitalization, the cultural value, economic value and ecological value of the courtyard in the public field of vision are reshaped.

3.3 Artistic and aesthetic characteristics under the spirit of the craftsman

The grape drying room, as the first productive building, has been expanded at many levels in the new era. Mazar village is the epitome of the typical oasis environment in Turpan oasis[9], which is adapted to the natural characteristics of the oasis environment.(See Figure 2).

Fig. 2. Architectural and aesthetic features in the environment.

4 Exploration: The ecological wisdom of the grape drying house under the endowment of regional resources

The difference and uniqueness of the endowment of regional resources are the foundation of ecological wisdom which is formed with distinct regional differences and environmental corresponding[10]. The feedback of harmonious ecological wisdom maintains the local ecological balance and harmonizes the relationship between man and nature[11]. ① The effective use of climatic resources, ② Ecological adaptation of man-land relationship(Multiple types of site selection, convenient and economical travel range， the diversity of land use),which map man-land relationship, and solve the effective production and life model of livelihood and ecological problems. Oasis residents in the residential roof and hillside facilitate the ventilation place to build a grape room. The special construction mode integrates the local production and life style with the natural conditions of climate.

4.1 Ecological wisdom of replying to climate characteristics

The ingenious combination of the grape drying room and residence, natural and artificial ecological air drying system save the cost of building the room, convenient production, practical life, They create a "livable, suitable" oasis human settlement.

Through the author's reference to the relevant academic achievements, in the light of its own from 2016 to 2017 for Grape barn of on-site investigation in Tuyugou Township, 2 rules of climate adaptability are summed up by the author: ① The terrain elevation is the main factor of the change of the microclimate of the Flaming Mountains, and it is also one of the important considerations for the construction of the site of the grape drying room. The results of this study are based on the analysis of Arc Gis10.0 geographic information software: The main range of the main elevation distribution in Mazha village and its periphery is 150m - 240m, and the number is 137, accounting for 65.35% of the total; In the range of 240 - 310m, the number is 52, accounting for 23.30%, and the above 310m above is less distributed here; In the relatively stable elevation, the grape drying room presents a cluster state, and the grape planting area and the distribution pattern of the living space are superimposed. ②The grape drying room follows the ecological law of the suitability of the change of surface temperature. The summer and autumn surface temperature in the distribution area of the grape drying room in the southwest part of Tuyugou area is 37.2~56.80 degrees, the average temperature is 45.98 degrees, and the surface temperature is 48~51 in the range of 48~51. As a whole, the most suitable surface temperature of the air room is 49~50 degrees, and the best temperature of the raisin drying in Shannan area is 49~50 [12].

4.2 Ecological wisdom based on the harmony of man-land relationship; and adjustment

Study of the morphological characteristics of a time and space settlement from the perspective of man-land relationship, the aim is to "explore the relationship between buildings (groups) and the natural environment and the spatial relationship between buildings", by analyzing the value of this special spatial location, we can establish a basic mode of "maximize the use of resources and minimize the cost of human labor"[13]. Therefore, based on the "time constraint mechanism" of production and life, and the allocation principle of land resource level, we discuss the driving factors of the distribution and utilization of grape drying room, and promote the production and life cost savings and the scientific and efficient allocation of land resources[14]. Mazai village is a typical sample to deal with the relationship between production and living land, and embodies the pattern of high efficient distribution of land resources.

4.2.1 The settlement level: the coupling relationship between the grape drying room and the grape planting land and the residence

The main agricultural planting structure of mazai village is made up of grapes, cotton, melon and other crops with low water consumption and low operating cost. Among them, the planting area of grape accounted for 68.52% of the ploughable area. In the agricultural income proportion, the grape planting crops were in the dominant economic position. The annual output of grapes used for processing raisins accounts for 37% - 45% of the total output[15], which is higher than the average level of 10% in Shanshan County. The planting area of Mazar village is relatively flat in the village area, which is conducive to the irrigation area. The middle part of the village is the core residential area, and the living and living land presents a dumbbell shape. Grape drying room is mainly distributed in the "Chuan fonts" river group houses two floors in the Gaogang zone along the hillside, with nearby areas, grape 3 spatial types (see Figure 3).

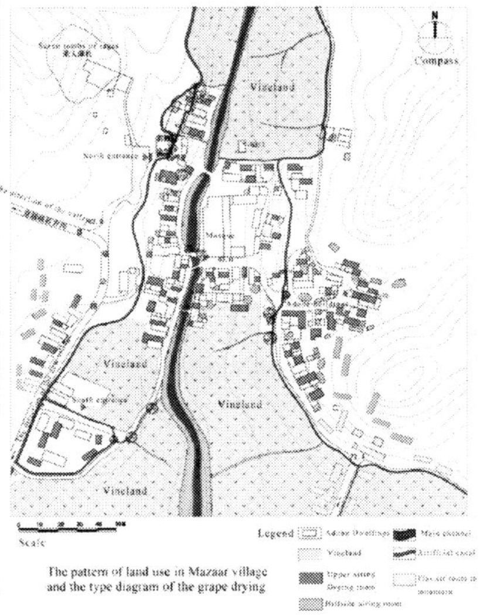

Fig. 3. Pattern of land use in mazaar village in Puyu gully and classification chart of grape drying room[1].

[1] Figure 3 is drawn by Yuehong Li, a member of the research group, and adapted by the author, Fuli Meng

The vertical characteristics of the settlement level in mazai village can be summarized as follows: in the field of land distribution and utilization, the grape drying room shows the vertical morphological characteristics of "low field of high field and upper air living". The distribution pattern of grape drying room in Mazar village is mainly composed of "living and production combination". The proportion of hillside potential type and flat front type is 30%, and the two type is 12%. The reason is that Mazar village is located on the top of the valley, which is impacted by the fan oasis. Considering the factors such as the topography, construction cost and technical conditions, it is relatively difficult to build a drying room on both sides along the slope. From the angle of ease of management to choose living and production space combination, this has become the main choice, it can be seen from the arcade barn, Tiaoyan barn. Based on the above analysis, Mazai village is not suitable for large-scale construction of airing room on the flat ground, so the vertical staggered roof air room is a typical local cultural landscape.

4.2.2 The courtyard level: the ecological wisdom of the organic compound of "production & life & ecology"

The organic compound of the living space and the airing space of "Upper air drying and lower living, ventilation and shelter", the vertical space of the airing room extends to the horizontal space.The compound residence and production space are of great significance for building the cool microclimate of residential space, saving the time cost of travel and realizing the transformation of multi-functional space: summer air house, autumn and winter storage, and intensive utilization of land resources.

Combination mechanism of "Upper air drying and lower living, ventilation and shelter": A layer of living space roof, the two layer is the grape drying room, and the thick and tall grape drying room plays the role of forming insulation space in summer, drying in autumn and keeping warm in winter. Home style grapes are mostly built at the top of the house to be ventilated. In general, residents of the building across the street or passageway upstairs upstairs, ventilation sun at the same time, saving land use effectively, increasing the shade area, reducing the courtyard and street temperature, the formation of small climate environment comfortable, not only meet the grape curing function and improving the living environment. Courtyard ecological model level: "level combination mode around the houses drying circle, attached ploughing and construction".

4.2.3 "Create wisdom of low technology, high involvement, ecological cycle"

Egional production space is used and inherited in a stable mode for a long time. The key factor is the richness of regional materials, the appropriateness of construction technology and the continuity of building subjects.

(1) Selection of materials of suit one's measures to local conditions, using material according to Low technology and high participation. These models can be presented in the construction and repair process of the grape drying room. From the aspects of site selection, climate adaptation and building interface processing, it can be concluded that the construction technology of grape room is much more complicated than that of residential buildings, and the production space plays a decisive role in the quality of family life. Therefore, the main body of Mazha village grape room is usually built on individual participation and collective participation. Though seemingly low-tech and co -construction activities play an important role in maintaining centripetal force.

(2) The unified coordination of 2 aspects of building space scale flexibly and masonry standards. ① the flexibility of building space scale: Usually the airing room is usually the East and west side length, the north and south side wide, is advantageous to the ventilation,

thus shortens the raisin drying time; The spatial scale of the construction depends on the planting area and family wishes of the grapes. ② standardization of construction technical scale: For the specification of adobe bricks commonly used construction of grape drying room is long 30cm, wide 15cm, high 7cm, the specification and construction of other building material specifications are basically the same, its main purpose is to use adobe bricks, cycle in different construction cost saving, convenient construction and repair for masonry.

5 Conclusion and Discussion

5.1 Conclusion

The study excavates and expounds its ecological wisdom from 3 aspects: settlement level, courtyard pattern and construction technology. This is an important basic work to achieve local localization cultural construction and ecological construction. 3 categories of summing up in multi scale: ① An ecological air - drying system combined with natural and artificial, ②Harmony and adjustment of Man-land relationship,③ Low technology, High involvement, ecological cycle,refined the ecological wisdom model of 7 small classes. This study considers that the traditional grape room under the background of traditional agriculture should be precisely protected and developed under the background of ecological civilization. We should regard the problems and values of dialectics as the dialectical development relationship, such as the recognition and respect of the relationship between man and land and the effective utilization of various resources. The traditional raw soil material has been largely replaced by brick of high energy consumption, low ecological; construction land has to be intensive, reducing the extremely fragile ecological zone of waste land; A new era of multi dimension value play to regional grape culture landscape resources, through the organic integration of traditional wisdom and the needs of the new era, and actively initiate long-term role in the minority villages develop specialty industries, promote the local characteristics of Rural Revitalization Strategy implementation, to achieve an affluent life, ecological environment and sustainable development.

5.2 Discussion

In this study, there is still room for further research on the development and conclusion of the research objects. The research samples are selected to have relevance, independence and typicality on the regional units. From the meso scale (traditional settlement level) to consider the representative, but because the selected unit (oasis area, courtyard settlement, etc.) the scale difference, studies the characteristics of homogeneous and heterogeneous research samples also showed different results, the influence of the mode of creating concise. Therefore, the universality and regularity of this research need further empirical analysis. We need more case studies to clarify the ecological wisdom of Turpan traditional grape drying house and adjust rules in the new era. This is also the focus of the follow-up research.

Acknowledgments

This study was supported by the National Social Science Foundation of China (14G126). In the process of writing, This paper could not be possible without the help of my Doctoral Supervisor Yi He as well as all the people from inside and outside Professor Weimin Que of city and the environment college of Peking University whose comments helped to improve the quality of a commanding height pointing, to express my sincere gratitude for them, and many thanks are due to the anonymous reviewers for their constructive comments and suggestions.

References

1. Zhang Yun. Analysis and Research on the elements of traditional humanities and natural native landscape in Turpan area . Beijing Forestry University. 2012,(06):51.
2. Lu Han, JiaYu. The characteristics of the traditional architecture of Xinjiang Uygur ethnic group, Journal of Architecture.1963, (01): 17-22.
3. Wang Yufen. Analysis of the artistic features of Turpan grape dry house architecture. Art education Research, 2012, (7)： 56.
4. Fuli Meng. Study on the spatial characteristics and types of Xinjiang oasis historical and cultural villages and their genetic mechanism [J], Journal of Guizhou National Research, 2018, (01):85.
5. Shan Jixiang. Into the world cultural landscape heritage M. Tianjin: Tianjin University press, 2010, (06): 55.
6. Lu Yinong, Wang Yuanxin. The grape drying room in Turpan - immortal earth architecture . Architecture and Culture, 2008, (02):82-83.
7. Li Chunjing. The arid climate environment of the local landscape design countermeasure research -- Taking Turpan mazarvillage and Yutian County old city for example, Xi'an University Of Architecture And Technology, 2011, (06): 65.
8. Yao Qian. Country culture is the soul of Rural Revitalization. Chinese culture newspaper, 2018, (02): (002).
9. Cai Wu Mei. Turpan area traditional residential space form of [D]. Shanghai. Shanghai Jiao Tong University School of media and design,.2011 (06):62.
10. Zhang Yingjie, Wang Wanjiang. The inheritance and protection of the architectural style of the grape dry room. Journal of Xi'an University Of Architecture And Technology (SOCIAL SCIENCE EDITION), 2015,34 (05):65-69.
11. Bai Baoli, Feng Kunsi. On the ethnic ecological ethics and the construction of harmonious society. Journal of Dalian University, 2007, (02):99-102.
12. Ma Jing. Study on the spatial pattern and dynamic changes of raisin processing air drying room in Shanshan oasis of Xinjiang [D]. Xinjiang Normal University, 2014, (06): 75.
13. Yue Bangrui, Wang Qingqing, Hou Quanhua. The perspective of man land relationship in Turpan mazarvillage oasis settlement pattern research of. Economic Geography,2011,31 (08),:1345-1350.
14. Li Yuehong. Research on the construction strategy of local settlement landscape under the constraints of water resources [D]. Xi'an University Of Architecture And Technology, 2011, (06): 51.
15. From 2010 to 2016 in Shanshan County, office of agricultural economic statistics report.

Using control charts for on-line video summarisation

Clare E. Matthews*, *Paria* Yousefi, and *Ludmila I.* Kuncheva

School of Computer Science, Bangor University, Bangor, Gwynedd, UK

Abstract. Many existing methods for video summarisation are not suitable for on-line applications, where computational and memory constraints mean that feature extraction and frame selection must be simple and efficient. Our proposed method uses RGB moments to represent frames, and a control-chart procedure to identify shots from which keyframes are then selected. The new method produces summaries of higher quality than two state-of-the-art on-line video summarisation methods identified as the best among nine such methods in our previous study. The summary quality is measured against an objective ideal for synthetic data sets, and compared to user-generated summaries of real videos.

1 Introduction

Lightweight, wearable devices allow consumers to capture a continuous stream of frames that provides a record of their daily activities [1]. Processing frames on-the-fly, to select a condensed set of frames that accurately represents the full content of the video, can greatly increase the duration over which such devices can operate. Methods for on-line video summarisation can be used for this process. As processing and memory resources are limited, the traditional high-level feature extraction from frames, e.g., through convolutional neural networks (CNN) [2], or methods requiring storage of all frames [3] may be infeasible. Similarly, elaborate summary selection methods may not be applicable on-line.

In our previous work [4], we proposed a taxonomy of on-line video summarisation methods. We described nine existing methods within the terms of the taxonomy, and compared them experimentally. These experiments highlighted the need for on-line methods to be robust to changes in parameter values. For example, parameters dependent on properties such as total video length are not suitable. The methods investigated are as follows: Shot boundary detection method (SBD) [5], Zero-mean normalised cross-correlation (ZNCC) [6], Diversity promotion (DIV) [2], Submodular convex optimisation (SCX) [7], Minimum sparse reconstruction (MSR) [8], Gaussian mixture model (GMM) [9], Histogram intersection (HIST) [10], Merged Gaussian mixture models (MGMM) [11], and Sufficient content change (SCC) [12]. We found that the SCX and MGMM methods consistently outperformed the others.

* Corresponding author: c.e.matthews@bangor.ac.uk

We use our previous findings to propose here a new on-line video summarisation method that meets the requirements of low computational complexity for feature extraction and summary selection, and with parameters that are relatively robust to different video type.

Despite the large number of video summarisation methods available, and the growing number of on-line methods, the evaluation, and therefore comparison, of methods remains a challenge. We compare our new method against the SCX and MGMM methods by running experiments on both synthetic and real data sets. For the synthetic data, an objectively "best" solution is available. For the real data, we choose a video data base where user-selected keyframe summaries are available, and can be used as ground truth.

The rest of the paper is organized as follows. Section 2 describes the classification system, and Section 3 introduces the new method. The experiments are presented in Section 4, and the conclusion, in Section 5.

2 Classification of methods

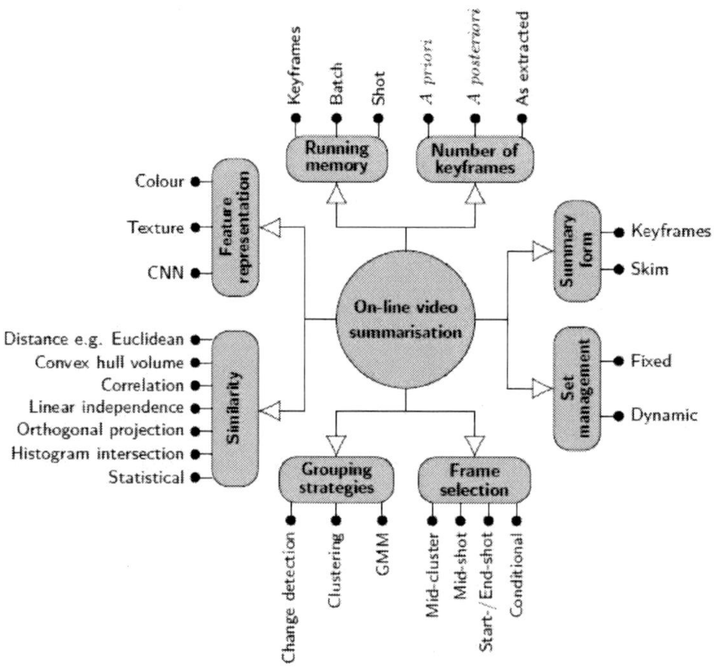

Fig. 1. A classification of on-line video summarisation methods.

The classification for on-line video summarisation methods proposed in [4] is reproduced in figure 1. This classification is adapted from that of Truong Venkatesh [12] for general video summarisation. Eight key components of an on-line summarisation method are identified:

- **Feature representation.** Each frame of the video is represented by an n-dimensional vector in some feature space, $x \in \Re^n$. Simple features may describe the colours within an image [5, 6, 9, 10], or the structure and texture [8]. Features derived from convolutional neural networks (CNN) [2, 7] are relatively complex.
- **Similarity.** A measure of similarity is used to compare the feature vectors of frames. Such measures include the distance between vectors in the feature space [7, 9], the volume of the convex hull of a set of frames [2], the degree of correlation

between frames [6], the intersection of feature vectors [10], the linear independence between batches of frames [5], and the orthogonal projection of a feature vector onto a set of existing features [8]. Statistical methods are also used, e.g. to test equivalence of sets of frames [11].

- **Grouping strategies.** Using the similarity measures described above, frames are grouped together. Groupings may be time-aware, e.g. time-contiguous shots identified using change-detection [5, 6, 10, 12], or independent of time, e.g. clustering of frames within the feature space [2, 7, 8, 13] or assigning frames to components of a Gaussian mixture model [9, 11].
- **Frame selection.** Different approaches exist for selecting a keyframe from a group of frames, e.g. choose the frame most central within a cluster [2, 7, 11], frames at the start / end [5, 12] or middle [6, 10] of a shot, or alternatively, frames that satisfy some condition, such as exceeding a ``difference" threshold [8, 10].
- **Set management.** The keyframe set may be fixed, i.e. once a keyframe is selected it cannot be removed from the set [5, 6, 7, 8, 9, 10, 12], or dynamic, i.e. a keyframe may be removed at some later point in the processing [2, 11].
- **Summary form.** Frames selected as a summary form either a static keyframe set [2, 5, 6, 7, 8, 10, 11, 12], or dynamic video skim [6, 9, 13, 3, 14].
- **Number of keyframes.** The number of keyframes in a summary is often variable and *as extracted*, determined by the algorithm and characteristics of the video [5, 6, 7, 8, 9, 10, 11, 12]. Alternatively, the number of frames can be defined *a priori* [2], or post-processing of the selected frames can reduce the set size *a posteriori*, to a pre-defined number of frames [2].
- **Running memory.** The memory required to run the summarisation is an important feature for on-line applications. Some methods need only keep the keyframe set in memory [2, 8, 9, 12], others process frames in batches and must therefore keep the full batch in memory [5, 7, 11]. Similarly, methods based on identifying shots within the video, may require a full shot to be stored in memory to then select the desired keyframe from the shot [6, 10].

3 Control-charts for on-line video summarisation

Here we propose a method that uses the statistical process of control-charts to identify shots from a streaming video. Control-charts [15] monitor a quantity of interest to detect when a process moves out of control. The mean, μ, of the quantity is used as a baseline value, and the process deemed to be "in control" while observations remain within a specified limit from the mean, typically three standard deviations, σ.

Algorithm 1: On-line control chart method

Input: Data stream $X = \{x_1,\dots,x_N\}$, $x_i \in \Re^L$, minimum shot length m, buffer size B, threshold for keyframe similarity θ.

Output: Selected set of keyframes $P \subset X$.

```
// Initialisation
1 P ← ∅
2 j ← 1 // Shot number
3 S_j ← {x_1,...,x_B} // First shot
4 for i ← [2,...,B] do
5     d_i ← d(x_i,x_{i-1}) // Euclidean distance
```

$\mu \leftarrow mean(d_2,\ldots,d_B)$
$\sigma \leftarrow std(d_2,\ldots,d_B)$

// Process video frame-by-frame
for $i \leftarrow [B+1,\ldots,N]$ **do**
  $d_i \leftarrow d(x_i,x_{i-1})$
  if $d_i < \mu + 3\sigma$ **then**
  // No new shot detected
    $[\mu,\sigma] \leftarrow$ update μ & σ with d_i
    $S_j \leftarrow S_j \cup X(i)$
  else
  // New shot detected
    if $|S_j| > m$
    // Shot sufficiently long
    $p_j \leftarrow \text{selectkeyframe}(S_j)$
    $\delta \leftarrow \text{keyframediff}(p_j,p_{j-1})$
    if $\delta < \theta$ **then**
      // Shots too similar: merge
      $S_j \leftarrow S_{j-1} \cup S_j$
      // Remove last keyframe from set
      $P \leftarrow P(1\text{:end}-1)$
      $p_j \leftarrow \text{selectkeyframe}(S_j)$
    $P \leftarrow P \cup p_j$
    $j \leftarrow j+1$
  else
    // Shot too short: ignore
    $S_j \leftarrow \emptyset$

Function $f = \text{selectkeyframe}(Y)$
// Select the frame closest to the mean
$f \leftarrow \underset{x\in Y}{\mathbf{argmin}}\, d(x,\bar{Y})$

Function $\delta = \text{keyframediff}(f_1,f_2)$
// Compare 16-bin Hue histograms of frames f_1 and f_2
$h_i = \text{hist16}(\text{hue}(f_i))$ // Normalised 16-bin Hue histogram
$\delta = \sum_{j=1}^{16} |h_1(j) - h_2(j)|$

3.1 Control-chart method (CC)

Assuming that each frame is represented as a point in some L-dimensional space, we take the Euclidean distance, d, between consecutive frames as the process to be monitored. A distance $d > \mu + 3\sigma$ defines a shot boundary. Once a full shot has been identified, a keyframe is selected as the frame closest to the centre of the cluster defined by the shot.

Potential issues with such a method are that: (1) consecutive shots identified by the algorithm may be too similar to warrant separate keyframes, and (2) short transitions may be identified as shots, but are not important to the summary. We address these issues as follows:

- Define a measure of similarity between frames, as follows [16]. Use the HSV representation of the frames to obtain 16-bin histograms of the hue value (H). If the Minkowski distance between the normalised histograms is less than a threshold of 0.5, the frames are similar.
- After identifying a shot and selecting the representative keyframe, we compare this frame with the previous keyframe (if available). If the two consecutive keyframes are similar according to the above measure, we assume that a shot boundary has been falsely identified. The boundary is removed, and the two shots are merged. A new keyframe is selected from the combined shot to replace the two keyframes from the individual shots.
- We define an empirical constant to state the minimum shot length. If a shot contains fewer frames, the shot is ignored and no keyframe is selected.

The CC method requires three parameters: a pre-defined threshold θ for classifying keyframes as similar, a minimum shot length m, and initial buffer size B for calculating the starting mean and standard deviation. If we assume that the number of frames per second will be constant across videos, and that the duration required for a shot to be of interest is largely independent of video content, the optimal value for m should be consistent across videos. We select two seconds to be the minimum duration of a shot for it to be of interest. The full control-chart method is given in Algorithm 1.

3.2 Feature representation

Our control-chart method may be used with any feature space. For an on-line application, feature extraction must be a computationally inexpensive process. Therefore, relatively complex features, such as those derived from CNN, are not feasible.

To select a feature space for testing the algorithm, we implement the extraction of a number of different features, including those used by existing on-line summarisation methods. Table 1 shows the time taken to extract the different features for a video containing 3,266 frames. The extraction time for the RGB moments is substantially shorter than HSV histograms, even when a relatively small number of bins are used for the histograms. We therefore select RGB moments as the feature space to use in the CC method.

The RGB moments are a 54-dimensional feature space; the mean and standard deviation of the three colour channels for the nine sub-images created from a uniform 3-by-3 grid.

Table 1. Average time to extract features for the VSUMM video #21, and methods that use the features.

Feature	Time to extract (s)
RGB moments of 9 blocks (CC)	25
HSV histogram - [8, 4, 4] bins for Hue, Saturation, Value (ZNCC, SBD, HIST)	85
CENTRIST 252-dimensional structural histogram (MSR) [17]	522
MPEG-7 colour layout descriptor (GMM) [18]	1,546
Penultimate layer of VGG CNN (DIV) [19]	> 1.5hr

4 Method testing and evaluation

Here we compare the results for the proposed CC method with the two existing methods, SCX and MGMM, found to perform best in our previous comparison study [4].

4.1 Synthetic data

We first consider the performance of the three methods on seven synthetic data sets. The first data set follows the example of Elhamifar et al. [20]. The data consists of three clusters in 2-dimensional space as illustrated in figure 2 (#1). Each point represents a frame in the video. The three clusters come in succession but the points within each cluster are generated independently from a standard normal distribution. The order of the points in the stream is indicated by a line joining every pair of consecutive points. The time tag is represented as the grey intensity. Earlier points are plotted with a lighter shade. The "ideal" selected set is shown with red target markers. In addition to the two dimensions plotted, two noise dimensions are added (from the distribution $N(0, 0.5)$). Data sets #2 - #5 are also shown in figure 2. Again, each data set contains an additional two noise dimensions. Data sets #6 and #7 follow a similar structure but with more dimensions, six and eight, respectively.

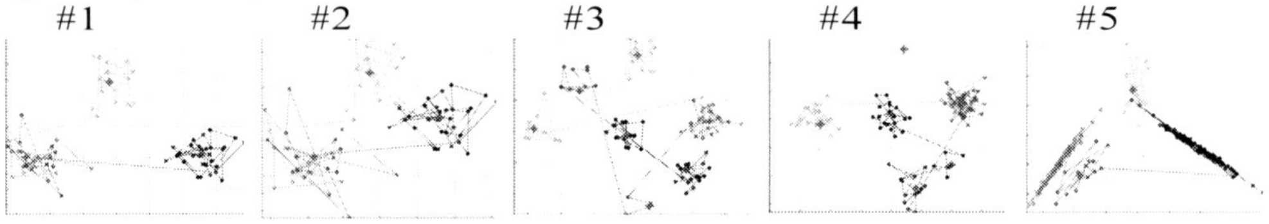

Fig. 2. Synthetic data sets #1 - #5. The time tag is represented as the grey intensity. Earlier points are plotted with a lighter shade. The "ideal" selected sets are shown with red target markers.

Using synthetic data allows an objective assessment of the summaries produced. If the video is already segmented into units (events, shots, scenes, etc.), the selected frames must allow for distinguishing between the units with the highest possible accuracy [21]. Therefore we use three complementary measures of the quality of the summary of synthetic data sets:

$$\text{Cardinality}: \quad K = |P| \tag{1}$$

$$\text{Approximation error}: \quad J = \sum_{i=1}^{N} d\left(x_i, p_i^*\right) \tag{2}$$

$$\text{Accuracy}: \quad A = 1 - nn(P) \tag{3}$$

where $X = (x_1,\ldots,x_N)$ is the sequence of video frames, N is the total number of frames in the video, $P = \{p_1,\ldots,p_K\}$ is the selected set of keyframes, p_i^* is the keyframe closest to frame x_i, d is the Euclidean distance, and $1 - nn(P)$ is the resubstitution classification accuracy in classifying X using P as the reference set. To obtain a good summary, we strive to maximise A while minimising J and K.

We train the method parameters on 50 randomly generated data sets following the distribution of data set #1. Solutions are evaluated as follows:

- Find the Pareto set for the three criteria A, K and J.
- Exclude any results in the Pareto set with $K > 10$. This step removes the solution that selects *all* frames as keyframes, giving perfect accuracy and no error.

- Select the summary with the best accuracy. Where multiple summaries tie, select that with the fewest frames, and use the approximation error to split any remaining ties.

Taking the 50 optimal parameter sets as a cluster, the set closest to the cluster centre is chosen as the tuned method parameters.

The methods are then tested on 300 randomly generated data sets, 50 from each of the remaining six data set patterns, using the parameters tuned on data set #1. For each data set the accuracy, cardinality and approximation error are calculated for each method. The methods are then ranked. Four paired-sample t-tests are performed, comparing the accuracy and approximation error for our proposed CC method against the two existing methods.

Table 2. Results of paired-sample t-tests comparing the accuracy (A) and approximation error (J) for the CC method summaries and the summaries generated by the MGMM and SCX methods. The confidence interval for the difference is shown for significant results (at the 0.05 significance level).

Method	Test	P-value	Confidence interval
MGMM	$A_{CC} - A_{MGMM}$	$1e^{-5}$	[0.02, 0.04]
	$J_{CC} - J_{MGMM}$	$6e^{-4}$	[-1.7, -0.4]
SCX	$A_{CC} - A_{SCX}$	0.7	-
	$J_{CC} - J_{SCX}$	$3e^{-23}$	[-4.0, -2.7]

Table 2 shows the results of the paired-sample t-tests. At the 0.05 level, there is no significant difference between the accuracy values for the CC and SCX methods (i.e. the difference has a zero mean). All other tests find a significant difference. The confidence intervals for the mean differences are less than zero for J, implying that the error tends to be less for the CC method, and greater than zero for A, implying that the accuracy tends to be greater for the CC method. The CC method summaries tend to rank best according to our three criteria; an average of 1.4 across the 300 experiments, compared to the existing methods that have average ranks of 2.2 and 2.3 for the MGMM and SCX methods, respectively.

4.2 VSUMM videos

The methods are tested on 50 real videos from the VSUMM collection[1] [16]. Whereas the summaries of the synthetic data can be assessed in relation to a "correct" result, there is no such objective assessment available for real videos; what constitutes a good summary is somewhat subjective. The VSUMM collection includes a database of five user-selected summaries for each video. These summaries can be used as a ground-truth, to compare method-generated summaries against. Following the approach of De Avila et. al [16], the match between two summaries is described by an F-measure calculated using the 16-bin histograms of the hue values of selected keyframes, as explained in Section 3.1.

Parameters for each method are tuned on video #21. We select the parameters that produce the summary with the highest average F-measure when compared with the five user ground-truth summaries. These parameters are used to run the methods on the other 49 videos.

Figure 3 shows the F-measure (averaged across the five user summaries) versus the number of keyframes selected by each method for the VSUMM videos. Each point on the plot corresponds to a video. The ideal summary has a high F-measure, and low number of frames. Points in the upper-left corner of the plots shown in figure 3 therefore represent the better summaries. The points for all methods are plotted with grey colour on all plots. The

[1] https://sites.google.com/site/vsummsite

points of the method in the title of the subplot are shown with black markers. The CC method generates a higher proportion of good summaries than the existing two methods. As an illustration of these results, figure 4 shows the summary of video #47 produced by the CC method, compared to the summary from user #1. All five frames in the user summary are matched in the CC method summary.

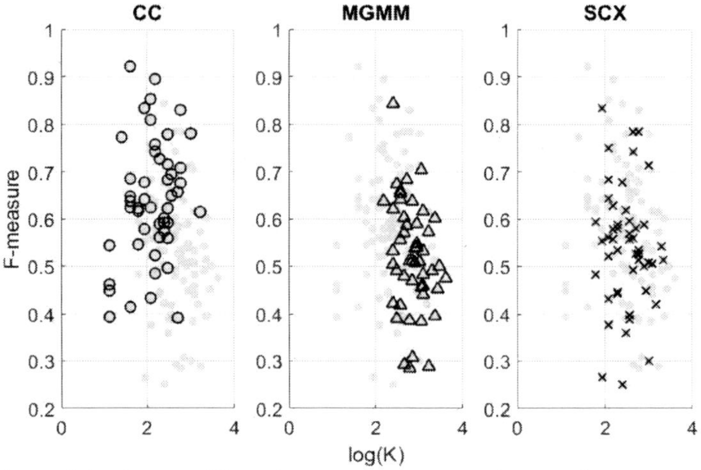

Fig. 3. Number of keyframes (K) and F-measure averaged over five user ground-truths, for summaries of the 50 VSUMM videos. Filled, grey circles show the results for all three methods, with the points for the named method highlighted in black.

Fig. 4. Comparison of VSUMM video #47 summaries from user #1 and the CC method. The matches have been calculated using the 16-bin histogram method with threshold 0.5 [16].

5 Conclusion

Control-charts are a simple and effective approach to on-line video summarisation. Our proposed CC method performs well in comparison to existing methods, both on small synthetic data sets and real videos. On-line methods require computationally inexpensive feature spaces. The CC method uses RGB moments, which are significantly faster to extract than the features used by some of the existing on-line methods. Feature extraction time can be improved further by working with compressed images. However, further work is required to assess the effect on summary quality.

The videos used for testing have well-defined shots, providing a relatively easy summarisation task. The performance of the new method may be different on other types of video, e.g. where the shots are less clearly defined or the variability within shots is greater. Examples of such type of data are egocentric videos and lifelogging photo streams. Performance on longer videos must also be considered. For the application of wearable devices, it may be necessary to introduce a restriction on the number of keyframes that can be selected.

Similarly, when shots can potentially become very long, or consecutive shots very similar, a more dynamic approach to sampling, and the shot detection and similarity thresholds may be beneficial, and will be investigated in future work.

Acknowledgment

This work was supported by project RPG-2015-188 funded by The Leverhulme Trust, UK

References

1. Betancourt A, Morerio P, Regazzoni C S and Rauterberg M 2014 CoRR **abs/1409.1484v1** (*Preprint* arXiv:1409.1484v1) URL https://arxiv.org/abs/1409.1484v1
2. Anirudh R, Masroor A and Turaga P 2016 *IEEE International Conference on Image Processing (ICIP2016)* pp 3329–3333
3. Lan S, Panda R, Zhu Q and Roy-Chowdhury A K 2018 *IEEE/CVF Computer Vision and Pattern Recognition*
4. Matthews C E, Kuncheva L I and Yousefi P 2018 SUBMITTED: *Machine Vision and Applications*
5. Abd-Almageed W 2008 *IEEE 15th International Conference on Image Processing (ICIP 2008)* pp 3200–3203
6. Almeida J, Leite N J and Torres R d S 2013 *Journal of Visual Communication and Image Representation* **24** 729–738 URL http://dx.doi.org/10.1016/j.jvcir.2012.01.009
7. Elhamifar E and Kaluza M C D P 2017 *IEEE Conference on Computer Vision and Pattern Recognition (CVPR2017)* pp 1818–1826
8. Mei S, Guan G, Wang Z, Wan S, He M and Feng D D 2015 *Pattern Recognition* **48** 522–533
9. Ou S H, Lee C H, Somayazulu V S, Chen Y K and Chien S Y 2015 *IEEE Journal of Selected Topics in Signal Processing* **9** 165–179
10. Rasheed Z and Shah M 2003 *Proceedings IEEE Computer Society Conference on Computer Vision and Pattern Recognition* **2** pp 343–343
11. Song M and Wang H 2005 SPIE 5803, *Intelligent Computing: Theory and Applications III* **5803** pp 174–184
12. Truong B T and Venkatesh S 2007 *ACM transactions on multimedia computing, communications, and applications (TOMM)* **3** 3 URL https://dl.acm.org/citation.cfm?doid=1198302.1198305
13. Zhao B and Xing E P 2014 *IEEE Conference on Computer Vision and Pattern Recognition*
14. Sharghi A, Gong B and Shah M 2016 *European Conference on Computer Vision*
15. Shewhart W A 1931 Economic control of quality of manufactured product (Van Nostrand Company)
16. de Avila S E F, Lopes A P B, da Luz Jr A and de Albuquerque Araújo A 2011 *Pattern Recognition Letters* **32** 56–68

17. Wu J and Rehg J M 2011 *IEEE Transactions on Pattern Analysis and Machine Intelligence* **33** 1489–1501
18. Kasutani E and Yamada A 2001 *IEEE International Conference on Image Processing* pp 674–677
19. Simonyan K and Zisserman A 2014 Very deep convolutional networks for large-scale image recognition arXiv:1409.1556
20. Elhamifar E, Sapiro G and Sastry S S *2016 IEEE transactions on pattern analysis and machine intelligence* **38** 2182–2197
21. Kuncheva L I, Yousefi P and Almeida J 2018 *Journal of Visual Communication and Image Representation* **52** 118–130 URL https://doi.org/10.1016/j.jvcir.2018.02.010

CNN and RNN mixed model for image classification

Qiwei Yin[1], *Ruixun* Zhang[2] , and *XiuLi* Shao[1,*]

[1]College of Computer and Control Engineering, Nankai University, Tianjin, China
[2]Laboratory for Financial Engineering, Massachusetts Institute of Technology, Cambridge, MA, USA

Abstract. In this paper, we propose a CNN(Convolutional neural networks) and RNN(recurrent neural networks) mixed model for image classification, the proposed network, called CNN-RNN model. Image data can be viewed as two-dimensional wave data, and convolution calculation is a filtering process. It can filter non-critical band information in an image, leaving behind important features of image information. The CNN-RNN model can use the RNN to Calculate the Dependency and Continuity Features of the Intermediate Layer Output of the CNN Model, connect the characteristics of these middle tiers to the final full-connection network for classification prediction, which will result in better classification accuracy. At the same time, in order to satisfy the restriction of the length of the input sequence by the RNN model and prevent the gradient explosion or gradient disappearing in the network, this paper combines the wavelet transform (WT) method in the Fourier transform to filter the input data. We will test the proposed CNN-RNN model on a widely-used datasets CIFAR-10. The results prove the proposed method has a better classification effect than the original CNN network, and that further investigation is needed.

1 Introduction

Current CNN [1-2] networks have become a standard method of machine learning for 'mesh data' (pictures, videos, etc.). Researchers have created many different structural network models on the basis of the CNN model, and have proved to be successful at a variety of image correlation problems including handwritten digital recognition [3], natural picture recognition, etc. Among them, [4] is one of the most classic networks currently used for image recognition. AlexNet [5] is also a neural network based on Convolution + Pool structure, It is one of the origins of deep learning. Starting from AlexNet. After that, there are a lot of CNN models appearing, ZF-Net [6] began to use the de-convolution to output the feature of the CNN network middle layer, and this feature enables people to more intuitively understand the working principle and network architecture within CNN. They discover the characteristics of edges, colors, etc. in shallow CNN detection images, and deeper CNN layer will detect shape features of objects to be identified. While the RNN has become the standard method for machine learning of sequence data (audio, natural

* Corresponding author: shaoxl@nankai.edu.cn

language, etc.). NLP [7] and machine translation [8-9] are the most current applications. Combining RNN processing sequence and CNN to process image data, the main research fields include image tagging [10], target detection [11], video screen behavior detection, etc. [12] innovatively proposed a network model ReNet for image classification. This network replaces the Convolution + Pooling layer in CNN using four common UNI-Dimensional RNNs layers. In this paper, we use a similar structure in the ReNet model, and propose our model CNN-RNN model for image classification. The model has some similarities with ResNet [13], this article is a reference to the jump connection method of ResNet model, using RNN Layer to Calculate Continuity Features in CNN Layer, and then connect these features to the final fully connected network for classification prediction. In this work, we will test the proposed CNN-RNN model on a widely used object recognition datasets CIFAR-10[14]. Experiments have shown that the CNN-RNN model has better accuracy and call more research in the future.

2 Model description

In this paper, we make some changes to the RNN layer in [12], and no longer calculate sequence results in four directions. We only calculate the sequence results in two directions, as shown in Figure 1 (example picture is for reference only, non-experiment dataset picture), it is a one layer structure of RNN calculations. It can be seen as a bidirectional RNN model. The RNN model can be LSTM (Long Short Term Memory) [15], GRU (Gated Recurrent Unit) [16], IRNN [17] or ConvLSTM (Convolutional LSTM Network) [18].

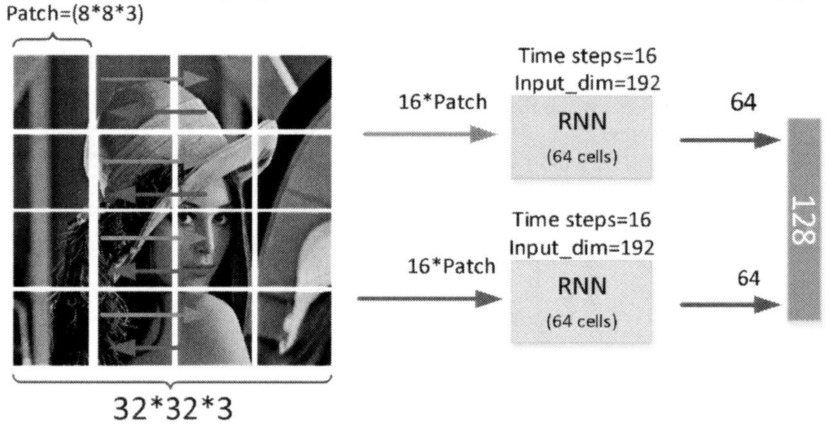

Fig. 1. One-Layer RNN.

Denote the input data of the RNN layer is $x \in R^{w \times h \times c}$, w and h denote the width, height of input x, and if x is the original picture, the c represents the number of color channels, other c represents the number of filters (or the number of convolution kernels). According to [6], the output of the convolutional layer after deconvolution operation is the result of the edge, texture, and shape of the image. Therefore, the output of the convolution layer is the original image after the filter processing, which has the original image part of the feature attributes. Define the identification area (Patch in the figure 1) $w_p \times h_p$, and we can split the input into a $M \times N$ patches $x_{m,n} \in R^{w_p \times h_p \times c}$, where $M = \frac{w}{w_p}, N = \frac{h}{h_p}$, Inputting $x_{m,n}$ into a bidirectional RNN network in a sequence:

$$S^F_{i,j} = f_{FWD}(S^F_{m,n-1}, x_{m,n}),\ for n = 1,...N \tag{1}$$

$$S^R_{i,j} = f_{REV}(S^R_{m,n+1}, x_{m,n}),\ for n = N,...,1 \tag{2}$$

where f_{FWD} and f_{REV} return forward sequence and reverse sequence recursively hide the results. If you need to add multiple layers of RNN, then return the entire sequence of $S_{i,j}^{F}$ and $S_{i,j}^{R}$, otherwise return the final state result of $S_{i,j}^{F}$ and $S_{i,j}^{R}$. The complete model diagram is shown in Figure 2. Both the CNN layer and the RNN layer can be added to multiple layers, and the RNN model can extract continuous dependency features from the output of the intermediate CNN layer.

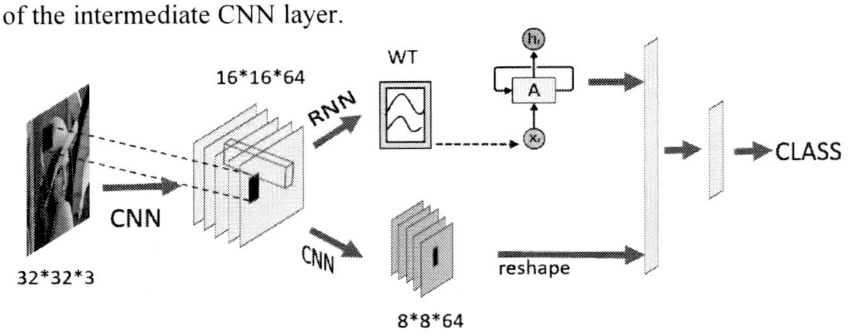

Fig. 2. CNN-RNN Model.

3 Model analysis

The filter is an operation in image processing. The specific operation of filtering is the sum of the product of the image pixels and the filter. Convolution and filtering are largely similar, but in fact it can be seen as a filtering process. The convolution operation can also be implemented using wave processing. The wave of the image is a non-periodic discrete signal in the time domain. Therefore, the Fourier transform can be used to operate the image wave. Convolution calculation is the most important in CNN. If the images in CNN are regarded as non-periodic discrete signals in time domain, the convolution of images is the process of taking waves in different frequency bands. The image can be viewed as a discrete two-dimensional function f(x,y), and the one-dimensional Fourier transform can be extended to two dimensions. The transformation from the spatial domain to the frequency domain is the Fourier transform:

$$F(u,v)=\int_{-\infty}^{\infty}\int_{-\infty}^{\infty}f(x,y)e^{-i(ux+uy)}dudv$$

The essence of a convolution kernel is a two-dimensional function, which has a corresponding spectral function. $F(u,v)$ in the formula is a certain kind of spectral function. The convolution of the image is actually taking the characteristics of different frequency bands in the image. The convolution operation in the time domain segment is actually equal to the multiplication operation in the frequency domain segment. However, the Fourier transform has limitations because it cannot characterize the local characteristics of the signal in the time domain, and does not work well for sudden and non-stationary signals. So using a wavelet transform is a better choice:

$$WT(a,\tau)=\frac{1}{\sqrt{a}}\int_{-\infty}^{\infty}f(t)*\psi(\frac{t-\tau}{a})dt$$

where ψ represents a wavelet basis, the difference from e^{ix} in the ordinary Fourier transform is that it is not a simple positive or cosine wave basis function but a wavelet basis function that satisfies certain conditions. Two-dimensional waves have not only local features but also sequence features. Therefore, this paper uses the RNN network to extract the sequence features of the two-dimensional wave, and also uses the sequence features as an important feature to judge the image classification.

ResNet's skip connection allows it to stack layers of the network deep without vanishing gradient problem [20], because of this hopping connection method, the depth of the ResNet model is not deep. The feature of the middle layer output can be directly connected to the final full-connection network. ResNet is actually a neural network structure composed of multiple shallow CNNs. The CNN-RNN method in this paper cannot achieve the same deep network structure as ResNet, but it can also use the different feature output of the middle layer to connect to the final full-connection layer.

4 Experiments

4.1 DataSets and Model Architectures

The CIFAR-10 dataset contains 60,000 images, and each of which belongs to one of ten categories; airplane, automobile, bird, truck, ship, horse, frog, dog, deer, and cat. The shape of each image is (32,32,3).

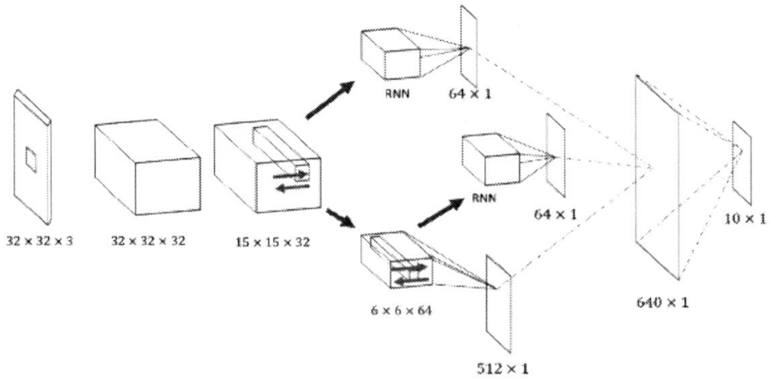

Fig. 3. CNN-RNN model used for CIFAR-10 classification.

ConvLSTM (Convolutional LSTM Network) was proposed by [18] to solve the problem of precipitation nowcasting. ConvLSTM is an improvement over full-length long-term memory networks (FC-LSTM). It is similar to an LSTM layer, but the input transformations and recurrent transformations are both convolutional. ConvLSTM not only has the timing modeling capabilities of LSTM, but also features local features like CNN. All ConvLSTM inputs $x_1,...,x_t$, Cell outputs $C_1,...,C_t$, hidden states $H_1,...,H_t$, and gates i_t, f_t, o_t are all 3D tensors:

$$
\begin{aligned}
& i_t = \sigma(W_{xi} * x_t + W_{hi} * H_{t-1} + W_{ci} \circ C_{t-1} + b_i) \\
& f_t = \sigma(W_{xf} * x_t + W_{hf} * H_{t-1} + W_{cf} \circ C_{t-1} + b_f) \\
& C_t = f_t * C_{t-1} + i_t \circ \tanh(W_{xC} * x_t + W_{hC} * H_{t-1} + b_C) \\
& o_t = \sigma(W_{xo} * x_t + W_{ho} * H_{t-1} + W_{co} \circ C_t + b_o) \\
& H_t = o_t \circ \tanh(C_t)
\end{aligned}
$$

where '*' indicates convolution operation and '∘' indicates Hadamard Product. For more details on ConvLSTM reference [18]. In this paper, we use various types of RNN models to compare and find that different types of RNNs have little effect on the results. Figure 3 shows the model structure of the CNN-RNN model trained on the CIFAR-10 data set proposed in this paper.

4.2 Training

The CNN-RNN model is also an End-to-End learning process that does not require additional preprocessing. To train the networks, we choose a wildly used adaptive learning rate algorithm, called Adam [21]. The CNN-RNN model will bring the problem of model overfitting due to the addition of an additional network structure. To reduce over-fitting problem. We will apply the normalization term imposed on the BatchNormalization [22] and RNN weights at the output of the RNN layer.

5 Results and Analysis

In the model training, we found that the standard VGG model will occur vanishing gradient problem, due to the network hierarchy being too deep, and the model hardly converged. Therefore, in this paper, we use a simplify VGG-9 model. We properly remove the two Pool layers to get this article's streamlined VGG-9. The learning rate in the model training process uses a fixed size of 0.001. The ResNet [13] used the variable learning rate in the original paper, so the results in this paper will be different. The training results of the model are shown in Table 1, including the accuracy of each model and the accuracy of the test.

Table 1. Model Test Results.

MODEL	TRAINING ACC	TEST ACC
CNN	0.932	0.772
CNN + RNN	**0.891**	**0.80**
VGGNET-9 [SIMONYAN K, ZISSERMAN A.]	0.943	0.769
VGGNET-9 + RNN	**0.929**	**0.782**
RESNET-20 [HE K, ZHANG X, REN S, ET AL.]	0.965	0.789
RESNET-20+ RNN	**0.961**	**0.812**

Figure 4 and Figure 5 show the accuracy of the original CNN model and CNN-RNN models modified by RNN (including ConvLSTM and RNN), respectively. It can be seen that the improved accuracy of the CNN-RNN model is generally higher than that of the original CNN model. The combination of CNN+ConvLSTM and CNN+IRNN does not differ greatly in accuracy. This shows that the feature of CNN output are not long-term dependencies. And there are only simple dependencies, so using an ordinary IRNN will satisfy the requirement. Therefore, the general RNN can meet the requirements.

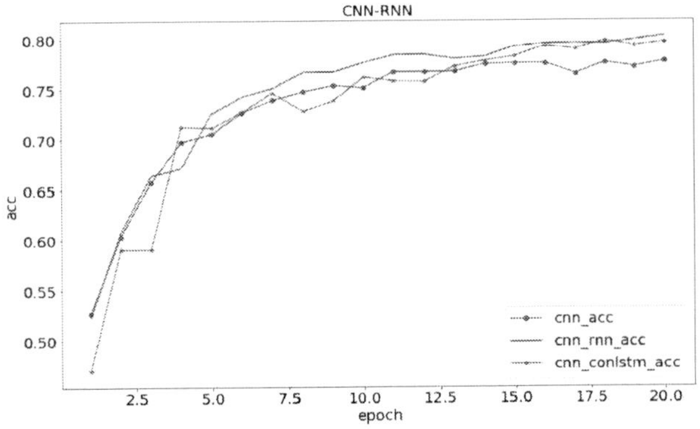

Fig. 4. CNN and CNN-RNN classification.

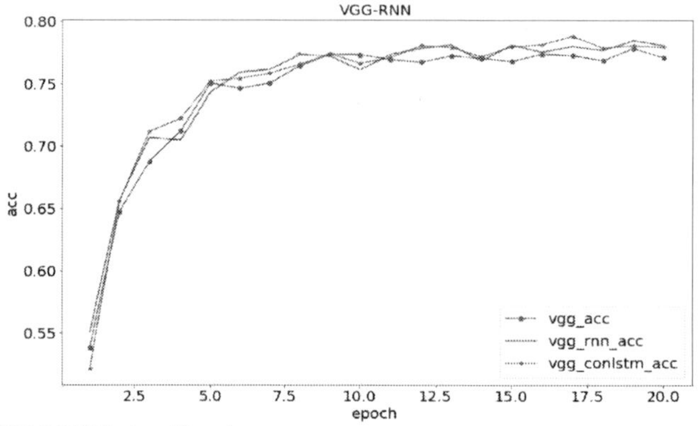

Fig. 5. VGG and CNN-RNN classification.

6 Conclusion and Future Work

In this paper, we propose an improved model CNN-RNN model, which has a certain improvement in the accuracy of image recognition problems. But it will increase the complexity of the model. It will reduce the training speed, and increase risk of overfitting. It need to add regularization items to limit the complexity of the model. In this paper, we simply extract feature from the CNN layer through the RNN network. There are many other extraction methods and combinations that need to be studied.

References

1. Fukushima K. Neocognitron: A self-organizing neural network model for a mechanism of pattern recognition unaffected by shift in position[J]. Biological Cybernetics, 1980, 36(4):193-202.
2. Lecun Y, Boser B, Denker J S, et al. Backpropagation Applied to Handwritten Zip Code Recognition[J]. Neural Computation, 1989, 1(4):541-551.
3. Cireşan, Dan, Meier U, Schmidhuber, Juergen. Multi-column Deep Neural Networks for Image Classification[J]. Eprint Arxiv, 2012, 157(10):3642-3649.
4. Simonyan K, Zisserman A. Very Deep Convolutional Networks for Large-Scale Image Recognition[J]. Computer Science, 2014.
5. Krizhevsky A, Sutskever I, Hinton G E. ImageNet classification with deep convolutional neural networks[C]// International Conference on Neural Information Processing Systems. Curran Associates Inc. 2012:1097-1105.
6. Zeiler M D, Fergus R. Visualizing and Understanding Convolutional Networks[C]// European Conference on Computer Vision. Springer, Cham, 2014:818-833.
7. Mikolov T A. Statistical Language Models Based on Neural Networks[J]. 2012.
8. Sutskever I, Vinyals O, Le Q V. Sequence to Sequence Learning with Neural Networks[J]. 2014, 4:3104-3112.
9. Bahdanau D, Cho K, Bengio Y. Neural Machine Translation by Jointly Learning to Align and Translate[J]. Computer Science, 2014
10. Mao J, Xu W, Yang Y, et al. Explain Images with Multimodal Recurrent Neural Networks[J]. Computer Science, 2014.

11. Ren S, He K, Girshick R, et al. Faster R-CNN: Towards Real-Time Object Detection with Region Proposal Networks[J]. IEEE Transactions on Pattern Analysis & Machine Intelligence, 2017, 39(6):1137-1149.
12. Visin F, Kastner K, Cho K, et al. ReNet: A Recurrent Neural Network Based Alternative to Convolutional Networks[J]. Computer Science, 2015, 25(7):2983-2996.
13. He K, Zhang X, Ren S, et al. Deep Residual Learning for Image Recognition[J]. 2015:770-778.
14. Krizhevsky A. Learning Multiple Layers of Features from Tiny Images[J]. 2009.
15. Surhone L M, Tennoe M T, Henssonow S F. Long Short Term Memory[J]. Betascript Publishing, 2010.
16. Cho K, Merrienboer B V, Gulcehre C, et al. Learning Phrase Representations using RNN Encoder-Decoder for Statistical Machine Translation[J]. Computer Science, 2014.
17. Le Q V, Jaitly N, Hinton G E. A Simple Way to Initialize Recurrent Networks of Rectified Linear Units[J]. Computer Science, 2015.
18. Shi X, Chen Z, Wang H, et al. Convolutional LSTM Network: A Machine Learning Approach for Precipitation Nowcasting[J]. 2015, 9199:802-810.
19. Y. LeCun, L. Bottou, Y. Bengio, and P. Haffner. Gradient-based learning applied to document recognition. Proceedings of the IEEE, 86(11):2278–2324, 1998.
20. Hochreiter S. Recurrent Neural Net Learning and Vanishing Gradient[J]. 1998.
21. Kingma D P, Ba J. Adam: A Method for Stochastic Optimization[J]. Computer Science, 2014.
22. Ioffe S, Szegedy C. Batch Normalization: Accelerating Deep Network Training by Reducing Internal Covariate Shift[J]. 2015:448-456.

MATEC Web of Conferences **277**, 02002 (2019) https://doi.org/10.1051/matecconf/201927702002
JCMME 2018

Ghost-removal image warping for optical flow estimation

Song Wang[1,2], and *Zengfu* Wang[1,2,*]

[1]Institute of Intelligent Machines, Chinese Academy of Sciences, Hefei 230031, Anhui, China
[2]University of Science and Technology of China, Hefei 230026, Anhui, China

Abstract. Traditional image warping methods used in optical flow estimation usually adopt simple interpolation strategies to obtain the warped images. But without considering the characteristic of occluded regions, the traditional methods may result in undesirable ghosting artifacts. To tackle this problem, in this paper we propose a novel image warping method to effectively remove ghosting artifacts. To be Specific, when given a warped image, the ghost regions are firstly discriminated using the optical flow information. Then, we use a new image compensation technique to eliminate the ghosting artifacts. The proposed method can avoid serious distortion in the warped images, therefore can prevent error propagation in the coarse-to-fine optical flow estimation schemes. Meanwhile, our approach can be easily integrated into various optical flow estimation methods. Experimental results on some popular datasets such as Flying Chairs and MPI-Sintel demonstrate that the proposed method can improve the performance of current optical flow estimation methods.

1 Introduction

The concept of optical flow, which describes the apparent motion between images or video frames, arises from the studies of biological visual systems. Optical flow has been widely used in many computer vision applications, such as video segmentation, three-dimensional reconstruction, and video compression. The first available optical flow estimation method was proposed by Horn and Schunck [1] in *1981*. Although there is abundant literature on this topic, accurate estimation of optical flow from real-world still remains a challenging problem.

The methods of optical flow estimation can be roughly divided into two categories. The first one is to estimate optical flow directly from the images in original size, the other category is based on the coarse-to-fine schemes. Both categories are popular and have massive fresh achievements. Full Flow [2] optimizes a classic optical flow objective over the full space of mappings between discrete grids without descriptor matching. Bailer et al. [3] proposed a dense correspondence field approach with much less outlier prone, which achieves the best results on MPI-Sintel dataset. Flownet, the first method， using convolutional neural networks (CNNs) for optical flow estimation was introduced by Dosovitskiy et al. [4]. The coarse-to-fine scheme is a classic strategy due to its capacity to

* Corresponding author: zfwang@ustc.edu.cn

handle large motions [5], and later it becomes popular again because it can combine naturally with deep learning [6].

Be it conventional methods or latest approaches based on convolutional networks, image warping is the essential ingredient in coarse-to-fine optical flow estimation schemes. The concept of image warping for optical flow estimation was proposed by Lucas and Kanade [7]. Sun et al. [8] verified that the spline-based bi-cubic interpolation is better than the MATLAB built-in function--*interp2* which is the widely used by many researchers. However, in the warped images, the existing work neglects the ghost regions, which may result in erroneous information propagated in the coarse-to-fine optical flow estimation scheme. For instance, we can clearly observe the ghosting artifacts in the examples shown in figure 1. Therefore, high-precision warped images without ghosting are necessary in the coarse-to-fine optical flow estimation schemes.

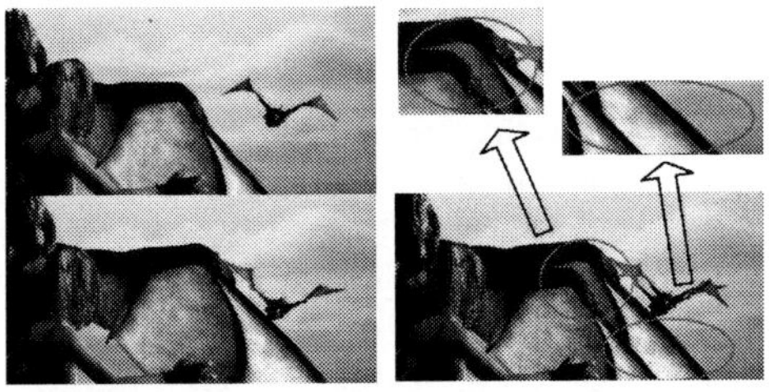

Fig. 1. Illustration of the ghosting artifacts. Left: image pair from MPI-Sintel dataset; Right: the warped image with ghosting.

In this paper, we propose a novel image warping strategy to handle the ghosting artifacts in the warped images. With the optical flow information, we first locate the ghost regions in the warped images. Then, the corresponding pixels in unwrapped images are determined according to the optical flow. Finally, the ghost effect is eliminated using an image compensation technique. In summary, the key contribution of this work are: 1) a ghost-removal image warping strategy which can be easily integrated into traditional and CNNs-based methods for optical flow estimation. 2) some insights about the importance of image warping for optical flow estimation.

2 Method

Image warping is an essential part of the coarse-to-fine spatial pyramidal structure for optical flow estimation. What's more, the warped images are directly related to the quality of optical flow estimation. We briefly introduce the coarse-to-fine optical flow estimation scheme in the first subsection. In the second subsection, we review the existing methods and propose our ghost-removal image warping method.

2.1 Coarse-to-fine optical flow estimation scheme

Optical flow is the pattern of apparent motion of objects in adjacent images, I_1 and I_2. In general, the first step is to construct the image spatial pyramids for both images respectively. We first estimated at the coarser resolution levels of the pyramids, is successively refined by taking into account the evidence of the finer resolution levels later. The overview of the coarse-to-fine optical flow estimation algorithm is presented in figure 2.

To be specific, we construct the spatial pyramids with n levels for both images firstly. We denote the k-th levels of pyramid of I_i as I_i^k, $i \in \{1,2\}$, $k \in \{1,2,...,n\}$. It should be point out that the bottom levels of pyramids, which labeled I_1^1 and I_2^1, are the images in original size. Then, we estimate the optical flow F_n at the coarsest resolution level using I_1^n and I_2^n. Then the warped image $\omega\,(I_2^{n-1},\ F_n)$ is obtained based on the optical flow F_n. Specifically, the result F_n needs to be interpolated before image warping. Next, with I_1^{n-1} and $\omega\,(I_2^{n-1},\ F_n)$, the optical flow F_{n-1} is estimated successively. Finally, the optical flow of every pixel in I_1 against I_2 can be effectively estimated after finite times of iteration.

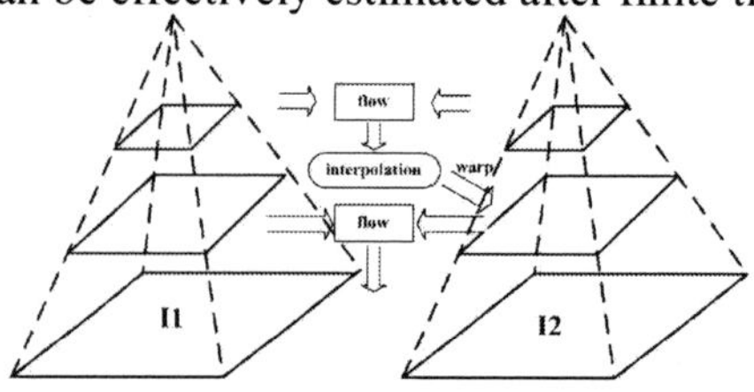

Fig. 2. Illustration of the coarse-to-fine spatial pyramidal optical flow estimation scheme.

2.2 Ghost-removal image warping method

Previous image warping method has neglected a disastrous problem—the ghosting artifacts in the warped images. The primary reason for ghosting is the warping image changed from I_1 to I_2. The occluded regions, which are covered by the occlusion regions, are virtually invisible in I_2. Without take into account this problem, the pixels of the occluded regions are replaced by those of the occlusion regions in the warped images wrongly. More specifically, the warped images record the moment before the motions, at which the occlusion regions and the occluded regions are separate. When the two regions coincide in I_2, only the occlusion regions are retained. However, the reason of ghosting in the warped images is that the occlusion regions and the occluded regions correspond to the same positions in I_2. Based on above descriptions and discussions, we propose our method which can discriminate and eliminate the ghosting artifacts in the warped images effectively.

Based on the statistics [9] of Flying Chairs and MPI-Sintel datasets, it is verified that the displacements of the foreground are larger than the background. Meanwhile, the occluded regions only belong to the background in most cases. As mentioned above, when the occlusion occurs, two or more positions in the warped image will trace back to the same pixel in I_2. According to the transformation matrix, it is easily to discriminate whether the positions in the warped image point to the same pixel. And the next step is to distinguish which position belongs to the foreground. In view of above-mentioned reasons, when the occlusion happened, only the position with larger displacement, meaning it belongs to the foreground, retains the corresponding pixel in I_2. Under this circumstance, the positions with small displacement are discriminated as the occluded regions, which are the ghost regions in most cases.

After the steps of mentioned above, we obtain the warped images in which the ghost regions are blank. The ways to fill the blank areas are closely related to the final quality of the warped images. Extensive experiments show that filling with *0* is not a wise choice. Since *0*, likes the value of the occlusion regions to some extent, is not the correct value of pixels in the blank areas. We propose two image compensation strategies for different situations. The first is utilize the neighbour pixels around the blank areas to filling the ghost regions, while the second one is filling the ghost regions with pixels in the corresponding positions in I_1. Details of the comparing experiments will be given in the next section, according to which filling with the pixels in I_1 is the optimal strategy for the optical flow estimation. It should be point out that I_1 is not the additional information for optical flow estimation since both images, I_1 and I_2, are required during computation.

3 Experiments

Image warping methods are the essential ingredient in the coarse-to-fine optical flow estimation schemes. Two optical flow estimation algorithms are used as the basic methods in this section. The first one is pyramidal LK optical flow (PLK) [10], and the second one is LDOF [5]. Replacing the traditional image warping methods with our ghost-removal image warping method are defined as our methods, called GrWarp-PLK and GrWarp-LDOF, respectively.

3.1 Image compensation strategies

In this subsection, we experimented with three image compensation strategies to filling the blank areas. One is filling with *0*, another is utilizing the neighbour pixels in I_2, and the third is using the corresponding pixels in I_1. For instance, the differences among three kinds of strategies mentioned above are shown in figure 3.

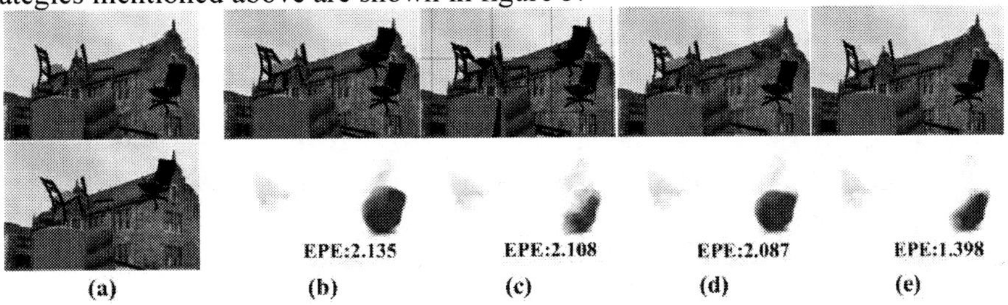

Fig. 3. Illustration of the warped images and the optical flow estimation. (a) image pair from Flying Chairs dataset; (b)the traditional method; (c) filling with *0*; (d) filling with the neighbour pixels in I_2; (e) filling with the corresponding pixels in I_1.

By comparing with the traditional method, the first strategy, which is filling with *0*, is not an acceptable choice, since *0* is not the correct value of pixels in the blank areas either. Filling with the neighbor pixels is similar to the image smoothing in some ways, which may reduce the possibility of mismatching in the ghost regions to some extent. The last strategy obtains the most accurate warped images, meanwhile achieving the best results. We can observe the significant differences on the borders of motion, which are the regions of ghosting, in the examples shown in figure 3. It is clearly that the ghosting artifacts directly influence the results. Meanwhile, our method eliminates the ghost regions effectively and achieves the best results.

3.2 Flying Chairs dataset

As discussed above, our method improves the results significantly. We have compared performance and quality with the basic methods in this subsection. Figure 4 illustrates the differences between the results produced by our method and those of the baseline method.

For each method, we calculate the average of endpoint error (AEE) for the test data. As of this writing, our method significantly reduces the endpoint error in both cases. Several examples of results from our method for this dataset are shown in figure 4. The comparison between the LDOF and the GrWarp-LDOF is used as examples in the following. The LDOF is almost fail to estimate optical flow of image pairs in the first row, while the GrWarp-LDOF achieves an acceptable result based on the ghost-removal strategy. It has a clearer contour of the chairs in optical flow image while using the GrWarp-LDOF in the second row, since our method can greatly reduce the mismatching on the borders of motion. It can be verified that the contour of the left chairs in the optical flow image in the third row shifted left obviously since the ghosting artifacts. Meanwhile, the optical flow image based on the GrWarp-LDOF can avoid this phenomenon successfully.

Fig. 4. Illustration of optical flow estimation on Flying Chairs dataset. In each row left to right: I_1, I_2, warped image based on the traditional method, optical flow estimation of LDOF, warped image based on our method, optical flow estimation of GrWarpLDOF. For all methods, the endpoint error is printed in the image.

3.3 MPI-Sintel dataset

Compared to the Flying Chairs dataset, the MPI-Sintel dataset is more difficult because of the complicated motion. Due to space limitations, we use temple 3, which including the self-occlusion and the occlusion, as examples in the following. The differences between the results produced by our method and those of the baseline method are shown in figure 5.

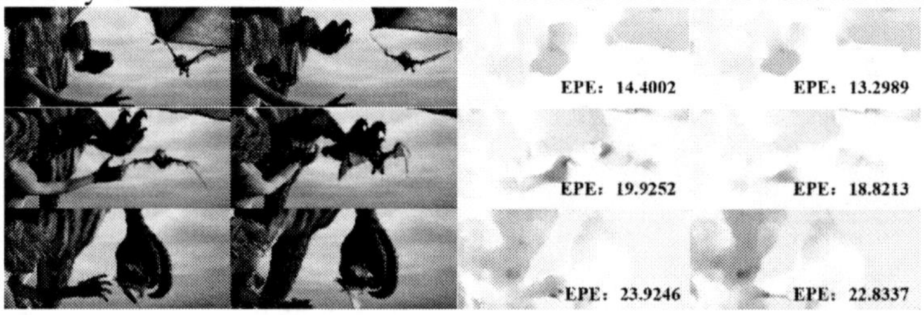

Fig. 5. Illustration of optical flow estimation on temple 3. In each row left to right: I_1, I_2, optical flow estimation of LDOF, optical flow estimation of GrWarp-LDOF. For all methods, the endpoint error is printed in the image.

As mentioned above this sequence is too difficult for the PLK. However, our method can reduce the endpoint error while using the rough optical flow estimation algorithm. It verifies our method still works at the coarse resolution levels of pyramid with the coarse optical flow results. As shown in figure 5, the GrWarp-LDOF has a great advantage in self-occlusion, such as the wings of bat in the first row and the clenched claws in the thrid row, and occlusion, such as the bat in the second row.

3.4 Supplementary details

We also compared performance and quality with the basic methods on the Middlebury and KITTI datasets and obtained the comparative results on both datasets. It is generally known that the Middlebury dataset is a classical optical flow benchmark, and it almost only contains small displacement motions. Without large displacement, our method does not present obvious improvement over the basic methods since the ghost regions are small. Due to space limitations, we will summary the results in table 1 without further details.

4 Conclusion

This paper presents an improved ghost-removal image warping method for optical flow estimation. We provide some insights about the importance of image warping, meanwhile, discriminating and eliminating the ghosting artifacts in the warped images effectively with utilizing the optical flow fully. Removing of the ambiguity of pixels of the ghost regions, improve the accuracy of optical flow estimation significantly. Furthermore, we hope that our ghost-removal image warping method, which can be easily integrated into the CNNs-based approaches, will support further advances.

Table 1. AEE of different methods.

dataset	PLK	GrWarp-PLK	LDOF	GrWarp-LDOF
Flying Chairs	11.8323	11.2567	4.4122	4.1783
Temple _3	41.3533	39.0473	12.3725	11.8819
KITTI(occ)	32.6077	32.3624	11.0514	10.9345
KITTI(noc)	25.4416	25.4497	5.4300	5.4402

Acknowledgments

This work is supported by National Natural Science Foundation of China (No:61472393).

References

1. Horn, B.K., Schunck, B.G.: Determining optical flow. 1981 *Artificial intelligence* **17** 185
2. Chen, Q., Koltun, V.: Full flow: Optical flow estimation by global optimizationover regular grids. 2016 *Proc. Int. Conf. on Computer Vision and Pattern Recognition.* (Las Vegas:IEEE) p 4706
3. Bailer, C., Taetz, B., Stricker, D.: Flow fields: Dense correspondence fields forhighly accurate large displacement optical flow estimation. 2015 *Proc. Int. Conf. on Computer Vision.*(Santiago:IEEE) p 4015
4. Dosovitskiy, A., Fischer, P., Ilg, E., Hausser, P., Hazirbas, C., Golkov, V., van derSmagt, P., Cremers, D., Brox, T.: Flownet: Learning optical flow with convolutional networks. 2015 *Proc. Int. Conf. on Computer Vision.*(Santiago:IEEE) p 2758
5. Brox, T., Malik, J.: Large displacement optical flow: descriptor matching in variational motion estimation. 2011 *IEEE Trans. on Pattern Analysis and Machine Intelligence* **33** 500
6. Ranjan, A., Black, M.J.: Optical flow estimation using a spatial pyramid network. 2017 *Proc. Int. Conf. on Computer Vision and Pattern Recognition.* vol 2 (Honolulu:IEEE)
7. Lucas, B.D., Kanade, T., et al.: An iterative image registration technique with anapplication to stereo vision 1981
8. Sun, D., Roth, S., Black, M.J.: A quantitative analysis of current practices inoptical flow estimation and the principles behind them. 2014 *Int. J. Computer Vision.* **106** 115
9. Ilg, E., Mayer, N., Saikia, T., Keuper, M., Dosovitskiy, A., Brox, T.: Flownet2.0: Evolution of optical flow estimation with deep networks. 2017 *Proc. Int. Conf. on Computer Vision and Pattern Recognition.* vol 2 (Honolulu:IEEE)
10. Bouguet, J.Y.: Pyramidal implementation of the affine lucas kanade feature trackerdescription of the algorithm. 2001 *Intel Corporation* **5** 4

MATEC Web of Conferences **277**, 02003 (2019) https://doi.org/10.1051/matecconf/201927702003
JCMME 2018

Content enrichment with expressive document modelling to leverage the understanding of unstructured data

Ganesh Selvaraj*, *Karla* Taboada, *Eloy* Gonzales, and *Habib* Baluwala

All in All Analytics Limited, New Zealand

Abstract. Most information in an enterprise is in the form of unstructured data which is usually managed using a document database. One of the key challenges is to define a generalized data model for this unstructured data and any information extracted from it using content enrichment algorithms. It is more challenging to incorporate provenance and temporal capabilities to such data models. Semantic databases use ontologies such as PROV-O to represent their provenance information expressively, and relational databases use for example Slowly Changing Dimensions (SCDs) concepts to represent temporal information. In this paper, we present a document model which has features inspired from Dublin core, PROV-O and temporal methodologies to generalize information extracted from unstructured data using content enrichment algorithms. Provenance information enables comparison of enrichment models, allows reproducibility and facilitates complex filtering on the enriched data. Temporal metadata helps in versioning the document and enables point-in-time and history queries conveniently.

1 Introduction

Most of the valuable enterprise data is either unstructured or semi structured [4, 8]. This data could be conversations, emails, notes, documents, logs, images, videos, audios, etc. From this data, content enrichment algorithms are used to extract information and features that have calculable value to the organization. Few of these algorithms are identification of parts of speech, entities (e.g. persons, organizations, locations, emails, URLs, dates, currencies, etc.), sentiments, keywords and relationships between entities. Complex data matching and identity resolution techniques can link the extracted information to the enterprise structured data. One of the key challenges here has been representing our content enrichment algorithm results efficiently in a document database using a generalized document model with temporal and provenance features. Using this generalized model, the application clients can use a single interface to parse all types of document. The document model concepts we used were inspired from Dublin Core metadata initiative [14].

* Corresponding author: allinallanalytics@gmail.com

Our method for provenance has been influenced by PROV-JSON [10] to some extent and on top of that we have tried to decouple data from provenance as much as possible.Data provenance and reproducibility of computations play vital role to achieve quality analytics. Real-time data, continuous or sliced snapshot data contains some temporal aspects of it [13]. Several researchers have focused on introducing time-based document models [12, 9]. However, most of the research has centred on relational and object-oriented models.

Incorporation of temporal and provenance features enabled us to perform model comparisons, complex filtration of data using trust as a parameter, temporal information retrieval, etc.Provenance also helps in achieving reproducible process and improves data exploration [5, 7].

2 Content Enrichment Pipeline Overview

Natural Language Processing (NLP) algorithms have enabled us to extract entities, topics, keywords, sentiments, relationships, etc. from the unstructured data e.g. text. Our enrichment pipeline helps us to transform the unstructured data into a valuable asset and interpret the data with less human intervention.

The proposed content enrichment pipeline deals with textual formats, audio files (e.g. call center conversations) and images (e.g. scanned documents) as of now. By integrating this content enrichment pipeline with an expressive document data model (discussed in section 3), understanding and using analyses results is made simpler, intuitive, and most importantly represented the concepts of domain knowledge to a well known common knowledge.

2.1 System Design

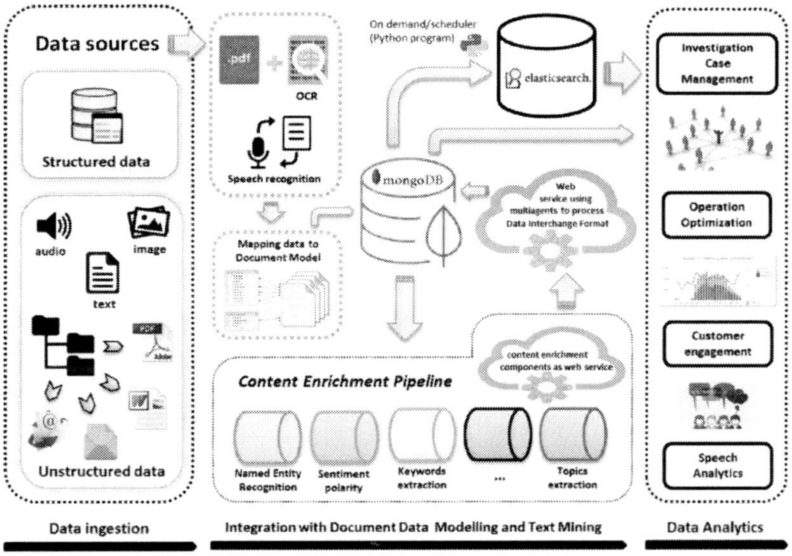

Fig. 1. Proposed System Architecture.

(i) **Data Collection:** The data collection is carried out in different data sources such as relational databases (e.g. Oracle), text files, emails, images, audio etc. The data collection process imports data from these sources and pass the data in a text format, i.e. audio is converted to text using speech translation techniques (e.g. CMUSphinx software[1]), images are processed by OCR tools(e.g. tesseract[2]). This data is then sent to a document modelling engine which is discussed in the next section.

(ii) **Mapping Data to Document Model:** The document model under discussion in this paper is JSON document format, and is schema free in nature. The reason for choosing JSON is to be aligned with the format accepted by popular document databases, in our case we are using MongoDB[3]. Mapping data to document model has two processes; initial import of document and adding enrichment metadata to the document. During initial import of documents, the modelling engine is made aware of the domain we are dealing with using configuration and mapping files. If the same document that is already present in MongoDB is passed through this engine, then it performs a logical merge of the documents in-memory, and inserts a new document with a different version.

(iii) **MongoDB:** The results from the content enrichment pipeline is persisted in MongoDB (version 3.4) using a document model discussed in section 3. MongoDB is our integration point i.e. after the documents are initially loaded into it, all applications read and write to MongoDB only. Having a database as an integration point also decoupled any contradicting technologies used in our applications e.g. JDK mismatches.

(iv) **Content Enrichment Pipeline:** In this section, we explain the different components we use to extract metadata from the content.

- Named entity recognition (NER) & Relationship Extraction: It is the process of identifying entities and relationships from text data. Our primary NER platform is General Architecture for Text Engineering (GATE version 8) [6]. Some of the entities are very domain specific which are not recognized by generic domain NER's, in those cases we use regular expression patterns, JAPE[4], etc.
- Gender Identification: For a given name (either available or extracted using NER), we identify the gender e.g. male or female. Stanford coreNLP (gender annotation for English names) and dictionaries of male and females names in English, Maori, Chinese, Indian, Spanish among others are used. Currently, dictionary approach is used if Stanford core is unable to identify gender.
- Sentiment Polarity Classification: It is the process of identifying the view or opinion associated to an information. We use Stanford CoreNLP libraries [11], which is based on a new type of Recursive Neural Network that builds on top of grammatical structures. Keyword Extraction: It is the process of extracting keywords (both generic and domain specific) from the unstructured data. We are using Knime's Keygraph (graph-based approach) and Chi-square keyword (co-occurrence statistics) extractor.
- Topic Extraction: It is the process of identifying topics a document could be related to. We are using Knime's Topic Extractor (Parallel LDA (Latent Dirichlet Allocation) statistical model [3]) node. The input is a collection of documents and the output is the document collection with topic assignments and the probability for each document to a certain topic.

(v) **Data Interchange Format to Document Model:** The content enrichment pipeline and its components can use different technologies, tools, programming languages, etc. Each

---

[1] https://cmusphinx.github.io/

[2] https://github.com/tesseract-ocr

[3] https://www.mongodb.com/

[4] https://gate.ac.uk/sale/tao/splitch8.html

of these technologies, for the same algorithm provides the output in a different format, and we do not want to implement our relatively complex document model mapping logic into all these components. So we came up with a lightweight data interchange format (DIF), which all these components will use to represent and exchange their results. We have a document modelling web service, which coordinates communication from the content enrichment components to MongoDB. This web service receives the DIF and converts it to the corresponding document model.

3 Proposed Document Model

A document can be defined as a collection of information (either core and/or meta) about something or even several things usually from a domain. This core or meta information can be either structured, semi-structured or unstructured. We use the content enrichment pipeline discussed earlier to extract more information from the unstructured or semi-structured part of the document. Our document model can handle unstructured data, other associated semi- structured metadata available from the domain (e.g. from legacy relational databases), and the extracted data from the enrichment pipeline. Along with this data, the document model features a temporal and provenance model as well. Documents are expressed in JSON as per requirement for document databases, hence the document model is also expressed in JSON to have a faithful representation.

3.1 Document Components

In this section, we describe the different components of our document model.

(i) **docId:** Every document will have an identifier in the form of a universal resource identifier (URI). This URI is unique to a document, but across versions of the same document, this is not unique. Each document has a version number, and along with this docId, this will create a compound key for unique identification of a document. If the base URI is like this http://www.xyz.nz/, then the document from domain retail, with id 1 will have an URI such as http://www.xyz.nz/retail?id=1. The domain is used as one of the fragments in the URI to prevent overlap of same identifiers across two domains.

(ii) **version:** Every document will have version information, which is of type number.

(iii) **textContent:** Any unstructured data of the document is stored in this field. When the textContent is cleansed e.g. boilerplate removal, the original text is retained and the cleansed text is added to the "cleansedTextContents" field.

(iv) **sentiments:** The sentiment value associated with a document is stored in this field. It will contain one of these values - very negative, negative, neutral, positive, very positive. Keywords extracted from the document also follows a similar model to the sentiment model.

(v) **parties:** Person or organization along with their roles e.g. customer, agent extracted from the text is generalized and stored in a field called "party". Based on the specialization, a party will have more or fewer fields. We use the "type" field to classify the party.

(vi) **names:** Any name (belonging to person or organization) mentioned/extracted from the text is stored in this field. In our current enrichment pipeline, we detect the gender according to the name, hence the gender model is also placed alongside the name model.

(vii) **addresses:** Any address information mentioned/extracted from the text is stored in this field. Each component of address can have coordinates information.

(viii) **topics:** The results of topic modelling are stored in this field. The "topicProbability" field contains the probability of the topic for that document. TopicWords has an array of pairs -"topicWord" and "topicWordWeightage" which are generated during topic modelling. Topic modelling is performed in a collection of documents, so whenever a topic modelling is performed, a corpusId is generated.

(ix) **metadate:** Date and time information is captured using "metadate" field. These meta dates can be attached to any part of the document.

(x) **prov:** Prov is a short hand for provenance. Provenance represents information such as who, how and when about a change. The prov model used in our document model is based on Prov-O ontological model [2]. Prov-O expressed the provenance using three concepts: entity, activity and agents. When modelling with documents, ideally any change will have only one prov, but a prov can have any number of activities. Nested activity can mean its a chain reaction (i.e. an activity influenced other). Adjacent activity means several activities (different ones) have said the same thing. The docId for prov and activities are randomly generated unique Id's to help transactional systems to easily locate a piece of information.

(xi) **deleted:** A boolean field which says if a document is deleted. We do not use physical delete, hence having a delete field to represent the deleted status of the document. By default deleted is not added to a document, and it is only added when explicitly specified.

(xii) **relationships:** Even though document model is not the most ideal for representing relationships, but we wanted to persist this information so when we build graphs later, we could make use of this information. Any relationships extracted from the unstructured data is represented using index free adjacency methodology, i.e. whenever a relationship is found for a document (e.g. party), a relationship document model is added for that part of the document. Every document and any embedded document has a docId field, and the relationship model makes use of this docId to refer the target document.

4 Example Queries

In this section, the expressiveness and usability of the proposed document model is described with help of few example queries. Consumer Compliant Dataset [1] has been used in this paper. It is a public dataset containing complaints received about financial products and services published by the US Consumer Financial Protection Bureau.

4.1 Information Retrieval using Queries

MongoDB is used to persist the dataset in our document model. For this experiment, a Mongo collection called "CIE" is created. The effectiveness of the proposed document model is described in this section, especially referred to write expressive queries using the MongoDB query language are used.

(i) **Query 1:** The query shown in Listing 1 finds all documents which are affected by any entity extraction algorithm. This is a very useful audit/profiling query to discover which documents have been used or missed out in NER process.

```
db.getCollection('CIE').find({"parties.prov.activity": { $elemMatch: { "name" : {
    $in: [ 'EntityExtraction' ]   }}}})
```

Listing 1. Query 1

(ii) **Query 2:** A common use case in machine learning is the evaluation or comparison of results by different algorithms. The proposed document model allows for comparison of results from different methods, e.g. in Listing 2 finds all documents with different sentiment values i.e. same document with negative and positive sentiment.

```
db.getCollection('CIE').find({"sentiments": {$elemMatch: {"sentiment": {$in: [ '
    NEGATIVE']}}},"sentiments": {$elemMatch: {"sentiment": {$in:[ 'POSITIVE
    ']}}}})
```

Listing 2. Query 2

(iii) **Query 3:** A more complex example is shown in Listing 3; a query to find all documents with negative sentiment and which have the key word extracted as *"child"*. From an operations point of view, a document involving a child and negative sentiment is a priority to look into it, and with our document model, such expressive queries can also be easily represented.

```
db.getCollection('CIE').find({$and:[  {"sentiments":  {$elemMatch: {"sentiment": {
    $in: [ 'NEGATIVE' ]  }}}},{"keywords":  {$elemMatch: {"keyword": { $in: [ 'child
    ']}}}}]})
```

Listing 3. Query 3

5 Conclusion

In this paper, we presented a document model to capture the enrichment results along with provenance and temporal metadata. The provenance feature based on PROV-O is added to the document model which enables content enrichment model comparison and reproducibility. Temporal capability in the document model allows for point in time and history queries. The document model is generalized for the most of the concepts. Hence applications can use a single interface to consume the document model for multiple domains as well. Our content enrichment pipeline comprises of different (and sometimes opposing) technologies, methodologies and tools, but they all work together by using a document database (MongoDB) as the integration point.

References

1. Consumer financial protection bureau. https://catalog.data.gov/dataset/consumer-complaint-database, [Online; accessed 12-May-2017]
2. Belhajjame, K., Cheney, J., Corsar, D., Garijo, D., Soiland-Reyes, S., Zednik, S., Zhao, J.: Provo: The prov ontology. w3c recommendation. Online]. http://www. w3. org/TR/provo (2013)
3. Blei, D.M., Ng, A.Y., Jordan, M.I.: Latent dirichlet allocation. Journal of machine Learning research 3(Jan), 993–1022 (2003)
4. Blumberg, R., Atre, S.: The problem with unstructured data. Dm Review 13(42-49), 62 (2003)
5. Cheah, Y.W., Plale, B.: Provenance analysis: Towards quality provenance. In: E-Science (e-Science), 2012 IEEE 8th International Conference on. pp. 1–8. IEEE (2012)

6. Cunningham, H., Wilks, Y., Gaizauskas, R.J.: Gate: a general architecture for text engineering. In: Proceedings of the 16th conference on Computational linguistics-Volume 2. pp. 1057–1060. Association for Computational Linguistics (1996)
7. Davidson, S.B., Freire, J.: Provenance and scientific workflows: challenges and opportunities. In: Proceedings of the 2008 ACM SIGMOD international conference on Management of data. pp. 1345–1350. ACM (2008)
8. Feldman, R., Sanger, J.: The text mining handbook: advanced approaches in analyzing unstructured data. Cambridge university press (2007)
9. Gregersen, H., Jensen, C.S.: Temporal entity-relationship models-a survey. IEEE Transactions on knowledge and data engineering 11(3), 464–497 (1999)
10. Huynh, T.D., Jewell, M.O., Sezavar Keshavarz, A., Michaelides, D.T., Yang, H., Moreau, L.: The prov-json serialization (2013)
11. Manning, C.D., Surdeanu, M., Bauer, J., Finkel, J.R., Bethard, S., McClosky, D.: The stanford corenlp natural language processing toolkit. In: ACL (System Demonstrations). pp. 55–60 (2014)
12. Ozsoyoglu, G., Snodgrass, R.T.: Temporal and real-time databases: A survey. IEEE Transactions on Knowledge and Data Engineering 7(4), 513–532 (1995)
13. Panigati, E., Schreiber, F.A., Zaniolo, C.: Data streams and data stream management systems and languages. In: Data Management in Pervasive Systems, pp. 93–111. Springer (2015)
14. Weibel, S., Kunze, J., Lagoze, C., Wolf, M.: Dublin core metadata for resource discovery. Tech. rep. (1998)

Machine translation using natural language processing

Middi Venkata Sai Rishita[1,*], *Middi Appala* Raju[2], and *Tanvir Ahmed* Harris[3]

[1]RNS Institute of Technology, India
[2]Christ University, India
[3]IIT Bombay, India

Abstract. Machine Translation is the translation of text or speech by a computer with no human involvement. It is a popular topic in research with different methods being created, like rule-based, statistical and example-based machine translation. Neural networks have made a leap forward to machine translation. This paper discusses the building of a deep neural network that functions as a part of end-to-end translation pipeline. The completed pipeline would accept English text as input and return the French Translation. The project has three main parts which are preprocessing, creation of models and Running the model on English Text.

1 Introduction

Machine translation (MT) is a domain of computational linguistics, which explores the use of software to translate text or speech from language to another. Machine translation simply performs substitution of words in one language for words in other, but that may not assure good translation. A more sophisticated method which is also a growing field used to address the issue of recognition of multiple phrases is with statistical and neural technique. In this translation of text from one language to another, there is no human involvement and it is the machine which performs the process of conversion. There are three types of machine translation system-rules based, statistical and neural. Rule based is a conventional method which is a combination of language and grammar and the support of dictionaries. This work focusses on building an end to end machine translation pipeline. We have discussed multiple existing architectures and finally proposed a hybrid model to achieve a more powerful system for machine translation from English to French.

2 Part 1: Dataset

The first step in the implementation of any Deep Learning project is the investigation of DataSet that would be used to train the pipeline as well as evaluate the pipeline. Most commonly, the datasets used for Machine Translation are from WMT (the website that is dedicated to research in statistical machine translation). Due to time constraints, the dataset contains small vocabulary. This facilitates the training in a reasonable time.

* Corresponding author: rishimiddi@gmail.com

The data located in a filepath is loaded. The file contains English sentences with their French translations. Load the data from these files. Print the first two lines from each file.

```
small_vocab_en Line 1:  new jersey is sometimes quiet during autumn , and it is snowy in april .
small_vocab_fr Line 1:  new jersey est parfois calme pendant l' automne , et il est neigeux en avril .
small_vocab_en Line 2:  the united states is usually chilly during july , and it is usually freezing in november .
small_vocab_fr Line 2:  les états-unis est généralement froid en juillet , et il gèle habituellement en novembre .
```

Fig. 1. First two lines of English and French from each file.

In the figure above, the punctuations have been delimited using spaces. The conversion to lowercase has been done. The complexity of vocabulary determines the complexity of the problem. The complexity of the dataset is given below.

```
1823250 English words.
227 unique English words.
10 Most common words in the English dataset:
"is" "," "." "in" "it" "during" "the" "but" "and" "sometimes"

1961295 French words.
355 unique French words.
10 Most common words in the French dataset:
"est" "." "," "en" "il" "les" "mais" "et" "la" "parfois"
```

Fig. 2. The complexity of dataset.

3 Part 2: Pre-process the data

The text data is not used as input to the model. The text is converted into sequences of integers using the pre-process methods which are Tokenization and addition of Padding.

3.1 Tokenize (Implementation)

The neural network has to comprehend the input data for it to predict on text data. The network can understand ASCII characters. The text data like "dog" is a sequence of ASCII character encodings. The operations performed on network include multiplication and addition operations, hence, text is converted to numbers.

Each character and word is assigned a number, conventionally called character IDs and word IDs, respectively. A word level model used word ids that generate text predictions for every word. This has lower complexity compared to character level model.

Keras has a utility class known as Tokenizer. This class allows to vectorize a text corpus, by turning each text into either a sequence of integers or into a vector where the coefficient for each token could be binary, based on word count. Run tokenize on sample data and sow the output fordebugging.

```
{'the': 1, 'quick': 2, 'a': 3, 'brown': 4, 'fox': 5, 'jumps': 6, 'over': 7, 'lazy': 8, 'dog': 9, 'by': 10, 'jove': 11
, 'my': 12, 'study': 13, 'of': 14, 'lexicography': 15, 'won': 16, 'prize': 17, 'this': 18, 'is': 19, 'short': 20, 'se
ntence': 21}

Sequence 1 in x
  Input:  The quick brown fox jumps over the lazy dog .
  Output: [1, 2, 4, 5, 6, 7, 1, 8, 9]
Sequence 2 in x
  Input:  By Jove , my quick study of lexicography won a prize .
  Output: [10, 11, 12, 2, 13, 14, 15, 16, 3, 17]
Sequence 3 in x
  Input:  This is a short sentence .
  Output: [18, 19, 3, 20, 21]
```

Fig. 3. Output of debugging.

3.2 Padding (Implementation)

When batching the sequence of word ids together, each sequence needs to be the same length. Since the sentences are dynamic in length, padding is done at the end of sentences to make them the same length. The function used to perform padding is present in Keras as pad_sequences. It takes three parameters which is x- List of sequences, length-Length to pad to and return padded numpy array of sequences. The output of padding is given below.

```
Sequence 1 in x
  Input:  [1 2 4 5 6 7 1 8 9]
  Output: [1 2 4 5 6 7 1 8 9 0]
Sequence 2 in x
  Input:  [10 11 12  2 13 14 15 16  3 17]
  Output: [10 11 12  2 13 14 15 16  3 17]
Sequence 3 in x
  Input:  [18 19  3 20 21]
  Output: [18 19  3 20 21  0  0  0  0  0]
```

Fig. 4. Output of padding.

3.3 Pre-process Pipeline

The Keras preprocess function is used to create preprocess pipeline. It takes two parameters x- Feature List of sequences, y- Label List of sentences and return a tuple of preprocessed x, preprocessed y, x tokenizer, y tokenizer. The Keras's sparse_categorical_crossentropy function requires labels to be in three dimensions.

4 Models

In this work, we have experimented on various neural network architectures which include simple Recurrent Neural Network, simple RNN with embedding, a Bidirectional RNN and an Encoder- Decoder RNN. We have compared the functioning of these four architectures. The final architecture was built to outperform all the four models discussed above.

The neural network needs to give the French Translation but it would be translating the input to word ids. The function logits_to_text will bridge the gab between the logits from the neural network to the French translation. It uses tokenizer to turn logits from a neural network to the French translation.

4.1 Simple RNN Implementation

Build and train a basic RNN on x and y. The parameters include a Tuple of input shape, length of output sequence, number of unique English words in the dataset and the number of unique French words in the dataset. It returns a built Keras model.

After this, train the neural network and reduce the batch size to 100 from 1024 and print the prediction. Prior to training the neural network, the layers are built and reshape the input to work with a basic Recurrent neural network.

The information about the layers and parameters are given below in the picture

```
Layer (type)                 Output Shape              Param #
=================================================================
gru_100 (GRU)                (None, 21, 128)           49920
_________________________________________________________________
time_distributed_116 (TimeDi (None, 21, 512)           66048
_________________________________________________________________
dropout_54 (Dropout)         (None, 21, 512)           0
_________________________________________________________________
time_distributed_117 (TimeDi (None, 21, 344)           176472
=================================================================
Total params: 292,440
Trainable params: 292,440
Non-trainable params: 0
_________________________________________________________________
```

Fig. 5. Architecture of a Simple RNN.

4.2 RNN with Embedding

An embedding is a vector representation of the word that is close to similar words in n-dimensional space, where the n represents the size of the embedding vectors. Instead of turning the words into ids, word embeddings are used. The input is reshaped before training the neural network. The passed index length is increased by 1 to avoid index error. After this, the batch size is reduced to 100. The number of epochs taken is 10 and the validation split is 0.2. The prediction is printed.

```
Input Shape (None, 21), Output Shape (None, 21, 344)
Input Shape (None, 21), Output Shape (None, 21, 345)
_________________________________________________________________
Layer (type)                 Output Shape              Param #
=================================================================
embedding_45 (Embedding)     (None, 21, 128)           25600
_________________________________________________________________
gru_102 (GRU)                (None, 21, 128)           98688
_________________________________________________________________
time_distributed_120 (TimeDi (None, 21, 512)           66048
_________________________________________________________________
dropout_56 (Dropout)         (None, 21, 512)           0
_________________________________________________________________
time_distributed_121 (TimeDi (None, 21, 345)           176985
=================================================================
Total params: 367,321
Trainable params: 367,321
Non-trainable params: 0
```

Fig. 6. Architecture of RNN with Embedding.

4.3 Bidirectional RNNs

The RNN outputs depend on present inputs and the memory of the previous outputs. But Recurrent Neural Networks don't see the future input. Thus, Bidirectional RNN are used. A Bidirectional RNN is a neural network in which the neurons of a regular RNN are split into two directions- positive time direction and negative time direction. For involving future information, delays need not be added when 2 time directions are incorporated. Implement the Bidirectional RNN by using the tanh activation function.

```
_________________________________________________________________
Layer (type)                 Output Shape              Param #
=================================================================
bidirectional_58 (Bidirectio (None, 21, 256)           99840
_________________________________________________________________
time_distributed_124 (TimeDi (None, 21, 512)           131584
_________________________________________________________________
dropout_58 (Dropout)         (None, 21, 512)           0
_________________________________________________________________
time_distributed_125 (TimeDi (None, 21, 345)           176985
=================================================================
Total params: 408,409
Trainable params: 408,409
Non-trainable params: 0
```

Fig. 7. Architecture of Bidirectional RNN.

4.4 Encoder-Decoder Models

The model consists is made up of an encoder and decoder. The encoder creates a matrix representation of the sentence. The decoder takes this matrix as input and predicts the translation as output. While implementing the encoder reverse the input sequence order for improved accuracy. Design an adapter to fit 2-dimensional output to required 3-dimensional input shape [samples, time steps, features]. The steps involved in encoder model are as follows:

1. Reverse input sequence order for improved accuracy.
2. Adapter to fit 2D output to required 3D input shape [samples, time steps, features]

The decoder is implemented after this. The neural network is trained and prediction is printed.

```
_________________________________________________________________
Layer (type)                 Output Shape              Param #
=================================================================
gru_107 (GRU)                (None, 128)               49920
_________________________________________________________________
repeat_vector_41 (RepeatVect (None, 21, 128)           0
_________________________________________________________________
gru_108 (GRU)                (None, 21, 128)           98688
_________________________________________________________________
time_distributed_128 (TimeDi (None, 21, 512)           66048
_________________________________________________________________
dropout_60 (Dropout)         (None, 21, 512)           0
_________________________________________________________________
time_distributed_129 (TimeDi (None, 21, 345)           176985
=================================================================
Total params: 391,641
Trainable params: 391,641
Non-trainable params: 0
```

Fig. 8. Architecture of encoder-decoder model.

4.5 Proposed model

In the proposed model, first embedding is done after which a bidirectional encoder is added. Add an adapter to fit 2D output to required GRU 3D input shape [samples, time steps, features].

Finally implement the decoder. The architecture is given below.

```
_________________________________________________________________
Layer (type)                 Output Shape              Param #
=================================================================
embedding_49 (Embedding)     (None, 15, 128)           25600
_________________________________________________________________
bidirectional_65 (Bidirectio (None, 256)               197376
_________________________________________________________________
repeat_vector_45 (RepeatVect (None, 21, 256)           0
_________________________________________________________________
bidirectional_66 (Bidirectio (None, 21, 256)           295680
_________________________________________________________________
time_distributed_136 (TimeDi (None, 21, 512)           131584
_________________________________________________________________
dropout_64 (Dropout)         (None, 21, 512)           0
_________________________________________________________________
time_distributed_137 (TimeDi (None, 21, 345)           176985
=================================================================
Total params: 827,225
Trainable params: 827,225
Non-trainable params: 0
```

Fig. 9. Architecture of the proposed model.

```
new jersey est parfois chaud au mois de il est est en est <PAD> <PAD> <PAD> <PAD> <PAD> <PAD> <PAD> <PAD>
Original text and translation:
['new jersey is sometimes quiet during autumn , and it is snowy in april .']
["new jersey est parfois calme pendant l' automne , et il est neigeux en avril ."]
```

Fig. 10. Original text and translation.

5 Results

The final sentence translations are returned in French with an appreciable accuracy of 96.71% The samples of French translations that were returned is given below;

```
Sample 1:
il a vu un vieux camion jaune <PAD> <PAD> <PAD> <PAD> <PAD> <PAD> <PAD> <PAD> <PAD> <PAD> <PAD> <PAD> <PAD> <PAD>
Il a vu un vieux camion jaune
Sample 2:
new jersey est parfois calme pendant l' automne et il est neigeux en avril <PAD> <PAD> <PAD> <PAD> <PAD> <PAD> <PAD>
new jersey est parfois calme pendant l' automne et il est neigeux en avril <PAD> <PAD> <PAD> <PAD> <PAD> <PAD> <PAD>
```

```
acc: 96.24%
acc: 96.90%
acc: 96.98%
96.71% (+/- 0.33%)
```

6 Conclusions

Thus, we have built a deep neural network that functions as part of an end-to-end machine translation pipeline to convert English text as input and return the French translation. The proposed network performed far better than the other architectures discussed previously in terms of the validation loss. The accuracy was 96.71%. A random seed was fixed for reproducibility, a 3-fold cross validation test harness was determined and the model was compiled and evaluated.

References

1. Keras documentation
2. SB Sall on "Example based machine translation using natural language processing" 2013
3. Gurleen Kaur sidhu on "Role of Machine Translation and Word sense Disambiguation in Natural Language Processing" 2013
4. MV Reddy on "NLP Challenges for machine translation from english to Indian languages"
5. Kehai Chen, Tiejun Zhao on "A neural Approach to source dependance based Context model for Statistical Machine Translation" 2018

DNS: A multi-scale deconvolution semantic segmentation network for joint detection and segmentation

Ning Feng[1], *Le* Dong[1,*], *Qianni* Zhang[2], *Ning* Zhang[3], *Xi* Wu[4] and *Jianwen* Chen[5]

[1]School of Computer Science and Engineering, University of Electronic Science and Technology of China, Chengdu, 611731, China
[2]School of Electronic Engineering and Computer Science, Queen Mary, University of London, London, E1 4NS, UK
[3]School of Information and Software Engineering, University of Electronic Science and Technology of China, Chengdu, 611731, China
[4]School of Computer Science, Chengdu University of Information Technology, Chengdu, 610103, China
[5]School of Information and Communication Engineering, University of Electronic Science and Technology of China, Chengdu, 611731, China

Abstract. Real-time semantic segmentation has become crucial in many applications such as medical image analysis and autonomous driving. In this paper, we introduce a single semantic segmentation network, called DNS, for joint object detection and segmentation task. We take advantage of multi-scale deconvolution mechanism to perform real time computations. To this goal, down-scale and up-scale streams are utilized to combine the multi-scale features for the final detection and segmentation task. By using the proposed DNS, not only the tradeoff between accuracy and cost but also the balance of detection and segmentation performance are settled. Experimental results for PASCAL VOC datasets show competitive performance for joint object detection and segmentation task.

1 Introduction

Semantic segmentation is a technique to assign semantic or object-class labels to individual pixels in images [1]. It is usually converted into pixel-wise classification problem, and focuses on the connection of semantics and location. The natural step in semantic segmentation is utilizing global information to resolve what while utilizing local information to resolve where [2].

Despite the attention it has received, global context and interplay between labelling and detection of object instance are still the restricted factors of semantic segmentation. Before the Simultaneous Detection and Segmentation (SDS) [3] were proposed, the detection and segmentation tasks are usually treated as individual one. Especially, most of the existing semantic segmentation approaches focus on single inference of net-design for higher accuracy, which are difficult to extend to incorporate other types of tasks. The effectiveness

* Corresponding author: ledong@uestc.edu.cn

of these convolutional nets largely depends on the sophisticated model design regarding depth and width, which has to involve many operations and parameters.

For the remarkable progress of recent deep convolutional neural nets, we in this paper resort to DCNN method similarly to manipulate semantic segmentation task. Besides, the problem of using a single network to handle multiple tasks has been repeatedly pursued in the stage of deep learning. In [4] the DCNN is used for recognition, localization and detection, while in [5] DCNN is trained for surface normal estimation, depth estimation and semantic segmentation, and [6] for joint detection, pose estimation and region proposal generation. Importantly, unlike the aforementioned holistic approaches, we are interested in exploiting semantic segmentation in order to improve both object detection and segmentation performance. Here, we take advantage of deconvolution mode with the sharing weight to combine these two tasks. Towards this goal, our convolutional network is trained under the supervision of bounding boxes and segmentation maps.

Meanwhile, real-time semantic segmentation has become crucial in many practical applications and brought with fundamental difficulty of reducing computation for pixel-wise label inference. As the prevalent image detection and segmentation pipelines, the approach remains expensive and relies on a region-based strategy that makes the network architecture inappropriate for semantic segmentation. Learning and inference in our model are efficient as we reason at the detection and segment level. We extensively evaluate the proposed model called 'DNS' on the PASCAL detection and segmentation benchmarks. What's more, the proposed method shows a graceful degradation compared with its counterpart.

Our main contributions are summarized below:

1. We focus on building a deep model for joint detection and semantic segmentation with a decent speed. To work out this problem, we introduce a multi-scale deconvolution mechanism which is a direct mode to perform easily. Specifically, we learn a multi-layer deconvolution network, which is composed of down-scale and up-scale stream. In this stream, we combine the multi-scale features instead of using the multi-scale inputs which has been demonstrated that outperforms average- and max-pooling, and can achieve excellent performance.

2. The trained 'DNS' network makes it possible to train a single net for multiple task (detection and segmentation). We achieved competitive advantages in PASCAL VOC benchmark. In addition to the trade-off between accuracy and inference cost, you will find that our DNS trained only on the PASCAL VOC dataset settle the balance of detection and segmentation performance.

The paper is organized as follows: Section 2 discusses related work; Section 3 presents our real-time multi-task framework. Finally, we devote to our experiments in Section 4, and Section 5 concludes the paper.

2 Related work

Before we introduce our approach, we now present techniques for both detection and semantic segmentation.

Representative methods [7-11] consider semantic segmentation task as simultaneous detection and segmentation (SDS), which in introduced in [7]. Semantic segmentation has recently witnessed rapid progress, but many leading methods are unable to identify object instances. To encourage the research on this problem, a Multi-task Network Cascades (MNCs) [11] for instance-aware semantic segmentation is proposed. This model consists of three networks, respectively differentiating instances, estimating masks, and categorizing objects. Although Multi-scale CNNs and their variants have made striking success for modelling the global scene structure for an image, they are limited in labelling fine-grained local structures like pixels and patches, since spatial contexts might be blindly mixed up without customizing their scales. Convolutional Feature Masking [10], connected Markov random field models [8], and Mask R-CNN [9] are also designed to address the issue of contexts of object labels.

In recent years, neural networks are driving advances in semantic segmentation, in which each pixel is labelled with the class of its enclosing region. Most of the convolutional versions of existing networks obtain precise segmentation from fixed-sized inputs in a particular dataset. These works [12, 3, 13], bring together DCNN methods and traditional computer vision algorithms for addressing pixel-wise segmentation problem.

Through the use of contextual information, 'deep CRFs' [14] is proposed to improve semantic segmentation, by combining the strengths of deep CNNs to learn powerful feature representations, with Conditional Random Fields (CRFs) which can capture contextual relation modelling. This method avoids repeated inference, and so is computationally tractable.

Incorporating multi-scale features in fully convolutional neural networks (FCNs) [2] has been a key element to achieving state-of-the-art performance on semantic segmentation. One common way to extract multi-scale features is to feed multiple resized input images to a shared deep network and then merge the resulting features for pixel-wise classification.
FCNs uses large receptive field and many pooling layers, both of which cause blurring and low spatial resolution in the deep layers. As a result, FCNs tends to produce segmentations that are poorly localized around object boundaries. Using a color-based CRF on top the FCN prediction [13] is one way that attempts to address this issue in post-processing steps.
Although post-processing the output of FCN with a fully-connected CRF can increase segmentation accuracy near object boundaries, mean-field inference in fully-connected CRF model is expensive in terms of both memory and CPU time. To this end, a task-specific edge detection model [15] using CNNs and a discriminatively trained domain transform is proposed. This domain transform can equivalently be seen as a recurrent neural network (RNN), and it is a special case of the recently proposed RNN with gated recurrent units.

In addition, using CRF on FCN require additional parameters and low-level features that are difficult to tune and integrate into the original network architecture. To overcome these problem, a Boundary Neural Field (BNF) [16] is proposed. It is a global energy model integrating FCN predictions with boundary cues. Further, some steer DNN architectures, like decoupled DNN [1], Deconvolution net [17] are designed to make precise per-pixel label prediction tasks.

While the discrete CRF is a natural _t for labelling tasks of semantic segmentation, a new end-to-end trainable deep network, referred to as Gaussian Mean Field (GMF) network [18], whose layers perform mean field inference over a Gaussian CRF, is proposed. The Gaussian CRF is composed of three sub-networks: a CNN-based unary network for generating unary potentials, a CNN-based pairwise network for generating pairwise potentials, and a GMF network for performing Gaussian CRF inference. This method outperforms various recent semantic segmentation approaches that combine CNNs with CRF models.

Meanwhile, some similar works set out to deal with both detection and semantic segmentation jointly [19-22]. Yao and Fidler [19] propose a traditional approach with high-order potentials to holistic scene understanding that reasons jointly about regions, location, class and spatial extent of objects. Fidler and Mottaghi [20] focus on how semantic segmentation can help object detection, and their model blends between the detector and the segmentation model, by boosting object hypotheses on the segments. Both of these two method neglect the strength of DCNN. Teichmann and Weber [21] introduce an approach (MultiNet) to joint classification, detection and semantic segmentation via a unified CNN architecture where the encoder is shared amongst the three tasks. However, the MultiNet trained and evaluated on KITTI dataset is limited, and mainly designed for autonomous driving. Kokkinos [22] introduce a convolutional neural network (CNN) that jointly handles low-, mid-, and high-level vision tasks in a unified architecture that is trained end-to-end. It is necessary to point out that UberNet initializes from a network that was trained with M-SCOCO data, which needs more dataset than our DNS.

3 Detection and semantic segmentation with DNS

In this paper, we propose an efficient and effective semantic segmentation architecture, called DNS, to jointly reason about object detection and semantic segmentation. Figure 1 presents DNS architecture. In this DNS, we add convolutional feature layers to the end of the truncated base net. These layers decrease in size progressively and allow predictions of detections at multiple scales. For object detection task, it is performed by a single convolutional layer that predicts the class and the coordinates of bounding box in the feature maps of the upscale stream. Similarly, we in the segmentation task upscale all the activations of the upscale stream and concatenate them to predict the pixel labels and produce segmentation maps.

In figure 1, the trained DNS is composed of two parts i.e. Convolution and Deconvolution networks. Firstly, the input image is pre-processed by a convolution network to produce a map with high-level features. We employ ResNet-50 or VGG-16 base net for convolutional part. Taking VGG-16 as the example, the convolution network discards the fully-connected softmax layers of VGG-16. We call this layer Conv5-3, following the deconvolution part. This part consists of Down-Scale and Up-Scale stream.

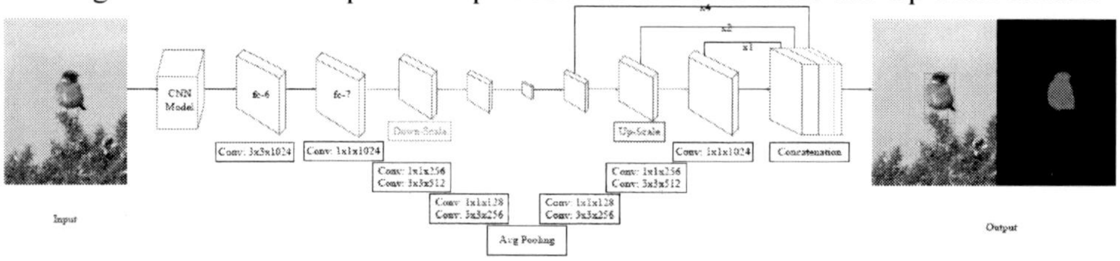

Fig. 1. The DNS architecture, which performs semantic segmentation with fully convolutional network. We adopt deconvolution layers to build the segmentation maps. Multi-upscale feature maps are utilized to make pixel prediction.

In the Down-Scale stream, we respectively use 3×3 and 1×1 convolutions to get the so-called 'fc-6' and 'fc-7' layer. Given the features produced by fc-7, we employ the similar block in each which includes the 1×1 and 3×3 convolutional layer as discussed in [23] followed by a pooling layer to produce more precise prediction. With three convolution block, we in the Up-Scale stream apply the same deconvolution pattern in [24] to skip the connections. This skip connection in our DNS is used likewise to prevent the gradient from affecting the backbone network too violently and ensure the stability of the network. With three deconvolution, the feature maps are concatenated in order to predict subsequently precise object masks and segmentation maps.

In the training stage, given training data annotated with bounding boxes and segmentation maps, we design the loss function which is simply the sum of two loss functions of these two task. Our training objective is expressed as:

$$\delta(w_0, det, seg) = B_{base}(w_0) + L(w_0, w_{det}, w_{seg}) \quad (1)$$

In equation (1), we use det and seg to the index two task; w_0 denotes the weights of the base net B_{base}, w_{det} and w_{seg} are task specific weights; B_{base} is the loss function of the base

CNN model(ResNet-50 or VGG-16); $L(w_0,w_{det},w_{seg})$ is the task specific loss function. This task-specific loss is written as follows:

$$L(w_0,w_{det},w_{seg}) = \frac{1}{N}\sum_{i-1}^{N} (L(f_{det}^i(w_0,w_{det}),y_{det}^i) + L(f_{seg}^i(w_0,w_{seg}),y_{seg}^i)) \quad (2)$$

where we use i to index training samples, denote by $f_{det,seg}^i$, $y_{det,seg}^i$ the task-specific network prediction and ground truth at the i-th example respectively, by $w_{det,seg}$ the task-specific network parameters.

To implement the detection operation, we follow the similar approach proposed in [23]. The objective function of detection task is to minimize error between ground-truth bounding boxes and the input image with anchor boxes. For segmentation task, the loss function is the cross-entropy between predicted and target class distribution of pixels. Specifically, we use a 1×1convolutional operation with 64 channels to map each layer of the upscale-stream to an intermediate representation. After this, each layer is up-scaled to the size of the last layer using bilinear interpolation and all maps are concatenated. This representation is mapped to c feature maps, where c is the number of classes, by using 3×3 convolutions to predict posterior class probabilities.

4 Experiments

We now present various experiments conducted on the Pascal VOC 2007 and 2012 datasets, for which both bounding box annotations and segmentation maps are available. Section 4.1 presents the datasets and the metrics in more details; Section 4.2 presents technical details which is important to make our work reproducible. The last section discusses the inference speed in the network architecture.

Our experiment has two objectives. The first one is to explore how DNS architecture addresses the two individual tasks. The second objective is how to settle the balance of detection and segmentation task in one single net (DNS). In order to examine this, we compare to the results in three aspects. Firstly, we contrast with the prevalent models only for the single detection task. Then the primary models merely used in segmentation task are compared. Finally, we contrast five representative approaches designed for joint detection and segmentation task.

4.1 Experimental setup

Datasets and Metrics: We use the PASCAL VOC07, and VOC12 datasets. All images in the VOC datasets are annotated with ground truth bounding boxes of objects. Both VOC07 and VOC12 consist of 20 foreground object classes and one background class. The VOC07 dataset is divided into 2 subsets, train-val (5011 images) and test (4952 images). The PASCAL VOC12-train subset contains 5717 images annotated for detection and 1464 of them have segmentation ground truth as well, while VOC12-val has 5823 images for detection and 1449 images for segmentation. We train our DNS on different subsets which consist of 'voc07-trainval-seg', 'voc12-train-seg', 'voc12-val', and 'voc12-val-seg'. In these four subsets, 'voc07-trainval-seg' subset includes 5011 segmentation images in PASCAL VOC07, while 'voc12-train-seg' 1464 segmentation images, 'voc12-val' 5823 images, and 'voc12-val-seg' 1449 segmentation images in PASCAL VOC12.

Optimization: Our DNS is coded in Python and TensorFlow. The experiments were conducted on a Tesla K40c GPU with 11439M memory. In all experiments, we use the Adam algorithm instead of SGD, with a mini-batch size of 32 images. The initial learning rate is set to 10^{-4} and decreased twice during training by a factor 10. We also use a weight decay parameter of 5×10^{-4} . As already mentioned, we use ResNet-50 as a feature extractor, 512 feature maps for each layer in down-scale and up-scale streams, 64 channels for intermediate representations in the segmentation branches. We evaluate our proposed methods on the PASCAL VOC 2007 and 2012 test set. We also compare our test set results with other competing methods.

4.2 Experimental evaluation

4.2.1 Individual object detection.

We start by verifying that diverse training PASCAL VOC dataset make much difference in our DNS. In the experiment, the default image size used to train our DNS is 300 × 300. The comparison of detection performance on different training PASCAL VOC datasets are reported in the Table 1. Table 2 and 3 respectively present the comparison of detection accuracy between DNS and state-of-the-art models on PASCAL VOC07 and VOC12.

Table 1. A Comparison of detection performance on Pascal VOC 2007 and 2012 test set. The model was trained on different VOC training datasets.

Network trained on different subsets	Base net	07mAP	12mAP
'voc07-trainval-seg', 'voc12-train-seg', 'voc12-val-seg'	ResNet-50	60.1	65.8
'voc07-trainval-seg', 'voc12-train-seg'	ResNet-50	62.8	68.4
'voc12-train-seg', 'voc12-val'	ResNet-50	64.2	70.6
'voc07-trainval-seg', 'voc12-train-seg', 'voc12-val'	**ResNet-50**	**69.0**	**73.3**
'voc07-trainval-seg', 'voc12-train-seg', 'voc12-val-seg'	VGG-16	63.4	67.6
'voc07-trainval-seg', 'voc12-train-seg'	VGG-16	65.9	69.4
'voc12-train-seg', 'voc12-val'	VGG-16	66.8	70.9
'voc07-trainval-seg', 'voc12-train-seg', 'voc12-val'	VGG-16	68.2	72.3

As shown in Table 1, our DNS result achieves the best results on the joint subset of voc07-trainval-seg, voc12-train-seg, and voc12-val, with 69.0 on 07mAP column and 73.3 on 12mAP column respectively. Note that the result on voc07-trainval-seg, voc12-train-seg, and voc12-val subset of ResNet-50 base net is better than each subset of VGG-16 base net. However, all the rest subset results on 07mAP and 12mAP column of base net of VGG-16 are better than that of ResNet-50.

Table 2. Comparison of detection performance on Pascal VOC 2007 test set. The models where trained on 'voc07-trainval-seg', 'voc12-train-seg', and 'voc12-val'.

Network	DNS	R-FCN[25]	Faster RCNN[26]	YOLO[27]	SSD[23]
Base net	ResNet-50	ResNet-101	ResNet-101	YOLO net	VGG-16
mAP	**69.0**	80.5	76.4	63.4	68.0
Aero	72.1	79.9	79.8		73.4
Bike	75.3	87.2	80.7		77.5
Bird	65.9	81.5	76.2		64.1
Boat	61.4	72.0	68.3		59.0
Bottle	30.6	69.8	55.9		38.9
Bus	77.7	86.8	85.1		75.2
Car	78.0	88.5	85.3		80.8
Cat	84.2	89.8	89.8		78.5
Chair	48.8	67.0	56.7		46.0
Cow	75.4	88.1	87.8		67.8

Table	67.0	74.5	69.4		69.2
Dog	77.5	89.8	88.3		76.6
Horse	81.6	90.6	88.9		82.1
Mbike	75.4	79.9	80.9		77.0
Person	71.2	81.2	78.4		72.5
Plant	41.0	53.7	41.7		41.2
Sheep	71.9	81.8	78.6		64.2
Sofa	70.8	81.5	79.8		69.1
Train	82.2	85.9	85.3		78.0
Tv	71.5	79.9	72.0		68.5

In Table 2, the results show that detection performance of DNS outperforms SSD and YOLO with a 69.0 mAP, while being a real time detector. Our DNS performs 5.6% better than YOLO and 1.0% than SSD for detection on Pascal VOC 2007 test set. We further improve the results by training for detection achieving 73.3 mAP on Pascal VOC 2012 test dataset in Table 3.

Table 3. Comparison of detection performance on Pascal VOC 2012 test set. The models where trained on 'voc07-trainval-seg', 'voc12-train-seg', and 'voc12-val'.

Network	DNS	R-FCN[25]	Faster RCNN[26]	YOLO[27]	SSD[23]
Base net	ResNet-50	ResNet-101	ResNet-101	YOLO net	VGG-16
mAP	**73.3**	77.6	73.8	57.9	72.4
Aero	82.6	86.9	86.5	77.0	85.6
Bike	80.4	83.4	81.6	67.2	80.1
Bird	75.8	81.5	77.2	57.7	70.5
Boat	55.9	63.8	58.0	38.3	57.6
Bottle	49.4	62.4	51.0	22.7	46.2
Bus	72.8	81.6	78.6	68.3	79.4
Car	77.1	81.1	76.6	55.9	76.1
Cat	90.5	93.1	93.2	81.4	89.2
Chair	57.2	58.0	48.6	36.2	53.0
Cow	79.6	83.8	80.4	60.8	77.0
Table	59.7	60.8	59.0	48.5	60.8
Dog	88.9	92.7	92.1	77.2	87.0
Horse	83.6	86.0	85.3	72.3	83.1
Mbike	82.0	84.6	84.8	71.3	82.3
Person	78.2	84.4	80.7	63.5	79.4
Plant	54.5	59.0	48.1	28.9	45.9
Sheep	76.5	80.8	77.3	52.2	75.9
Sofa	69.0	68.6	66.5	54.8	69.5
Train	83.7	86.1	84.7	73.9	81.9
Tv	68.4	72.9	65.6	50.8	67.5

As Table 3 shows, detection on VOC12 improves by more than 4.3% on VOC07. Meanwhile, our detection results are similar to Faster RCNN on VOC12, which is also better than YOLO and SSD. We argue than this result could be competitive even though it is still 3.3% less than R-FCN.

4.2.2 Individual semantic segmentation

The second task that we have tried is semantic segmentation. Even though a broad range of techniques designed for this problem, we compare to the state-of-the-art methods. Table 4 presents the comparison of different segmentation evaluation results on PASCAL VOC12 test.

Table 4. Semantic segmentation evaluation results on PASCAL VOC 2012 test set.

Method	Mean IOU(%)
SDS[7]	51.6
FCN-8s[2]	62.2
DeconvNet[17]	69.6
DeepLab-v2[28]	79.7
PSPNet[29]	82.6
RefineNet[30]	83.4
DeepLab-v3[31]	85.7
DNS	**69.8**

From Table 4, we can see that our DNS yields 69.8 on the mean IOU metric evaluation, which improves over SDS (the primogenitor of simultaneous detection and segmentation) by a large 18.2% margin, while improving 7.6% with respect to FCN-8s. The performance of DeconvNet is competitive to our DNS, about 0.2% less than ours. Note that RefineNet and DeepLab-v3 are the most representative works in semantic segmentation which focus on segmentation thus neglecting detection, which is also the goal of our pursuit.

4.2.3 Joint detection and semantic segmentation

Motivated by the empirical results in the previous paragraphs, we have explored the ability of how DNS architecture addresses the two individual tasks. Now for the joint task, we contrast to five representative methods, which similarly make use of a single net to address joint detection and segmentation (final goal) in Table 5. We also shows the visualization results of detection and segmentation on PASCAL VOC 2007 test images in figure 2.

From table 5, we can observe that very little of works address this two task jointly. Among the rest method which conduct on PASCAL VOC Dataset, our DNS is largely better than SDS, Holistic Scene Understanding and segDPM method. What's important is that our DNS achieves second good performance for joint detection and segmentation task.

Table 5. Comparison of Jointly Detection and Segmentation Evaluation.

Method	Dataset	Mean mAP (%)	Mean IOU (%)
SDS[7]	VOC12Det+Seg	53.9	51.6
HSU[19]	MSRC-21Det+VOC10 Seg	49.3	31.2
segDPM[20]	VOC10Det+Seg	61.4	34.8
DAG[32]	VOC07Det+VOC10Seg	67.09	72.07
UberNet[22]	VOC07Det+VOC12Seg	78.8	71.1
DNS	VOC12Det+Seg	73.3	69.8

Compared with DAG which is a little bit less than our performance, the detection performance of DNS achieves 73.3 with about 6.2 mAP growth, while segmentation performance 69.8 with about 2.3 mean IOU decline. The main reason is that the object detection architecture of DAG is based on the Faster-RCNN, which is a little bit better than ours, and the semantic segmentation architecture of DAG is based on FCN, that is a little bit worse than ours. Especially the accuracy of DAG drops (both mAP and mean IOU) significantly after the adversarial perturbations are added.

Compared with UberNet which achieves the best performance, in conjunction with Faster-RCNN, UberNet makes use of the VOC 2007 dataset to fine-tune the MS-COCO pretrained network for detection task. For segmentation task, UberNet deviates from the Deeplab-FOV architecture by using linear operations on top of skip layers to reach the similar result of DeepLab-v2. In spite of this, the mAP accuracy of our DNS is about 5.5 under UberNet, while the mean IOU accuracy of DNS just 1.3 under UberNet.

4.3 Speed comparison

To strike the balance between accuracy and inference cost, we report speed comparison to other state-of-the-art pipelines in figure 3. Our approach is the most accurate among these five detectors working 24 frames per second (FPS) and in the setting close to real time (19 FPS), it can provide the real-time detections among the counterparts, while also providing semantic segmentation mask.

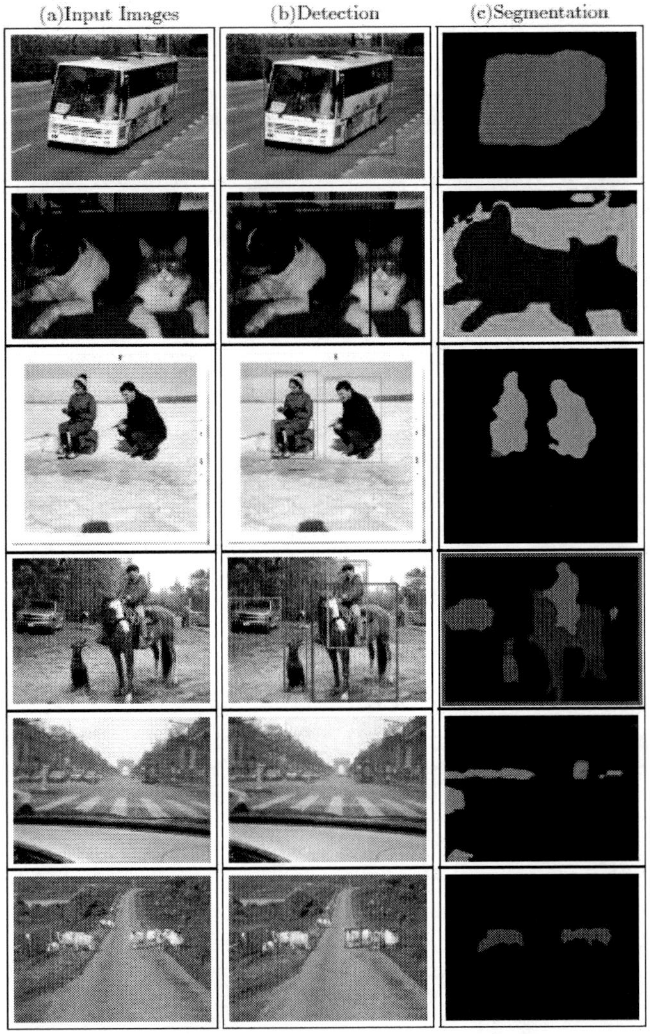

Fig. 2. Example results of several PASCAL VOC 2007 test images based on DNS models. (a) Input Images. (b) Detection results in VOC07. (c) Segmentation results in VOC07.

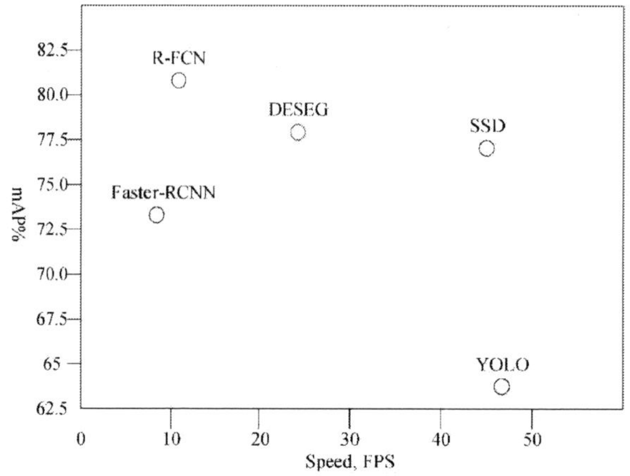

Fig. 3. Speed comparison. X-coordinate is their speed, in FPS. Y-axis is the detection accuracy of measured in mAP.

5 Conclusions

This paper adapts a deep deconvolution semantic segmentation model (DNS) to handle both detection and segmentation task. Experiments on PASCAL VOC dataset have shown that: (1) Our DNS network based on weight-sharing is advantageous to both detection and segmentation task. (2) Merging the down-scale and up-scale features not only improves the performance over deconvolution baselines, but also allows us to fast the detection speed. (3) Our network demonstrates the competitive performance in PASCAL VOC detection and segmentation benchmark.

Acknowledgments

This work was supported in part by the National Natural Science Foundation of China under Grant 61772114, in part by the Fundamental Research Funds for the Central Universities (No. ZYGX2016J077), and in part by the program of China Scholarships Council (No. 201606070055).

References

1. Hong S, Noh H and Han B 2015 Decoupled deep neural network for semi-supervised semantic segmentation *In: Advances in Neural Information Processing Systems*
2. Long J, Shelhamer E and Darrell T 2014 Fully convolutional networks for semantic segmentation. *IEEE Transactions on Pattern Analysis and Machine Intelligence* **39** 640-51
3. Hariharan B, Arbelaez P and Girshick R 2015 Hypercolumns for object segmentation and _ne-grained localization *In: The IEEE Conference on Computer Vision and Pattern Recognition*
4. Sermanet P, Eigen D, Zhang X, Mathieu M, Fergus R and LeCun Y 2014 Overfeat: Integrated recognition, localization and detection using convolutional networks *Preprint arXiv:1312.6229*

5. Eigen D and Fergus R 2015 Predicting depth, surface normals and semantic labels with a common multi-scale convolutional architecture *In: The IEEE International Conference on Computer Vision*
6. Gkioxari G, Girshick R and Malik J 2015 Contextual action recognition with r*cnn *In: The IEEE International Conference on Computer Vision*
7. Hariharan B, Arbelaez P, Girshick R and Malik J 2015 Simultaneous detection and segmentation *In: European Conference on Computer Vision*
8. Zhang Z, Fidler S and Urtasun R 2016 Instance-level segmentation for autonomous driving with deep densely connected mrfs *In: The IEEE Conference on Computer Vision and Pattern Recognition*
9. He K, Gkioxari G, Dollr P and Girshick R 2017 Mask r-cnn *In: The IEEE International Conference on Computer Vision*
10. Dai J, He K and Sun J 2015 Convolutional feature masking for joint object and stuff segmentation *In: The IEEE Conference on Computer Vision and Pattern Recognition*
11. Dai J, He K and Sun J 2015 Convolutional feature masking for joint object and stuff segmentation *In: The IEEE Conference on Computer Vision and Pattern Recognition*
12. Mostajabi M, Yadollahpour P and Shakhnarovich G 2015 Convolutional feature masking for joint object and stuff segmentation *In: The IEEE Conference on Computer Vision and Pattern Recognition*
13. Chen C, Papandreou G, Kokkinos I, Murphy K and Yuille L 2015 Semantic image segmentation with deep convolutional nets and fully connected crfs. *In: International Conference on Learning Representations*
14. Lin G, Shen C, Reid I and Hengel A 2016 Efficient piecewise training of deep structured models for semantic segmentation *In: The IEEE Conference on Computer Vision and Pattern Recognition*
15. Chen C, Barron T, Papandreou G, Murphy K and Yuille L 2016 Semantic image segmentation with task-speci_c edge detection using cnns and a discriminatively trained domain transform *In: The IEEE Conference on Computer Vision and Pattern Recognition*
16. Bertasius G, Shi J and Torresani L 2016 Semantic segmentation with boundary neural fields *In: The IEEE Conference on Computer Vision and Pattern Recognition*
17. Noh H, Hong S and Han B 2015 Learning deconvolution network for semantic segmentation *In: The IEEE International Conference on Computer Vision*
18. Vemulapalli R, Tuzel O, Liu Y and Chellappa R 2016 Gaussian conditional random field network for semantic segmentation *In: The IEEE Conference on Computer Vision and Pattern Recognition*
19. Yao J, Fidler S and Urtasun R 2012 Describing the scene as a whole: Joint object detection, scene classification and semantic segmentation *In: The IEEE Conference on Computer Vision and Pattern Recognition*
20. Fidler S, Mottaghi R, Yuille A and Urtasun R 2013 Bottom-up segmentation for top-down detection *In: The IEEE Conference on Computer Vision and Pattern Recognition*
21. Teichmann M, Weber M, Zoellner M, Cipolla R and Urtasun R 2016 Multinet: Real-time joint semantic reasoning for autonomous driving *Preprint arXiv: 1612.07695*
22. Kokkinos I 2017 Training a `universal' convolutional neural network for low-, mid-, and high-level vision using diverse datasets and limited memory *In: The IEEE Conference on Computer Vision and Pattern Recognition*

23. Liu W, Anguelov D, Erhan D, Szegedy C, Reed S, Fu Y and Berg C 2016 Ssd: Single shot multibox detector *In: European Conference on Computer Vision*
24. Fu Y, Liu W, Ranga A, Tyagi A and Berg C 2017 Dssd : Deconvolutional single shot detector *Preprint arXiv: 1701.06659*
25. Dai J, Li Y, He K and Sun J 2016 R-fcn: Object detection via region-based fully convolutional networks *In: Advances in Neural Information Processing Systems*
26. Ren S, He K, Girshick R and Sun J 2017 Faster r-cnn: Towards real-time object detection with region proposal networks *IEEE Transactions on Pattern Analysis and Machine Intelligence* **39** 1137-49
27. Redmon J, Divvala S, Girshick R and Farhadi A 2016 You only look once: Unified, real-time object detection *In: The IEEE Conference on Computer Vision and Pattern Recognition*
28. Chen C, Papandreou G, Kokkinos I, Murphy K and Yuille L 2018 Deeplab: Semantic image segmentation with deep convolutional nets, atrous convolution, and fully connected crfs *IEEE Transactions on Pattern Analysis and Machine Intelligence* **40** 834-48
29. Zhao H, Shi J, Qi X, Wang X and Jia J 2017 Pyramid scene parsing network *In: The IEEE Conference on Computer Vision and Pattern Recognition*
30. Lin G, Milan A, Shen C and Reid I 2017 Refinenet: Multi-path refinement networks for high-resolution semantic segmentation *In: The IEEE Conference on Computer Vision and Pattern Recognition*
31. Chen C, Papandreou G, Schroff F and Adam H 2017 Rethinking atrous convolution for semantic image segmentation *Preprint arXiv: 1706.05587*
32. Xie C, Wang J, Zhang Z, Zhou Y, Xie L and Yuille A 2017 Adversarial examples for semantic segmentation and object detection *In: The IEEE International Conference on Computer Vision*

Semantic representation for visual reasoning

Xubin Ni[1], *Lirong* Yin[2], *Xiaobing* Chen[1], *Shan* Liu[1,3], *Bo* Yang[1], and *Wenfeng* Zheng[1,*]

[1]School of Automation, University of Electronic Science and Technology of China, Chengdu, 610054, China
[2]Geographical & Sustainability Sciences Department, University of Iowa, Iowa City, IA, 52242, USA
[3]Department of Modelling, Simulation, and Visualization Engineering, Old Dominion University, Norfolk, VA 23529, USA

Abstract. In the field of visual reasoning, image features are widely used as the input of neural networks to get answers. However, image features are too redundant to learn accurate characterizations for regular networks. While in human reasoning, abstract description is usually constructed to avoid irrelevant details. Inspired by this, a higher-level representation named semantic representation is introduced in this paper to make visual reasoning more efficient. The idea of the Gram matrix used in the neural style transfer research is transferred here to build a relation matrix which enables the related information between objects to be better represented. The model using semantic representation as input outperforms the same model using image features as input which verifies that more accurate results can be obtained through the introduction of high-level semantic representation in the field of visual reasoning.

1 Introduction

Visual Question Answering (VQA) is a technology that combines natural language processing with digital image processing. The general process for solving a VQA problem is to take the image and the corresponding question in natural language as input and finally get the answer. If the question involves reasoning, it is called visual reasoning. The issues studied by visual reasoning are similar to VQA but may require more interdependent inference steps to solve the problem.

The models used in VQA can be divided into classical machine learning models and deep learning models. Most classical machine learning models are based on Bayesian theory. [1] proposed a Bayesian framework for VQA, predicting the type of answer to a question and using it to generate an answer. [2] combined the semantic tree obtained from the semantic analyser and image to construct the SWQA model to predict the corresponding answer. After CLEVR dataset was proposed [3], those classical machine learning models performed poorly on this dataset compared with some deep learning models [4-7]. The workflow of the existing deep learning models can be roughly divided into three steps: extracting features from the question sentence, extracting features from the image, and combining the question embedding and image features to generate answers. For question

* Corresponding author: wenfeng.zheng.cn@gmail.com

embedding, techniques such as Long Short-Term Memory (LSTM) can be used [8, 9]. Regarding image features, it is generally considered to extract image features using a convolutional neural network (CNN). The iBOWIMG model used a pre-trained GoogleNet image classification model to extract image features, using the word embedding of each word in the question as a text feature. After stitching the image features and text features, the answer was obtained by Softmax regression [4]. [5, 6] continuously generated a neural network for each image and problem selected from various problem-based sub-modules and combined them to generate a neural network. The relational network made it possible to grasp the key of relational reasoning by constraining the structure of the function and obtained the state-of-art result [7].

Although deep learning models have made significant progress over classical machine learning models, it still has a large gap with the level of human reasoning on complex issues. In this research it is planned to use the semantic representation of the image as input in the visual reasoning task, instead of directly using the pixel or the image features extracted by CNN as input, to explore whether the result of introducing high-level semantic representation can be better. If better results can be gained, this idea can be transferred to other areas of computer vision, even other areas of deep learning.

2 Method

2.1 Dataset: Enhanced Sort-of-CLEVR(ESOC)

Most current visual reasoning works are based on the CLEVR dataset [3]. To simplify the image processing part and the natural language understanding part, the ESOC dataset based on Sort-of-Clevr dataset [7] was built in this research. ESOC dataset is similar to Sort-of-Clevr dataset but the scenes of ESOC dataset are more detailed and complex.

ESOC contains images, several questions for the image and corresponding answers. Each image contains several 2D geometric objects. The property of each object is randomly chosen from Figure 1. The questions are divided into relational questions (RQ) and non-relational questions (NRQ). The non-relational question involves only one object in the image, querying its shape, position or size. The relational question involves the positional relationship between two objects. The entire dataset is divided into three groups which are 2-shapes dataset, 4-shapes dataset, and 6-shapes dataset to test the performance of the model on different complexity scenes. N-shapes means that there are n objects in each scene.

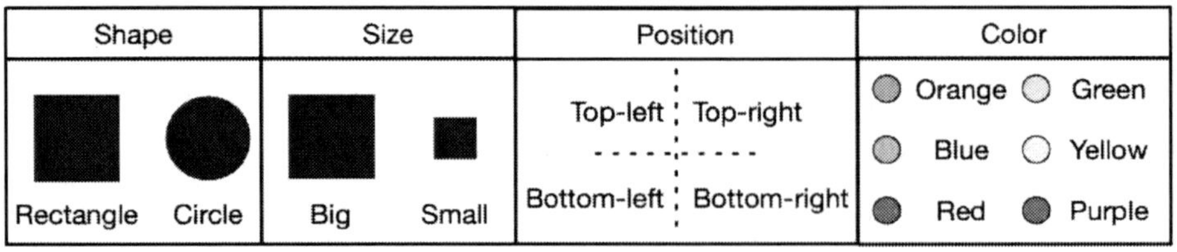

Fig. 1. The elemental composition of the ESOC dataset.

2.2 Process of extracting semantic representation

Assuming that the input image is x, the question for the image is q, the answer is a. The purpose of visual reasoning task is to find a system H to get the answer $a = H(x,q)$. In this research, a semantic network capable of characterizing enough information of the image is used as input rather than image features extracted by CNN. The process is shown in Figure 2. Suppose O is the set of objects contained in the image and $O = \{O_1, O_2, \cdots, O_n\}$. Image segmentation technology is needed for extracting the collection O. For simple images, it can be solved according to edge segmentation. If the content of the dataset is a complex image, such as CLEVR dataset, which contains 3D objects that may overlap each other, instance segmentation techniques such as Mask R-CNN [10] and SSD [11] may be employed. For each object, the set of its attributes is $P = \{P_1, P_2, \cdots, P_m\}$. In this research, these attributes may include shape, size, color, and location. Modules M to extract these properties can be designed separately, where $M = \{M_1, M_2, \cdots, M_m\}$. The module $M_i \in M$ for extracting image semantic representation has no fixed requirements and is selected according to specific tasks. For example, the shape of the target object can be obtained by CNN. Assuming that the value of the attribute P_i is V_i. The value V_i of the attribute P_i of the object O_k can be obtained by $V_i = M_i(O_k)$. Thus, a set of triples describing the entire image can be obtained which is $S = \{S_{O_1}, S_{O_2}, \cdots, S_{O_n}\}$, where

$$S_{O_i} = \{(O_i, P_1, V_1), (O_i, P_2, V_2), \cdots, (O_i, P_m, V_m)\} \tag{1}$$

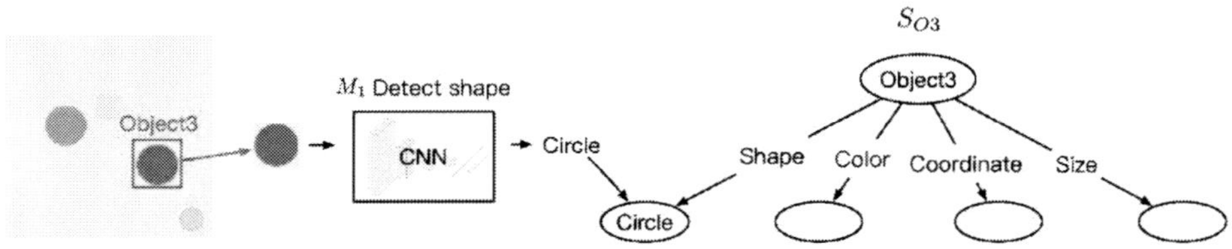

Fig. 2. The baseline of extracting a semantic net form the image.

Natural language cannot be understood by computers directly, so questions need to be pre-processed first. It can be done using LSTM, like the method in machine translation called Seq2Seq [12], to map the question to a vector space. Suppose the question is q, the encoder is L, and the question embedding is characterized by $Q = L(q)$. Finally, $a = H(S, Q)$.

Each sample in ESOC dataset contains (x, S, Q, a), where x is the original image; S is the semantic representation of the original image; Q is the question embedding; a is the ground-truth answer. (x, Q, a) and (S, Q, a) were trained in the neural network of the same architecture to compare the results of the two. Adam optimizer was used for gradient descent training and dropout mechanism was used to prevent overfitting [13]. The termination of training was judged by the early stopping method [14].

2.3 Relation matrix

At the beginning of the research, the way to organize semantic representation was to directly add the semantic representation of each object to a list to form a multidimensional vector. However, in practical experiments, such effects were found to be limited. The probable reason is that the network only captures the individual information of each object, but does not learn a way to characterize the relative relationship between different objects. A similar situation was also found in the study of style transfer. The visual style is an overall intuitive perception. When comparing the style differences between the two images, it is inefficient to take the feature layers of the two images directly because each feature map can only reflect some of its own rather than the overall feature. Similarly, the semantic vectors here were independent of each other, and the model did not have a good understanding of the relationship between them and the relationship between the individual and the whole scene. In [15], a Gram matrix of feature layers was constructed to describe the style of an image. The reason why the Gram matrix is valid in that research is that it can be used to measure the characteristics of each dimension and the relationship between different dimensions. In the multi-scale matrix obtained after the inner product, the diagonal elements provide information about the different feature maps themselves, and the remaining elements provide relevant information between the different feature maps. Such a matrix can not only reflect the characteristics of the corresponding image but also reflect the closeness of different features.

Using a similar idea, a relation matrix was constructed to describe each object's own information and the relationship between different objects. Suppose that the semantic representations of object O_1 and object O_2 are S_{O_1} and S_{O_2}, and the question embedding is Q. The network F consisting of multiple fully connected layers is used to extract the relationship between O_1 and O_2 which is $F(S_{O_1}, S_{O_2}, Q)$. Assuming that there are n objects in the graph, the size of the relationship matrix is $n \times n$, where the elements of the i_{th} row and the j_{th} column is $R(i,j) = F(S_{O_i}, S_{O_j}, Q)$. The relation matrix is:

$$R = \begin{bmatrix} R(1,1) & R(1,2) & \cdots & R(1,n) \\ R(2,1) & R(2,2) & \cdots & R(2,n) \\ \vdots & \vdots & \ddots & \vdots \\ R(n,1) & R(n,2) & \cdots & R(n,n) \end{bmatrix} \tag{2}$$

2.4 Model

The model of this research consists of two parts which are feature extraction layer and inference layer, as shown in Figure 3. Feature extraction mainly involves image features extraction, semantic representation extraction and question features extraction. The structure of the inference layer of the image-feature-based model and semantic-representation-based model is the same which is multi-layer fully connected neural networks. The convolutional neural network for extracting image features is composed of four layers of convolutional layers.

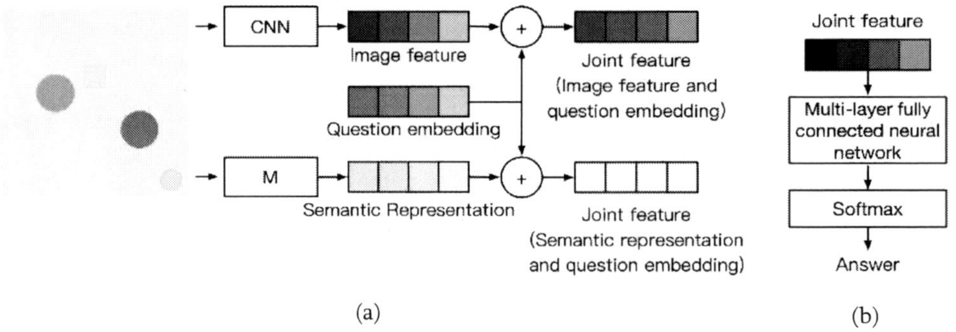

Fig. 3. Two parts of the model in this research. (a) Feature extraction layer. (b) Inference layer.

For image-feature-based model, the features of the image are first extracted by the convolutional layer. Then the image features and question embedding are jointly embedded as the input to inference layer, and the answer is finally obtained. For semantic-representation-based model, the feature extraction method is as described in section 2.2 and 2.3. After that, the semantic representation vector and the question embedding can be jointly embedded as the input to inference layer.

3 Result

The experiment was based on the ESOC dataset and the model described in section 2.2. The main task was to compare the difference in results with semantic representation as input and image features as input. At first, semantic vector was directly arranged into a large-size vector as input. The result is shown in Table 1. The accuracy promotion after replacing image features with semantic representation as input is shown in Figure 4.

Table 1. The accuracy (%) of image-feature-based (IF) models, semantic-representation-based (SR) models and semantic-representation-based models with relation matrix.

Model Type	IF			SR			SR + relation matrix		
Object numbers	2	4	6	2	4	6	2	4	6
Accuracy of NRQ	69.64	59.39	60.14	99.03	97.81	95.77	99.75	99.72	99.61
Accuracy of RQ	69.33	64.10	55.61	99.06	66.79	57.90	99.76	94.91	93.28

It can be seen that the accuracy is improved on three datasets after the semantic representation is used as the input. Among them, on the non-relational question, semantic-representation-based model has a significant effect on the improvement of accuracy, but on the relational question, the accuracy promotion of the 4-shapes dataset and 6-shapes dataset is not apparent, only about 4%. The reason why the accuracy of the 2-shapes dataset on the relational question is as high as that on the non-relational question is that when there are only two objects in the scene, the relational question and the non-relational question are equivalent.

In order to further improve the results, the relation matrix was used in the data pre-processing part, and other parts did not change. The results are shown in Table 1 and Figure 5. It can be found that after using the relation matrix, the accuracy of the relational question is also significantly improved compared to the image-feature-based model. For 4-shapes dataset, the accuracy promotion increases from 4.20% to 48.07%, and for the 6-shapes dataset it is from 4.12% to 67.74%. The improvement is undeniable. It can be observed that a simple model can complete complex reasoning tasks as long as the semantic representation is processed slightly.

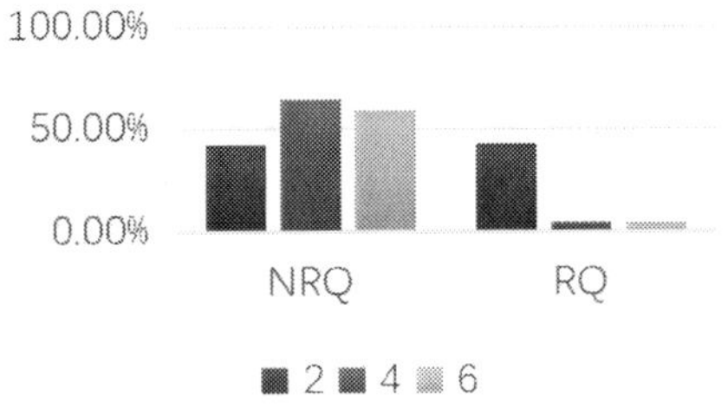

Fig. 4. The promotion of accuracy after using semantic representation as input directly.

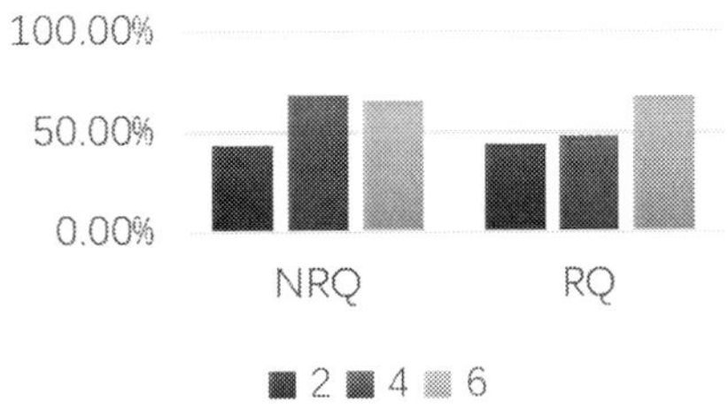

Fig. 5. The promotion of accuracy after using semantic representation with relation matrix.

4 Conclusion

This paper demonstrates how the semantic representation of an image can be used as the input to a visual reasoning system and verifies that changing the representation of the image can further improve system performance. After replacing the visual features with unprocessed semantic representation vectors, the accuracy of the model on non-relational questions was significantly improved, but the accuracy on the relational questions was only slightly improved. Then the semantic vector was pre-processed by constructing a relation matrix. This simple operation made the model have a 43% to 67% improvement on relational and non-relational questions. It can be seen that the effect of semantic representation itself is not worse than image features. And semantic representation is simple and easy to carry out other processing. After this, experiment results can be further improved. In this experiment, the baseline model was fundamental, with only a convolutional neural network and a multi-layer fully connected layer. If the semantic representation is more complex and more specific in a particular field, it could be expected that the introduction of higher-level representation will make results more accurate, which is one of the possible future research directions.

The problem with the current work is that although the final accuracy is improved, the work of extracting semantic information requires manual intervention to select appropriate semantic features, which undoubtedly increases system complexity and additional time overhead compared to the end-to-end system. So, the next work will focus on how to make system extract useful semantic information automatically and reduce manual intervention.

Acknowledgments

Funds for International S&T Cooperation and Exchange R&D Project of Sichuan Province (Grant No. 2017HH0054).

References

1. Kafle, K. and C. Kanan. *Answer-Type Prediction for Visual Question Answering*. in *Computer Vision and Pattern Recognition*. (2016).
2. Malinowski, M. and M. Fritz. *A multi-world approach to question answering about real-world scenes based on uncertain input*. in *Advances in Neural Information Processing Systems*. (2014).
3. Johnson, J., et al., *CLEVR: A Diagnostic Dataset for Compositional Language and Elementary Visual Reasoning*. (2017).
4. Zhou, B., et al., *Simple Baseline for Visual Question Answering*. Computer Science, (2015).
5. Andreas, J., et al. *Neural Module Networks*. in *IEEE Conference on Computer Vision and Pattern Recognition*. (2016).
6. Johnson, J., et al., *Inferring and Executing Programs for Visual Reasoning*. (2017).
7. Santoro, A., et al., *A simple neural network module for relational reasoning*. (2017).
8. Sak, H., A. Senior, and F. Beaufays, *Long Short-Term Memory Based Recurrent Neural Network Architectures for Large Vocabulary Speech Recognition*. Computer Science, (2014): p. 338-342.
9. Andreas, J., et al., *Deep Compositional Question Answering with Neural Module Networks*. Computer Science, (2015). **27**: p. 55-56.
10. He, K., et al., *Mask R-CNN*. IEEE Transactions on Pattern Analysis & Machine Intelligence, (2017). **PP**(99): p. 1-1.
11. Liu, W., et al. *SSD: Single Shot MultiBox Detector*. in *European Conference on Computer Vision*. (2016).
12. Sutskever, I., O. Vinyals, and Q.V. Le, *Sequence to Sequence Learning with Neural Networks*. (2014). **4**: p. 3104-3112.
13. Srivastava, N., et al., *Dropout: a simple way to prevent neural networks from overfitting*. Journal of Machine Learning Research, (2014). **15**(1): p. 1929-1958.
14. Yao, Y., L. Rosasco, and A. Caponnetto, *On Early Stopping in Gradient Descent Learning*. Constructive Approximation, (2007). **26**(2): p. 289-315.
15. Gatys, L.A., A.S. Ecker, and M. Bethge, *A Neural Algorithm of Artistic Style*. Computer Science, (2015).

Rehabilitation recognition skeleton data depth learning based on RNN

Qingzhi Zhang*, *Panfeng* Wu, *Xiaohui* Du, *Hualiang* Sun, and *Lijia* Yu

Shandong Institute of Space Electronics Technology, China

Abstract. With the extensive application of deep learning in the field of human rehabilitation, skeleton based rehabilitation recognition is becoming more and more concerned with large-scale bone data sets. The key factor of this task is the two intra frame representations of the combined co-and the inter-frame. In this paper, an inter frame representation method based on RNN is proposed. Pointtion of each joint is joint-coded they are assembled into semantic both spatial and temporal domains.we introduce a global spatial aggregation which is able to learn superior joint co features over local aggregation.

1 Introduction

Human rehabilitation recognition is the basic scenario of AI application. The description of human actions to joints is the best description of human rehabilitation. Firstly, compared with other optical flow and RGB, the data amount is very small, and the target is clear. Secondly, the joint data has the best robustness to the interference of the background noise. These two characteristics make skeleton based models have the inherent advantages of lightweight and high recognition rate.

In this paper, we focus on the problem of Workflow for Rehabilitation Recognition Skeleton (Figure 1).

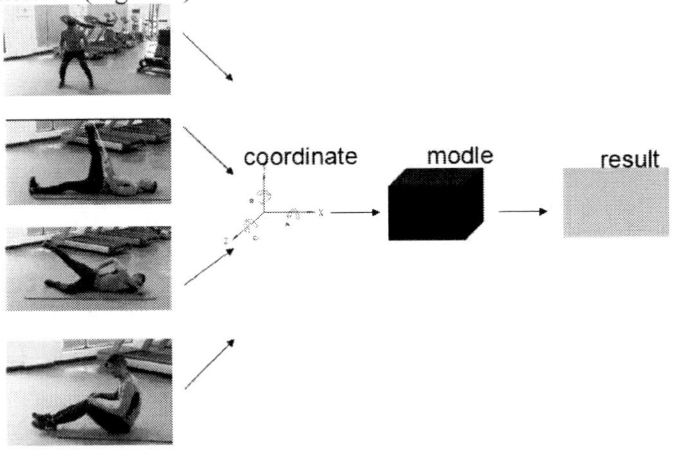

Fig. 1. Workflow for Rehabilitation Recognition Skeleton.

* Corresponding author: Qzzhang28@gmail.com

2 Related Work

The RNN network is designed to address the entire process simulation of exercise rehabilitation. The RNN network is a natural choice that has been used on a large scale to complete deep learning from the skeleton sequence. With the widespread use of deep learning, more and more literature uses RNN to learn skeleton features and complete large-scale citations in various scenarios.

3 Methods

3.1 Co-occurrence Feature Learning with RNN

RNN is one of the most powerful and successful neural network models, which has been widely used in image classification, object detection, video classification and so on. Compared with RNN and other sequence structures, RNN can utilize historical information, take action sequence into account, and use time and space to encode asynchronously. Using the convolution characteristics of RNN, the human body's rehabilitation action is decomposed into two steps .that is, the action characteristics of the cross space domain, including the motion position and the range of motion, and the motion feature fitting in the Ag and the whole motion process. Finally, the results obtained from soft max can give the accuracy of the rehabilitation result to the posture and range of movement. It indicates that T is a D1 D2 D3 3D tensor flow. In the convolution process, the array values of motion posture and range of motion can be transposed to meet different rehabilitation needs. All tensors of dimension DI can be aggregated in whole process. If the key actions are specified in the action process, the two preceding and subsequent actions of the key actions can be used as key convolution frames.

3.2 Explicit Skeleton Motion

In addition to the co-occurrence of movement, a group of multiple time movement of joints is the key frame of rehabilitation action. Therefore, in human rehabilitation, skeleton action is taken as the key frame and the convolution network of RNN is introduced.

We formulate it as Kt ={D1t; D2t; : : : ; DNt}, where N is the number of joint and D = (x; y; z) is a 3D joint coordinate. The skeleton motion is defined as the temporal difference of each joint between two consecutive frames:

Zt = Kt+1-Kt
={D1t+1-D1t; D2t+1_D2t; ::::; DNt+1-DNt }:

The original skeleton coordinate K and skeleton motion D are sent into convolution respectively by spatial coordinates and time coordinates. In the subsequent calculation, the two are combined to merge.

		Temporal difference	
conv1	1x1x64	1x1x64	Point-level feature learning
conv2	3x1x32	3x1x32	
	(0,2,1)	(0,2,1)	
conv3	3x3x32, /2	3x3x32, /2	
conv4	3x3x64, /2	3x3x64, /2	
	Concat		Co-occurrence feature learning
conv5	3x3x128, /2		
conv6	3x3x256, /2		
	Flatten		Convolution
fc7	256		Transpose
			FC
fc8	class		

Fig. 2. Overview of the proposed hierarchical co-occurrence.

3.3 Hierarchical Co-occurrence Network

In this section, the detailed Rehabilitation recognition base on RNN will be described. The network architecture is shown in Figure 3. The joint tensor sequence X can be represented by Z G W tensor, Z represents the number of frames, G represents the number of joints in the skeleton, W represents coordinates. Skeletal movement is involved in network feedback in the same way as X. They are as the two input traffic. The two networks interact in the same system and their parameters. Their characteristics are data fusion after unified convolution. After a given state and motion requirement of a given motion skeleton, hierarchical learning is performed. In the first calculation, the point level features are computed by convolution with 1 (1) and 1 (2). The size of the body is fixed, so the independent three-dimensional coordinate system of each joint is fixed, and the feature of point level is added to this joint coordinate system. After that, we use RNN to transform the 3D coordinates of the joints into the moving channels. In the second calculation, all the global features will be classified by two fully linked layers of recovery action and rehabilitation time.

In the three converged calculations, the best accuracy occurs at the most operational time. The results of the test are as shown in Table 1.

Table 1. Performance of different fusion methods for multi-person feature on the NTU RGB+D dataset in the cross-subject setting.

Method		Accuracy (%)
Early fusion		84.7
Late fusion	Mean	88.3
	Concat	88.4
	Max	88.9

4 Experiments

The NTU RGB+D data set is so far the largest skeleton-based human action recognition dataset. It contains 56880 skeleton sequences, which are annotated as one of 60 action classes.

There are two recommended evaluation protocols, i.e. Cross Subject (CS) and Cross-View (CV).In the training process, according to the international standard method, the random subsequence is extracted, and the extraction rate is uniformly distributed in [0.5, 1]. In the calculation process, the ratio of skeletal sequence is 0.9. The time of the sample is different. We normalize the sample and interpolate the sample with insufficient time. The total input training model of the algorithm is 100k, each sequence is 1024, the initial learning ability is 0.0001, the exponential decay is 0.999 per 10K step.

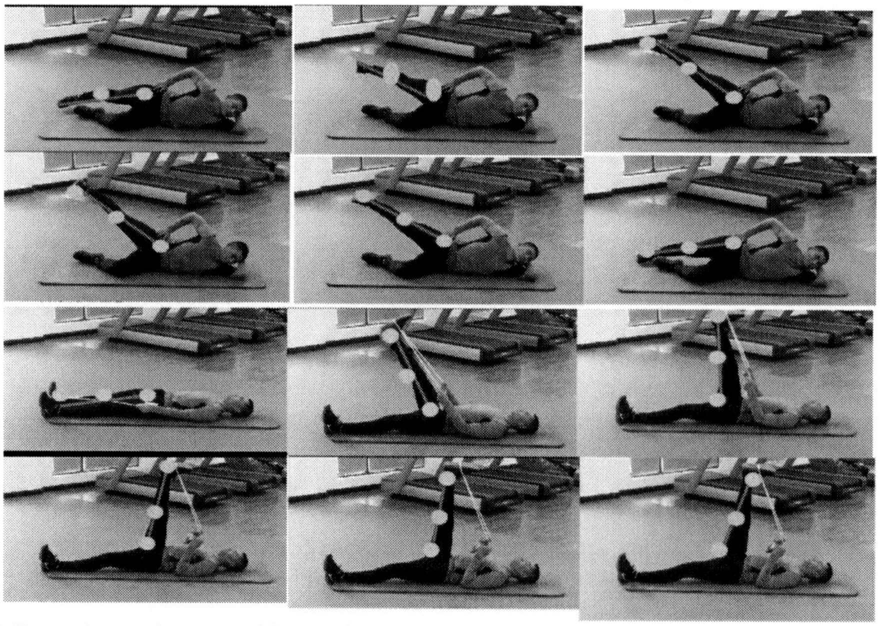

Fig. 3. Exemplary action recognition results.

5 Conclusions

We present a deep learning framework for skeleton rehabilitation based on RNN for human rehabilitation. By using the convolution network of RNN, a three-dimensional coordinate based on the body skeleton is established between the confusions and the time of rehabilitation, and then the characteristics of the current skeleton and the time of motion are fused. The experimental results show that it can obviously improve the recognition of rehabilitation actions and improve the accuracy of evaluation of rehabilitation actions.

References

1. [Du et al., 2017] Chao Li, Qiaoyong Zhong, Di Xie, Shiliang Pu. Co-occurrence Feature Learning from Skeleton Data for Action Recognition and Detection with Hierarchical Aggregation, pages 579–583, 2016.
2. [Girshick et al., 2014] Ross B Girshick, Jeff Donahue, Trevor Darrell, and Jitendra Malik. Rich feature hierar-chies for accurate object detection and semantic segmentation. CVPR, pages 580–587, 2014.
3. [He et al., 2016] Kaiming He, Xiangyu Zhang, Shaoqing Ren, and Jian Sun. Deep residual learning for image recog-nition. In CVPR, pages 770–778, 2016.
4. [Hussein et al., 2013] Mohamed E. Hussein, Marwan Torki, Mohammad A. Gowayyed, and Motaz El-Saban. Human action recognition using a temporal hierarchy of covariance descriptors on 3d joint locations. In IJCAI, pages 639–44, 2013.
5. [Ji et al., 2014] Yanli Ji, Guo Ye, and Hong Cheng. Inter-active body part contrast mining for human interaction recognition. In ICMEW, pages 1–6, 2014.
6. [Jin and Choi, 2012] Sou Young Jin and Ho Jin Choi. Essen-tial body-joint and atomic action detection for human ac-tivity recognition using longest common subsequence algorithm. In ICCV, pages 148–159, 2012.
7. [Ke et al., 2017] Qiuhong Ke, Mohammed Bennamoun, Senjian An, Ferdous Sohel, and Farid Boussaid. A New Representation of Skeleton Sequences for 3D Action Recognition. In CVPR, July 2017.
8. [Kingma and Ba, 2015] Diederik P Kingma and Jimmy Lei Ba. Adam: A method for stochastic optimization. ICLR, 2015.
9. [Li et al., 2016] Yanghao Li, Cuiling Lan, Junliang Xing, Wenjun Zeng, Chunfeng Yuan, and Jiaying Liu. On-line human action detection using joint classification-regression recurrent neural networks. ECCV, pages 203– 220, 2016.
10. [Li et al., 2017a] Bo Li, Huahui Chen, Yucheng Chen, Yuchao Dai, and Mingyi He. Skeleton boxes: Solving skeleton based action detection with a single deep convolutional neural network. In ICMEW, pages 613–616, July 2017.
11. [Li et al., 2017b] Chao Li, Qiaoyong Zhong, Di Xie, and Shiliang Pu. Skeleton-based action recognition with con-volutional neural networks. In ICMEW, pages 597–600, July 2017.
12. [Liu et al., 2016] Jun Liu, Amir Shahroudy, Dong Xu, and Gang Wang. Spatio-temporal lstm with trust gates for 3d human action recognition. In ECCV, pages 816–833, 2016.
13. [Liu et al., 2017] Chunhui Liu, Yueyu Hu, Yanghao Li, Si-jie Song, and Jiaying Liu. PKU-MMD: A large scale benchmark for continuous multi-modal human action understanding. ACM Multimedia workshop, 2017.

Prognostics for an actuator with the combination of support vector regression and particle filter

Runxia Guo*, *Jianfei* Sui, and *Jiusheng* Chen

School of Electronic Information and Automation, Civil Aviation University of China, Jin North Road No. 2898, Dongli District, Tianjin 300300, China

Abstract. The accurate prognostics for actuator malfunctions is a challenging task. Developing reliable prognostic methods is vital for providing reasonable preventive maintenance schedules and preventing unexpected failures. Particle filter has been proved to be a traditional approach to deal with actuator prognostic problems. However, the measurement function in the particle filter algorithm cannot be obtained in the prediction process, this paper presents a hybrid framework combining support vector regression (SVR) and particle filter (PF). The SVR output prediction results are employed as the "measurements" for the subsequent PF algorithm. To accomplish the accurate prognostics for actuator fault of civil aircraft, an improved PF based on Kendall correlation coefficient is put forward to solve the problem of particles' degeneracy. The experimental results are presented, demonstrating that the SVR-PF hybrid approach has satisfactory performance with better prognostics accuracy and higher fault resolution than traditional approaches.

1 Introduction

Developing a satisfactory fault prognostics technique for the aircraft automatic flight control system (AFCS) is of great importance for the civil aviation safety. Generally, the actuator is strongly coupled, severely nonlinear and electro-hydraulic, which greatly increases the difficulty of prognostics. In particular, with the development of AFCS, the focus of the contradiction is gradually shifting to the actuators. How to ensure the smooth running of the actuator is of great interest in both the aerospace industry and the theoretical research.

The essential fault prognostics approaches can be categorized into model-based methods and data-driven methods. A typical model-based fault prognostics approach generally starts with the internal working mechanism of the object system, and predicts the trend of development by using an analytical model that can reflect the physical laws of the system. However, in most practical applications, the establishment of models from the internal working mechanism of the target system or the component requires a lot of expert knowledge, especially for complex electronic systems (actuator, autopilot, etc.), building

* Corresponding author: rxguoblp@163.com

effective physical models are usually not possible. Data-driven fault prognostics technology does not require the prior knowledge of the target system (analytical model or expert experience), while directly putting the test or state monitoring data as an object to estimate the future evolution of the target system's trend, thus the drawback of model-based fault prognostics techniques can be avoided. Data-driven methods use several instruments, most of which are originated from artificial intelligence or statistical fields, and consist of neural networks (NN), support vector machine (SVM) and particle filter (PF) etc. SVR is a generalization of SVM in the case of regression; see [1]. SVR can get smaller errors based on limited training samples, and also ensure that small errors are maintained on independent test sets. A direct remaining useful life estimation method based on SVR is proposed in [2]. Recently, nonlinear filtering has become an increasingly active subject. The PF algorithm has been proven to be an effective approach for solving nonlinear and non-Gaussian problems [3]. PF is capable of achieving state estimation that can reflect more detailed and sufficient information of the target component when compared to the SVR's pure output prediction (Typically, SVR can only embody the fitting of the input and output relations, and can't learn the state of the system.). Therefore, the prognostics method based on PF has higher resolution than the output prediction based on SVR, i.e., the prognostics results may contain more abundant information. Unfortunately, the measurement function in the PF algorithm might be inaccessible in the long-term prediction process. It doesn't have the ability to predict the future state accurately in the absence of the latest measurement value. Moreover, a common phenomenon of PF is the sample degeneracy. Resampling is the main technique for solving this problem, which will always cause another noticeable problem: the loss of particle diversity.

The main contributions of this paper lie in two aspects: 1) From the perspective of data-driven approach, a novel framework combining SVR and PF is proposed to accurately estimate the state variables inside the actuator that are not visible (can't be directly measured), while is capable of achieving fault prognostics based on state variables with higher resolution; 2) An improved method for PF is developed to alleviate the problem of particle degeneracy to achieve better prediction accuracy.

The structure of this paper is organized as follows. In Section 2, the basic principle of PF is introduced. The proposed SVR-PF method is also described in detail in Section 2, where we present an improved PF algorithm. Section 3 illustrates the experiment and results analysis. Conclusions and future works are summarized in Section 4.

2 Prognostic method

2.1 Classical particle filter

Consider a discrete-time nonlinear state-space model with additive noise, given by:

$$x_k = f\left(x_{k-1}, v_{k-1}\right) \tag{1}$$

$$z_k = h\left(x_k, w_k\right) \tag{2}$$

where $x_k \in \mathbb{R}^{n_x}$ and $z_k \in \mathbb{R}^{n_z}$ are the state vector and the measurement vector, respectively. $f(\cdot)$ and $h(\cdot)$ are nonlinear functions that describe the system and measurement models, respectively. v_{k-1} is the process noise and w_k is the measurement noise.

To approximate the true posterior probability density function (PDF) $p\left(x_k \middle| z_{1:k}\right)$, new samples and their corresponding weights are generated from the Sequential Importance

Sampling (SIS) approach. Firstly, drawing samples x_k^i from the importance density $q\left(x_k \middle| x_{k-1}, z_{1:k}\right)$ for every $i = 1,\ldots,N$. Secondly, using the latest measurement z_k ,the weights ω_k^i are updated by the Bayes rule

$$\omega_k^i = \omega_{k-1}^i \cdot p\left(z_k \middle| x_k^i\right) p\left(x_k^i \middle| x_{k-1}^i\right)\left[q\left(x_k^i \middle| x_{k-1}^i, z_{1:k}\right)\right]^{-1} \quad (3)$$

to approximate the target distribution and normalized

$$\omega_k^i = \omega_k^i \cdot \left(\sum_{j=1}^{N} \omega_k^j\right)^{-1} \quad (4)$$

The posterior PDF is conventionally represented by a set of weighted samples, denoted by:

$$p\left(x_k \middle| z_{1:k}\right) = \sum_{i=1}^{N} \omega_k^i \delta\left(x_k - x_k^i\right) \quad (5)$$

where δ denotes the Dirac delta function. When the importance density is selected as $p\left(x_k \middle| x_{k-1}\right)$, the formula (3) changes to the form as follows:

$$\omega_k^i = \omega_{k-1}^i g_k\left(z_k \middle| x_k^i\right) \quad (6)$$

In addition, the PF monitors the degeneracy phenomenon by using effective sample N_{eff} . Resampling step is triggered if $N_{eff} \le N_{thres}$,with N_{thres} a pre-defined threshold.

2.2 Improved particle filter based on Kendall correlation coefficient

Clearly, the standard PF algorithm avoids degeneracy phenomenon by eliminating small weight particles and retaining large weight particles, but this can also cause loss of particle diversity. Choosing the appropriate importance density is a way to solve the problem of samples' degeneracy. The selection of the importance density function has a very obvious influence on the efficiency of the algorithm and the rate of weight degradation. When the likelihood is too narrow or it is located in the tails of the prior distribution, most particles which extracted from the prior distribution have small weights, and the estimation accuracy can be significantly decreased. Due to the neglect of the latest measurement information, the importance density only contains the model information, failed to overcome the problem of sample weight degeneracy. To conquer the mentioned drawback, a novel technique of sampling from prior importance density is proposed. Then particles are pushed in to the regions of highly likelihood by using the measurement information. We use the Kendall correlation coefficient (KCC) incorporated into the standard particle filter as a criterion to determine whether the particle is similar to the real state and then decide that the particle should be ignored or should be considered in the next step of the process.

2.2.1Kendall correlation coefficient

The Kendall correlation coefficient (KCC) is a non-parametric statistic used to measure the relevance of two random variables (real-valued vectors), and it is one of the most widely used methods for similarity measurement. More detailed explanations about this topic can be found in [4]. The Kendall correlation coefficient τ can be calculated as follows:

$$\tau = 2\left(N_c - N_d\right) \cdot N_b\left(N_b - 1\right)^{-1} \quad (7)$$

where $-1 \le \tau \le 1$, N_c represents the logarithm of elements that satisfy consistency and N_d represents the logarithm of elements that don't satisfy consistency. The coefficient τ ranges from -1 to 1 and it is invariant to linear transformations of either variables.

2.2.2 KCC-based particle filter

Suppose that a particle's measurement path is close to the measurement path of the true state, then its path is more likely to be close to the path of the true state. The degree of similarity between the paths of the two measurements can be computed by the coefficient τ . When the degree of similarity is high, it indicates that the particle is closer to the real state. On the contrary, the similarity between the particle and the real state of the system is low. The particles are potentially pushed into the significant areas of the state space depend upon the KCC correction term.

Let $X_k^* = \left\{ \left[x_j^*(i) \right]_{j=k-L+1}^{k} \right\}_{i=1}^{N}$ represent all the N particles from time $k-L+1$ to time k ,where $1 < L \le k$ is a constant that is pre-determined, and these particles are sampled from the importance density. $Z_k^* = \left\{ \left[z_j^*(i) \right]_{j=k-L+1}^{k} \right\}_{i=1}^{N}$ represent the corresponding measurement paths of the N particles, where $z_j^*(i)$ can be computed by using the measurement model (2), denoted by $z_j^*(i) = h\left(x_j^*(i), 0\right)$.

Let $Z_k = \left[z_j \right]_{j=k-L+1}^{k}$ represent the measurement path of the system state that is actually measured from time $k-L+1$ to k , and the latest observations are provided by the SVR output prediction. In this case, an exponential function that measures the degree of similarity between the two measured values can be defined as:

$$\beta_k^i = \mathcal{J}\left(\left[z_j^*(i) \right]_{j=k-L+1}^{k}, \left[z_j \right]_{j=k-L+1}^{k} \right) = e^{(\alpha \times \tau)} \tag{8}$$

where $\beta_k^i > 0$ represents the degree of similarity between the measured value's path of a particle and the measured value path of the true system state from time $k-L+1$ to time k . τ can be computed by the formula (7). Here, $\alpha > 0$ is a scaling factor, which needs to be set before the algorithm runs. The value α can be adjusted according to the magnitude of the measurement noise to change the relative size of the value τ corresponding to the different value β_k^i , so as to improve the estimation accuracy. For instance, when the measurement noise is small, the value α can be appropriately increased. When the measurement noise is large, the value α can be appropriately reduced. It is noted that we select the particles that is the most similar to the real state as the final particles x_k^i to perform state estimation. Unlike standard particle filter, we pick out the particles that are closest to the true states in the prediction step.

2.3 The hybrid framework of SVR-PF

Our aim is to combine the improved SVR and the improved PF algorithm in depth to achieve actuator state prediction. On the one hand, owing to the monitoring data generated from the actuator is available, while the actuator is operating, the proposed SVR-PF method could be used to achieve the prognostics for the actuator. On the other hand, due to the real measurement value might be inaccessible, the output prediction value could be used as the approximated value of the real value of the actuator. Therefore, the output prediction from the improved SVR time series model could be used as the state measurement equation to improve the precision of prognosis. Meanwhile, the improved PF algorithm could make further fine estimate of the invisible future state of the system, and obtain the uncertainty representation. Firstly, the improved SVR model is established to achieve output prediction of the actuator data, its output is named "*SVRpredict(k)*." This prediction result is used as the measurement value for the state updating equation of the improved PF algorithm. The final state prediction result is output by the improved PF algorithm, denoted as "*PFpredict(k)*." Furthermore, the uncertainty representation can be obtained by the improved PF algorithm. If the "*PFpredict(k)*" reaches the fault threshold x_{thr} (x_{thr} from the component maintenance manual (CMM) of the actuator.), the fault warning is performed. If the fault threshold is not reached, the model is updated and the forward prediction is continued.

This hybrid framework is used to emphasize the fact that the SVR algorithm and the PF algorithm are complementary. The output of the SVR algorithm is the input of the PF algorithm, which provides the "data excitation" for the subsequent state estimation of the PF algorithm, while the PF algorithm is a further enrichment and refinement of the SVR algorithm, i.e., on the basis of obtained the prediction values of the actuator output, the prediction values of the state variables which are not visible in the actuator are further obtained.

3 Experiment and results analysis

The actuator is a mechanical part that connects the actuator body with the corresponding rudder surface. When the internal oil leakage of the actuator occurs, due to the insufficient of oil supply pressure, the actuator will not carry out normal piston movement. Thus the position of the actuator cylinder is selected as the variable to predict. To fully validate the performance of the designed prognostics approach, we choose an actuator cylinder movement model as the research object, and its parameters are analyzed as the following expression:

$$\begin{cases} x_k = x_{k-1} + \Delta T \phi\left(x_{k-1}\right) + \nu_{k-1} \\ z_k = x_k + w_k \end{cases} \tag{9}$$

where x_k denotes the actual actuator cylinder position and

$$\phi(x_{k-1}) = K_{ci} K\left(u_{k-1} - x_{k-1}\right) \left\{ \frac{\Delta P - \upsilon F_{aero} S^{-1}}{\Delta P_{ref} + K_a \left[K_{ci} K\left(u_{k-1} - x_{k-1}\right)\right]^2 S^{-1}} \right\}^{1/2} \tag{10}$$

The process noise $\nu_{k-1} \sim N\left(0, 0.2\right)$ and measurement noise $w_k \sim N\left(0, 0.4\right)$ are assumed to be Gaussian white noise.

The actuator is a fast identification system. According to the actual data sampling rate of the airborne equipment, the airborne data acquisition module (SDCU) extracts a running data at intervals of 250 milliseconds. In addition, in order to improve the computational efficiency and satisfy the requirements of practical applications, we properly control the size of the data set. Therefore, the sampling time interval is set to 250 milliseconds. In other words, four data are collected within one second. The flight control input u_k is set to a sine function, and its form is $u_k = A\sin\left(2\pi\varsigma k\right)$ .The experimental time is set to 20 seconds and the time interval ΔT between two consecutive measurement value is 250 milliseconds. The oscillation frequency ς is 1.2Hz.The rest of the selected initial parameters in the experiments are listed in Table 1. We use the root mean square error (RMSE) and mean absolute percentage error (MAPE) to evaluate the performance of the proposed methods.

Table 1. List of parameter values.

Term	Physical significance	Model value
K_{ci}	A simple or double slope gain	2.326
K	Servo control gain	1.085
ΔP	Hydraulic pressure delivered to actuator	3045.79psi
F_{aero}	Aerodynamic force applying on the control surface	≈ 0
K_a	Actuator damping coefficient with maximum rod speed	0.8
ΔP_{ref}	Constant differential pressure	2581.67psi
S	Actuator piston surface area	4.38sq.in
A	Amplitude of the actuator rod oscillation	16.68in
υ	Tuning parameter	1.645

[a] It is worth noting that psi is acronym for pounds per square inch.

[b] F_{aero} is set to zero, which means that the aerodynamic forces exerted on the control surface is negligible.

The performance of the proposed SVR-PF approach will be evaluated, and also compares with the standard PF, the radial basis function particle filter (RBF-PF) [5] and the particle filtering technique combined with a kernel smoothing method (PF-KS) [6]. In the improved PF algorithm, due to the particles with the most similarity of the true state are chosen as the final particles according to the KCC, and the distribution of these particles are adjusted by the improved SVR. Both the particle degeneracy problem and the computational complexity reduction problem have been solved. Therefore, the proposed SVR-PF method can improve the efficiency of calculation and reduce the burden of calculation. We compare the SVR-PF approach with standard PF, RBF-PF, and PF-KS. The sample number is set to be 200 particles. As shown in Figure.1, the introduced SVR-PF derives a satisfactory prediction result. Table 2 gives a comparison of the prediction performance of standard PF, RBF-PF, PF-KS, and SVR-PF in terms of RMSE and MAPE. It can be found from Figure.1 that the tracking performance of the proposed SVR-PF approach is better than other approaches. Therefore, the proposed SVR-PF can be considered as an excellent tool to be applied to the state prediction of the actuator.

Table 2. Comparison for the four methods.

Algorithm	Time steps	RMSE	MAPE
PF (200)	120s	5.2713	12.83%
RBF-PF (200)	120s	4.6942	8.71%
PF-KS (200)	120s	2.5211	6.04%
SVR-PF (200)	120s	1.1875	1.32%

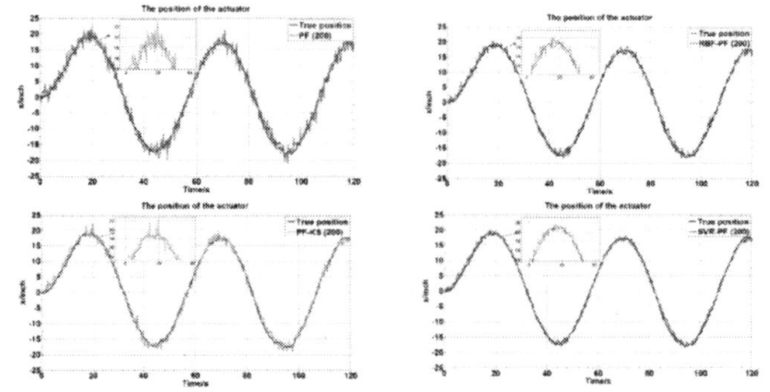

Fig. 1. Comparison of state tracking capabilities.

4 Conclusion

In this paper, starting from a data-driven perspective, we explored a set of new deep hybrid fault prediction method for the actuator. Satisfactory experimental results have been achieved by using the real actuator's data, and the SVR-PF demonstrated improved prediction capability over the standard PF. The main contributions of this paper are summarized as: 1) A novel state prediction approach of actuator is proposed, which provides a good foundation for multi-step forward prediction; 2) The improved PF algorithm based on Kendall correlation coefficient effectively alleviates the degeneracy problem. Future work is devoted to reducing the computational complexity of the proposed SVR-PF algorithm and further improving the prediction accuracy of the SVR algorithm.

References

1. V N Vapnik 2002 The nature of statistical learning theory *IEEE Trans. Neural Netw* **8** 1564
2. Khelif, Racha, Chebel-Morello, Brigitte, Malinowski and Simon 2017 Direct remaining useful life estimation based on support vector regression *IEEE Trans. Ind. Electron* **64** 2276-85
3. N J Gordon and A F M Smith 1993 Novel approach to nonlinear/non-Gaussian Bayesian state estimation *IEE Proceedings-F* **140** 107-113
4. Kendall M 1938 A new measure of rank correlation *Biometrika* **30** 81-89
5. C Sbarufatti, M Corbetta, M Giglio and F Cadini 2017 Adaptive prognosis of lithium-ion batteries based on the combination of particle filters and radial basis function neural networks *J. Power Sources* **344** 128-140
6. H Yang, P Baraldi, FD Maio and E Zio 2015 A particle filtering and kernel smoothing-based approach for new design component prognostics *Reliab Eng Syst Safe* **134** 19-31

Remaining useful life prognostics for the electro-hydraulic actuator using relevance vector machine and optimized on-line incremental learning

Jiankang Dong[1], *Jiaqi* Wang[2], and *Runxia* Guo[1,*]

[1]School of Electronic Information and Automation, Civil Aviation University of China, No. 2898 Jin North Road, Tianjin, 300300, China
[2]School of Computer science and technology,Civil Aviation University of China, No. 2898 Jin North Road, Tianjin, 300300, China

Abstract. The electro-hydraulic actuator plays a significant role in the automatic flight control system, so it is vital to predict the remaining useful life (RUL) for the electro-hydraulic actuators. Relevance vector machine (RVM) is flourishing in the field of RUL prognostics and gradually applied to the prediction of complex systems or components, but the general RVM cannot achieve on-line prediction efficiently due to its high computational complexity, besides, the sparse RVM model which is only based on historical data set could cause a large prediction error in the long term. To deal with these plights, an optimized incremental learning algorithm based on RVM is presented taking full advantage of the on-line updating samples to improve the precision of prognostics.

1 Introduction

Electro-hydraulic actuator is a core component in the Automatic Flight Control System (AFCS) of civil aircraft. Failures in electro-hydraulic actuator could lead to loss of operation, functional degradation, machine halt and even catastrophic failure [1]. Therefore, the reliability of electro-hydraulic actuator becomes a subject of great interest to the civil aircraft avionics system and predict the remaining useful life (RUL) in an effective way are imperative [2]. Figure 1 shows a typical simplified diagram of an electro-hydraulic actuator.

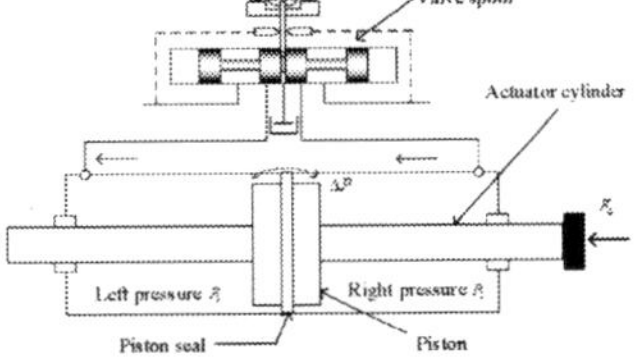

Fig. 1. Schematic of the electro-hydraulic actuator.

* Corresponding author: rxguoblp@163.com

In the existing literature, the prognostics approaches for RUL can be broadly divided into model-based approaches and data-driven approaches. Model-based approaches like Kalman filtering (KF) and particle filtering (PF) typically involve the establishment of a physical model to describe the mechanism of the system degradation and failure evolvement. Therefore, they may not be suitable for many practical applications [3]. Contrarily, data-driven approaches which are used to capture the inherent relationships could avoid the dependence on the system's analytical model, thus they can extract essential information from historical and monitoring data without modelling. Therefore, data-driven approaches have gradually became the primary methodology in the field of prognostics due to the easy-to-use and flexible modelling characteristics [4].

Thus far, the prognostics based on RVM algorithm is generally off-line which does not possess dynamic updating capability of the model and has low prediction precision in long-term [5]. Particularly, the RVM algorithm equipped with the characteristic of small sample size still has no efficacious on-line training technology that can be actualized to process on-line modelling and prognostics [6]. To overcome these defects, this paper presents an optimized on-line incremental learning algorithm based on RVM to realize electro-hydraulic actuator's RUL prognostics. In the proposed method, the RVM algorithm is herein applied to predict the RUL of electro-hydraulic actuator by capturing the correspondence between the hydraulic difference and sample entropy, in addition, the uncertainty of the predictive results is also characterized.

2 Reviews of related theories

2.1 Relevance vector machine

The RVM introduces a prior to the weight of the model and the most probable value is iteratively estimated from the data set [7]. Given a set of examples of input vectors $\{x_i\}_{i=1}^{N}$ and their corresponding target vectors $\{t_i\}_{i=1}^{N}$. The purpose is to learn a dependency model between the targets and inputs vectors for the sake of making precise prognostics value of t_i for unseen value of x_i [8]. Generally, the predictions are based on a function $y(x)$ defined over the input space and the function $y(x)$ can be defined as below:

$$y(x_i;\omega)=\sum_{i=0}^{N}\omega_i K(x,x_i) \tag{1}$$

where $\omega=[\omega_0,\ \omega_1,\cdots,\omega_N]^T$ denotes the weight vector, $K(x,x_i)$ is a kernel function.

RVM seems to predict target t_i according to

$$t_i=y(x_i;\omega)+\varepsilon_i \tag{2}$$

where ε_i is an independent sample from noise process with mean 0 and variance σ^2 .

The likelihood estimation of all the sampled data sets can be written as (3)

$$P(t|\omega,\sigma^2)=(2\pi\sigma^2)^{-N/2}\exp\left\{-(2\sigma^2)^{-1}\|t-\Phi\omega\|^2\right\} \tag{3}$$

where $\Phi=[\varphi(x_1),\varphi(x_2),\ldots,\varphi(x_N)]^T$ is an $N\times(N+1)$ design matrix, $\varphi(x_i)=[1,K(x_i,x_1),K(x_i,x_2),\ldots,K(x_i,x_N)]^T$.

Given a new input x^* , its corresponding output probability distribution is

$$P(t^*|t,\alpha_{MP},\sigma_{MP}^2)=N(t^*|y^*,\sigma_*^2) \tag{4}$$

where the mean and the variance are respectively $y^* = \mu^T \phi(x^*)$, $\sigma_*^2 = \sigma_{MP}^2 + \phi(x^*)^T \Sigma \phi(x^*)$.

2.2 Sample entropy

In this paper, the sample entropy is represented by a symbol E_{samp}. Here, using E_{samp} to process the original voltage sequence can greatly reduce the training samples, which provide potential for high computational efficiency of on-line prediction.

For a given N_c total number of data points, we form the $N_c - m + 1$ vectors $x_m(p)$ as

$$x_m(p) = [x(p), x(p+1), \ldots, x(p+m-1)], 1 \le p \le N_c - m + 1 \tag{5}$$

where m is the length of a subseries of the dataset N_c.

The distance between two such vectors and is defined as follow

$$d_m[x_m(p), x_m(q)] = max[\|x_m(p+k) - x_m(q+k)\|], 0 \le k \le m-1 \tag{6}$$

The E_{samp} is ultimately estimated by

$$E_{samp} = -ln\left[\frac{A^m(r)}{B^m(r)}\right] \tag{7}$$

where $B^m(r)$ and $A^m(r)$ represent the probabilities that two sequences match for m and $m+1$ points, respectively.

3 Method for the RUL prognostics of electro-hydraulic actuator

3.1 Data preprocessing

Sampled N total number cycles of voltage sequence $\{x_m(p)\}_{p=1}^{N_c}$ from an electro-hydraulic actuator and each cycle contains N_c data points, then compute the E_{samp} of voltage sequence for each cycle. Then the new data set $\{E_{samp}(i)\}_{i=1}^{N}$ representing a sample entropy vector is employed as data input and the corresponding hydraulic difference vector $\{\Delta P(i)\}_{i=1}^{N}$ as target vector of RVM, respectively.

$$\Delta P(i) = \|P_1(i) - P_2(i)\| \tag{8}$$

where $P_1(i)$ represents the pressure on the left of the cylinder, $P_2(i)$ stands for the pressure on the right of the cylinder in Figure 1.

3.2 Training algorithm for RVM

In this section, an RVM prognostics model is built by training the newly obtained sample entropy data set. The steps for training RVM model are as follows:

Step 1. Initialization: initialize the weight ω and variance σ^2.

Step 2. Expectation step: utilize $\omega^{(\kappa)}$ and $(\sigma^2)^{\kappa}$ to estimate the $(\kappa+1)th$ iteration and $E(\omega\omega^T)$. $\omega^{(\kappa+1)} = (\sigma^{-2})^{(\kappa)}\left(\Psi^{(\kappa)} - \Psi^{(\kappa)}\Phi^T\Lambda^{-1}\Phi\Psi^{(\kappa)}\right)\Phi^T t$,

$$E(\omega\omega^T) = \left(\Psi^{(\kappa)} - \Psi^{(\kappa)}\Phi^T\Lambda^{-1}\Phi\Psi^{(\kappa)}\right) + \omega\omega^T \tag{9}$$

where $\Lambda = \Phi\Psi^{(\kappa)}\Phi^{T} + (\sigma^2)I$, $\Psi^{(\kappa)} = diag\left[\left|\omega_0^{(\kappa)}\right|^2, \left|\omega_1^{(\kappa)}\right|^2, \ldots, \left|\omega_N^{(\kappa)}\right|^2\right]$ and $\omega^{(\kappa)}$, $\omega^{(\kappa+1)}$ represent the weights at iterations κ and $\kappa+1$.

Step 3. Maximization step: calculate the new variance according to $\omega^{(\kappa+1)}$ obtained in Step 2.

Step 4. Convergence judgment: stop the iteration if $\left\|\omega^{(\kappa+1)} - \omega^{(\kappa)}\right\| / \left\|\omega^{(\kappa)}\right\| < \delta$ or the algorithm reaches the maximum iteration number, or else go to Step 2 to start a new iteration.

3.3 Updating on-line model with optimized incremental learning algorithm

The incremental learning algorithm is adopted in this paper with the merit of learning new knowledge without forgetting previously learned. However, with the on-line new data updating, the size of on-line training samples will increase gradually, which could lead to low computational efficiency. Therefore, a new optimized algorithm is proposed in this paper to improve the calculation efficiency of on-line prognostics. Detailed steps are shown in Figure 2.

Take τ_0 cycle as the start point of the prognostics, predict the hydraulic difference $\Delta P(i)$ using the initial model until the predicted value reaches the failure threshold U , then the real hydraulic difference values will obtain one by one during this process. After a l-step prediction stage, the prognostics precision should be checked. Adopt the real hydraulic difference values of the l-step (from cycle τ_0+1 to τ_0+l) to update the corresponding predictive values and add them to the previous training samples. Set an updating threshold T to determine whether to update the current model or not. Compute the correlation coefficient (C) of the two time series before and after the updating, if the result outweighs the updating threshold, it indicates that the newly added samples have little influence on the current model and there is a good accordance between the predictive and actual values, so the model does not need to be updated. Continue predicting and adding the next l-step's real hydraulic difference values, compute the correlation coefficient until the calculation result is less than the updating threshold, then update the current model using the latest training samples and continue to predict with the new model at cycle τ_1 . Repeat the above steps.

The correlation coefficient (C) is defined as follow:

$$C = \frac{\sum_{i=1}^{l}(y_i - \bar{y})(\hat{y}_i - \bar{\hat{y}})}{\sqrt{\sum_{i=1}^{l}(y_i - \bar{y})^2}\sqrt{\sum_{i=1}^{l}(\hat{y}_i - \bar{\hat{y}})^2}} \tag{10}$$

where y_i is the actual value, $\hat{y}_i$ is the predictive value, $\bar{y}$ is the mean of actual values, and $\bar{\hat{y}}$ is the mean of predictive values, l is the length of a prediction stage.

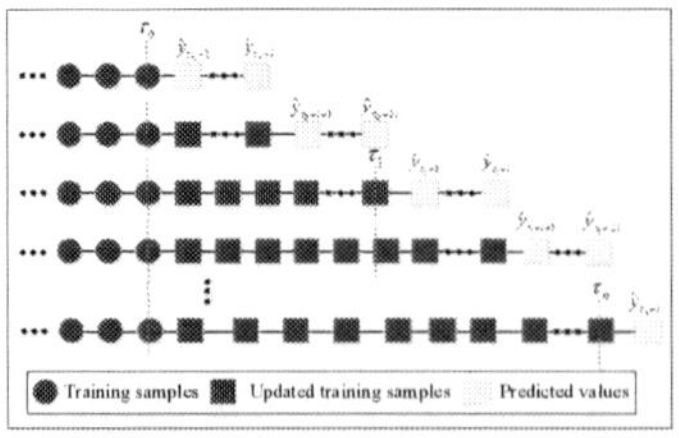

Fig. 2. Basic schematic of online model updating.

3.4 RUL

The prognostics stops and the remaining useful life (RUL) is calculated at cycle τ_{end} when the failure threshold of the electro-hydraulic actuator met. The failure threshold can be considered to occur when the predictive hydraulic difference degradation value drops to 70% of its initial value in actual application, and the RUL at the current cycle can be obtained by measuring the distance between the cycle τ_{end} and the cycle τ_0 at which the prediction starts. The predicted RUL is given as:

$$RUL = \tau_{end} - \tau_0 \tag{11}$$

where τ_{end} is the intersection of the prognostics curve and the failure threshold line.

4 Experiment and analysis

In order to further verify the performance of the proposed on-line RUL prognostics algorithm, we conduct an experiment against the internal leakage fault of an electro-hydraulic actuator. The voltage data set is selected from an electro-hydraulic actuator, which has nominal voltage of 115V with 400HZ and the maximum hydraulic difference of 2200psi. The actual life of the electro-hydraulic actuator is about one million cycles. In order to accelerate the results of the experiment, the electro-hydraulic actuator experiment is conducted with the 720 times' acceleration to simulate run-to-failure test. The experiment performed on the data set at three different prognostics start cycles, which are 30% end of life (EOL), 45% EOL and 60% EOL, respectively. EOL represents the failure cycle where the hydraulic difference reaches the failure threshold defined as 800psi in this study. Under the 720 times simulation accelerated run-to-failure test, the actual cycle life of the electro-hydraulic actuator is 1400 cycles, and a cycle refers to the time length of an actuator cylinder swinging back and forth.

4.1 Case study of the proposed RUL prognostics method

In this study, the start points of the data set at 30% EOL, 45% EOL and 60% EOL are 420 cycles, 630 cycles and 840 cycles, respectively. The failure threshold takes at 1400 cycles and the time interval ΔT of a cycle is set 2.89s. l represented the length of a prediction stage is set through the experience and the suitable one namely 30 is chosen to conduct this experiment.

In order to fully validate the superiority of the proposed optimized incremental RVM (optimized I-RVM) prognostics algorithm, Figure 3 shows the process of updating the prognostics model constantly and the 13 different RUL results predicted by different updated models indicate that the more times the model is updated, the higher the prognostics accuracy is.

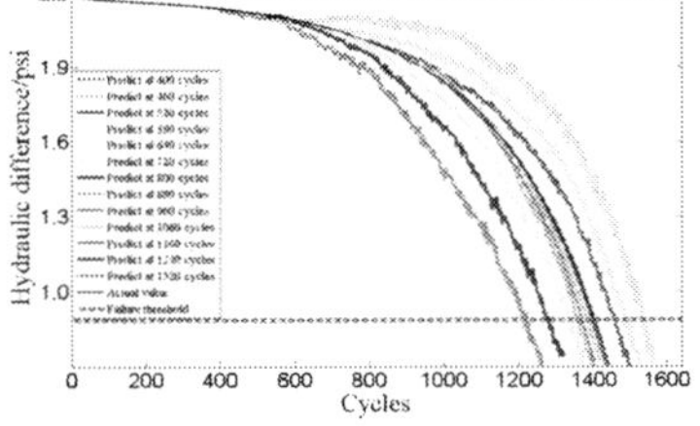

Fig. 3. The process of updating the prognostics model.

The tracks of the RUL predicted by original RVM, retraining RVM (R-RVM), incremental RVM (I-RVM) and optimized incremental RVM (optimized I-RVM) at 30% EOL, 45% EOL and 60% EOL are shown in Figure 4, respectively.

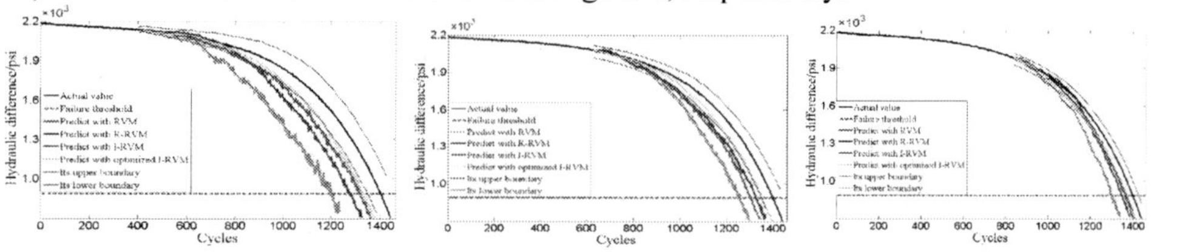

(a) (b) (c)

Fig. 4. The comparison results of the proposed different methods at different start cycles (a): at 30% EOL (b): at 45% EOL (c): at 60% EOL.

Table 1 shows the predict performance comparison of the three methods.

Table 1. Predict performance comparison of the three methods.

Algorithm	Start points	RUL error (cycle)	Time consumption (s)
RVM	30% EOL	189	1.465±0.043
R-RVM	30% EOL	161	7.812±0.121
I-RVM	30% EOL	140	5.657±0.052
optimized I-RVM	30% EOL	132	4.579±0.040
RVM	45% EOL	165	1.960±0.037
R-RVM	45% EOL	149	7.281±0.156
I-RVM	45% EOL	118	5.268±0.049
optimized I-RVM	45% EOL	76	4.023±0.038
RVM	60% EOL	101	2.595±0.045
R-RVM	60% EOL	63	6.915±0.146
I-RVM	60% EOL	45	5.024±0.036
optimized I-RVM	60% EOL	36	3.652±0.035

It can be seen from Figure 4 and Table 1 that the predicted RUL results with the proposed optimized I-RVM prognostics mothed are almost close to the actual values and have the minimum error value all the time.

5 Conclusion

In this paper, a novel on-line RUL prognostics algorithm for electro-hydraulic actuator has been proposed in view of the problem that RVM can not realize on-line prediction. The incremental learning algorithm realizes the on-line prognostics and improves the performance of RVM long-time prediction. The optimized I-RVM method has greatly reduced the training samples and improved the computational efficiency by using sample entropy to process the original voltage data set. The experimental comparison results show a satisfactory performance of the optimized I-RVM on-line prognostics method in terms of prediction efficiency accuracy and computational efficiency, which may act as a further reference for other similar domains related to remaining useful life prognostics and real time condition monitoring.

Acknowledgements

This work was jointly supported by the National Natural Science Foundation of China (Grant 61603395); the National Natural Science Foundation of China and the Civil

Aviation Administration of China (Grant U1433103); the National Key R & D Project of China (Grant 2016YFB0502405); the Special Program of Talents Development for Excellent Youth Scholars in Tianjin.

References

1. Li H, Pan D and Chen C L P 2014 Intelligent Prognostics for Battery Health Monitoring Using the Mean Entropy and Relevance Vector Machine *IEEE Trans. Systems Man & Cybernetics Systems*. **44** 851-862
2. Hu X, Jiang J, Cao D and Bo E 2016 Battery Health Prognosis for Electric Vehicles Using Sample Entropy and Sparse Bayesian Predictive Modeling *IRE Trans. Ind. Electron.* **63** 2645-56.
3. Guo R, Wang J, Zhang N and Dong J 2018 State prediction for the actuators of civil aircraft based on a fusion framework of relevance vector machine and autoregressive integrated moving average *J. Syst. Control Eng.* 1-13
4. Liu D, Wang H, Peng Y, Xie W and Liao H 2013 Satellite Lithium-Ion Battery Remaining Cycle Life Prediction with Novel Indirect Health Indicator Extraction *Energies*. **6** 3654-68
5. Liu D, Peng Y, Li J and Peng X 2013 Multiple optimized online support vector regression for adaptive time series prediction *Measurement*. **46** 2391–404
6. Chen C, Kunche S, and Pecht M 2013 Incremental learning approach for improved prediction *IEEE Int. Conf. Prognostics and Health Management*. pp 1–7
7. Lou J, Jiang Y, Shen Q, Shen Z and Wang Z 2015 Software reliability prediction via relevance vector regression *Neurocomputing*. 66-73
8. Widodo A, Shim M C, Caesarendra W and Yang B S 2011 Intelligent prognostics for battery health monitoring based on sample entropy *Expert Systems with Applications* **38** 11763-69

An operational effectiveness evaluation method of the swarming UAVs air combat system

Niping Jia, *Zhiwei* Yang*, and *Kewei* Yang

College of System Engineering, National University of Defense Technology Changsha, Hunan, P.R. China

Abstract. In this paper, we propose an operational effectiveness evaluation method of the swarming UAVs air combat system. The system network model is firstly established before the evaluation. We divide the edges in the network into nine types according to the Generalized Operation Loop and give the edge attribute of operational information entropy based on node attributes. When calculating the system operational effectiveness, we focus on the striking ability and propose the concept of Operation Loop Capability to represent the target-attack effectiveness of a certain operational loop, which can be calculated from the edge operational information entropies. We can get the combat effectiveness to a certain target from the Operation Loop Capability, and the combat effectiveness of the swarming UAVs air combat system is represented by the sum of the striking capabilities to all targets. Finally, the scene of swarming UAVs air combat is taken as a case study to apply and verify our evaluation method. This paper gives a quantitative perspective on swarming UAVs combat system assessment and can help to analyze the contribution of UAVs and other weapons in the system.

1 Introduction

Recently, the new battlefield concept of swarming unmanned aerial vehicles (UAVs) combat is getting world-wide attention. It is derived from the concept of the "distributed killing" capability, which is listed as one of the most significant abilities to be developed in the US "third offset strategy"[1], and it means decomposing the functions of large equipment by multiple small unmanned combat platforms.

At present, many scholars have introduced and analyzed the swarming UAVs air combat system [2][3]. But most of these studies are descriptive, lacking quantitative modeling and assessment of the swarming UAVs combat system, which is of great significance for UAVs development planning and weapon system requirement analyzation. However, the swarming UAV air combat system is also a typical military system, and there have been many literatures about the system modeling and evaluation issue. One important modeling and evaluation method is the network theory. American scholar Jeffery R. Cares proposed the IACM model and divided the operational nodes in the weapon network into

* Corresponding author: zhwyang88@hotmail.com

four types [4][5]. The research team in National University of Defense Technology proposed a network description and evaluation method of the weapon system called Operation Loop (OL) [6]. They divided the system weapons into sensors, decision-makers, attackers and targets, which provided a good perspective for combat system modeling and evaluation.

Based on the present studies on military systems, in this paper, we propose an operational effectiveness evaluation method of the swarming UAVs air combat system through network modeling. This work gives a new perspective for modeling and assessing the swarming UAVs air combat system quantitatively.

2 Network modeling of the swarming UAVs air combat system based on Generalized Operation Loop

2.1 The concept of Generalized Operation Loop (GOL)

The concept of Generalized Operation Loop (GOL) is based on Operation Loop (OL). Tan Yuejin firstly proposed OL theory and defined the combat process as a closed loop consisting of one or more sensors, deciders, attack weapons and targets [6]. In the swarming UAVs combat system, in addition to the basic relationships in OL, there may be more relationships among the same kind of weapons. We add these relationships to the standard OL theory and define it as Generalized Operation Loop.

Definition 1: Generalized Operation Loop (GOL) is a modified closed loop based on traditional Operation Loop (OL), besides the cycle of sensors (S), decision-makers (D), attackers (A) and targets (T), it contains more weapon relationships, including the information sharing relationships between sensors, the coordination of command and control relationships between decision-makers, and the autonomous coordinated attack relationships between attackers.

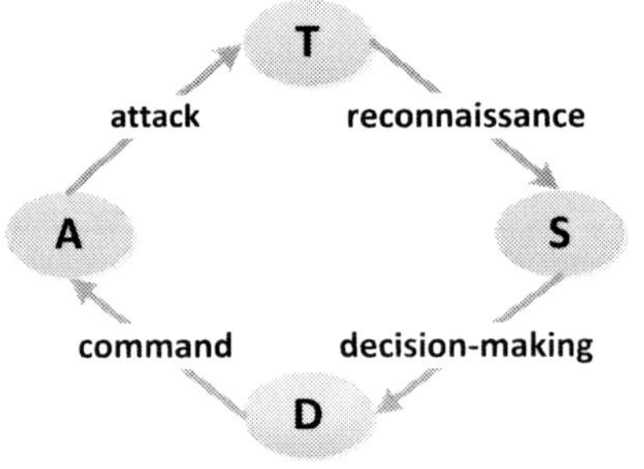

Fig. 1. Standard Operation Loop (OL).

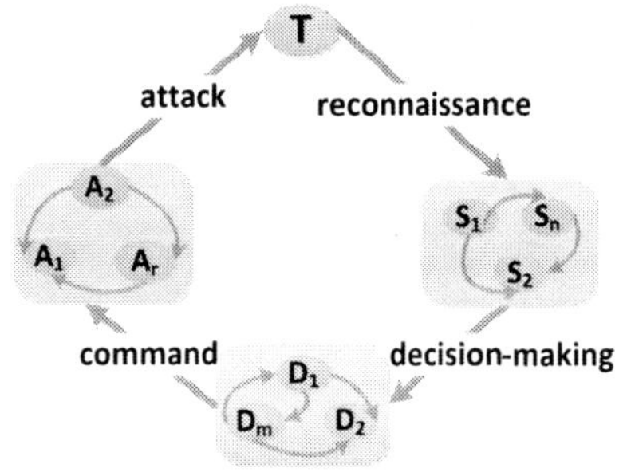

Fig. 2. Generalized Operation Loop (GOL).

2.2 Nodes definition

In the system network, the weapons are abstracted as nodes and can be classified as $V=\{S,D,A,T\}$. Where, S, D, A, T represents the weapon of sensor, decider, attacker and target entity respectively. The attributes of the four types of nodes are set in figure 3-6.

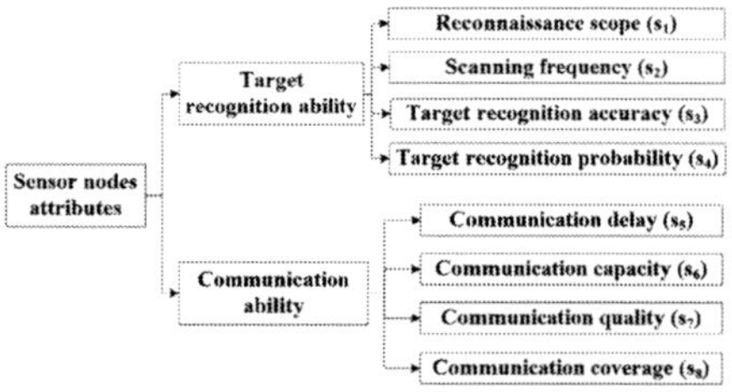

Fig. 3. Attributes of sensor nodes.

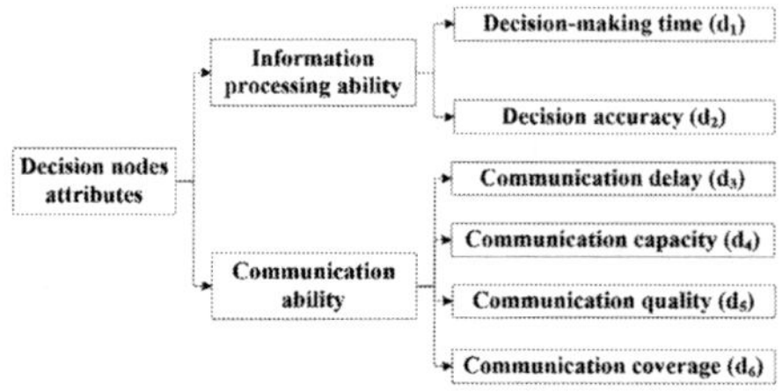

Fig. 4. Attributes of decision nodes.

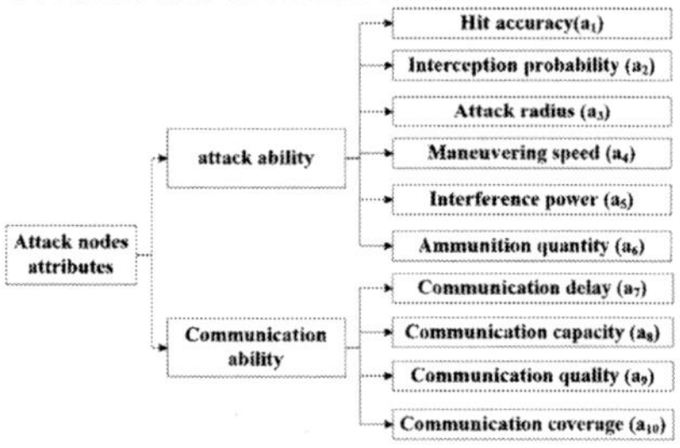

Fig. 5. Attributes of attack nodes.

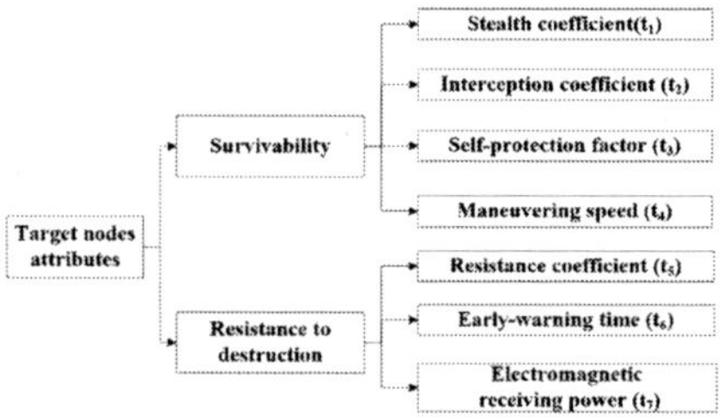

Fig. 6. Attributes of target nodes.

2.3 Edges definition

The types of edges represent different relationships between S, D, A and T nodes. This part aims to find out all the possible relationships (edge types) among the system weapons and give corresponding edge attributes for further effectiveness evaluation. During the modeling process, we mainly consider the information sharing and the complex communication relationships between UAVs.

2.3.1 Edge types

As shown in table 1, we carefully analyzed the relationships between S, D, A, and T, and the edges with practical meanings are listed.

Table 1. Edge classifications and meanings.

Edge meanings	Edges	Edge direction	Interpretation
Command edges	$D \to A$ $D \to S$	One-way Directed	The command nodes can control the actions of attack weapons and sensor devices.
Intelligence edges	$S \to D$ $A \to D$	One-way Directed	This edge type represents the feedbacks of battlefield information to the decision nodes.
Attack edge	$A \to T$	One-way Directed	It represents the attack task to the target.
Reconnaissance edge	$T \to S$	One-way Directed	The information of the targets can be detected, and it means the information flow from targets to sensors.

Command coordination edge	$D \leftrightarrow D$	Bidirectional	This edge means the command cooperation between the decision - makers.
Information sharing edge	$S \leftrightarrow S$	Bidirectional	It mainly represents the information sharing between the reconnaissance UAVs in the combat process.
Attack coordination edge	$A \leftrightarrow A$	Bidirectional	It mainly represents the attack cooperation between the attack UAVs in the combat process.

2.3.2 Edge attribute

For the combat effectiveness evaluation, we define the edge attribute of the operational information entropy using the nodes attributes data of S, D, A and T. In 1948, Shannon proposed the concept of "information entropy" [7] and he described the information entropy as

$$H = -\sum_{x \in U} P(x) \lg P(x) \tag{1}$$

In the combat process, the low operational information uncertainty (also information entropy) means the better knowledge of the battle situation, which will help to get better combat effectiveness. Thus, we will use the information entropy to make the combat effectiveness evaluation and give the following modified edge operational information entropy definition.

Definition 2: Edge operational information entropy is used to reflect the information uncertainty of the information flow in the combat network, signifying the control and knowledge of battlefield situation, and is defined as h_e .

$$h_e = -\sum_{i=1}^{n} w_i \lg f_i \tag{2}$$

$$f_i = 1 - \frac{2}{e^{x_i} + e^{-x_i}} \tag{3}$$

Supposing that the edge e connects the nodes of S_1 and S_2, then $x_i (i = 1, 2, ..., n)$ represents the comprehensive index calculated from the attributes of S_1 and S_2. n is the number of x_i and w_i is the weights of x_i . f_i is a membership function to guarantee $h_e > 0$ (that is, $\log f_i < 0$, $f_i \in [0,1]$). For different types of edges, there is a subtle difference in the number of x_i and its calculation method, thus the edge operational information entropies will also be different slightly. Here we just take the $S \leftrightarrow S$ edge for example and give its edge operational information entropy calculation method. Supposing that there are two reconnaissance UAVs S_1 and S_2, which are all sensor nodes (S) and their attributes are shown in figure 3. x_1 is set to mark whether there is a communication connect between the two UAVs, the information capacity of the communication link is set to be x_2 , the information quality of the communication link is x_3 , and x_4 means the information transmission delay, then

$$x_1 = \begin{cases} 1, d_{S_1 S_2} \le \min(s_8^1, s_8^2) \\ 0, d_{S_1 S_2} \le \min(s_8^1, s_8^2) \end{cases} \tag{4}$$

$$x_2 = \min(s_6^1, s_6^2), x_3 = s_7^1 \times s_7^2, x_4 = s_5^1 + s_5^2 \tag{5}$$

$$h_{e,S_1S_2} = -\sum_{i=1}^{4} w_i \lg(1 - \frac{2}{e^{x_i} + e^{-x_i}}) \quad (6)$$

where, $d_{S_1S_2}$ means the distance of UAV S_1 and S_2, s_j^1 and s_j^2 represent the attributes of node S_1 and S_2, as figure 3 shows. w_i is the weight of x_i . h_{e,S_1S_2} is the operational information entropy of the edge $S \leftrightarrow S$, meaning the information uncertainty and reflecting the combat effectiveness.

3 The operational effectiveness evaluation for the swarming UAVs air combat system

Based on the above network model, we put forward the concept of Operation Loop Capability and evaluate the target-attack effectiveness of the swarming UAVs air combat system.

Definition 3: Operation Loop Capability is an index to quantify the impact of a certain Operation Loop (both standard and generalized loop) to a particular target in the combat system network.

As is mentioned in 2.4.2, the edge operational information entropy is a measure of the combat information uncertainty. A low operational information entropy means smaller information uncertainty of the target and will lead to stronger combat capability. Therefore, there is an inverse relationship between the information entropy and the combat capability. Hence, we define

$$C_k = e^{-h_{e,k}} (k = 1,2,...,m) \quad (7)$$

where m is the number of edges of one Operation Loop, C_k represents the capability of the k_{th} edge in the Operation Loop, $h_{e,k}$ is the operational information entropy of the k_{th} edge in the Operation Loop. Then the Operation Loop Capability (C_{OL})can be calculated by:

$$C_{OL} = \prod_{k}^{m} C_k \quad (8)$$

Supposing that an attack is launched to target p , the number of participated operational loops is r , then the striking effectiveness to p is equal to the sum of the combat capabilities of all the corresponding operational loops. Namely,

$$C_{T_p} = \sum_{j=1}^{r} C_{OLj} \quad (9)$$

Definition 4: The target-attack effectiveness of the swarming UAVs air combat system (C) is the sum of the striking capabilities to all targets.

$$C = \sum_{p=1}^{l} w_p C_{T_p} \quad (10)$$

where l is the number of the targets in the system, $w_p \in [0,1]$ is the weights of the targets based on target threat degree ranking. w_p can be calculated *TOPSIS* method [8] based on target attributes data.

4 Case study

In this part, we take the attack to enemy ground command centers and infrastructures as an example.

4.1 Network model

To build the system network model, we have to figure out all the combat weapons and clarify their types. The weapons in the swarming UAVs air combat scene are listed in table 2.

Table 2. The corresponding weapons in the swarming UAVs air combat system.

Entities	Nodes type	Weapons and quantities
Sensors	S	Ground radar station (1); Electronic reconnaissance aircraft (Y-8, 1); Reconnaissance UAVs (7)
Deciders	D	Early warning aircraft (KJ-2000,1)
Attackers	A	Attack UAVs (8); Fighter plane (J-10, 3); Bombers (H-6, 2); Attack plane (Q-5, 1); Air-to-ground guided missile (1); Cruise missile (1); Aerial artillery (1); Airborne antimissile system (1)
Targets	T	Fighter plane (S-30,1); Fighter plane (F-15,1); Interceptor plane (F-106,1), Enemy command post (1); Enemy airport (1); Detection radar (1); Guidance radar (1); Missile launch location (1); Antiaircraft gun launch location (1); Air-to-ground guided missile (1); Antiaircraft gun (1); Interceptor missile (Patriot 3, 1)

The second step of the network model is to determine edges among the above weapons. The attack weapons undertake the combat mission to the enemy and the strike relationships are as follows:

Table 3. Attack relationships.

Weapons	Enemy targets	Weapons	Enemy targets
J-10	F-15	Cruise missile	S-30; F-106; F-15; Command post; Airport
J-10	S-30	Air-to-ground guided missile	Airport; Command post; Missile launch location; Antiaircraft gun launch location
J-10	F-106	Reconnaissan ce UAVs	S-30; Command post; Airport; Missile launch location; Antiaircraft gun launch location
H-6	Command post; Airport; Antiaircraft gun launch location	Attack UAVs	S-30; F-15; F-106; Command post; Airport; Antiaircraft gun; Air-to-ground guided missile; Antiaircraft gun launch location; Missile launch location; Patriot 3
H-6	Command post; Airport; Missile launch location	Aerial artillery	S-30; F-106; F-15; Command post
Q-5	Missile launch location; Antiaircraft gun launch location; Detection radar; Guidance radar	Airborne antimissile system	Patriot 3; Antiaircraft gun; Air-to-ground guided missile

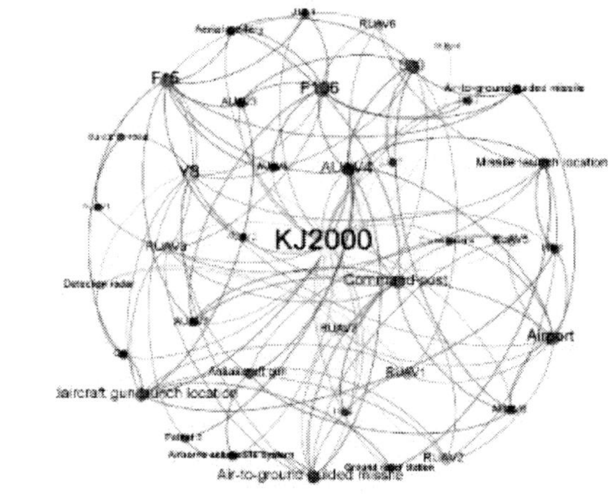

Fig. 7. Due to the large number of nodes, we merely listed the attack relationships in table 3, and other relationships are shown as the edges in the left operational network. The green, yellow, blue and red nodes represent sensors, deciders, attackers and targets respectively. All the edges in the network are divided into nine edge types, as table 1 shows.

4.2 Operational effectiveness evaluation

The effectiveness of the swarming UAVs air combat system (C) is defined as the sum of the striking capabilities to all targets, so we have to firstly calculate the operational effectiveness to a certain target. Just take the target of "Antiaircraft gun launch location" as an example, table 4 shows its attribute data. Notably, all the weapon attributes' (as figure 3-6) data is from relevant literatures or Internet, and due to the space limitation, we didn't list all the data.

Table 4. Data of "Antiaircraft gun launch location".

Attributes	t_1	t_2	t_3	t_4	t_5	t_6	t_7
Values	3	4	3	0	6	0.5	4

Aimed at the target of "Antiaircraft gun launch location", one of the operational loops in the network is "Antiaircraft gun launch position → Y-8 → KJ-2000 → H-6 → Antiaircraft gun launch position". In this loop, the edge types including $S \to D, D \to A, A \to T$ and $T \to S$. According to the attribute data of these weapons, the operational information entropies of the four edges are calculated.

$$h_{e,TS} = x_0 \times [-\sum_{i=1}^{7} \frac{1}{7} \lg(1 - \frac{2}{e^{x_i} + e^{-x_i}})] = 0.18033 \tag{11}$$

$$h_{e,SD} = -\sum_{i=1}^{4} 0.25 \lg(1 - \frac{2}{e^{x_i} + e^{-x_i}}) = 0.88768 \tag{12}$$

$$h_{e,DA} = -\sum_{i=1}^{4} 0.25 \lg(1 - \frac{2}{e^{x_i} + e^{-x_i}}) = 0.79973 \tag{13}$$

$$h_{e,AT} = x_0 \times [-\sum_{i=1}^{5} 0.2 \lg(1 - \frac{2}{e^{x_i} + e^{-x_i}})] = 0.15888 \tag{14}$$

Then the capability of this Operation Loop is

$$C_{OL} = \prod_{k}^{m} C_k = \prod_{k}^{m} e^{-h_{e,k}} = e^{-(h_{e,TS}+h_{e,SD}+h_{e,DA}+h_{e,AT})} = 0.1318 \quad (15)$$

After a series of calculations of all the Operation Loops including the target "Antiaircraft gun launch location", the operational effectiveness to this target C_{T_p} is summed to be 211.3263.

Table 5. The operational effectiveness to all targets.

Targets	Antiaircraft gun launch location	Missile launch location	S-30	F-106	F-15	Comman d post
w_p	0.0441	0.0441	0.0871	0.0954	0.0911	0.1231
C_{T_p}	211.3263	146.6069	175.8427	284.4926	367.9419	234.7211
Targets	Airport	Detection radar	Guidance radar	Air-to-ground guided missile	Antiaircraf t gun	Patriot 3
w_p	0.0986	0.1380	0.1454	0.0472	0.0392	0.0472
C_{T_p}	403.524	27.63429	8.97049	104.0399	135.9472	61.2898

Table 5 illustrates the operational effectiveness to all targets in the swarming UAVs air combat system and the system combat effectiveness C can be certified as

$$C = \sum_{p=1}^{l} w_p C_{T_p} = 178.6749 \quad (16)$$

Although the proposed system combat effectiveness index is a figure and sometimes cannot be certified, it provides a quantitative method of assessing the swarming UAVs combat scene. Besides, it may live a foundation for weapon contribution or importance evaluation by changing node quantities.

5 Conclusion

In this paper, we propose an operational effectiveness evaluation method of the swarming UAVs air combat system. Before the effectiveness evaluation, the system network model is firstly established. Considering the unique weapon relationships in UAVs system, the concept of Generalized Operation Loop (GOL) is put forward and the edge operational information entropy is proposed to reflect the information uncertainty and combat effectiveness based on the network nodes attributes. As for the system operational effectiveness evaluation, we mainly focus on the striking capability and give the concept of Operation Loop Capability to evaluate the system target-attack effectiveness, which can be calculated from the edge operational information entropies. The combat effectiveness of the swarming UAVs air combat system is represented by the sum of the striking capabilities to all targets. Finally, the scene of swarming UAVs air combat is taken as a case study to apply and verify the proposed operational effectiveness evaluation method. It provides a good quantitative perspective for swarming UAVs system performance assessment and can help to evaluate weapon contribution or importance.

In the future works, we will make a further analyzation of the equipment importance in the system and their influence on the network structure invulnerability.

Acknowledgments

The work was supported by the National Natural Science Foundation of China under Grant 71571185; and the Research Project of National University of Defense Technology under Grant 15GJ003-278.

References

1. Bitzinger R. US-China Competition, the Third Offset Strategy, and Implications for the Global Arms Industry[J]. Institute on Global Conflict & Cooperation Working Paper, 2017.
2. Shen Chao, Wu Kunlin and Song Yiran. Key developments in the development of drone bee colony operations[J]. Aerospace Missile, 2016(11): 28-33.
3. Shen Taoran and Sang Yiyong. Overview of the development of foreign UAV equipment and typical combat modes [J]. New Industrialization, 2018(5).
4. Cares J R, Incorporated A. An information age combat model [J]. Paper for Iccrts, 2004.
5. Cares J R. LCS Study - design principles of distributed, networked forces[J]. 2005.
6. Tan Yuejin. Principles of systems engineering (Revised edition) [M]. Beijing: Science Press, 2010. 6-22.
7. Shannon C E. A mathematical theory of communication[J]. Bell Labs Technical Journal, 1948, 27(4):379-23.
8. Deng H, Yeh C H, Willis R J. Inter-company comparison using modified TOPSIS with objective weights[J]. Computers & Operations Research, 2000, 27(10):963-973.

Analysis on the endogenous mechanism of big data and tower management framework in intelligent manufacturing enterprises

Mingwei Sun[1], *Feng* Wang[2,*], *Haihua* Li[3], *Ying* Liu[4], and *Shuli* Wang [5]

[1]Public Teach. and Education Dep., Changchun Automobile Industry Institute, China
[2]Big Data Research Centre, School of Management, Jilin University, China
[3]School of M-Science and Info. Engineering, Jilin University of Finance and Economics, China
[4]Aviation University of Airforce, Flight Training Base, Changchun 130000, China
[5]Computer Teaching and Research Center Jilin University, China

Abstract. This paper makes an in-depth comparative analysis of years of experience in intelligent manufacturing projects and literature research related to Big Data. The 4.0 value chain model and concept are put forward to carry out the logical analysis of the endogenous relationship drive and endogenous management mechanism of intelligent manufacturing. The intelligent manufacturing business management process under the 4.0 value chain is established, and the tower Big Data management framework of intelligent manufacturing enterprises is innovatively proposed. This paper discusses the connotation, elements and drive relationship of enterprise Big Data from three dimensions of business operation, information drive and management policies. The hierarchical structure and related connotation of Big Data are revealed, and the basic characteristics of intelligent manufacturing enterprises Big Data are analyzed. The purpose of this paper is to clarify the difference of the concept between enterprise Big Data and mass data and open the Big Data fundamental research driven by digitization management. It provides basic innovative ideas and scientific research methods for the new generation of digital virtual simulation, digital factory construction and industrial chain management.

1 Introduction

Under the requirement of development of the new era, Big Data has attracted the attention of enterprises, countries, academia, scientific and technological circles on its research and application. Industrial Big Data is the core technology of industrial internet and industry 4.0. The manufacturing industry needs to open up industrialization with new Big Data thinking. It is urgently needed to integrate Big Data with business management and production operation activities. To promote the intelligence transformation and upgrading of manufacturing enterprises, especially to open up the basic cognition of Big Data and re-examine it with scientific thinking and management methods, it is necessary and important

* Corresponding author: Wf64@163.com

to build the connotation, elements, management mechanism and framework of scientific Big Data and take charge of its innovative development.

2 Current situation of the research and application of Big Data

The futurist and social thinker Alvin Toffler (1980) first proposed the concept of "Big Data" in *The Third Wave. Demand Survey of the Application and Control of Data Visualization* first proposed the terminology "Big Data" [1]. *Big Data Era* firstly proposed 4 v features (Volume, Velocity, Variety and Value) widely as the definition and concept of Big Data [2]. In 2008, *Nature* special issue *Big Data* introduces the challenges brought by mass data from the aspects of internet technology, network economics, supercomputing, environmental science and biomedicine, etc. In 2010, Apache Hadoop defines Big Data as: "large data sets ordinary computer software can't seize, manage or process within an acceptable time". *Big Data: The Next Frontier of Innovation, Competition and Productivity* proposes that "Big Data requires new processing modes to have stronger insight, decision-making and process optimization abilities". The concept and potential of Big Data are profoundly discussed by EMC/IDC in *Extracting Value from Chaos*. In 2011, IDC also defined Big Data: "Big Data technology describes the new-generation technology and framework system, and extracts the economic value of various mass data through high-speed collection, discovery or analysis". In April 2012, the research journal of European Association of Informatics and Mathematics, ERCIM News special issue, *Big Data*, discussed the intensive form of research on the innovation technology of Big Data and other issues. "OODA cycle model is considered the main model of Big Data supply chain", "which further extended to a Big Data driven proactive manufacturing architecture".

Technology and Framework of Industrial Big Data proposed "industrial Big Data is the general name of all kinds of data and related technologies and applications generated at each stage of the whole product life cycle surrounding the typical intelligent manufacturing model". Industrial Big Data can be subdivided into "product data, production data, value chain data and external data". According to *Technical Review of Industrial Big Data*, industrial Big Data mainly includes "enterprise information data, industrial IoT data and cross-border data". It is proposed that small data and mass data have strong correlative and structural relationship with industrial Big Data. "Big Data + industry" establishes a data-driving value chain. The new space for the development of manufacturing industry is generated through the interaction of external data and internal data.

GE construction APM cloud platform helps enterprises to create and develop industrial internet applications based on Predix standards. At Hanover industrial show in April 2017, Siemens presented the core content of MindSphere. French company Schneider integrated Schneider electric industrial interconnecting products with the application of edge control and launched the EcoStruxure platform. In April 2015, three major associations of German BITKOM, VDMA and ZVEI jointly released and proposed three core dimensional industry 4.0 basic framework and RAMI 4.0 system. In addition, domestic and foreign ERP enterprises Oracle, SAP, Sagesoft, Microsoft, Inspur, UFIDA and Kingdee all join in the development and extension of Big Data products, and launch enterprise cloud level industrial Big Data framework and technology management platform.

2.1 Research status of the application and management of Big Data

So far, problems with Big Data and even industrial Big Data are as follows:

2.1.1 Firstly, accurate definition, basic research on Big Data and clarification of concept boundaries are lacked

Up to now, no research institution or scholar has given a precise definition of Big Data. The concept of Big Data is usually confused with that of mass data. The mass data research is regarded as Big Data research and the definition of industrial Big Data and industrial enterprise Big Data is unclear. Quantized model, hierarchical elements [3], value standards, and digital-driving framework are lacked and the associated structure of Big Data and the essence of digital virtual simulation management are ignored. Keen on technology and lack of basic management research, this kind of research will inevitably lead to Big Data deviating from the basic needs and goals of enterprises and society, causing the failure of 85% Big Data projects with capital scale of millions and even tens of millions.

2.1.2 Secondly, Big Data belongs to the product of technology and the category of management

Manufacturing enterprises Big Data is the integrated product of the frontier interdiscipline. Developers of Big Data research mostly start from the category of Big Data technology and excavate the commercial value of the mass data. They try to study Big Data as a "Data nature" inclined to the natural science concept of "data boundary" [4]. It was ignored that the mainstream hot spots of Big Data should return to the high-level management application fields of the society, enterprises. The management and control driven by the innovation of Big Data should be enhanced with the return of more industrial and commercial value. For this reason, Nick Heudecker, CEO of Gantt, tweeted: "The key of Big Data isn't about technology."

2.1.3 Thirdly, not realizing to get industrial, financial and market Big Data causes enterprises to lose business opportunities

Most enterprises lack internet business awareness, mobile finance and business development thinking, and lose mass valuable data information and Big Data (mobile payment, customer finance, market sales, product performance) with the most commercial value in the market. Or it will be pushed into the hands of "third-party payment" e-commerce enterprises, thus losing or losing the main business role of e-commerce channels in the market of enterprises (important Big Data creators).

3 Endogenous management mechanism and framework of Big Data in intelligent manufacturing enterprises

3.1 Intelligent manufacturing enterprise is the creator of Big Data

Industrial manufacturing is the biggest contributor and motivator of e-commerce mass data. IT integrates industrialization and informatization together. Many industrial enterprises have crossed the boundaries to link all fields, so the integration of internal and external industries of manufacturing and production has become the direct factor affecting the Big Data ecology. Since the industrial revolution from Watt, the production efficiency and speed of machines far exceed that of manual workshops. Big Data have entered into a rapid and vigorous development condition, especially after the third industrial revolution. Since IT appeared, a vast amount of information was produced by the interconnection of all things. Data management systems such as CMIS, AIS, MRP, MES, PDM, CRM and ERP store high-density integrated data of industries, supply chains and industrial chains. Think Big, a Big Data consulting company, ranks IoT and machine data as the top priority. In particular, industrial robots in the 4.0 era are continuously creating industrial Big Data in a high-speed, precise and large-scale way. The whole world has entered to a mode with the explosive growth of data volume.

3.2 Endogenous management and interactive mechanism of intelligent enterprise Big Data

3.2.1 4.0 Value chain of intelligent manufacturing enterprises

All enterprises should have 4 basic endogenous functions: value-added operation; profits from sales; financial analysis and control optimization. The four endogenous functional factors of Big Data related to intelligent manufacturing enterprises are in the charge of the enterprise value ecosystem: value driving chain. Michael Porter put forward the value chain theory in 1985 delineating the cost pattern, business process and value-added model created by enterprise [5]. It clarifies that the business process is created by product value and describes the drive relationship between common business elements in an enterprise. It has the following defects: firstly, in Porter Value Chain it is difficult to accurately explain and evaluate the corresponding digital value of intelligent manufacturing, intelligent logistics, intelligent finance and intelligent sales. Digital factories need more accurate digital value chain to simulate the value form of intelligent manufacturing enterprises; secondly, the endogenous drive mechanism of enterprise digital resources and automated activities of "purchase, production and sale" cannot be constructed and analyzed [6]; therefore, in today's advanced manufacturing industry 4.0, digital enterprises need to reconstruct the value chain theory [7].

Use industry 4.0 and new generation IT technology, according to the ecological reality created by value, management drive and restraint mechanism to analyze specific input elements related to the value created by enterprise products (intelligent CNC processing center, combination process, resource consumption, finance expenses and other factors) and to reveal the production cycle of enterprise value creation. The digital management simulation value chain about the overall process from the purchase of raw materials to the production of ordered goods by intelligent automated processing center is called the intelligent manufacturing 4.0 value chain. （figure 1）

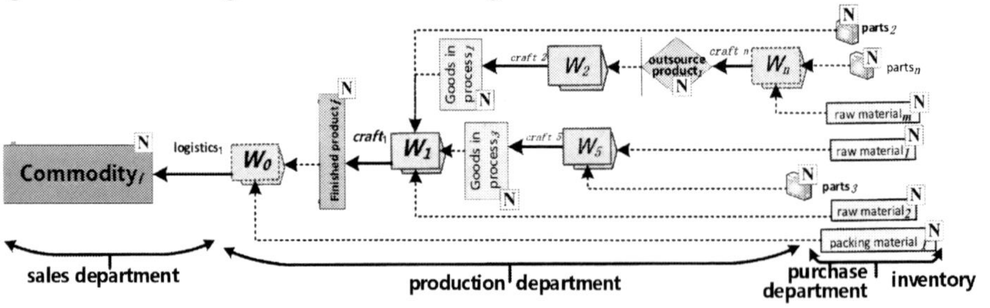

Fig. 1. The 4.0 value chain of intelligent manufacturing enterprises.

Intelligent 4.0 value chain illustrations: [N] represents the material number ; W_n represents the N th MES process automated manufacturing center ; ← represents the required process and processing lead time; the dotted arrow represents the material needed to supply the manufacturing equipment.

3.2.2 The internal drive management mechanism and process simulation of enterprises under the 4.0 value chain

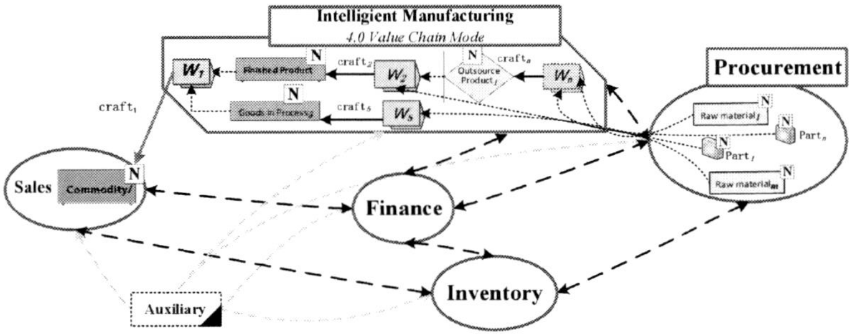

Fig. 2. Intelligent manufacturing enterprise business-driven mechanism based on 4.0 value chain.

Under the mode of industry 4.0, restrained by market size, production planning, contract orders, cost, profit, and social services, etc., the intelligent finance under E-payment can be regarded as the stimulation point of intelligent sales of E-commerce, which can trigger value creation and operation activities of "production, storage and purchase" of enterprises including process scheduling, production of goods, warehousing of supplies and procurement, synergy and outsourcing, and other main business activities. This activity also needs a certain degree of support: product research and development, investment and financing, fixed assets procurement and maintenance [8], staff recruitment, quality service, quality inspection and after-sales activities. Certain production activities will inevitably lead to the consumption of the internal and external resources of the enterprise, and thus cause financial capital management, control and ABC management of activity cost. And after the enterprise continuing to go through one or several cycles of cycle operation and accumulation and generating several business information data sets, enterprises can collect, manage and optimize data through business intelligence. Therefore, the 4.0 value chain causes the enterprise: from intelligent finance → intelligent sales → intelligent manufacturing → intelligent procurement → intelligent inventory logistics to the formation of the drive process of the internal business management relationship within the enterprise. (figure 2)

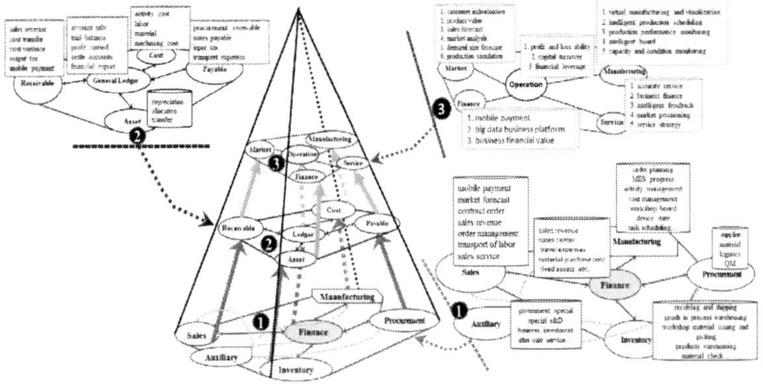

Fig. 3. Enterprise tower Big Data management framework and components.

Demand - based production pulling the interaction between logistics and capital leads to profit and loss analysis on product costs and related financial and management accounting, etc. (enterprise accounting information). Enterprises use business to drive information data storage and accumulation. This provides a deep level optimization of business decision-making, control and execution through business intelligence [9]. Therefore, the digital drive mode from below has formed the manufacturing enterprise tower Big Data management framework (figure 3). This helps enterprises to realize the intellectualization of whole-process operation of Big Data system [10], and lay a foundation for enterprise upgrading, transformation, global layout, and capability improvement.

3.2.3 Reference conditions and rules of integrated virtual operation and management under Big Data mode

The reference conditions of digital management in enterprises virtually driven by Big Data: The enterprise has the open interface for bank E- payment, which is used for the verification, payment and authorization of the order transaction of intelligent finance; individual users need Big Data authentication system (electronic identity and security authorization authentication interface) of government human resources; the client enterprise needs the electronic authentication (safety credit) of industrial and commercial bureau and the electronic transaction e-authorization interface of bank. There are two ways of Big Data driving digital management in enterprise: order sales and planned production.

4 Analysis of the definition, classification and connotation of enterprise big data

The intelligent manufacturing enterprise Big Data consists of 3 layers and 5 levels (figure 4).

4.1 Level 1：intelligent manufacturing enterprise Big Data

Intelligent manufacturing enterprise Big Data is defined: using artificial intelligence and CPS technology, the integrated operation process of "purchase, sales and storage" driven by intelligent manufacturing in the ecological process created by enterprise value actuates the disclosure of financial accounting information and the intelligentized management and control of profit and loss analysis. The three - dimensional integration of digital data collection composed of business operation, information management and decision-making control are formed, which is called intelligent manufacturing enterprise Big Data.

Characteristics of intelligent manufacturing enterprises Big Data: authority, reliability, security credibility, value, strong correlation, interoperability, integrity, virtuality, management complexity, strong correlation, dynamic, and openness (enterprise, industry).

4.2 The second level of Big Data is divided into three layers: business, information, and BI

There are significant logical associations and drive relationship between them. (see figure 3)

4.2.1 Business Big Data (secondary data)

Based on the 4.0 value chain theory, economic activities will lead to changes in enterprise resources or values within the enterprise, from which business relationship data are generated. Business Big Data can be divided into main business / non-main business / auxiliary Big Data.

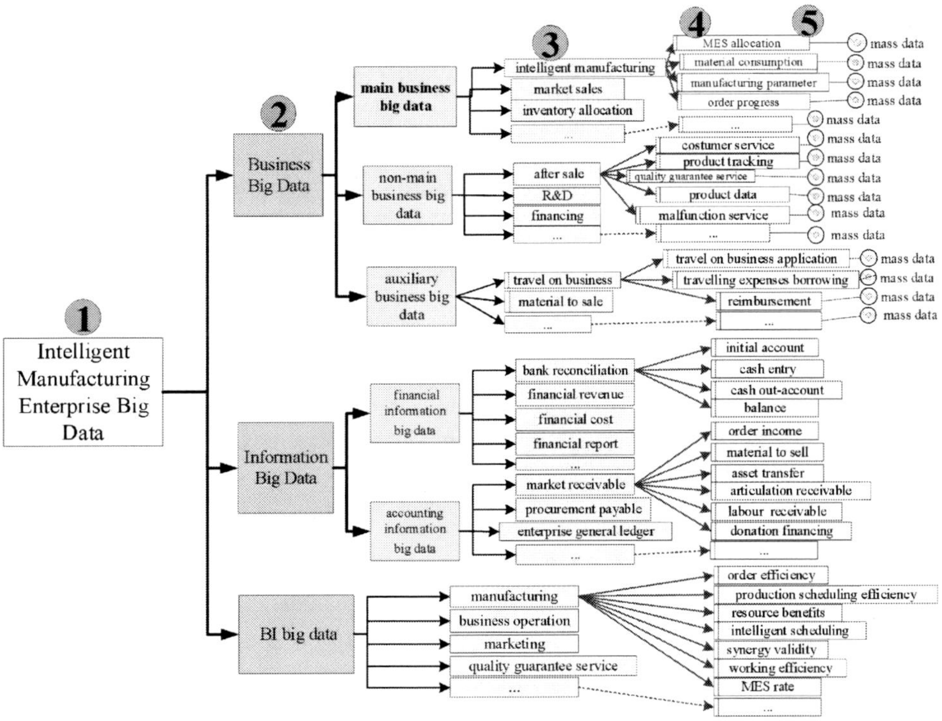

Fig. 4. Schematic diagram of intelligent manufacturing enterprise Big Data.

4.2.2 Information Big Data (secondary data)

The change of enterprise value and capital expenses caused by business operations leads to the information collection about enterprise financial payment, financial revenue and expenditure, cost accounting, profit and loss, accounting information disclosure and other information, which is called information Big Data.

4.2.3 BI Big Data (secondary data)

Enterprise Big Data BI decision layer is a large, intelligent, continuously improved, management, analysis, control and decision-making system which aims to improve and obtain the best operating efficiency of the enterprise in at least two ways: one is the industry operation data experience from many years' business management and the other is the ability of managing and controlling business information from the external competitive environment or whole business resource allocation in industrial chain network. After automated analysis, the operation parameters can be improved and optimized.

4.3 Level 3: functional Big Data

With the attribute of department management, it aims at division of labor, efficient management for enterprises and occupational division of enterprises' organizations, usually including intelligent manufacturing, financial management, inventory, intelligent logistics and market sales, etc. Although different enterprises have different functions, they are inseparable from the three basic functions of "purchase, sales and storage".

4.4 Level 4: enterprise metadata

Enterprise metadata is the smallest data unit in enterprise Big Data structure; it is also the smallest data unit in enterprise Big Data management level, which is used to constitute the smallest function and transmission unit of intelligent Big Data. Multiple metadata are used to structure and interpret the management characteristics of functional Big Data. Different from ordinary data or numbers, enterprise metadata usually has the following 6 characteristics: business subjectivity; value correlation; management drive; objectivity; safety credibility and emulation virtuality.

4.5 Level 5: mass data

Mass data is a kind of special auxiliary metadata. Information related to equipment, products, components, users and other related auxiliary characteristic data which occur on a node in a detailed field of an enterprise many times continually is called mass data. Mass process data which include details of angle, length, chip error, voltage, environmental humidity, temperature, etc. of CNC machine tools are used to reflect the working condition and equipment reliability. Characteristics of mass data: non-business subjectivity (auxiliary); legal relativity; privacy; commerce; indirection; altruistic business values.

5 Conclusion

Intelligent manufacturing enterprise Big Data is a very important concept in Big Data family, and also the foundation of simulation management such as cloud system, digital intelligent enterprise, etc. This paper starts the basic research on the definition, connotation, mechanism, research framework of intelligent manufacturing enterprise Big Data. The research on 4.0 value chain, manufacturing enterprise tower management framework, endogenous business-driven mechanism and process logic, etc. will open a new chapter of industrial digitalization and the transformation and upgrading of traditional manufacturing.

Acknowledgment

Fund support project (2017ZZ042): Research on Reconstruction of Management Platform for Intelligent Manufacturing ERP Platform under Big Data Mode, supported by "The Fundamental Research Funds for the Central Universities".

References

1. Guidelines for the Construction of National Intelligent Manufacturing Standard System, Ministry of Industry and Information Technology, National Standardization Management Committee, 2015.
2. Liu qiang, Qin Sizhao. Research Prospect of Large Data Modeling in Process Industry [J]. Journal of Automation, 2016(2):161-171.
3. Li Guojie, Scientific Value of Big Data Research. Chinese Computer Society Communication, 2012,8(9): 8. 15.
4. Zhou Jiajun, Yao Xifan. Advanced Manufacturing Technology and New Industrial Revolution [J]. Computer Integrated Manufacturing Systems, 2015, 21(8):1963-1978.
5. Zhou Ji. Intelligent Manufacturing-main direction of "Made in China 2025" [J].China Mechanical Engineering, 2015, 26 (17):2273-2284 (in Chinese).

6. Yao Xifan, Zhang Jianming, LIN Y Z.The Basic Theory and Technical Framework for Wisdom manufacturing systems [J].Systems Engineering-Theory&Practice,2016,36(10):2699-2711 (China)
7. Yao Xifan, Zhou Jiajun, ZHANG Cunji, et al. Proactive Manufacturing-a Big Data Driven Emerging Manufacturing Paradigm [J]. Computer Integrated Manufacture Systems, 2017, 23 (1) :172-185 (in Chinese) .
8. LEE J. Industrial Big Data: the Revolutionary Transformation and Value Creation in Industry 4.0era[M].Beijing: China Machine Press, China, 2015 (in Chinese)
9. Yao Xifan, Jin Hong, Xu Chuan, et al. Virtualization and Servitization of Cloud Manufacturing Resources[J].Journal of South China University of Technology: Natural Science Edition, 2013, 41 (3) :1-7 (in Chinese)
10. Myriam Jahn. Industry 4.0: Transformation of Business Mode of Intelligent Manufacturing and Marketing in the Future [M],2017:108-127

Early and late rice identification from Tiangong-2 wide band images based on CNN

Zhuang Zhou[1,2*], *Shengyang* Li[1,2], and *Yuyang* Shao[1,2]

[1]Technology and Engineering Center for Space Utilization, Chinese Academy of Sciences, China
[2]Key Laboratory of Space Utilization, Chinese Academy of Sciences, China

Abstract. The wide band images acquired from the Tiangong-2 space laboratory covers many spectral bands such as visible light, shortwave infrared and thermal infrared. These high-quality images can be used for space science experiments such as earth observation. In this paper, we use CNN (convolutional neural networks) to extract the spectral features of different landcover from the wide band images, then identify the early rice and the late rice accurately in Huarong County, Hunan Province, China. With advanced techniques such as deep learning, the spatial distribution information of crops can be effectively obtained from the wide band images which can provide data services for agricultural production management.

1 Introduction

Tiangong-2 space laboratory was successfully launched on 15th September 2016. After the successful mission of Tiangong-1 space laboratory, Tiangong-2 is the second space laboratory emission launched by China. Tiangong-2 has carried out a number of new space applied load equipment and more than ten applications and experiments, which relate to earth observation and space science. The wide band imaging spectrometer is one of the playload in Tiangong-2 which has wide FOV (Field of View) and combination of image and spectrum.

It is the first sensor which integrates visible and near infrared, short wavelength infrared and thermal infrared spectrums in one instrument in the world. It is a push-broom sensor that is channel programmable in the visible near infrared spectrums and multi-spectral detection in short wavelength and thermal infrared spectrum. There are 14 band channels with a spatial resolution of 100 meters which is mainly aimed at medium resolution and large-scale object monitoring. It is suitable for the study of landcover classification, as well as for ocean and coastal zone water color and water temperature observation.

Numerous traditional landcover classification methods have been applied in the field of earth observation, such as k-nearest-neighbors, minimum distance, support vector machines (SVM), random forest and logistic regression [1-3]. Recently, some more effective feature extraction methods as well as advanced classifiers were proposed, such as deep learning. In the past few years' deep learning are the most popular and efficient approaches for

* Corresponding author: zhouzhuang@csu.ac.cn

landcover classification of remote sensing [4]. More and more application results show that the methods based on deep learning are superior to SVM. As a powerful machine learning method, deep learning can be applied in solving various tasks in the fields of image classification, speech recognition and natural language processing. The main idea of deep learning is to simulate the human visual system and extract effective features from the information in a hierarchical manner. In the field of image classification, a large number of deep learning models, frameworks and standard data sets are available for reference [5].

Convolutional neural networks (CNN) is one of the most popular models of deep learning, which is widely used in dealing with image vision problems. The weight-sharing structure of CNN makes it similar to the biological neural network which can reduce the complexity of the model and has prominent advantages in two-dimensional image processing. The idea of CNN was first proposed in [6], further developed in [7], and simplified and refined in [8]. CNN has been demonstrated to provide excellent classification performance than traditional classifiers in earth observation area. However, the research of remote sensing classification based on deep learning mainly for high-resolution or hyperspectral remote sensing images which are difficult to obtain. To explore the feasibility of crops classification and recognition from wide band images based on deep learning is of great significance.

2 The study area and data

2.1 The study area

The purpose of this paper is to achieve the landcover and crops classification from Tiangong-2 wide band images for Huarong county, which located in the northeast of Hunan province, China, with an area of around 1642 square km (112.31 ~113.03°E; 29.17~29.81°N). The climate of the study area is subtropical humid with an annual rainfall of 1304mm and average temperature of 17.2 °C. The superior geographical location and natural conditions make Huarong County an important rice producing area.

The crops planting structure of this region is stability which are mainly composed of early rice and late rice from May to October every year. Early rice is transplanted in mid-April and harvested in July, late rice is transplanted in July and harvested in October. The main landcover of the study area is classified into five types such as early rice, late rice, forest, buildings and water. The false color composite image of the study area from Tiangong-2 is shown in figure1.

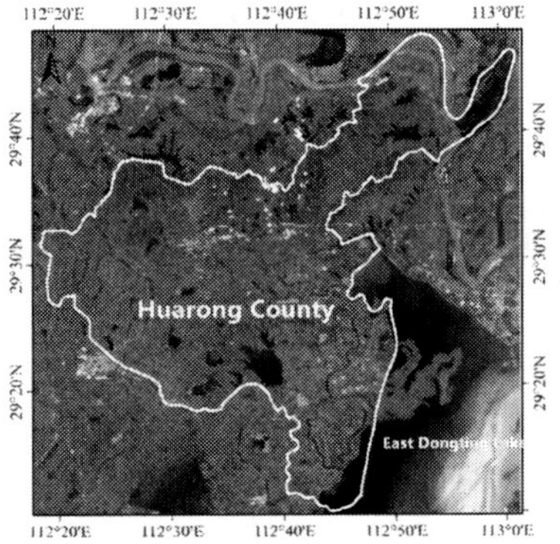

Fig. 1. The false color composite image of study area from Tiangong-2.

2.2 Data

The wide band images from Tiangong-2 we selected were acquired in July 25, 2017, which contains 14 bands of channels and has a spatial resolution of 100 meters. Abundant spectral information provides effective data source for rice feature extraction and identification. In order to obtain accurate sample label data, we selected the Sentinel-2A high-resolution remote sensing images (transiting time of July 17, 2017, spatial resolution of 10m) in similar period for visual interpretation and collected the main five types of ground object samples. Through the spatial position matching, the corresponding spectral vectors of the samples in the wide band images of Tiangong-2 are extracted, resampled and transformed into label images. A total 179 different types of polygons were collected by artificial visual interpretation which divided into training and testing data sets to train and validate the classifier, respectively. All of the labeled pixel data we collected were randomly divided into training set and verification set according to the proportion of 1:1.

3 The research method

3.1 CNN

CNN is a feed-forward neural network which composed of input layer, convolutional layer, pooling layer, fully connected layer and output layer. With local connections and weights sharing, the number of parameters can be drastically reduced. In addition, the risk of over-fitting can also be decreased. A normal CNN network usually consists of several pairs of convolutional layer, max pooling layer and ends with a full connection layer. The architecture of CNN classifier constructed in this paper is shown in figure 2.

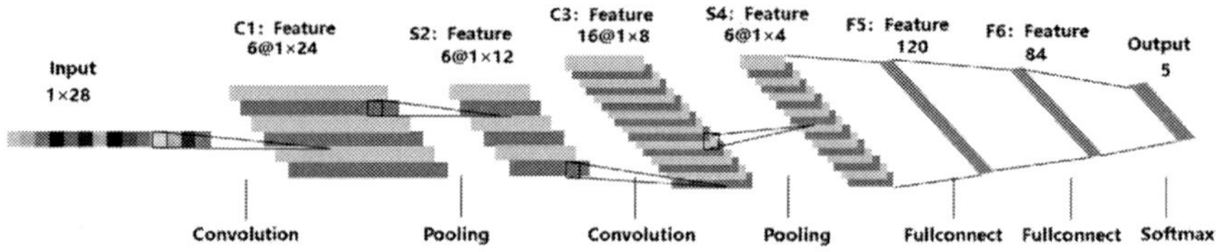

Fig. 2. The architecture of the proposed CNN classifier.

In a regular neural network, every neuron was connected to all neurons in the next layer. In contrast, the neurons on the convolutional layer of CNN are sparsely connected to the neurons in the max pooling layer based on their relative location. Neurons that belong to the same layer share the same weights which reduce the number of model parameters and make the model training process more efficient. The max pooling layer is generally attached behind the convolutional layer.

Max pooling partitions the input data into a set of nonoverlapping windows which can select the maximum value for each sub region and reduces the calculation complexity of model calculations and enhance the invariance of feature translation. In order to classify, the computation chain of the CNN ends with a fully connected network, which integrates all the feature maps of different locations.

3.2 CNN-based classification

The hierarchical structure of CNN has proved to be the most effective way to learn image representation in visual field. Spectral vectors of 14 bands from the wide band images are resampled and converted to grayscale images with dimension size 28 in figure 3.

As illustrated in figure 2, the CNN network constructed in this paper consists of two convolutional layers, each of which is connected to a max pooling layer. The last two layers of the model are fully connected layers. The rectified linear unit (ReLU) function was chosen as the activation function which is also the most popular and effective activation function in deep learning field. Compared to the sigmoid function, the advantages of using ReLU enables more efficient computations and faster gradient diffusion for training CNN.

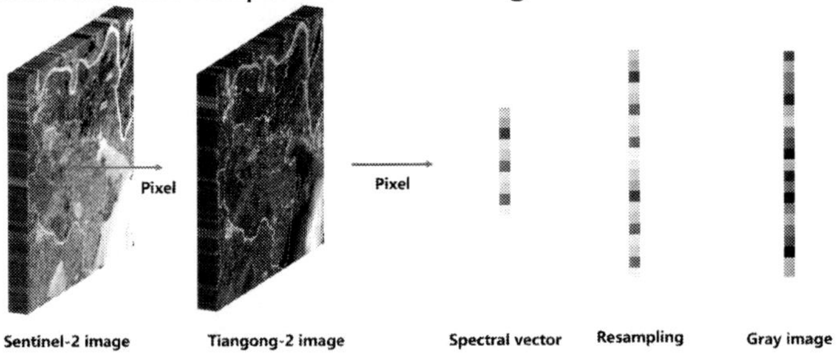

Fig. 3. The gray image processing.

The difference of CNN network structure mainly lies in the number of network layers and the training mode of network. As shown in Figure 2, the network consists of eight different layers, including the input layer, two convolution layers, two maximum pool layers, two full connection layers and an output layer.

In wide band images, each pixel sample can be viewed as a two-dimensional gray image with a height of 1. Therefore, the size of the input layer is just ⌈1×28⌋, and 28 is the resampled to 2 times number of bands. The ⌈1×28⌋ input data was filtered by the first hidden convolutional layer with 6 kernels of size ⌈1×5⌋.

The first hidden convolutional layer contains ⌈6×1×24⌋ nodes, and ⌈24=28-5+1⌋. The number of trainable parameters between the first two layers are ⌈6×(5+1)⌋. The first max pooling layer is the second hidden layer, and the kernel size is ⌈1×2⌋. This layer contains ⌈6×1×12⌋ nodes, and ⌈12=24/2⌋.

The second convolutional layer contains ⌈16×1×8⌋ nodes, and ⌈8=12-5+1⌋. The kernel size of the second max pooling layer is ⌈1×2⌋. This layer contains ⌈6×1×4⌋ nodes, and ⌈4=8/2⌋. All the fully connected layers connected with the softmax classifier where output number is 5.

According to the corresponding CNN and the afore mentioned parameters, classifying a specified pixel of wide band images can be realized. In our architecture, convolutional layers and max pooling layers can be viewed as a trainable feature extractor to the input spectral information of wide band images, and the second fully connected layer is trainable classifier to the feature extractor. The output of sampled is the actual characteristics of raw data. The CNN model designed in this paper can finally extract 120 kinds of features from original images.

All the trainable parameters in our CNN should be initialized to be a random value between -0.05 and 0.05. The first step of training process is forward propagation which aims to compute the actual classification result of the input data with current parameters. The second training process step is back propagation which is employed to update the trainable parameters in order to make the discrepancy between the actual classification output and the desired classification output as small as possible.

Because the structure and all the relevant weight values are specified, we can build the CNN classifier and reload the saved parameters classifying wide band images, the classification process is just like the forward propagation step.

4 Experiment

All the programs are implemented using Caffe which is a CNN framework that makes us easily define, optimize, and evaluate mathematical expressions involving multidimensional arrays efficiently and conveniently on GPUs. The results are generated on a PC equipped with an Intel Core i7 with 2.6 GHz and Nvidia GeForce GTX 530 graphics card.

4.1 Date preprocessing

The main five types of ground object samples were collected from the Sentinel-2A high-resolution images by visual interpretation. Then, through the spatial position matching, the corresponding spectral vectors of the samples in the wide band images of Tiangong-2 are extracted, resampled and transformed into label images. Available training data are divided into training and testing samples for training the parameters and calculating the model loss of the proposed CNN classifier. Moreover, each pixel is scaled to [0,1] uniformly.

4.2 Results and comparisons

Figure 4 illustrate the distributions of early rice and late rice of Huarong County in 2017 from Tiangong-2 wide band images by CNN. The results showed that the early rice mainly concentrated in the west of Huarong County, showing a continuous and massive distribution, indicating that the early maturing rice in the western region had not yet been harvested. The distribution of early rice and late rice in the eastern and central regions was rather fragmented, reflecting that the harvesting and transplanting alternating features.

To demonstrate the relationship between accuracies and the differences species of landcover, we present the detail accuracies of our proposed CNN classifiers in Table 1. The overall accuracy of corresponding classification is 95.96% and kappa coefficient is 0.95. The accuracies more than 90% for all classes of major agricultural crops (early rice and late rice).

The excellent performance verifies that the proposed CNN classifier has discriminative capability to extract subtle visual features, which is even superior to human vision for classifying complex curve shapes.

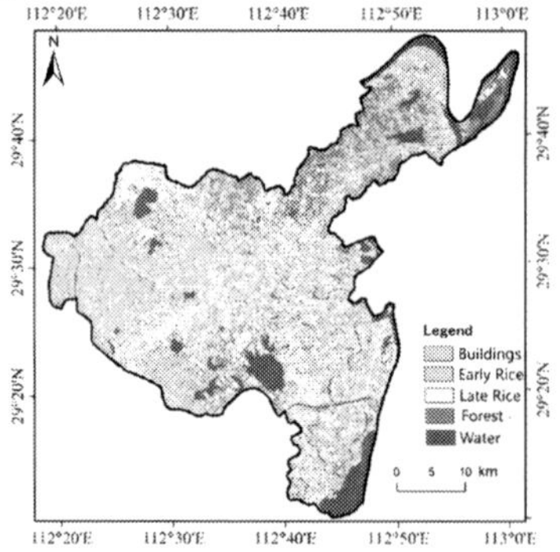

Fig. 4. The spatial distributions of the peanut in Huarong.

Table 1. The classification accuracy of Huarong.

	Buildings	Water	Forest	Early Rice	Late Rice	Total	UA
Buildings	97	0	0	2	2	101	96.04%
Water	0	286	0	0	0	286	100.00%
Forest	0	0	87	1	1	89	97.75%
Early Rice	0	6	0	173	4	183	94.54%

Late Rice	0	0	9	5	150	164	91.46%
Total	97	292	96	181	157	823	
PA	100.00%	97.95%	90.63%	95.58%	95.54%		
OA	95.96%	Kappa	0.95				

5 Conclusion

In this paper, we proposed a CNN-based method for early and late rice planting area extract on wide band images from Tiangong-2. Our work is an exploration of using CNN for crops recognition and achieves excellent performance. The architecture of our proposed CNN classifier contains two convolutional layers, two max pooling layers and two fully connected layers, due to the small number of training samples. In the feature, some techniques, such as Dropout, can be used to alleviate the overfitting problem caused by limited training samples. Furthermore, only spectral information of crops was considered in our work, we should pay more attention to multi-features such as time phases information of crops, to improve the result of classification.

In addition, recent researches in deep learning have indicated that unsupervised learning can be employed to train CNN, reducing the requirement of labeled samples significantly. Deep learning, especially deep CNN, should have great potentiality for remote sensing classification in the feature. Moreover, in the current work, we do not consider the spatial correlation and only concentrate on the spectral signatures. We believe that some spatial-spectral techniques also can applied to further improve the CNN based classification.

References

1. Brown J C, Kastens J H, Coutinho A C, et al. Classifying multiyear agricultural land use data from Mato Grosso using time-series MODIS vegetation index data. Remote Sensing of Environment, 2013, 130: 39-50.
2. Li X, Xu X, Wang J, et al. Crop classification recognition based on time-series images from HJ satellite. Transactions of the Chinese Society of Agricultural Engineering, 2013, 29(2): 169-176. (in Chinese with English Abstract)
3. Khatami, Reza, Giorgos Mountrakis, and Stephen V. Stehman. "A meta-analysis of remote sensing research on supervised pixel-based land-cover image classification processes: General guidelines for practitioners and future research." Remote Sensing of Environment 177 (2016): 89-100.
4. Chen, Yushi, et al. "Deep learning-based classification of hyperspectral data." IEEE Journal of Selected topics in applied earth observations and remote sensing 7.6 (2014): 2094-2107.5.
5. Kussul, Nataliia, et al. "Deep Learning Classification of Land Cover and Crop Types Using Remote Sensing Data." IEEE Geoscience and Remote Sensing Letters 14.5 (2017): 778-782.
6. Fukushima K. Neocognitron: A hierarchical neural network capable of visual pattern recognition. Neural networks, 1988, 1(2): 119-130.
7. LeCun Y, Bottou L, Bengio Y, et al. Gradient-based learning applied to document recognition. Proceedings of the IEEE, 1998, 86(11): 2278-2324.
8. Simard P Y, Steinkraus D, Platt J C. Best practices for convolutional neural networks applied to visual document analysis. //ICDAR. 2003, 3: 958-962.

Reduction of interference using Orthogonal Vertical Beamforming in an indoor communication

Paleerat Wongchampa*

Department of Electronics Engineering Technology, College of Industrial Technology, King Mongkut's University of Technology North Bangkok, Thailand

Abstract. The interference between users is one of the factors that limit the achievable user throughput in the current indoor communication. In this paper, an evaluation of transmit beamforming contribution is analyzed in the context of an indoor environment. The technique of the Orthogonal Vertical Beamforming (OVB) has been proposed to reduce interference in an indoor communication. The proposed concept is validated through computer simulation in term of Signal Interference plus Noise Ratio (SINR). The obtained results show that the OVB provides higher performance over the conventional vertical beamforming and Orthogonal Beamforming (OBFM) in an only vertical plane.

1 Introduction

The wireless communication in indoor mainly used the wireless local area network (WLANs). WLANs has been very popular [1] because a lot of its utilities. WLANs does not only give us the convenience and freedom for the usage but also to set up the WLANs network could possibly connect in house or office computer system. Moreover, there's no need to use uncomfortable signal cable for connecting the network. Portable PC and notebooks could connect freely in the signal-range area to each other or the network from other distances. The IEEE 802.11ac was the new WLANs which would be targeted to replace IEEE 802.11n. The Multiple Input Multiple Output (MIMO) technology has used in IEEE 802.11ac. The MIMO technique needs the feedback information to the system. The beamforming is the technique which without the feedback information. So, the beamforming is less complex than MIMO when using in an indoor communication.

From the work presented in [2], the Orthogonal Beamforming (OBFM) mainly considered the gain provided by the horizontal plane while cannot considering the gain by the vertical plane shown in Fig. 1. The figure 1 (a) and (b) show the top view and side view of the scenario of users stay at the same direction (angle in horizontal plane) but they are located in different distances. In this case, the proposed OBFM cannot handle as it can only the horizontally form one beam to one particular user. The results of conventional vertical beamforming concept [3-4] are low Signal to Interference plus Noise Ratio (SINR) because the interference from other users shown in figure 2. Therefore, this paper proposes

* Corresponding author: paleerat.w@cit.kmutnb.ac.th

the Orthogonal Vertical Beamforming (OVB) to reduce interference in an indoor environment in vertical plane.

The rest of this paper is as follows. After the brief introduction, a brief concept of the OVB is described in Section 2. Then, the computer simulation is performed to show the performance of proposed concept in Section 3. Finally, Section 4 concludes the paper.

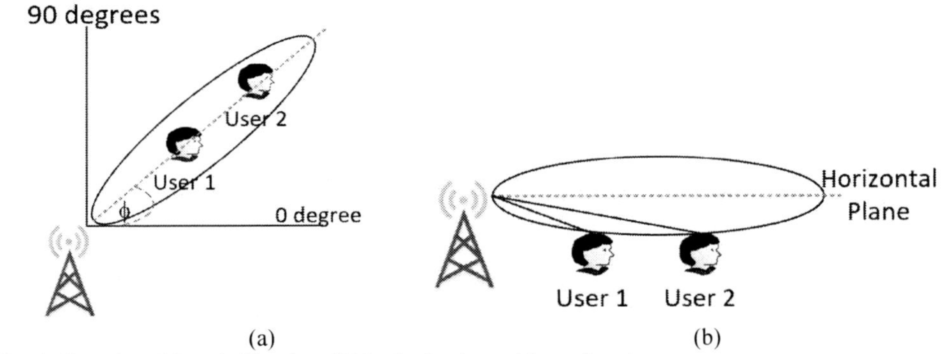

Fig. 1. Top view (a) and side view (b) in the horizontal beamforming.

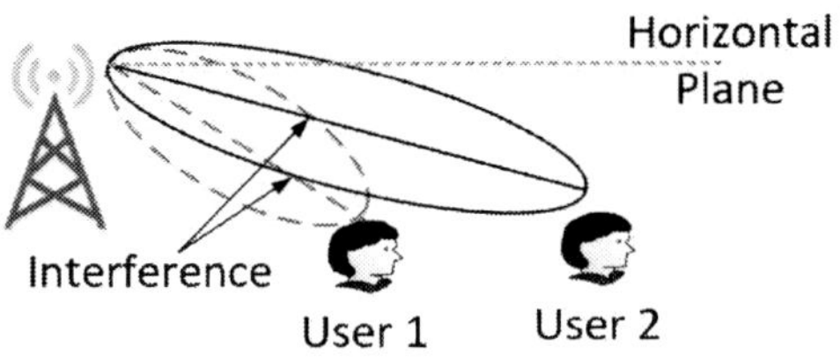

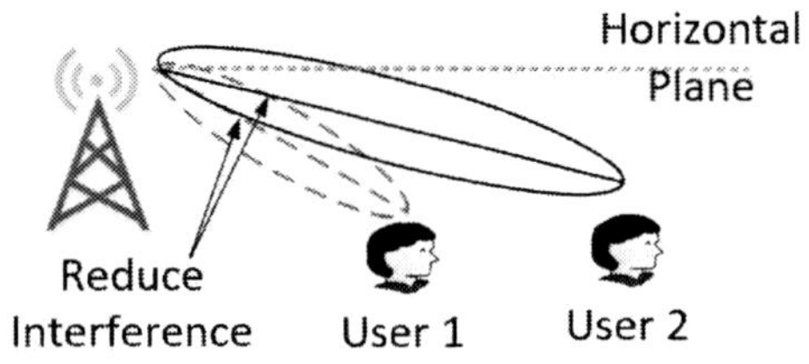

Fig. 2. Conventional vertical beamforming. **Fig. 3.** Orthogonal vertical beamforming (OVB) concept.

2 Concept of Orthogonal Vertical Beamforming

The concept of OVB in this paper is depicted in figure 3. As we can see, the users stay at the same angle in the horizontal plane but they are located at different distances. So, the beam must be able to tilt along with orthogonal in the vertical plane to each user. This paper focuses on a propagation scenario in multi-user communications an indoor environment. Consider a downlink system consisting of an Access Point (AP) consists of N linear array antennas and M users with a single antenna. The vertical antenna pattern relating to AP and the users can be expressed in dB scale as [5]

$$A_{V_M} = min\left(12\left(\frac{\theta\_\theta_{AP}}{\theta_{3dB}}\right)^2, A_m\right) \quad (1)$$

where θ_{3dB} is the vertical 3 dB beamwidth of the AP antenna. θ_{AP} is the vertical boresight angles of the AP antenna array. θ is the vertical angle from AP to the user. A_m is the vertical sidelobe attenuation. The characteristic of beam steering for linear array [6] is given in (2).

$$AF = \sum_{n=1}^{N} e^{jkd_x sin\theta cos\ + \beta_x} \quad (2)$$

where N is the number of antenna elements in the x-direction. The antenna elements along the x-axis spaced by d_x. β_x is phase delays in antenna elements. θ and are directions of the direction of arrival in vertical and horizontal plane respectively.

In computer simulation for this paper, considering the scenario where the users are at the same but the different θ. Figure 4 shows the system model for Orthogonal Vertical Beamforming with M users. So, the received signal from the AP can be written as

$$y = \sqrt{P}\boldsymbol{wHx} + \boldsymbol{n} \quad (3)$$

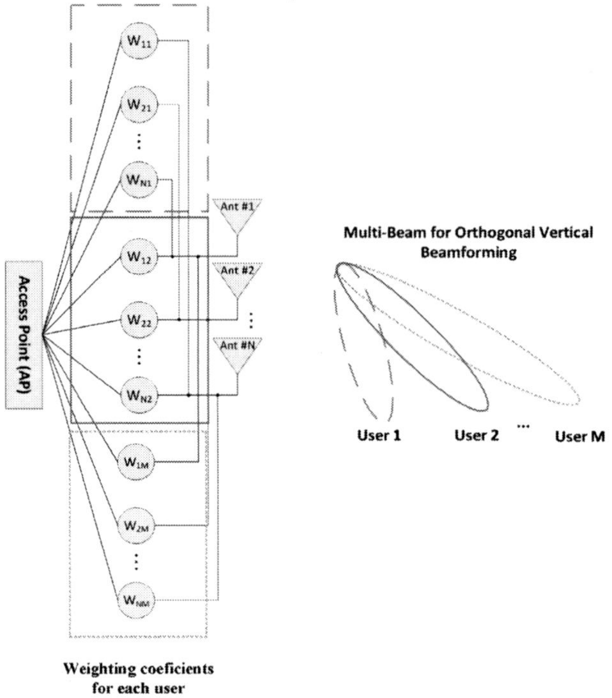

Fig. 4. System model for Orthogonal Vertical Beamforming (OVB).

where $\boldsymbol{y}$ is the received signal vector at AP after performing beam formation and P is the transmitted power of each user, $\boldsymbol{w}$ is weighting coefficient vector of each antenna element at AP, $\boldsymbol{x}$ is the $M\times 1$ signal vector containing the user data $x_1, x_2, \ldots, x_m$, $\boldsymbol{n}$ is AWGN noise and $\boldsymbol{H}$ is a communication channel for indoor environment include the line of sight, non-line of sight signals and AF of linear array antennas that can be formulated as

$$\boldsymbol{H} = \boldsymbol{S}_{LOS} + \boldsymbol{S}_{NLOS} \quad (4)$$

$\boldsymbol{S}_{LOS}$ is the Line-Of-Sight (LOS) signal vector multiplied with the steering vector of the LOS (AF_{LOS}) signal, $\boldsymbol{S}_{NLOS}$ is the Non-Line-Of-Sight (NLOS) signal vector multiplied with the steering vector of the NLOS (AF_{NLOS}) signal in equation (2). The system considers in an indoor environment so, each user transmits a signal to the AP that including LOS and NLOS signals. Note that, the power level for both LOS and NLOS components is given by (5) as it is the path loss calculation for indoor environment [6].

$$path\ loss = 37 - 30\log d + 13.8n^{\left(\frac{n+2}{n+1}-0.46\right)} + X_\sigma \quad (5)$$

where d is the distance between the BS and user, n is the number of floors. Also, X_σ is zero mean Gaussian distributed random variable (in dB) with standard deviation σ. So, the received signal can be expressed by

$$y = \sqrt{P}\boldsymbol{wS}_{LOS}\boldsymbol{x} + \sum_{l=1}^{L}\sqrt{P}\boldsymbol{wS}_{NLOS}\boldsymbol{x}_l + \boldsymbol{n} \quad (6)$$

$$y = \sqrt{P}\boldsymbol{w}(PL_{LOS}\boldsymbol{AF}_{LOS})\boldsymbol{x} + \sum_{l=1}^{L}\sqrt{P}\boldsymbol{w}(PL_{NLOS,l}\boldsymbol{AF}_{NLOS,l})\boldsymbol{x}_l + \boldsymbol{n} \quad (7)$$

where PL_{LOS} is the LOS signal coming from user can be modelled using a simplified path loss. $PL_{NLOS,l}$ is the NLOS signals coming from user and l path can be modelled using Rayleigh fading channel.

The beamforming must be calculate the weighting coefficients for turning the main beam to the users. The weighting coefficients consider in vertical plane. So, assume the users are at the same $\varnothing$ and diferent θ. Then, the $\boldsymbol{w}$ must be inverse of AF_{LOS} signal.

$$\boldsymbol{w} = \left(\boldsymbol{AF}_{LOS}\right)^{-1} \tag{8}$$

$$\boldsymbol{w} = [\boldsymbol{AF}(\theta_1) \quad \boldsymbol{AF}(\theta_2) \quad \cdots \quad \boldsymbol{AF}(\theta_M)]_{N\times M}{}^{-1} \tag{9}$$

$$\boldsymbol{w} = \begin{bmatrix} 1 & 1 & \cdots & 1 \\ e^{jkd_x sin\theta_1 cos\varnothing+\beta_x} & e^{jkd_x sin\theta_2 cos\varnothing+\beta_x} & \cdots & e^{jkd_x sin\theta_M cos\varnothing+\beta_x} \\ \vdots & \cdots & \ddots & \vdots \\ e^{jkd_x(N-1)sin\theta_1 cos\varnothing+\beta_x} & e^{jkd_x(N-1)sin\ _2 cos\varnothing+\beta_x} & \cdots & e^{jkd_x(N-1)sin\theta_M cos\varnothing+\beta_x} \end{bmatrix}^{-1} \tag{10}$$

The proposed concept is to find an appropriate weighting coefficients $\boldsymbol{w}$ to avoid interference in the main beam direction as all beams are launched at the same time and the same frequency in the vertical plane. When the system uses the weighting coefficients that the received signal will independent relation between weighting coefficients and steering vectors, so-called orthogonal property. As all beams are orthogonal to one another in the vertical plane, users in the vertical plane are not interference to themselves anymore.

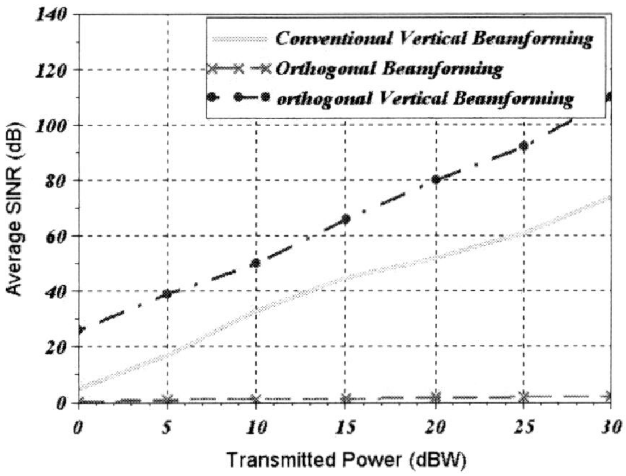

Fig. 5. Average SINR vs. Transmitted power for 4 users that located in vertical angle (θ) at 20, 40, 60 and 80 degrees.

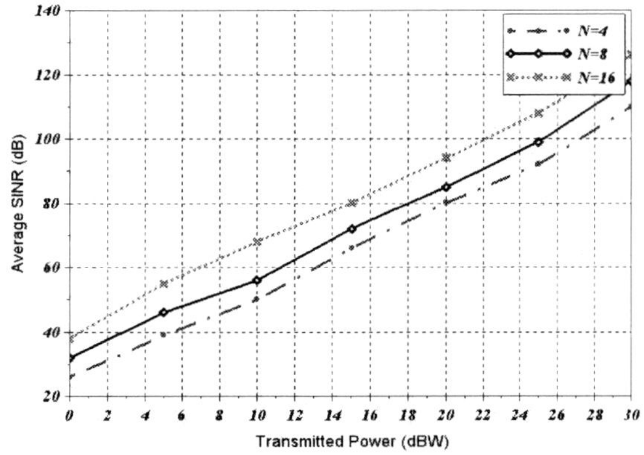

Fig. 6. Average SINR vs. Transmitted power for orthogonal vertical beamforming when N = 4, N=8 and N = 16.

3 Simmulation results

In this section, we present simulation results to show the performance of proposed Orthogonal Vertical Beamforming (OVB) in indoor communications compared with conventional vertical beamforming and Orthogonal Beamforming (OBFM). We show the impact of the system SINR on the transmitted power. Fig. 5 shows the plots of simulated average SINR vs. transmitted power where the AP is equipped with 4 omnidirectional antennas. The 4-users are located in vertical angle (θ) at 20, 40, 60 and 80 degrees when all users are an equally horizontal plane (∅) given at 30 degrees. For the simulation, we assume a practical indoor scenario with the path-loss exponent of 1.4. We substitute parameters and vary the transmitted power of user from 0 to 30 dBW. Then, the average SINR increases when the transmitted power of user increases.

Moreover, the proposed orthogonal vertical beamforming provides a higher average SINR comparing to conventional vertical beamforming and orthogonal beamforming. The OBFM provides the lowest SINR because of all users at the same horizontal plane (∅). So, the interference cannot be effectively rejected as they are all users stay in the region of the main beam. The result confirms that the proposed orthogonal vertical beamforming can increase the performance in indoor communications in case of the users are in the same horizontal angle but different the vertical angle.

For the transmitted power between average SINR at the number of antennas is 4, 8 and 16 elements of OVB technique shown in Fig. 6. At the result of OVB technique simulation at four users, the number of antennas is 4 in blue dash line which has lower average SINR than the number of 8 and 16 elements. The increased number of antennas resulted in an increased gain and narrower beam width. So, the interference between users decreased and SINR increased.

4 Conclusion

This paper has proposed an Orthogonal Vertical Beamforming (OVB) concept to reduce interference in an indoor environment for users are in the same horizontal angle. The results of the system from the simulation that provides higher performance in term of SINR over the conventional vertical beamforming and Orthogonal Beamforming (OBFM) in the vertical plane. The future work will combine the OBFM in the horizontal plane with OVB in the vertical plane for improving the performance in term of SINR and capacity in an indoor environment.

Acknowledgment

This research was funded by College of Industrial Technology, King Mongkut's University of Technology North Bangkok (Grant No Res-CIT*0320*/2018)

References

1. Wireless LAN medium access control (MAC) and physical layer (PHY) specifications, *IEEE802.11Std.*,Aug.1999.
2. Paleerat Wongchampa and Monthippa Uthansakul, "Orthogonal Beamforming for Multiuser Wireless Communications", *IEEE Antennas&Propagation magazine*, 2017, pp. 38-49.
3. Wookbong Lee et al, "Downlink Vertical Beamforming Designs for Multi-User MISO Systems", *IEEE conference* 2015.

4. Jiancun Fan et al, "Vertical beamforming with downtilt optimization in downlink cellular networks", *IEEE conference* 2015.
5. Radiocommunication Sector of International Telecommunication Union, "Report M.2135-1: Guidelines for Evaluation of Radio Interface Technologies for IMT-Advanced," *ITU-R M Series*, October 2009.
6. ETSI TR 136 931, "*LTE; Evolved Universal Terrestrial Radio Access (E-UTRA);Radio Frequency (RF) requirements for LTE Pico Node B*" 3GPP TR 36.931 version 9.0.0 Release 9.

Analysis of spatial and environmental justice in three Mexican cities

L Araujo-Alvarado[1], *C Y Ortega*-Montoya[2*], and *A Ávila*-Galarza[1]

[1]Facultad de Ingenieria, Universidad Autonoma de San Luis Potosí, Av. Manuel Nava #8, Zona Universitaria, 78290, San Luis Potosi, Mexico
[2]Escuela de Humanidades y Educacion. Tecnologico de Monterrey, Av. Pedro Coronel No.16, Dependencias Federales, 98600, Guadalupe, Zacatecas, Mexico

Abstract. The accelerated growth of cities in developing countries poses a challenge for urban planning such as guaranteeing the safety of all inhabitants from natural and anthropogenic threats. The objective of this study is to determine the urban areas and social sectors of three medium sized Mexican cities that are exposed to high risk activities. The results indicate that for the metropolitan area of San Luis Potosi, 57.6% of high risk activities are within the urban sprawl, while 100% are in the cities of Zacatecas and Fresnillo. In all three case studies, most of the activities studied are in areas with low and medium rates of marginalization. In San Luis Potosi, 4.1% are in areas with high and very high population density but none of the case studies showed a conglomeration of these types of risks in a determined area of the cities. A database of companies with high risk activities, regulated at a federal level, public became public in Mexico in 2014. This information is of great value at a local level to define the strategies of land use planning that guarantee access to environmental justice as well as effective prevention strategies and responses to chemical emergencies.

1 Introduction

The United Nations Development Programme (UNDP) has presented within its sustainable development objectives the construction of sustainable cities and communities, therefore recognizing that currently half of the population lives in urban areas and that by the year 2050 it will increase to two thirds. Because of this, it is of great importance to transform and efficiently manage urban spaces.

A problem related to the accelerated growth of urban areas is the formation of poverty belts around cities where the poorest inhabitants tend to settle erratically. The urban space is therefore spatially configured in relation to the social sectors that inhabit the different areas of the city. One of the challenges to urban management is the prevention, preparation and response to environmental risks of not only natural but also anthropogenic origin.

*Corresponding author: yazmin.ortega@tec.mx

Cannon [1] states that social processes create an uneven exposure to risk, thus making some sectors of the population more susceptible to suffering disasters than others.

Chemical risks are relevant because industrial spaces for the transformation of consumer goods are grouped within the city. The analysis of risk scenarios is important because it allows us to identify the most susceptible groups of a population or ecosystem that can be exposed during a possible disaster. An adequate management of risk is the only guarantee in facing the future with a solid and secure foundation [2]. For this reason, since 1992, the United Nations has recognized the importance of chemical risks in urban environments through the APELL program, which establishes that a better understanding of risk scenarios and the mechanisms for prevention and protection in case of emergency are vital for the construction of capabilities within a society and the agencies in charge of public safety.

According to these scenarios, it is very important that the instruments for land use planning, such as the case of the ecological and environmental management, incorporate into their diagnostic, prognostic and resulting programs, the consideration of risk analysis for the different existing hazards, thus identifying the level of vulnerability of the population for each one of them [3]. Recognizing the conditions of vulnerability as risk generators, and therefore as generators of disaster, has a political dimension. It implies recognizing that in situations of risk and disaster there is social and political responsibility thus referring to notions of environmental justice [4].

While people have the right to a safe and clean environment, research in environmental justice has shown that lower and middle class communities have a greater exposure to higher levels of risk than other groups in society [5]. The environmental justice concept was developed in the United States due to a movement that sought equality in the distribution of environmental risks among ethnic and racial groups. This is to say; all people have the right to be protected from pollution and to live and enjoy a safe and clean environment [6].

Environmental justice establishes that groups with a high concentration of low-income families or racial and ethnic minorities are more exposed to a series of higher environmental risks and problems. These groups also do not equally benefit from the positive implications derived from environmental policies and regulation [7]. According to environmental justice studies, the most common sources of risk tend to be facilities with toxic emissions, landfills and industrial zones. Several studies have shown that a significant relation exists between the proximity of environmental risks and health problems such as respiratory conditions, pregnancy problems (malformations, injury to the central nervous system of the fetus), child cancer, leukemia and kidney disease, among others. Although the population that is close to these sources of risk seems to be more likely to have these diseases or health problems, the proximity is not equivalent to the individual exposure and the negative impact is not the same for all inhabitants [8].

In addition to this, in their research, Cole, Elliott and Khemmarat [9] found that ethnic division and polarization of communities have a positive and statistically significant correlation with toxic releases. This suggests that, under equal conditions, the greater the ethnic divisions within an area, the greater the level of toxic emissions will be. Another important aspect in case studies of environmental justice is the use of Geographic Information Systems (GIS) as a way of exploring spatial relations between factors that can be associated with the locations of risk facilities. Population density, land use and employment are factors that can be used to predict the installation of emission sources [10].

Ash et al. [11] indicates that while it is true that a larger number of cases where there is a lack of environmental justice have been present in minorities such as ethnic groups, low-income populations and immigrants (some authors even include pregnant women, children and senior citizens in this term), this is a problem that encompasses all society. They suggest that urban areas that contain minorities experiment exposure to pollution but in the end, this is a problem that affects all subgroups of a population. Their conclusion is that access to environmental justice is good for everyone, not just for minority groups.

1.1 Regulation of risk activities in Mexico

In Mexico, high risk activities are defined as service, commercial or industrial sites where materials with explosive, toxic or flammable characteristics, that exceed the quantities published in the First and Second High Risk Activities List, are generated or handled. These activities are regulated at a federal level.

The storage of these hazardous substances represents a chemical threat that can affect the environment or the population´s health if an accidental release of one of these substances were to occur. The analysis of high risk activities in Mexico was possible due to the approval of the General Act of Transparency and Access to Public Information and consequently has opened up information at a federal level. Thanks to this law and through a request for public information it was possible to learn about the high risk activities that are present in the national territory and that previously were considered classified information that could not be accessed.

2 Case studies

The metropolitan area of San Luis Potosi-Soledad de Graciano Sanchez is considered the 11th largest in the country (1,040,443 people) and one of the most important in the state since a third of the services and population are centered there. The urban area of San Luis Potosi has gained economic, political and social importance due to its proximity to the three main population centers in the country (Mexico City, Guadalajara and Monterrey). Likewise, the city is part of two important trade corridors in the country: the road and rail network that connect the central-northern part of the country (the core of NAFTA) and the network that connects the Gulf of Mexico and Pacific Ocean seaports. This confirms its favored position as a cultural, industrial, commercial and logistical zone.

The metropolitan area of Zacatecas (275,027 people) is experiencing an important population growth and currently ranks 43 out of 56 in the national urban system with a little more than 270 thousand inhabitants. On the other hand, the city of Fresnillo has been the most populated municipality in the state of Zacatecas for 50 years (230,865 people). The main problems of this city are centered on a deficient urban structure that is due to the disorderly growth of the city and to an indiscriminate mixture of land use that presents a marked incompatibility.

The case studies and their population are shown in figure 1.

Fig. 1. Location of case studies.

3 Methodology

This research was developed following the subsequent stages:

3.1 Consultation of the National Register of high risk activities

The National Register of companies with high risk activities was obtained through a public information request that was submitted to the Secretariat of Environment and Natural Resources (SEMARNAT) in 2014.

3.2 Update of the National Register of high risk activities

The addresses and main information of companies with high risk activities (such as business name, address and state) were verified through a consultation of the National Institute of Statistics and Geography´s [12] National Directory of Economic Units (DENUE). Said directory offers the identification details, location, economic activity and size of the companies active in the national territory.

3.3 Geographic positioning of risk activities

The high risk activities were geographically arranged using the information from DENUE and the geographic information system ArcMap 10.1.

3.4 Delimitation of industrial areas

Industrial zones were delimitated within the cities studied using Google Earth. In the case of San Luis Potosi, the secondary zoning plan of the 2003 Strategic Population Central Plan of San Luis Potosi-Soledad de Graciano was used. For Zacatecas, the current urban land use plan of the 2016 Urban Development Plan of Zacatecas-Guadalupe was used. In the case of the city of Fresnillo, the zones were defined using the Google Earth tool, considering the connectivity and aggregation level of the industrial infrastructure.

3.5 Consultation of social development benchmarks

The 2010 Urban Marginalization Index was consulted for the areas studied at the level of a Basic Geostatistical Area (AGEB). This index is developed by the National Population Council [13] and is included in the INEGI´s Census Information Consult System (SCINCE).

3.6 Consultation of population density

The population density of the AGEB´s studied was obtained from the total population variables available in the 2010 Housing and Population Census [14]. The ArcMap tool "calculate geometry" was used to obtain the area of the AGEB´s in km^2 units.

$$\text{Population Density}=\frac{\text{Population (Inhab)}}{\text{Surface (km}^2\text{)}}$$

3.7 Environmental justice analysis

The level of social marginalization and population density of the basic geostatistical areas (AGEB) where the companies with high environmental risk are located were analyzed. An AGEB is a polygon that groups between 25 and 50 city blocks.

The population density and social marginalization information are important variables for the analysis of risk scenarios since they allow us to identify the characteristics of the systems potentially affected and the conditions of social vulnerability.

4 Results

According to the National Register of High Risk Activities, in San Luis Potosi there are 73 high risk activities. 57.6% of companies considered as high risk are located within the industrial zone of the city, while 42.4% are dispersed throughout the urban and suburban areas (Figure 2a). In the case study of the Zacatecas metropolitan area, there are 6 companies with high risk activities. None of these companies are located within the industrial polygon, but rather are located in areas destined for commercial and residential use in the center and outskirts of the metropolitan area (Figure 2b). The same occurs for Fresnillo where the 11 companies with high risk activities are located within the urban sprawl (Figure 2c.)

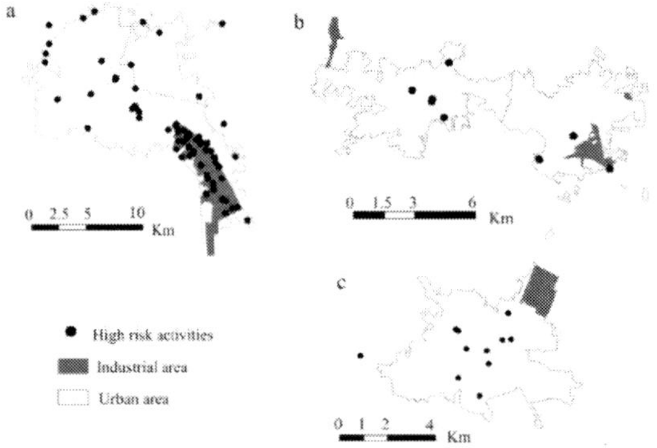

Fig. 2. Companies with high environmental risk in the metropolitan areas of a) San Luis Potosi, b)Zacatecas and c) Fresnillo Source:[12].

In the three case studies the high risk activities are randomly distributed among the different sectors of the cities without a particular pattern for these type of activities.

It was found that the most frequent high risk activities in the capital of San Luis Potosi are: the production of chemical products and the storage and sale of LP gas. While in Zacatecas it is the sale of LP gas and the production of ice. Finally, a mining company with high risk activities is located within the metropolitan area of Zacatecas (Table 1).

Table 1. Main activities of the companies with high risk activities[a]

Main Activities	Number of companies	
	San Luis Potosi	Zacatecas
Treatment plants and hazardous waste treatment	7	-
LP and butane gas storage and distribution	9	5
Pemex	3	1
Manufacturing of chemical products	12	-
Ice Manufacturing	4	4
Automotive companies	7	3
Food industry	10	2
Mining	-	1
Others	21	2

[a] Source: National Directory of Economic Units (DENUE).

4.1 Social vulnerability

The location of the companies with high risk activities was correlated to the rate of marginalization of the basic geostatistical areas. In the case of the metropolitan area of San Luis Potosi the results obtained were: 28.8% of the companies are located in areas with low and very low rate of marginalization, the majority of these are located near the industrial zone and the center of the urban sprawl, 9.6% in areas of medium marginalization and only 1.3% in areas with a high rate of marginalization, primarily in the outskirts of the city (Figure 3a). On another note, 60.3% of said companies are found in areas where the degree of marginalization is unknown.

In the metropolitan area of Zacatecas (Figure 3b), 50% of the high risk activities studied are located in areas classified with a low rate of marginalization, 33.3% in areas with a very low rate and 16.7% in areas without data. In the case of the city of Fresnillo (Figure 3c), 63.6% of the companies with high risk activities are located in areas with a low rate of marginalization, while only 27.3% are situated in areas with a medium rate and 9.1% in areas without data.

It is important to note that the areas where information on the level of marginalization does not exist tend to be areas without human settlements or non-urban towns with a population of less than 2,500 inhabitants.

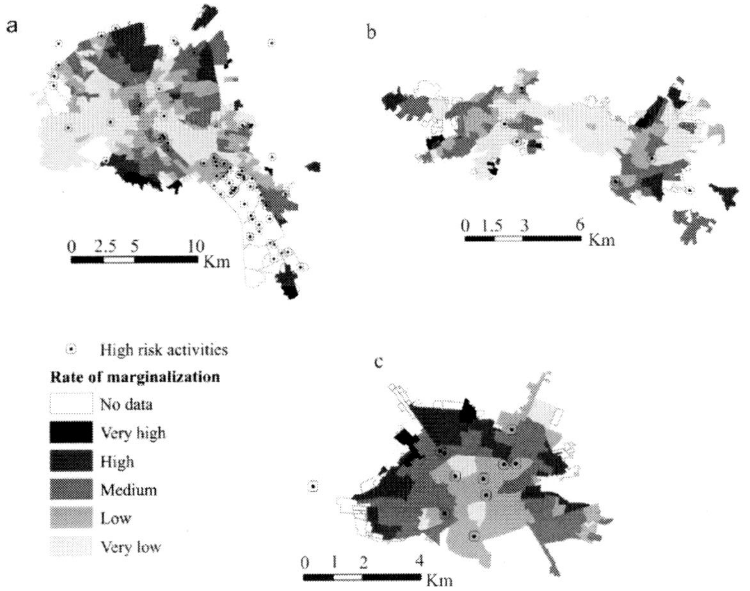

Fig. 3. Marginalization rate in the case studies a) San Luis Potosi, b) Zacatecas and c) Fresnillo. Source [12][13].

4.2 Population density

The analysis of the population density by AGEB in the urban area of San Luis Potosi shows that 48% of companies with high risk activities are located in AGEBs with very low population density. As we can observe in figure 4a, 16.4% of said companies are in areas with low and very low density, 13.7% present no data, 11% with medium density, 6.8% with hih density and 4.1% of them are in AGEBS with very high density, especially in the eastern part of the city.

In the case of the urban area of Zacatecas, 50%% of high risk activities are located to the north of the city in areas with high population density. 16.6% show no data. The three remaining companies (50%) are in medium population density areas as shown in figure 4b.

In the city of Fresnillo, 54.5% of high risk companies are in areas with medium and low population density. 36.4% are located in areas with high density and 9.1% show no data (Figure 4c).

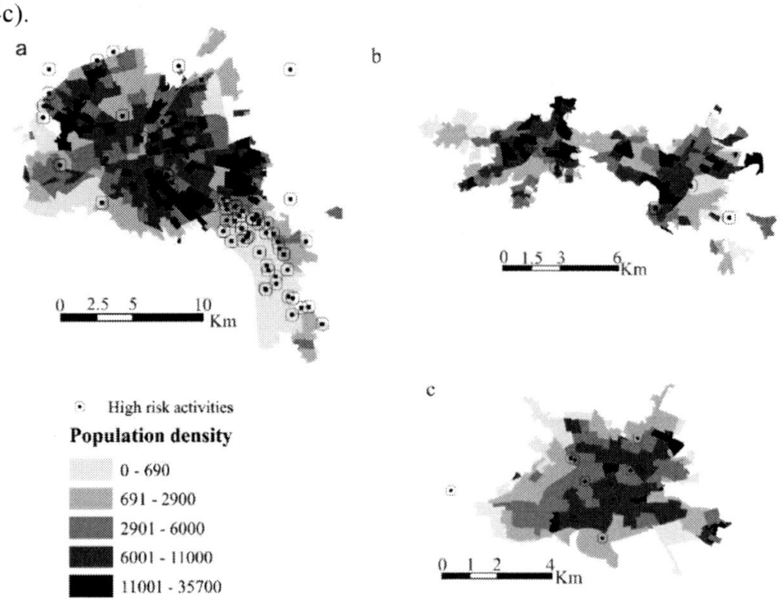

Fig. 4. Population density in the case studies a) San Luis Potosi, b)Zacatecas and c) Fresnillo. Source [12][14]

5 Discussion

In the case of San Luis Potosi it was found that 57% of the companies are located in areas with industrial land use vocation, while the rest of the companies are located in the interior of the urban sprawl. This is in contrast to the Population Central Plan of San Luis Potosi-Soledad de Graciano which establishes that all types of industry or large-scale storage that have high risk and/or pollution characteristics must be located in industrial areas or corridors designed for that purpose.

The analysis of the degree of marginalization per urban AGEB indicates that in San Luis Potosi only one high risk company exists in an AGEB with a very high degree of marginalization (in the outskirts of the city). However, a large number of companies are located in the surroundings of the urban sprawl, in areas without data on social marginalization. In areas considered non-urban (with less than 2,500 inhabitants) this variable can have greater relevance in the study of risk scenarios. Ortega [15] indicates that rural settlements are not included in urban marginalization rates, but that one would expect greater vulnerability from this scenario due to its poverty conditions which represents a lower response in an emergency.

In the three case studies of companies with high risk activities, the number of inhabitants near these activities turned out to be the variable with the most critical aspect. This is due to the fact that the companies are located in urban areas where the population density is high and job opportunities reduce the rate of marginalization.

With respect to the relationship between the degree of marginalization and a greater exposure to risks caused by companies with environmental hazards, this study found no correlation. This was mentioned by Mennis and Jordan [10] who claimed that while high risk facilities are associated with minority groups in society, there is not a clear pattern in all areas of study. It is important to note however that in other parts of the country this situation is in fact related to environmental justice, where low income populations are the ones that are most exposed to sources of danger due to high risk activities.

6 Conclusion

The analysis of the social marginalization and population density variables in the urban areas of San Luis Potosi, Zacatecas and Fresnillo were useful in identifying the distribution of chemical and technological risks. Thus confirming the presence of high risk activities not in industrial zones, but rather in areas with commercial and residential use within the urban sprawl.

The program that governs the city´s growth in the metropolitan area of San Luis Potosi has not been updated in more than 14 years. Another important point is that none of the cities studied has spatial planning.

In the three cases studied, most of the companies were in areas with low and medium rates of marginalization. There is only one case in the northern part of San Luis Potosi where a company exists in an AGEB with a very high level of marginalization. In San Luis Potosi there is a greater proximity between high risk activities and areas with high population density. Particularly interesting is where people are living close to a source of danger with the risk of an accident happening that can affect the health of the population and the environment. The dissemination of information regarding the risk of living near a high risk activity and training the population on what measures to take in case of a leak, fire or explosion, are effective strategies that can be used to decrease their vulnerability and increase their responsiveness in an emergency.

State and municipal governments in the areas studied must develop (or update) their urban planning instruments based on adequate ecological and spatial planning. The proposal and execution of pertinent corrective measures that contribute to the decrease of high risk activities in commercial and residential areas is also necessary.

The dissemination and access to the information gathered in this type of research project is important as it leads to the development of better urban planning policies, results in actions that decrease the vulnerability of people and establishes proof and precedents useful in improving access to environmental justice.

Acknowledgments

To the Engineering Department at the Autonomous University of San Luis Potosi and the Urban Development Secretariat of San Luis Potosi and Zacatecas.

References

1. Cannon T 2000 Vulnerability Analisys and Disaster *Flood* Ed Parker D J 45-55 (Oxford: Routledge)
2. Maturana P A 2011 Evaluación de riesgos y gestión en desastres. 10 preguntas para la década actual *Rev Méd Clín Las Condes*. 22 545-55
3. Rivera E, Palacio G, Villalobos G, Silva R and Afonso P 2004 *Evaluación de daños en las zonas costeras de la península de Yucatán por el huracán "Isidoro". Desarrollo de Propuestas de Investigación y Mitigación en Manejo Integrado de Recursos Costeros. Sección Campeche,* (Campeche: Universidad Autónoma de Campeche)

4. Campos M, Toscana A and Alanís J 2015 Riesgos socionaturales: vulnerabilidad socioeconómica, justicia ambiental y justicia espacial. *Cuad Geogr-Rev Col Geogr* ***24*** 53-69
5. EPA 1992 Environmental Equity. Reducing risk for all communities. (Washington: Environmental Protection Agency)
6. Hervé D 2010 Noción y elementos de la justicia ambiental: directrices para su aplicación en la planificación territorial y en la evaluación ambiental estratégica. *Rev der* **23** 9- 36. https://dx.doi.org/10.4067/S0718-09502010000100001
7. Arriaga A and Pardo M 2011 Justicia ambiental: estado de la cuestion. *Rev Int Sociol* Revista v Internacional de Sociología **69** 627-48. DOI: 10.3989/2009.12.210
8. Maantay J, Chakraborty J and Brender J 2011 Proximity to Enviromental Hazards: Environmental Justice and Adverse Health Outcomes. *Strengthening Enviromental Justice Research and Decision Making: A Symposium on the Science of Disproportionate Enviromental Health Impacts* (Washington: Environmental Protection Agency)
9. Cole M A, Elliott R J and Khemmarat K 2013 Local exposure to toxic releases: Examining the role of ethnic fractionalization and polarisation *Ecol econ,* ***93*** 249-259.
10. Mennis J and Jordan L 2005 The Distribution of Enviromental Equity: Exploring Spatial Nonstrationarity in Multivariate Models of Air Toxic Releases *Ann Am Assoc Geogr* **95** 249- 68
11. Ash M, Boyce J K, Chang G, Pastor M, Scoggins J. and Tran J 2009 *Justice in the Air: Tracking Toxic Pollution from America's Industries and Companies to Our States, Cities, and Neighborhoods* (*Massachusetts;* Political Economy Research Institute)
12. INEGI 2014 *Directorio Estadístico Nacional de Unidades Económicas.* (Aguascalientes: Instituto Nacional de Estadística y Geografía)
13. CONAPO 2010 *Indice de Marginación Urbana 2010. Base AGEB Urbanas y Población.* (Mexico City: Consejo Nacional Población)
14. INEGI 2010 *Censo de Población y Vivienda del año 2010.* (Aguascalientes: Instituto Nacional de Estadística y Geografía)
15. Ortega-Montoya C Y 2015 *Propuesta Metodológica para la Evaluación del Riesgo Ambiental Debido al Manejo de Materiales Peligroso en el Estado de San Luis Potosí* (Mexico City: Repositorio Nacional CONACYT)

MATEC Web of Conferences **277**, 02015 (2019) https://doi.org/10.1051/matecconf/201927702015
JCMME 2018

Analysis of shrinkage and creep effect for bridge widening based on TB-FEM method and experimental research

Keke Peng[1,*], and *Fangzhen* Wen[2]

[1]Transportation and civil buildings college, Foshan University, Foshan 528000, China
[2]Guangdong Highway Construction Co., Ltd., Guangzhou 510627, China

Abstract. Based on TB-FEM method and laboratory test, the influence of shrinkage and creep on the structure of widened bridge is studied. The shrinkage test uses C50 concrete and C50 double-mixture (steel polyester fiber) concrete to prepare the standard prism specimen, and the creep test uses the 15m hollow slab which exactly has the same material, size, pre-stress and construction method of a real bridge as the creep test specimen. The strain-time curve obtained by shrinkage and creep is used to define the shrinkage and creep characteristics of the material, and Trost-Bazant is combined with finite element method to analyze the mechanical response of the new and old bridge splicing structure caused by shrinkage and creep. The result shows that under the condition of rigid splicing, the new bridge should be supported by the simply supported structure, and it is not suitable to use the same simply supported and continuous hollow slab as the original bridge. Theoretical analysis and example study show that combination of laboratory test and TB-FEM method can reasonably analyze the effect of shrinkage and creep on bridge widening, and provide a theoretical basis for the type selection of widening structure.

1 Introduction

Adopting of appropriate widening measures can improve the traffic capacity of old bridge and meet the needs of modem transportation. In the form of splice structure, there are three commonly used methods at present: (1) the deck width and bearing capacity of old bridge are insufficient. After a new half of bridge is built, the traffic will be diverted to this new half and the old bridge will be removed, and when the whole new bridge is built to the full width, the traffic will be diverted to the whole bridge. All traffic will be undertaken by the new bridge after reconstruction. (2) The deck width of the old bridge is insufficient. The old bridge structure is basically unchanged during the widening and the new bridge will be built on one side or both sides of the old bridge, while the old bridge continues to bear part of the traffic. After the reconstruction, the new bridge structure is independent from the old bridge structure. (3) The deck width of old bridge is insufficient, and the bearing capacity and stress in service stage meet the requirements of specification. After widening, the deck pavement of new and old bridge is spread continuously and jointly; in order to ensure

* Corresponding author: Pengkeke8481@126.com

safety and comfort of driving, the compatibility of deformation of new and old bridge must be strictly guaranteed. For the case (1) and (2), the new and old bridge will not be under stress jointly, so the shrinkage and creep will not affect the widened structure; for the case (3), the new and old bridge will be under stress jointly and the shrinkage and creep of old bridge are basically completed, while the shrinkage and creep of new bridge are in progress, so the impact of shrinkage and creep on the new and old bridge widening structure must be considered[1-3].

Creep and shrinkage are among the basic characteristics of concrete which is a kind of viscous-elastic material, and they have a great influence on the bridge structure. They also have a long duration and complicated changing process, so it is not easy to control them [4-5]. In practical concrete structures, creep and shrinkage are mixed with temperature strain. In the analysis and calculation, temperature stress and temperature strain are often considered separately. Creep and shrinkage can be considered together [6]. There are three methods to analyze the influence of creep and shrinkage on structural internal force and component section stress: (1) Differential equation solution method [7]. (2) Algebraic equation solution method [8] and (3) stepwise analysis of finite element [9]. The first two methods are the solutions using the finite element stepwise analysis method. In this method, the time-varying nonlinear stress-strain relationship is used to consider the shrinkage and creep and it is necessary to solve the coupled-differential equations or algebraic equation; these equations are difficult to solve and the analytical solutions are long, so it is inconvenient to use them. Based on the experimental study, the shrinkage and creep characteristics of the material are obtained, and the characteristics are putted into the finite element model to analyze the influence of shrinkage and creep on the splicing structure of the new and old bridge.

2 Research on the finite element calculation method for shrinkage and creep based on the Trost-Bazant-finite element method (TB-FEM method)

The relationship between stress increment and strain increment caused by creep and shrinkage can be expressed by the Dischinger method [10] and Trost-Bazant method [11]. This paper is based on the TB method. The relationship between stress and strain increment is expressed in a more precise form as follows by taking t_i as the calculation time:

$$\begin{aligned}\Delta\varepsilon_{cs}(t_i,t_{i-1}) &= \frac{\Delta\sigma_{cs}(t_i,t_{i-1})}{E(t_{i-1})}[1+\chi(t_i,t_{i-1})\varphi(t_i,t_{i-1})] \\ &+\left\{\sum_{j=1}^{i-1}\frac{\Delta\sigma(t_j)}{E(t_j)}[\varphi(t_i,t_j)-\varphi(t_{i-1},t_j)]+\Delta\varepsilon_s(t_i,t_{i-1})\right\}\end{aligned} \quad (1)$$

where: $\Delta\varepsilon_{cs}(t_i,t_{i-1})$ and $\Delta\sigma_{cs}(t_i,t_{i-1})$ - strain increment and stress increment caused by the creep and shrinkage in the period from t_{i-1} to t_i; $\Delta\sigma(t_j)$ - stress increment at the time t_j; $\Delta\varepsilon_s(t_i,t_{i-1})$ - shrinkage strain increment occurred in the period from t_{i-1} to t_i; $E(t_j)$ - the modulus of elasticity at the time t_j.

In the finite element analysis of the creep and shrinkage of concrete bridges, the process from the beginning of construction to completion and until the completion of shrinkage and creep can be divided into several calculation stages, and each calculation stage can be divided into several appropriate time intervals. The structures built in each computation stage are divided into several units, so that the concrete of each unit has the uniform creep and shrinkage characteristics. The data simulation test shows that if the concrete creep coefficient and shrinkage strain are simulated as the multiple exponential coefficient expressions by taking e as the base, the memory consumption of computer will be greatly reduced and the calculation speed will be quicker when the finite element analysis is used for calculation. The above theory is the basic idea of the TB-FEM method. In this paper, the Workbench program in ANSYS12 software is used to conduct the iterative analysis on the influence of shrinkage and creep on the widening bridge by finite element method.

3 Project background

The old bridge used as an example adopts the $10 \times 15 = 150m$ cross-structure continuous hollow slab, which has a bridge width of 157.75m and cantilever of 37.5cm; it is composed of15slabs, as shown in the figure 1.

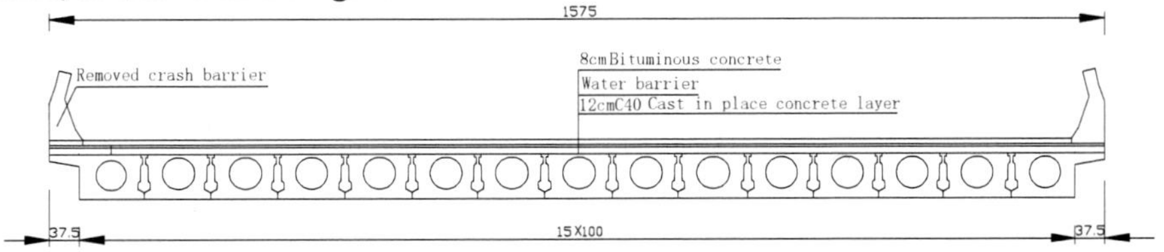

Fig. 1. Cross Section of Old Bridge.

Three longitudinal widening structure schemes are given for the original structure of the bridge used as example. Scheme I: the simply supported structure is used for the new bridge to realize rigid splicing with the original bridge and end cross beam is arranged at the support location; scheme II: the simply supported structure is used for the new bridge to realize rigid splicing with the original bridge and end cross beam is not arranged at the support location; Scheme III: rigid splicing of new bridge is implemented by using the same structural continuous hollow slab with the original bridge, and end cross beam is arranged at the support location. Finite element model is established for each of the above schemes. The new bridge takes account of the shrinkage and creep of concrete, and the old bridge does not consider the shrinkage and creep of concrete to compare and select the optimum scheme.

4 Concrete shrinkage and creep test

4.1 Concrete shrinkage test

According to the requirements of the specification GB/T50082-2009[12], this test uses the C50 concrete and C50 double-mixture concrete (steel fiber and polyester fiber), and 2 groups of specimen are made for each type of concrete for purpose of test. Each group consists of 3specimen which are 100mmx100mmx515mm prism concrete specimen, and these specimens are used for testing of the length shrinkage of concrete specimen not caused by external force under the standard temperature and moisture condition. The specimen will be removed from the standard curing room at the age of 3 days (starting from water adding into concrete in the mixing station) and immediately moved into the constant temperature and humidity chamber at a temperature of (20 ± 2) ℃ and relative humidity (60 ± 5) % to test the initial length. After that, measure the strain readings (start from the date of moving into the constant temperature and humidity chamber, and the age shall be determined according to the actual situation) at the specified time interval (1d, 3d,7d, 14d, 28d, 45d, 60d, 90d, 120d, 150d, 180d, 240d, 300d and 360d). The curve for the change of shrinkage strain of each concrete specimen with time can be obtained according to the measured shrinkage strain data of concrete specimen, as shown in the figure 2.

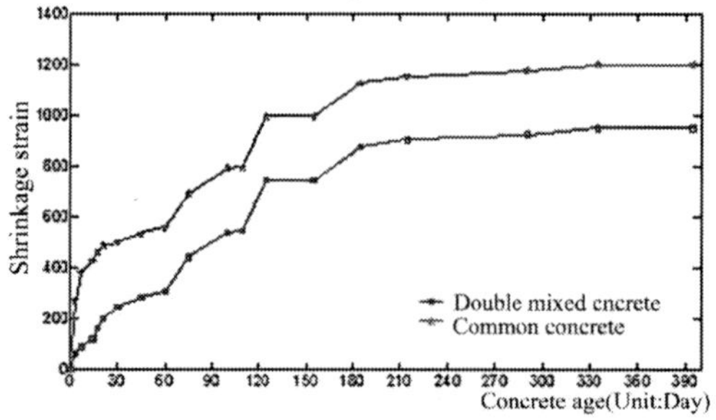

Fig. 2. Concrete Shrinkage Strain Time Curve.

Through the analysis of data and shrinkage strain-time curve obtained from the concrete shrinkage test, it is found that the shrinkage strain of concrete increases with time, and the shrinkage deformation of concrete specimen is faster in the early stage: the shrinkage deformation of most specimen accounts for a large percentage up to 70% in the total shrinkage deformation in 1 month after the preparation of specimen. The concrete shrinkage can be effectively reduced by adding a certain amount of steel fiber and polyester fiber in concrete.

4.2 Concrete creep test

According to the test method specified in the specification GB/T50082-2009 [12], creep and shrinkage tests of concrete belong to the category of long-term durability of concrete. The creep test of concrete is performed by using the 15m hollow slab (a piece of middle plate) which exactly has the same material, size, pre-stress and construction method with the real bridge as the creep test specimen. The test beam is designed according to the partial pre-stressed A-class components and constructed by using the pre-tensioning method; the concrete is C50 concrete; the pre-stressed steel strand has a nominal diameter of $\phi^s 15.2$ mm, nominal area 140mm^2, tensile strength standard value of $f_{pk} = 1860MPa$ and modulus of elasticity $E_p = 1.95\times10^5 MPa$. The common steel bar is hot rolled HRB335 steel. Electronic level is used for the specimen to measure the displacement of the inverted arch in the hollow slab, and displacement strain sensor is used to measure the creep strain changes of the main girder in the mid-span.

Take the four angle points at the upper left comer, lower left comer, upper right comer and lower right comer on the cross section of the mid-span as the measuring points. In order to intuitively reflect the difference between the actual creep values and the theoretical values of the test beams at each measurement point, the actual creep strain data of the test beam and the theoretical values are plotted in a time history curve for purpose of comparison and analysis, as shown in figure 3.

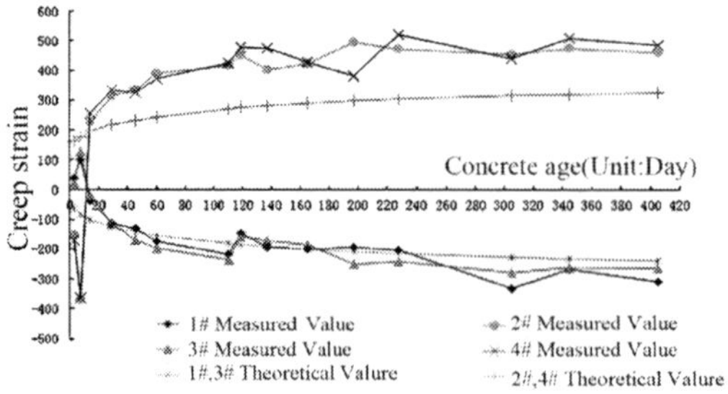

Fig. 3. Creep Strain-Time Curve of Test Beam.

4.3 Conclusion of shrinkage and creep test

(1) The shrinkage deformation of concrete specimen is faster in the early stage, and the shrinkage deformation of most specimen accounts for a large percentage up to 70% in the total shrinkage deformation in 1 month after the preparation of specimen. (2) Due to the existence of pre-stress, the creep strain of test beam shows a trend of gradual increase in the initial period and reversed development in the later period, and the actual creep development curve is different from the theoretical creep strain calculated by the creep coefficient stipulated in the JTG D62-2004. (3) At the $\varepsilon(t)$ moment, the strain of the concrete member can be expressed as follows:

$$\varepsilon(t)=\varepsilon_i(\tau)+\varepsilon_c(t)+\varepsilon_s(t)+\varepsilon_T(t)=\varepsilon_\sigma(t)+\varepsilon_n(t) \quad (2)$$

where, $\varepsilon_i(\tau)$ is the initial strain at loading time; $\varepsilon_c(t)$ is creep strain at the $t>\tau$ time, $\varepsilon_s(t)$ is the shrinkage strain, and $\varepsilon_T(t)$ is the temperature strain.

$\varepsilon_\sigma(t)=\varepsilon_i(\tau)+\varepsilon_c(t)$ is the strain generated by the stress; $\varepsilon_n(t)=\varepsilon_s(t)+\varepsilon_T(t)$ is the non-stress strain.

After entering the shrinkage and creep strain relationship; $\varepsilon_{s,c}(t)=\varepsilon_s(t)+\varepsilon_c(t)$, obtained from test into the finite element model, the related shrinkage and creep parameters in the theoretical calculation can be corrected to obtain the theoretical model which is closer to the actual working state, which makes the comparison analysis between the theoretical calculation and measured date become more reasonable.

5 ANSYS simulation analysis result of shrinkage and creep

5.1 Finite element model

The automatic grid division method is used to divide into different units and these units are the tetrahedron units without intermediate node. The number of units and nodes of the whole bridge is shown in the table 1, while the entity model is shown in the figure 4 (here only list the finite element model for the Scheme I due to the space limit).

Table 1. Number of Units and Nodes of each Scheme.

Scheme	Nodes number	Elements number
Scheme I	367029	159050
Scheme II	344045	147833
Scheme III	367994	159782

Fig. 4. Finite Element Mesh Division of Entity Model of Scheme I.

5.2 Simulation analysis result

The TB-FEM method is used to analyze the influence of shrinkage and creep on each scheme and the main calculation results are extracted as figure 5.

The three schemes in aspects of the maximum shrinkage, creep strain and its changes along the beam length when t=10 years. As shown in figure 5, in term of the maximum shrinkage and creep stress of the cross section in the 10th year, the Scheme III has the maximum stress, Scheme I has the medium stress and scheme II has the minimum stress; Scheme I and 2 have the similar stress. The shrinkage and creep stress values of scheme III are much larger than those of the other two schemes. The stress value of scheme III has a trend of gradual increase from middle to both ends, while the stress values of scheme I and 2 are relatively unstable, which means that the use of simply supported structure for bridge expansion can effectively reduce the spatial transfer of shrinkage and creep and decrease the stress value of shrinkage and creep. The stress values of the three schemes are basically consistent at the ends, which is because the scalability at ends can mitigate the influence of shrinkage and creep.

Figure 6 shows the stress comparison of the new and old bridge splicing belts in the 3 schemes. As shown in the figure 6, the scheme III has the maximum stress at the splicing belt, scheme II has the medium stress and Scheme I has the minimum stress. The effect of shrinkage and creep is larger in the initial period of bridge completion, and its growth rate gradually slows down with time. The stress values of splicing belts in the Scheme I, 2 and 4 increases slightly with time, but the growth rate is decreased with time. The influence of shrinkage and creep difference on the splicing belts of bridge is higher than that on the new bridge.

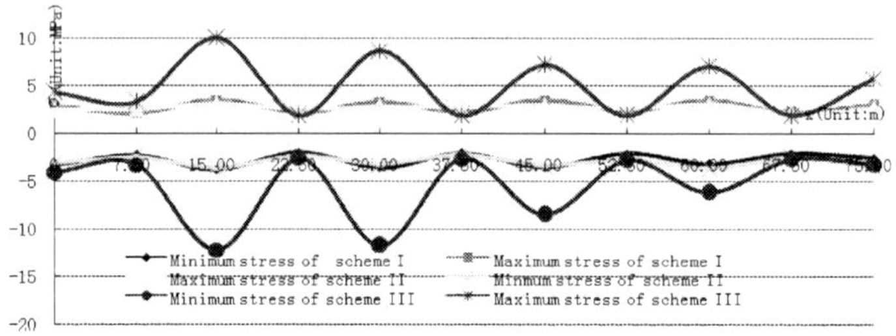

Fig. 5. Distribution of Maximum Stress along the Beam Length (10 years).

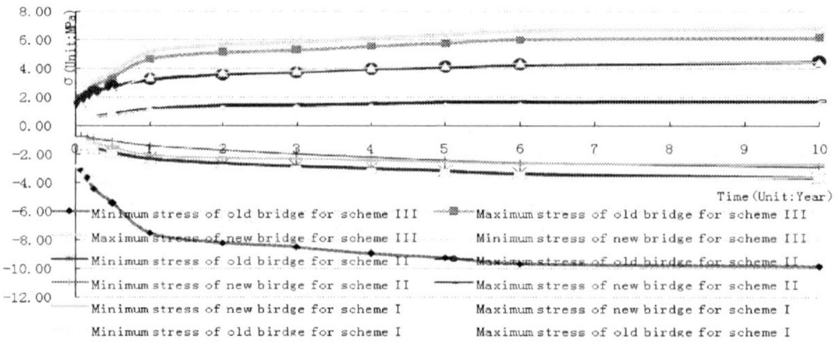

Fig. 6. Time-varying Diagram for the Maximum Stress at the Splicing Belts (10 years).

It can be seen from the above calculation that the stress value generated by the shrinkage in the Scheme III is too large, which will cause damage especially to the integrated layer; according to the calculation and comparison, the Scheme III cannot be considered as a rigid splicing scheme for the continuous hollow slab beam of a long joint structure.

6 Conclusion and prospect

The shrinkage test uses C50 concrete and C50 double-mixture concrete to make the standard prism concrete specimen; the creep test uses the 15m hollow slab which has the same material size, pre-stress and construction method with the real bridge. The strain-time curve obtained by shrinkage and creep is used to define the shrinkage and creep's characteristics of the material, and the TB method and finite element method are combined to analyze the mechanical response of the new and old bridge splicing structure caused by shrinkage and creep. The research shows that under the condition of rigid splicing, the new bridge should be supported by the simply supported structure, and it is not suitable to use the simply supported and continuous hollow slab. This study can provide a theoretical basis for the splicing time of new and old bridge as well as the type selection of new bridge structure. As the new and old bridge splicing scheme includes the type selection of new bridge structure, comparison and selection of transverse splicing scheme, comparison and selection of splicing time, etc., the proposed method in this project only studies the problem of type selection of new bridge structure under the condition of transverse rigid splicing of new and old bridge after considering the factor of shrinkage and creep difference. The identification of other factors and the decision of splicing scheme need further study.

Acknowledgments

The author would like to acknowledge the support from the Natural Science Foundation of China (Project number: 51608119）.

References

1. Du Huidong 2017 *Shrinkage and Creep & Foundation Settlement Theory based Research for Bridge Expansion Program* (Kumming: Thesis of Kunming University of Science and Technology).
2. TuBing, Fagzhi 2016 *Analysis of Shrinkage and Creep Effects on Newly-built and Existing Concrete Slab Bridge after Lateral Jointing* (China: China Journal of Highway and Transportation) Vol 29 No10 pp 66-76.
3. Wang P, Hu W, Zhang C. 2015 *Method to Analyze the Effect of Concrete Shrinkage and Creep on Widened T-Beam Bridges* International Conference on Intelligent Transportation, Big Data and Smart City.(Las Palmas de Gran Canaria) pp 511-514.
4. Tu B, Fang Z, Dong Y, et al. 2017 *Time-variant reliability analysis of widened deteriorating pre-stressed concrete bridges considering shrinkage and creep* (England: Engineering Structures) Vol 153 pp 1-16.
5. Sagara A, Pane I. 2015 *A Study on Effects of Creep and Shrinkage in High Strength Concrete Bridges* (England: Procedia Engineering) Vol 125 pp 1087-1093.
6. Wang J, Yao Y, Su M.2017 *Coupling Effect between Concrete Creep, Shrinkage and Temperature on Long Span Box Girder* (Shijiazhuang: Bridge Journal of Shijiazhuang Tiedao University) Vol 30 No1 pp 1-7.
7. Yang Yingen. 2016 *Differential Equation Method for Effects of Creep and Shrinkage in Steel-Concrete Composite Beam* (Kumming: Thesis of Kunming University of Science and Technology).
8. Han Chunxiu, Zhou Donghua, Yang Yanhua, Chen Xu,Yao Kaicheng 2016 *Use of an algebraic constitutive equation to calculate the stress redistribution due to creep and shrinkage of composite beams* (Harbin: Journal of Harbin Engineering University) Vol 37 No 8 pp 1041-1049.
9. Huang Ying, Gao Jie 2016 *Computational Analysis and Finite Element Simulation Study on Shrinkage and Creep of Pre-stressed Concrete Structures* (Sichuan: Journal

of Sichuan University of Science & Engineering (Natural Science Edition)) Vol 29 No 3 pp 75-80.

10. Zdeněk P. Bažant, Liisa Panula 1978 *A note of amelioration of the creep function for "improved dischinger method"* (England: Cement & Concrete Research) Vol 8 No 3 pp 381-386.
11. Zdeněk P. Bažant 2016 *Random Creep and Shrinkage in Structures: Ampling* Journal of (England: Structural Engineering) Vol 111 No 5 pp 1113-1134.
12. Ministry of Housing and Urban-Rural Development of the People's Republic of China (MOHURD) General Administration of Quality Inspection and Quarantine of the People ' s Republic of China 2009 *Standard for test methods of long-term performance and durability of ordinary concrete* (State Standard of the People's Republic of China : 2009-11-30 Published, 2010-07-01 Put into effect).

Preliminary evaluation of indoor thermal comfort in Malaysia heritage mosque

W F M Yusoff[1,2,*], and *N H* Ja'afar[1,2]

[1]Centre for Innovative Architecture and Built Environment, Faculty of Engineering and Built Environment, Universiti Kebangsaan Malaysia, Malaysia
[2]Secretariat for National Identity Architecture, Faculty of Engineering and Built Environment, Universiti Kebangsaan Malaysia, Malaysia

Abstract. Malaysia heritage buildings are the country's architectural gems. Among them is the religious building such as mosque. The Malaysia heritage mosque is normally built in response to the local culture and environment. Unlike most of Malaysia modern mosques where air conditioning is opted as ventilation mode, the heritage mosques achieve indoor thermal comfort via the passive architectural design. Due to that, this study was executed with the purpose of investigating the indoor thermal comfort of a Malaysia heritage mosque located at traditional street in Melaka, namely Masjid Tanah. The methodology used in the study was field measurement, in which the parameters measured were the indoor and outdoor air temperatures. The field measurement was conducted for two days in the end of March and early of April 2018, from 9 am to 4 pm. The indoor predicted comfort temperature was derived using the adaptive model. The findings indicated that the measured indoor air temperatures were lower than the indoor predicted comfort temperatures at most of the time. Therefore, it shows that the passive architectural design adopted by the Malaysia heritage mosque, which is Masjid Tanah, is able to provide indoor thermal comfort to the users.

1 Introduction

The Oxford Dictionary defines heritage as anything that is inherited, while for a heritage building, it is referred as a nation's historic building that is worth to be preserved [1]. However, according to Wilkinson and Remoy [2] , there is no explicit definition of heritage building, but the criteria represented by it consist of cultural, historical, inherent and symbolic values, as well as having relationship with the location. Meanwhile, in the Malaysia National Heritage Act 2005 [3], the word heritage is defined as a generic national heritage such as sites, objects or cultural heritage, whether it is registered or not in the National Heritage List. Therefore, from those definitions, heritage building can be summarized as any building that has heritage value to the people, place or nation. Traditional buildings are not necessarily indicated as heritage buildings as they are normally buildings of custom that represent the people, place and culture. Heritage

* Corresponding author: wardahyusoff@ukm.edu.my

buildings encompass many functions such as transportation buildings, factories, religious buildings, offices, apartments and many more [4].

Heritage buildings are normally situated at the traditional streets. This is because they normally possess strong relationship with the locations and context. Even, some of the buildings are the main factor that contribute to the growth of the population and township where they are located. During their early period of existence, they may become the centre for people's activities, such as the religious and communal buildings. These functions may remain till nowadays or may also change according to the needs and suitability. Traditional streets are the streets that are located in an old town that has historical value. Other characteristics that made a street becomes traditional street are the surrounding buildings and activities which have historical values, as well as the surrounding topography such as the presence of river, which once played significant role in the development of the area [5]. In Malaysia, the traditional streets usually characterized by the presence of old shophouses that encompass public walkway, which is also called five-foot walkway. The activities along the street are normally dominated by informal trading or retailing, which sometimes also occur at the five-foot walkway [6].

The studies of heritage buildings have been conducted since long time ago as people value the existence of such buildings in documenting the history of one civilization. Among the scope of studies are the building functions [7;4], the energy performance [8], the building structures [9], the indoor environment [10] and many more. A comfort indoor environment of a heritage building is of paramount importance as it affects the people's activities inside the building. Indoor thermal comfort is one of the indoor environmental parameters as it relates to the thermal condition of the indoor spaces, such as the hot and cold conditions.

People are stated to be in thermal comfort conditions when they are satisfied with the thermal environment. Though satisfaction is very subjective and vary among different people, efforts of identifying thermal comfort conditions are still ongoing till nowadays, using various thermal comfort models. During the initial study of thermal comfort, people used the PMV model developed by Fanger. This model was initially developed for temperate climate by conducting experiment in a controlled environment. With that controlled environment, this model is stated to be suitable for a steady-state environment only. It involves six parameters namely air temperature, relative humidity, mean radiant temperature, air movement, metabolic rate and clothing insulation [11]. Through time, people realize that humans have the ability to adapt themselves to the changes in the environment. With that awareness, another model was developed for thermal comfort study which is adaptive model.

The adaptive model was initially developed by Nicol and Humphreys with the realization that in actual condition, people adapt to larger range of temperature compared to the range predicted by PMV model. The ability to adapt depends on many factors such as climatic conditions, cultural background, as well as physiological and psychological aspects. In the adaptive model, the outdoor air temperature is indicated to have significant role in influencing the indoor comfort temperature [12]. Hence, in the adaptive model, people with different backgrounds and climatic conditions may have different ranges of thermal comfort.

The ability of people to adapt to the surrounding climate is proven via previous studies of thermal comfort such as by Caetano et al. [13] and Yusoff [14]. The findings of these studies indicated that people in hot and humid climate tend to accept higher air temperature as their comfort temperature. Caetano et al. [13] investigated office buildings in Brazil, where though they were air-conditioned, the workers still had freedom to control their indoor environment by opening the windows. The study adopted PMV model, in which based on the model, the derived indoor comfort temperatures were between 22.5 °C to 25.5 °C. Nevertheless, the study concluded that people in hot and humid climate able to tolerate and adapt to higher air temperature. This is due to the interesting findings where there were respondents who felt slightly cold or neutral at the indoor air temperature of above 26 °C. Meanwhile, in the study by Yusoff [14] which was executed in Malaysia, the number of respondents who felt neutral or slightly warm at 33 °C were equal, in which these numbers were the majority. In addition, majority of the respondents were also satisfied with such indoor air temperature condition.

As the PMV model was developed under a controlled environment, it is more appropriate for buildings that are air-conditioned [13]. On contrary, the adaptive model was stated to be more suitable for the naturally ventilated building which has a continuously changing environment [13;15]. Therefore, the selection of appropriate thermal comfort model depends on the condition of the investigated indoor environment. In this study, the adaptive model was employed instead of the PMV model as the investigated heritage mosque, which was Masjid Tanah, was designed to utilize natural ventilation.

In term of the equation, the PMV model is much more complex in comparison to the adaptive model. The adaptive model equation just requires the outdoor air temperature value only in deriving the neutral temperature or indoor predicted comfort temperature [11]. Among the adaptive model equation that can be utilized is the one that was developed by Humphreys et al. [16]:

$$Tn = 13.8 + 0.53\ (\pm 0.02)\ To \qquad (1)$$

where Tn is the neutral temperature or indoor predicted comfort temperature, and To is the average outdoor air temperature. All the temperatures are in Degree Celsius unit. In this study, the adaptive model equation developed by Humphreys et al. [16] was used to search the indoor predicted comfort temperature. This temperature is essential in having initial idea on the appropriate indoor temperature that provides comfort to the users.

For a heritage building, it is necessary to evaluate its present environmental performance. This is because the environmental performance of a building consists of parameters that respond directly to the current situation. The changes in the climatic and surrounding elements will influence the environmental performance. For example, the surrounding development such as urbanization will alter the surrounding context where the building is located, hence affects the climatic elements such as the wind and air temperature profiles of the area. Basically the environmental performance and energy usage of a building are not associated with the time it is constructed, but more to the architectural and technological solutions that are implemented in it [8]. Therefore, this study intends to examine the environmental performance of a heritage building by focusing on the indoor thermal comfort. The selected heritage building in this study is a mosque due to its significant function in a community. The findings from this study provide initial idea whether the architectural and technological solutions adopted by the heritage mosque have good respond to the local climate and environment throughout the ages.

2 The description of Masjid Tanah

Masjid Tanah was selected for this study due to its uniqueness in term of its history, location, simple layout plan, construction materials and three-tier pitched roof. Masjid Tanah, located in the district of Alor Gajah, Melaka, is one of the traditional mosques in Melaka that has been gazetted as Melaka heritage building by Melaka State Museum Corporation. It was originally built in 1800s by a Muslim scholar from Gujerat. However, in 1951 it had undergone some renovation to fulfil the needs and convenience of the users.

Its strategic location which is at the junction that connects the traditional streets, as well as besides a river has made this mosque one of a significant landmark in that area (figure 1). Moreover, the town where it is situated is also named after the mosque. Therefore, when people refer to Masjid Tanah, it can be either the town, which name is also Masjid Tanah, or the mosque itself. The mosque is also surrounded by old shophouses that accommodate heritage businesses such as jewellery shops, restaurants, traditional attire shops and many more. The streets at the area are called traditional streets as they are located in an old town that has historical value.

The word '*tanah*' in Malay language means soil. The mosque had been named as Masjid Tanah due to its walls that were constructed using mixture of egg white, sand and laterite. Hence, the overall thickness of the wall is 250mm, which is thicker than the normal brick construction nowadays. With all the uniqueness mentioned above, it is undeniable that Masjid Tanah is a valuable heritage to the country.

Fig. 1. The location of Masjid Tanah.

Masjid Tanah is considered a mosque with a simple layout plan compared to the modern Malaysian mosque layout nowadays. It has a praying hall area with the size of approximately 61 m^2, and surrounded by veranda (figures 2a). At the end of the praying hall is a '*mihrab'*, which is a small area that is usually designated for '*imam'* or leader of the prayers. The roof of the mosque is also unique as it is composed of three-tier pitched roof, made of clay roof tiles (figure 2b). This configuration of roof, which is the three-tier pitched roof, is very synonym with the traditional mosques in Melaka.

Among the passive design characteristics adopted by the mosque are the three-tier pitched roof that provides appropriate height between floor and roof level for good natural ventilation, as well as wide roof overhang, which is about 1 m length, that protects the indoor area from heavy rain and direct solar radiation. Moreover, the veranda all around the mosque also functions as buffer zone that prohibits the direct penetration of solar radiation into the praying hall (figure 2a). In addition, the presence of openings at all walls also allow cross ventilation to occur.

(a) (b)

Fig. 2. The veranda (a), and the three-tier pitched roof (b) of Masjid Tanah.

3 Research methodology

The research methodology employed in this study was field measurement, in which the parameters measured were outdoor and indoor air temperatures. Since the study was a preliminary evaluation, hence the measurement was executed for short duration only which was two days. The measurement was conducted on 31st March and 1st April 2018, from 9 am to 4 pm. The measuring tools used for the measurement were two sets of multiparameter ventilation meters namely TSI Velocicalc 9565 series (figure 3). The measuring tools were able to measure and record various parameters such as air temperature, relative humidity, air pressure and air velocity. In this study, the parameter measured was air temperature. The accuracy for air temperature measurement was ±0.3°C.

Prior to actual measurement, the measuring tools were calibrated by placing them in parallel, and at the same height, which was about 1 m above the floor level. The process of calibration was executed for 5 hours. The results of calibration showed that the differences in the air temperature readings between the two measuring tools were less than 0.5%.

During the actual measurement, the measuring tools were placed at the indoor and outdoor area as shown in figure 4. The locations were selected with the consideration of the safety of the measuring tools, as well as the reliability of the measured data. The measuring tools were located at 1.2 m height from the floor level, which was within the normal height of a human. The air temperature data was recorded every five minutes interval.

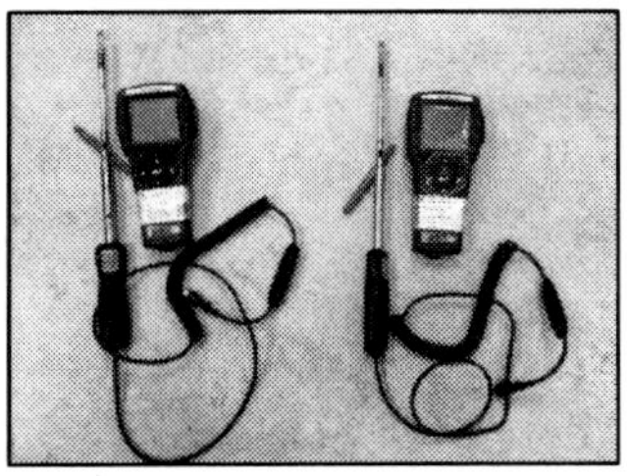

Fig. 3. The two sets of multiparameter ventilation meters namely TSI Velocicalc 9565 series.

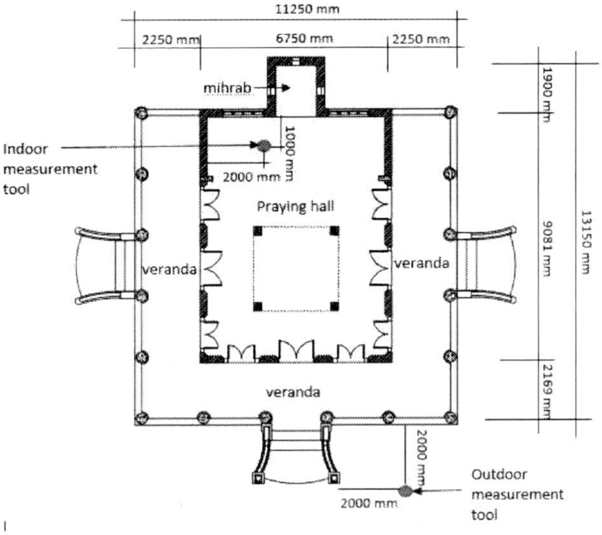

Fig. 4. The locations of the indoor and outdoor measuring tools.

4 Results and discussion

The average outdoor and indoor air temperatures for two days measurement were presented in table 1. The data tabulated in table 1 shows that the air temperature differences between https://doi.org/10.1051/matecconf/201927702016

the outdoor (To) and indoor (Ta) were less than 7 °C at all time. The highest air temperature difference was 6.8 °C, which was at 11 am, whilst the lowest was at 4 pm, which was 0.8 °C. Hence, the data indicated that the heat accumulated inside the building through time, where in the late afternoon, the indoor air temperature was almost similar to the outdoor air temperature. Nevertheless, the indoor air temperature was still lower than the outdoor air temperature at all time.

Table 1. The air temperature differences between the outdoor (To) and indoor (Ta).

	Outdoor air temperature, To (degC)	Indoor air temperature, Ta (degC)	Differences (To – Ta) (degC)
9 am	35.9	29.5	6.4
10 am	36.1	30.2	5.8
11 am	37.8	30.9	6.8
12 pm	37.3	31.6	5.7
1 pm	35.8	31.9	3.9
2 pm	34.9	32.1	2.8
3 pm	33.8	32.2	1.6
4 pm	32.9	32.1	0.8

The results depicted in figure 5 were derived using the adaptive model equation developed by Humphreys et al. [16]. The results indicated that during the morning hours and until 1 pm, the measured indoor air temperatures (Ta) were lower than the indoor predicted comfort temperatures (Tn). However, starting from 2 pm and above, the results were vice versa.

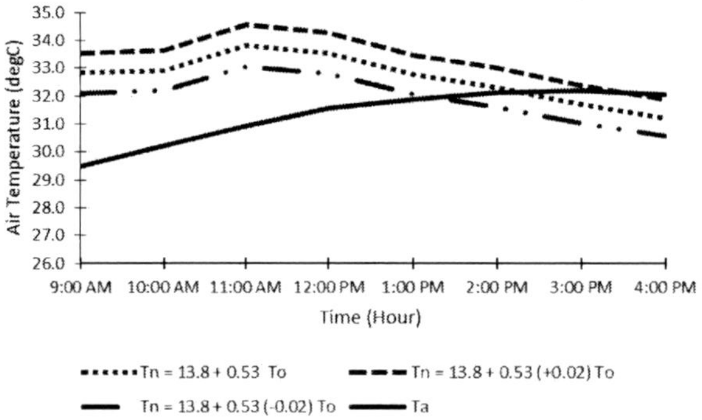

Fig. 5. The comparison between the indoor predicted comfort temperature (Tn) and measured indoor air temperature (Ta).

Therefore, the results depicted in table 1 and figure 5 show that thermal comfort was able to be achieved for the indoor environment of Masjid Tanah, especially during the morning hours. Though the outdoor air temperatures demonstrated high values during the morning hours, which were more than 35 °C, the indoor environment of Masjid Tanah was still comfortable for the users. The highest measured indoor air temperature recorded during the study was at 3 pm, which was 32.2 °C. This value is still lower than the accepted comfort temperature by the respondents in the study by Yusoff [14], which was 33 °C.

The comfortable indoor environment of Masjid Tanah is due to the presence of veranda all around that prevents direct penetration of solar radiation into the praying hall. However, the accumulation of heat through time had made the indoor environment slightly warmer in the afternoon, though there was reduction of outdoor air temperature compared to the morning hours.

5 Conclusion

The study shows that heritage building is among the best example to learn about local architectural characteristics that have good respond to the local climate and environment. The passive design characteristics of Masjid Tanah that respond to the local tropical climate of Malaysia have enabled the indoor environment of the praying hall to be in thermal comfort condition most the time. Its simple layout plan with no indoor partition, as well as the presence of many openings also allow good cross ventilation to occur. The existence of veranda also prohibits the penetration of direct solar radiation into the praying hall. With the evolution of mosque design nowadays, it is important to note that the shading of the facades and indoor spaces from direct solar radiation, as well as effective cross ventilation must be highly considered in achieving indoor thermal comfort in hot and humid climate. The globalization and modernization have made these two aspects sometimes ignored in the building design nowadays.

It is also important to note that this study is just a preliminary evaluation of thermal comfort condition in a heritage mosque via field measurement and adaptive model equation. Regarding to the positive results derived from this study, it is suggested in the future to extend this study to a more detail evaluation by prolonging the duration of the study, as well as conducting questionnaire survey with the users. The comparative analysis between the field measurement and the questionnaire survey will provide more solid findings in determining the thermal comfort condition inside a heritage mosque.

The findings from this study is also hoped to become motivation to investigate in detail the indoor environment of heritage building. Though it is already known that heritage buildings normally have good respond to the local climate and environment, the detail investigation using measurement tools will enhance the understanding of the actual condition. With deep understanding, it will help the effective implementation of heritage building characteristics at modern buildings.

Acknowledgements

The authors would like to express great gratitude to Ministry of Education Malaysia and Universiti Kebangsaan Malaysia for funding this study through research grant FRGS/1/2015/SSI11/UKM/02/2 and GGP-2017-046, respectively. The authors would also like to thank Center of Excellence for Engineering and Technology (CREATE), Public Works Department of Malaysia.

References

1. Thompson D 1996 *The Concise Oxford Dictionary of Current English* Fifth ed. (London: BCA) p 634
2. Wilkinson S, Remøy H 2017 Heritage building preservation vs sustainability? Conflict isn't inevitable [cited 2018 Jul 25]. Available from: http://theconversation.com/heritage-building-preservation-vs-sustainability-conflict-isnt-inevitable-83973
3. Law of Malaysia 2006 *National Heritage Act 2005* Percetakan Nasional Malaysia Bhd.
4. Darmawan E and Enis A S 2016 The changing function of heritage building for roling tourism planning progress *Proceedia-Social Behav Sci* **227** 712–9
5. Ja'afar N H, Sulaiman A B and Shamsuddin S 2012 Traditional street activities in Kuala Lumpur city centre *Int J Multidiscip Thoughts* **2** 93–105
6. Shamsuddin S and Ujang N 2008 Making places: The role of attachment in creating the sense of place for traditional streets in Malaysia *Habitat Int.* **32** 399–409

7. MIslrlIsoy D and Günçe K 2016 Adaptive reuse strategies for heritage buildings: A holistic approach. *Sustain Cities Soc.* **26** 91–8
8. Fabbri K, Zuppiroli M and Ambrogio K 2012 Heritage buildings and energy performance: Mapping with GIS tools. *Energy Build* **48** 137–45
9. Llorens J and Zanelli A 2016 Structural Membranes for Refurbishment of the Architectural Heritage. *In: Procedia Engineering* **155** 18 – 27
10. Martinez-Molina A, Boarin P, Tort-Ausina I and Vivancos J L 2018 Assessing visitors' thermal comfort in historic museum buildings: Results from a Post-Occupancy Evaluation on a case study *Build Environ.* **132** 291–302
11. Djamila H 2017 Indoor thermal comfort predictions: Selected issues and trends *Renew Sustain Energy Rev* **74** 569–80
12. Carlucci S, Bai L, de Dear R and Yang L 2018 Review of adaptive thermal comfort models in built environmental regulatory documents *Build Environ* **137** 73–89
13. Caetano D S, Kalz D E, Lomardo L L B and Rosa L P 2017 Evaluation of thermal comfort and occupant satisfaction in office buildings in hot and humid climate regions by means of field surveys *Energy Procedia* **115** 183–94
14. Yusoff W F M 2017 Thermal comfort at Semi Open Plaza in Commercial Building: The Case Study of Bangi Gateway Shopping Complex. In: *Putrajaya International Built Environment, Technology and Engineering Conference (PIBEC3)* Putrajaya p 30–8
15. He Y, Liu M, Kvan T and Peng S 2017 An enthalpy-based energy savings estimation method targeting thermal comfort level in naturally ventilated buildings in hot-humid summer zones *Appl Energy* **187** 717–31
16. Humphreys M A, Rijal H B and Nicol J F 2013 Updating the adaptive relation between climate and comfort indoors; new insights and an extended database *Build Environ* **63** 40–55

MATEC Web of Conferences **277**, 02017 (2019) https://doi.org/10.1051/matecconf/201927702017
JCMME 2018

Planning and formulation of environmental management system for Indian infrastructure projects

C Akin[1,*], *R Samuel* Devadoss[2], and *V Vandhana* Devi[3]

[1]Assistant Professor, Department of Civil Engineering, KCG College of Technology, Chennai-600097, India
[2]Associate Professor, Department of Civil Engineering, Hindustan Institute of Technology and Science, Chennai-603103, India
[3]Associate Professor and Head of Department, Department of Civil Engineering, KCG College of Technology, Chennai-600097, India

Abstract. Every year the impact on natural environment is more due to consumption of natural resources for large infrastructure and construction projects. Even though by implementing modern construction techniques and low cost effective materials the volume of impact reduction is not efficient. India is a country which has enormous natural resources and it has been utilized for large infrastructure and construction projects which makes an environmental impact. This present paper focuses on planning and formulation of Environmental Management System (EMS) to reduce the environmental impact due to large infrastructure and construction projects. The planning of EMS is done based on the environmental impact checklist provided by Ministry of Environmental and Forest Climate change (MoEF&CC) under Government of India and the Formulation is done through PERT and CPM techniques which is to implement EMS in short duration to reduce environmental impact through large infrastructure and construction projects. Finally, the Planning and Formulation of EMS gives solution to reduce more volume of environmental impacts due to large infrastructure and construction projects.

1 Introduction

The Natural resources are the root cause for the aesthetics and sustainability of the earth. The basic needs of the living beings have to be fulfilled by the consumption of natural resources. As the modern era arrives more infrastructure development has been taking place all over the world in terms of Transportation systems, Tunnels, building constructions, etc. The infrastructure projects are the important consumers of natural resources especially through deforestation, Quarrying of huge rocks and mountains, Ground water consumption and finally it creates Environmental impact to Land, Air and Water Environment through Soil erosion, Air pollution and Carbon footprints which causes Global warming and Hazards. Every year the environmental impact increases due to consumption of more

*Corresponding author: akin.akin766@gmail.com

natural resources for large infrastructure and Construction Projects. To reduce environmental impact Modern Construction techniques such as grouting and guniting techniques have evolved and low cost effective materials such as fly-ash, slag in replacement of cement has not proved its overall efficiency in reducing Environmental Impact. India is a Developing Country which has enormous amount of natural resources covering its huge mass of land area and with huge population. Infrastructure projects are an important source for the development of Nations growth through development of Roads, Highways, Dams and Tall and Multi storied buildings.

Therefore the utilization of natural resources is in need for the development of Infrastructure projects. Government of India has taken initiatives to mitigate environmental impact and provided guidelines through various research activities [1], [2]. Ministry of Environment, Forest and Climate Change (MoEF&CC), Government of India has provided environmental impact checklist for reducing the environmental impact for large construction projects in many aspects considering the Land environment, Water environment, Air Environment, Aesthetics and Energy conservation techniques. The Environmental impact checklist of MoEF &CC can be followed systematically by applying Environmental Management System (EMS) in which it saves time and cost of the project simultaneously [1], [2], [3]. EMS is an ISO 14001 concept which consists of Plan, Do, Check, Act also known as PDCA cycle which is used to save time, cost and energy [2], [4], [5]. In this Present Paper, Planning of Environmental Management System for infrastructure projects is done based on the Environmental impact checklist given by MoEF&CC at Annexure 2 of form 1A for construction projects listed under item 8 of the schedule and Formulation of EMS is done through Programme Evaluation Review Technique (PERT) and Critical Path Method (CPM) to predict duration, variance, deviation and critical path which is used to implement EMS in short duration to reduce environmental impact through large infrastructure and construction projects [1], [2], [3]. Finally, the environmental impact checklist of MoEF & CC with respect to proper planning and formulation of EMS the time, cost and energy can be saved and simultaneously it helps to reduce more volume of environmental impact for large infrastructure and construction projects in India.

2 Methodology

The methodology used here for planning and formulation of EMS in Indian Infrastructure and construction project has been classified typically into two steps; one is by planning and other by formulating. The first step is by Planning the stages of environmental impact due to an infrastructure and construction project with steps taken to mitigate by applying the EMS concept of Plan, do, check and act process. The stages of Environmental impact were obtained from the checklist provided in MoEF & CC at Annexure 2 of form 1A for construction projects listed under item 8 of the schedule [1], [2], [6]. The Second step is by formulating the planned activities of the first step through PERT and CPM to implement the EMS process of reducing environmental impact within the estimated time duration and to predict the variance, standard deviations and the critical path of the overall process of applying EMS [2], [7].

3 Results and discussion

Based on the environmental impact checklist the first step planning was done to mitigate the impact due to a large infrastructure and construction project by applying the EMS process. Table 1 shows the method of planning based on environmental impact checklist by applying EMS. The stages of Environmental impact were obtained from the checklist provided in MoEF & CC at Annexure 2 of form 1A for construction projects listed under item 8 of the schedule.

The Second step is by formulating the planned activities of the first step through PERT and CPM to implement the EMS process of reducing environmental impact within the estimated time duration and to predict the variance, standard deviations and the critical path of the overall process of applying EMS [2], [3], [8]. Table 2 shows the estimated time duration in days for implementing EMS. Where the optimistic time estimate was denoted as 'to', most likely time estimate was denoted as 'tm' and pessimistic time estimate was denoted as 'tp'. Table 3 shows the expected duration in days which was denoted as 'te' and the expected variance was denoted as 'σ' for applying EMS. Figure 1 shows the network diagram of the activities in the environmental impact checklist for the EMS process to reduce the environmental impact caused by a large infrastructure and construction project within the estimated time duration.

Table 1. Planning method based on Environmental impact checklist by EMS for construction projects.

Activity	Environmental Impact Check list	Plan	Do	Check	Act
1	Land Environment	Land use	Earthwork	Disturbances	Plant trees/ Solar
2	Water Environment	Water Requirement	Waste Water treatment	Water balance	Water Recharge
3	Vegetation	Local ecosystem	Site Modification	Threats	Tree Plantation/ Create Water bodies
4	Fauna	Displacement of fauna	Create barriers	Impacts	Corridors/ fish ladders
5	Air Environment	Atmospheric concentration	Transport infrastructure	Movement Patterns	Traffic Management
6	Aesthetics	Vicinity	Urban design	Obstruction	Landscaping
7	Socio-Economic Aspects	Local population	Social infrastructure	Adverse effect	Safeguard local community
8	Building materials	Embodied Energy	Energy Efficient process	Pollution	Recycle materials
9	Energy Conservation	Power requirement	Renewable Energy consumption	Energy balance	Non conventional energy technologies
10	Environmental Management Plan	Construction Phase	Operation Phase	Environmental Regulations	Mitigation Measures

Table 2. Estimated time duration (in days) for implementing EMS.

Activity	to	tm	tp
1-2	3	4	5
2-3	1	2	3
2-4	2	3	4
3-5	3	4	5
4-5	1	3	5
4-6	3	5	7
5-7	4	5	6
6-7	6	7	8
7-8	2	4	6
7-9	1	2	3
8-10	4	6	8
9-10	3	5	7

Table 3. Expected duration in days and Expected Variance for implementation of EMS.

Activity	to	tm	tp	Expected duration $te = \frac{to + 4tm + tp}{6}$	Expected Variance $\sigma^2 = (\frac{tp-to}{6})^2$
1-2	3	4	5	4	0.11
2-3	1	2	3	2	0.11
2-4	2	3	4	3	0.11
3-5	3	4	5	4	0.11
4-5	1	3	5	3	0.44
4-6	3	5	7	5	0.44
5-7	4	5	6	5	0.11
6-7	6	7	8	7	0.11
7-8	2	4	6	4	0.44
7-9	1	2	3	2	0.11
8-10	4	6	8	6	0.44
9-10	3	5	7	5	0.44

From table 3, it is observed that the expected duration of the activities in environmental impact checklist for EMS process implementation is 29 days which saves time and cost and simultaneously it reduces environmental impact for large infrastructure and construction projects.

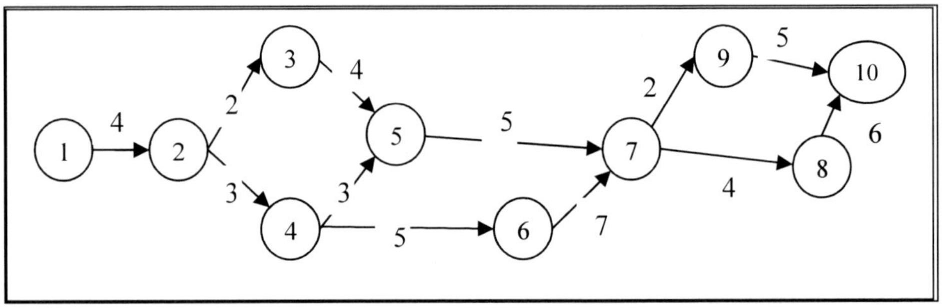

Fig. 1. Network diagram of the activities in environmental impact checklist for EMS process within the estimated time duration.

Therefore from the expected duration, it is observed that the critical path of the overall EMS process as provided in the network diagram at figure 1 is 1-2-4-6-7-8-10. The expected variance of the length of the EMS process will be the sum of the expected variances of all the critical activities in environmental impact checklist for EMS process. Therefore, the expected variance of the length for implementation of EMS process = 0.11+0.11+0.44+0.11+0.44+0.44 =1.65 and the expected standard deviation of the completion time of the EMS process = $\sqrt{1.65}$ = 1.29

4 Conclusion

The present paper helps to reduce environmental impact for Indian infrastructure and construction projects. The planning and formulation technique is applied to implement EMS process for environmental impact checklist provided by MoEF & CC in which each and every stages of impact are mitigated through Plan, do, check and act process and the EMS process was formulated using PERT and CPM techniques and it was identified that the expected time duration for implementation of EMS process for Indian large infrastructure and construction project was 29 days and approximately 30 days. The expected variance and standard deviation was also identified successfully. Hence, Planning and formulation of Environmental Management System for Indian Infrastructure project can reduce major environmental impact and also it can save time, cost and energy of any infrastructure and construction project.

Acknowledgement

The environment impact checklist was obtained from Manual on norms and standards for environment clearance of large construction projects. The Authors wish to thank Ministry of Environmental and forest climate change, Government of India for contributing the manual to the Nation's development in reducing the environmental impact due to large construction projects. The Author Akin C wish to thank Mr. Pitchi Rajan M, Assistant Professor and Mr. Hameedul Irshad, Assistant Professor, KCG College of Technology, Chennai, India for their friendly support.

References

1. Manual on norms and standards for environment clearance of large construction projects, (2006) New Delhi, India.
2. Sundaresan, V, Ganapathy Subramanian, K.S, Ganesan, K. (2000), "Resource Management Techniques", A.R. Publications, Tamil Nadu, India.
3. Gwen Christini, Michael Fetsko, and Chris Hendrickson, (2004), "Environmental Management Systems and ISO 14001 Certification for construction firms", Journal of construction engineering and Management, DOI: 10.1061/ (ASCE)0733-9364(2004)130:3(330).
4. Mohamed Marouen Amiri, Hédi Noubbigh, Kamel Naoui and Nouha Choura, (2015), "Environmental Management System: Environmental Impacts and Productivity", International Journal of Business and Management, doi:10.5539/ijbm.v10n11p107.
5. Aydemir Akyürek and Osman Nuri Ağdağ, (2017), "Evaluation of Environmental Management System Implementation in Construction Projects", European Scientific Journal, July 2017 /SPECIAL/ edition.
6. Report of the comptroller and Auditor general of India on environmental clearance and post clearance monitoring, (2016), Indian Audit and Accounts department, Ministry of Environment, forest and climate change, Government of India, Report No. 39 of 2016.
7. Xianbo Zhao and Natee Singhaputtangkul, (2015), "Effects of Firm Characteristics on Enterprise Risk Management: Case Study of Chinese Construction Firms Operating in Singapore", Journal of Management in Engineering, DOI: 10.1061/(ASCE)ME.1943-5479.0000434.
8. S. X. Zeng, C. M. Tam, Z. M. Deng and Vivian W. Y. Tam, (2003), "ISO 14001 and the construction industry: Survey in China", Journal of Management in Engineering, DOI: 10.1061/ (ASCE)0742-597X(2003)19:3(107).

Numerical limit analysis of the ultimate load bearing capacity of a pier foundation under complex load

H F Wang*, and *X H* Cheng

Department of Civil Engineering, Tsinghua University, Beijing 100084, China

Abstract. A pier foundation has plenty of advantages as the foundation form for large onshore wind turbines in the terrain of the Gobi desert and collapsible loess areas. The ultimate load bearing capacity design, as an important part in the design phase of this foundation form, is not sufficiently in terms of design theory and the checking method, especially for application in collapsible loess areas. In this paper, numerical limit analysis has been employed to analyze the load bearing capacity for the scheme of selection of a pier foundation of a 100 MW wind farm in certain collapsible loess geology in Ningxia. The results were then compared with the empirical formulas, the limit equilibrium solutions and the finite element solutions to verify the accuracy of the results. It has been demonstrated that under the ultimate engineering load, the pier foundation can meet the stability requirement. In the ultimate state, whether the surrounding soil of the foundation falls in plasticity is associated with the ratio of the depth to the diameter of the foundation. The numerical limit analysis method can automatically determine the failure surface.

1 Introduction

Due to advantages like rich resources, a good economic effect and low pollution, wind power has gained extensive attention worldwide.

According to the forces acting on it, the conventional foundation of a wind turbine includes a gravity type foundation, a pier foundation and a pile foundation. The gravity type foundation counteracts the external forces and bending moments acting on the wind turbine using the weight of the concrete and backfill and the corresponding couples. A pile foundation can be further classified as an ordinary reinforced concrete pile and rock micropile foundation, relying on the weight of the foundation's platform together with the pile soil to counteract the external loads. These two foundation forms are common and have been widely applied with relatively complete design methods. As shown in Fig.1,the pier foundation originated in America [1, 2], with a general burial depth of 8 m to 11 m and a top width of the foundation between 5 m and 9 m. The overall geometry is an inverted cylinder, relying mainly on the force couple caused by the interaction of the soil pressures of the in-situ soil and the foundation's concrete to counteract the external loads.

* Corresponding author: timwang517@tsinghua.edu.cn

Large-scale development of onshore wind power has occurred relatively late in China, and imported technology has played the main role in its development. In regions with good soil properties like Xinjiang, the pier foundation has broad applications. However, as for the collapsible loess areas, the pier foundation should be applied with caution due to the mechanical mechanisms and limitations in the calculation methods.

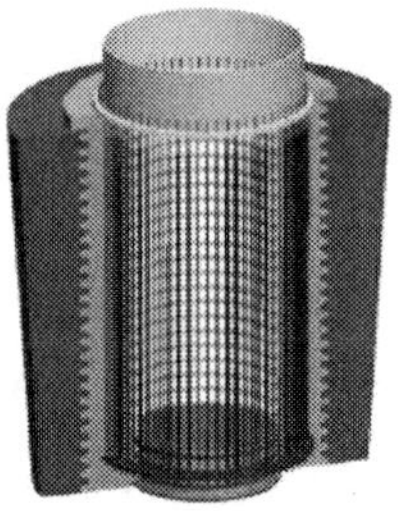

Fig. 1. Typical situation of a pier foundation for a wind turbine.

At present, the analysis methods for the ultimate anti-overturning load bearing capacity of a pier foundation mainly include the limit equilibrium method [3, 4], the plastic upper limit analysis [5, 6] and the finite element method [7-10]. Generally, the limit equilibrium method considers the foundation and the soil mass inside it as a rigid entity and performs a series of approximation and assumptions relating to key parameters such as soil pressure distribution and depth of the center of rotation. Engineering experience indicates that this method can reasonably estimate the foundation's stability, but the accuracy cannot be guaranteed theoretically. The plastic upper limit analysis employs the kinematically admissible velocity field to satisfy the constraints, including the strain compatibility, the velocity boundary conditions and the associative flow rules, and the theoretical basis is strict. Nevertheless, the analytical capability of this method is limited since it is difficult to perform load bearing capacity analysis for complex foundations or load conditions. The finite element method has distinct advantages in dealing with complex problems, which is one of the major methods used for the performance analysis of a pier foundation. However, in the design phase, an error range between the finite element calculation value and the actual engineering load bearing capacity is not foreseeable, and it is difficult to properly select the constitutive model and parameters relating to the soil mass for specific engineering projects.

Based upon the conventional limit analysis theorem, the numerical limit analysis is a numerical method that utilizes finite element discretization technology to discretize the soil mass and the structure, it establishes the kinematically admissible velocity field (upper limit) and the static admissible stress field (lower limit), and applies a planning algorithm to obtain the maximum (upper limit) and minimum (lower limit) ultimate load. This method has the following advantages: (1) without assumption of the failure mode, it automatically acquires the failure mechanism, the plastic zone of the soil and the stress distribution through program optimization; (2) it has the capability of dealing with complex loads and complex geometric conditions with the finite element method, and has a strict theoretical basis for limit analysis; (3) it envelops the true results by analysis of the upper and lower limits and is capable of providing the error range.

Numerical element limit analysis concept was proposed by Lysmer in 1970 [11]. Sloan introduced the steepest edge active set algorithm in 1988, so that the efficiency of the algorithm was greatly improved [12]. Sloan [13], as well as Sloan and Kleeman [14], calculated the undrained vertical load bearing capacity of a strip foundation by applying upper limit and lower limit analyses of the finite element limit, respectively. Ukritchon et al. [15, 16] calculated the undrained ultimate load bearing capacity and stability of a foundation pit support for a strip foundation under complex force conditions by applying the numerical limit analysis. Theoretical study and the engineering application of numerical limit analysis started late in China. Yang Feng et al. [17] analyzed the stability of a shallow tunnel by upper limit analysis of the numerical limit. Huang et al. [18, 19], as well as Li and Wang [20] calculated the load bearing capacity and stability of a side slope for a strip foundation by using the lower limit analysis of the finite element limit, respectively.

The correct treatment of the contact relationship (tangential and normal phases) on the bucket-soil interface is one of the key factors in the accuracy of the simulation. The friction strength of the bucket-soil interface is associated with the material properties of the bucket and the soil, which can be measured using the model test. In the present work, the friction strength of the bucket-soil interface and the undrained shear strength were assumed to be different.

Nowadays, researchers at home and abroad have developed three-dimensional numerical limit analysis technology [19, 21], but the analyses are limited when dealing with simple problems like a strip foundation and a homogeneous soil layer due to limited processing capacity. A relatively mature two-dimensional plane strain program can analyze more complicated problems with relatively illustrative results. Some previous investigations have proved that there is a slight difference between the load bearing capacity of a complex load of a shallow round foundation acquired from three-dimensional analysis, and that acquired from a plane strain analysis [10]. Therefore, the work in this paper employed the plane strain numerical limit analysis.

2 A brief introduction to the principles of numerical limit analysis

The principles of numerical limit analysis fall into two parts: the finite element discretization method and the limit analysis solution. A brief introduction to the principles of both the lower limit analysis and the upper limit analysis has been given below.

The simplified analysis model for calculating the ultimate load bearing capacity of a pier foundation by using numerical limit analysis has been shown in figure 2(a). The top of the foundation was aligned with the ground, and the foundation was in the soil. The center of the top of the bucket was subject to gravity (V), the horizontal force (H) and the overturning moment (M). A rigid-plastic assumption was used for the soil. Since the analysis mainly considered the interaction between the soil and the structure rather than the bending of the steel structure itself, the foundation was assumed to be rigid, that is, 1) a single structural element is rigid without internal strain; 2) the adjacent structural elements share the same angle of rotation and displacement at the same node coordinate without considering a plastic hinge.

2.1 Principles of the lower limit analysis

In the lower limit analysis, the soil was discretized as 3-node constant-strain-rate triangle elements. As shown in figure 2(a), each node had three stress components, and each stress component was assumed to be linear within the element. This was different from the conventional finite element, in finite element limit analysis each node belongs and only belongs to one element (i.e., multiple nodes may share the same coordinates), so that the stress discontinuity is caused at the boundary of the adjacent elements. A suction caisson was discretized as 2-node beam elements and a single-node plastic hinge element (without real dimensions); each node belonged to one element and had three degrees of freedom (two acting forces and one bending moment). It was supposed that the stress was linearly distributed along the structural elements. The configuration of the plastic hinge unit in the upper and lower limit analyses was there to simulate a rigid-plastic structure. As the foundation was assumed to be rigid, no plastic hinge did arise.

In the lower limit analysis, the static admissible stress field was established, and it could be taken as a linear planning model to solve the objective function (see table 1). The constitutive relationship of soil used the Tresca yield criterion. Since a linear planning algorithm as used, the linearization treatment of the yield function as performed using the inscribed regular 24-gon (lower limit) and the circumscribed regular 24-gon (upper limit) (figure 2(c)). For the detailed formulas and analysis procedures, refer to Sloan [13].

The objective function of the lower limit analysis is the maximum external load that satisfies the conditions of the static admissible stress field. Regarding the pier foundation, the maximum external load was from the transverse force and the overturning torque was transferred from the upside of the wind turbine and gantry, and both resulted from the effect of the wind on the structures.

2.2 Principles of the upper limit analysis

In the upper limit analysis, the discretization forms of the soil and structure were the same as those in the lower limit analysis (figure 2(b)). Each node of the soil element and beam element had two velocity components, and the node of the plastic element had three velocity components. The distribution of the velocity field was linear in element. Each node independently belonged to one element, and thus a velocity discontinuity was allowed at the boundary of the adjacent elements.

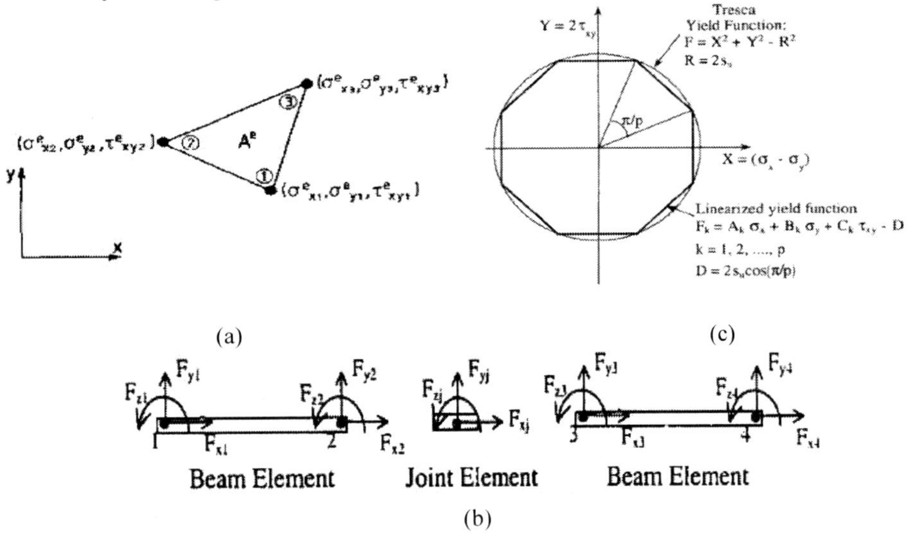

Fig. 2. Problem discretization for caisson foundation and summary of plain-strain elements in numerical limit analyses: (a) Lower bound discretization for soil elements, (b) Lower bound discretization for structural elements, (c) Tresca yield criterion for lower bound analyses.

In the upper limit analysis, a kinematically admissible velocity field was established, and a minimal external force meeting the requirements was obtained after the planning (see table 1). For detailed formulas and procedures, refer to Sloan and Kleeman [14].

The objective function of the upper limit analysis is obtained by equalizing the work of the external and internal forces. Especially for the problems analyzed in this paper, the external forces could be classified as: (1) weight of the soil and the bucket, (2) horizontal load and overturning torque at the top of the bucket; and the internal forces could be classified as: (1) plastic deformation inside the soil element and between elements, (2) slip and cracks at the soil-structure interface.

Table 1. Solution of the numerical limit analyses.

Limit analyses	Lower	Upper
Objective function	$\mathrm{MAX.}\left\{\mathbf{C}_1^{\mathrm{T}}\boldsymbol{\sigma}\right\}$	$\mathrm{MIN.}\left\{\mathbf{C}_2^{\mathrm{T}}\dot{\boldsymbol{\lambda}} + \mathbf{C}_3^{\mathrm{T}}\mathbf{U}^{\pm}\right\}$
Constraint function	$\begin{bmatrix} \frac{\partial}{\partial x} & \frac{\partial}{\partial y} \\ \frac{\partial}{\partial y} & \frac{\partial}{\partial x} \end{bmatrix} \begin{bmatrix} \sigma_x & \tau_{xy} \\ \sigma_y & \tau_{xy} \end{bmatrix} = \begin{bmatrix} 0 \\ \gamma \end{bmatrix}$	$\dot{\varepsilon}_x = -\frac{\partial u}{\partial x} = \dot{\lambda}\frac{\partial F}{\partial \sigma_x}, \dot{\varepsilon}_y = -\frac{\partial v}{\partial y} = \dot{\lambda}\frac{\partial F}{\partial \sigma_y}$ $\dot{\varepsilon}_{xy} = -\left(\frac{\partial v}{\partial x} + \frac{\partial u}{\partial y}\right) = \dot{\lambda}\frac{\partial F}{\partial \tau_{xy}}, \dot{\lambda} \geq 0$ $F = X^2 + Y^2 - R^2 = 0$
	$\sigma_{n1} = \sigma_{n2}, \tau_1 = \tau_2$	$\Delta u_n = \lvert\Delta u_t\rvert \tan\phi, \lvert\Delta u_t\rvert = u_t^+ + u_t^-$ $\Delta u_t = u_t^+ - u_t^-, u_t^+ \geq 0, u_t^- \geq 0$
	$F = X^2 + Y^2 - R^2 \leq 0$	$u_1 = d_1, v_1 = p_1$
	$\sigma_{n1} = q_1, \tau_1 = t_1$	$W_{\mathrm{ext}} = \int_L \sigma_n v_n dL + \int_A \gamma v dA = W_{\mathrm{int}}$

3 Engineering example

3.1 Overview of the engineering and numerical simulation

A wind farm has been proposed for the Ningxia Hui Autonomous Region, China, with a scale of 100 MW. The stratum is loess-like soil (Q41). The loess-like silt is soil with medium-high compressibility, and it is characterized by its collapsibility, and is classified as Class-I (slight) non-weight collapsible loess to Class-II (medium) self-weight collapsible loess. The stratum is continuously distributed in the field with a thickness of 0.5 - 10.5 m. The main physical and mechanical indicators of each specific layer of rock and soil have been listed in table 2.

For the foundation of a wind turbine, the vertical load is its own weight, and the horizontal load and the overturning torque is caused by the forces from the wind acting on the structure. As provided by the supplier of the wind turbine, the weight of the wind turbine and the associated structure was 3,891 kN, and the ultimate horizontal load and the ultimate overturning load at the bottom of the structure were 854 kN and 71,413 kN.m, respectively (combined and corrected according to the norm [22]).

Table 2. Parameters of the soil strata.

Name of soil layer	Coefficient of compressibility a_{1-2}	Volume weight g/(kN·m^{-3})	Saturated quick shear test		Modulus of compressibility /MPa
			Cohesion c_q/kPa	Angle of internal friction j/(°)	
Loess-like silt -1	0.39	15.3	7	24.1	5.04
Loess-like silt -2	0.218	17.7	8.4	21.7	8

This paper has proposed a standard selection of the type of foundation and has conducted three transformation modes for different foundation dimensions according to other methods, with consideration of the requirements of penetrating the wet limit layer, so as to analyze the effects of increasing the diameter and rim grout on the foundation's load bearing capacity. One of the main advantages of numerical limit analysis is that a true value can be enveloped within the upper and lower limit solutions.

Model-1: D_1 (upside diameter) = 8 m, D_2 (lower diameter) = 7 m, D (mean diameter) = 7.45 m, L (foundation depth) = 11 m;

Model-2: the mean diameter of the foundation D was extended by 1 m, 1.5 m and 2 m in the same direction (i.e., increasing the C20 grouting range outside), and the corresponding mean outer diameters of the bucket were 8.45 m, 8.95 m and 9.45 m.

Model-3: kept the upper diameter of the bucket D_1 and the outer side slope constant, and the burial depth of the foundation was increased to 11 m, 12 m and 13 m, respectively.

Model-4: the C20 grouting range was locally extended in the upper part of the bucket, an extension was conducted at a burial depth of 4 m and the extension's diameter was 11 m, and then it was gradually extended to a diameter of 12 m at the top.

3.2 Factor of safety and method validation

Factor of safety (*FS*) is the ratio of the foundation's ultimate load bearing capacity to its ultimate load under certain conditions of its own weight and *M*/*H*. All models *FS* results are shown in table 3. Considering the compatibility of safety and economy, the follow-up scheme used the data results of Model-1. Therefore, a *FS* check was performed on Model-1 using the soil stratification, the method of limit equilibrium method, the coefficient of subgrade reaction and the ultimate bearing capacity method in table 4.

Table 3. Factors of safety of the pier foundations.

Model types	Model-1	Model -2			Model -3			Model -4
		D=8.45	D=8.95	D=9.45	L=11	L=12	L=13	
FS	1.31	1.46	1.64	1.82	1.24	1.44	1.68	1.65

Table 4. Comparison of the different method's FS for Model-1.

Validating method	Soil stratification method	Coefficient of subgrade reaction	Ultimate bearing capacity method
FS	1.37	1.29	1.40

3.3 Failure mechanism

The failure mechanisms (i.e., the deformation failure modes) and the soil plastic zone distribution that were solved by the upper limit solution of the finite element limit analysis have been shown in figure 3 and figure 4. Obviously, under all load conditions, the basic failure mechanism was dominated by rotation around a point in the foundation. In all cases, the soil around the bucket would be destroyed. However, in Model-4, the proportion of the damaged area would be large. In fact, the assumption of the rigidity of the soil in the bucket was used in the case of $L/D>2$ at first. As to the foundation of an onshore wind turbine, since the main effect comes from the overturning moment rather than from the horizontal force or the uplift force, the value of L/D is considerably reduced and less than 2 in general, and can be close to 1 especially in Model-4. In this case, the assumption of the rigidity of the soil in the bucket was no longer true. Nonlinear finite element analysis requires optimization of the constitutive equations to reflect the soil plasticity caused by the rotation of the principal stress so as to improve the estimation accuracy of the plastic zone.

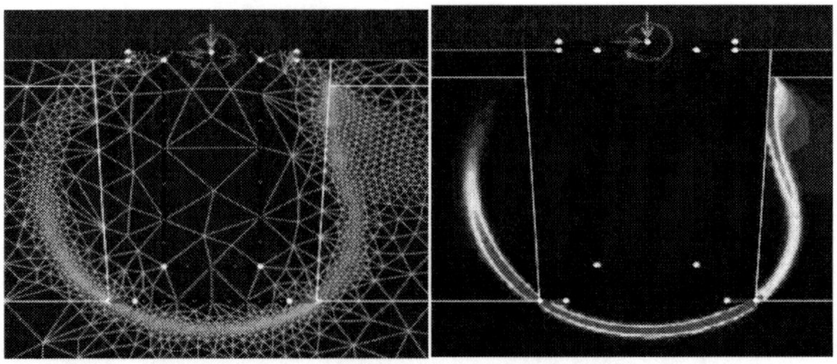

Fig. 3. Contours of the numerical limit analyse.

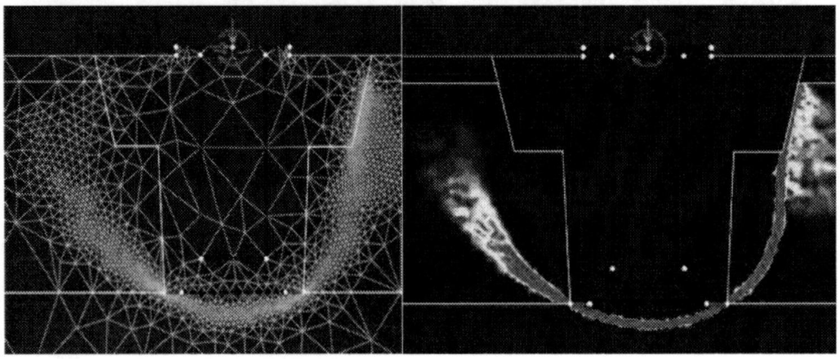

Fig. 4. Contours of the numerical limit analyses for Model-4.

4 Conclusions and suggestions

This paper has used numerical limit analysis to study the ultimate load bearing capacity of the pier foundation of a 2-MW wind turbine in a wind farm in Ningxia under the combined effect of gravity, horizontal loads and overturning moments. Assumption of the failure modes was not needed in the analysis, and the basic failure mechanisms, the soil plastic zone and the soil stress distribution could be automatically obtained through optimization. The main conclusions and suggestions were as follows:

(1) Concerning pier foundation type selection, the factor of safety of the ultimate load bearing capacity for different dimensions have been provided and the factor of safety of the schemes were compared. This indicated that the basic type selection with a diameter of 7.45 m and a depth of 11 m was the optimal scheme.

(2) Based on the comparison between the empirical formulas, the limit equilibrium method, the method of the coefficient of subgrade reaction and the finite element solution, it can be seen that the numerical limit analysis was characterized by a strict theorem and the calculation results thereof were reliable and accurate.

(3) With these address parameters, increasing the foundation depth and diameter could dramatically enhance the ultimate load bearing capacity, and the effect of increasing the diameter was superior to that of increasing the foundation depth in increasing the anti-overturning load bearing capacity. This requires further analysis to find out whether this conclusion also applies to other soil properties.

(4) Numerical limit analysis could automatically distinguish the development of a crack at the soil-structure interface and the calculation results were more reasonable.

References

1. Henderson A P and Patrick M B 1996 Tensionless pier foundation *EP, US5586417*

2. Henderson A P and Patrick M B 1998 Pier foundation under high unit compression *US, US5826387*
3. Sun X Y, Luan M T and Tang X W 2010 Study of horizontal bearing capacity of bucket foundation on saturated soft clay ground *Rock Soil Mech.* **31** 667-72
4. Clukey E C and Morrison M J 1993 A centrifuge and analytical study to evaluate suction caissons for TLP applications in the Gulf of Mexico *Design and Performance of Deep Foundations: Piles and Piers in Soil and Soft Rock* (ASCE) pp 141-56
5. Aubeny C and Murff J D 2005 Simplified limit solutions for the capacity of suction anchors under undrained conditions *Ocean Eng.* **32** 864-77
6. Zhang J H, Chen Z Y and Li F 2010 Three dimensional limit analysis of suction bucket foundations *Ocean Eng.* **37** 790-9
7. Li D Y, Feng L Y, Guo Y X and Cao L X 2013 Numerical analysis of lateral bearing behaviors of skirted suction caissons *Chin. J. Geo. Eng.* **35** 33-8
8. Wang Z Y, Luan M T, Yang Q, Wang Z T and Zhang Q Y 2009 Numerical analysis of ultimated bearing capacity behavior of suction caisson foundations under cyclic loading *J. Dalian Univ. Techno.* **49** 414-8
9. Randolph M and House A 2002 Analysis of suction caisson capacity in clay *Offshore Technology Conf.*
10. Bransby M F and Yun G J 2009 The undrained capacity of skirted strip foundations under combined loading *Geotechnique* **59** 115-25
11. Lysmer J 1970 Limit analysis of plane problems in soil mechanics *Journal of Soil Mechanics & Foundations Div* **96** 1311-34
12. Sloan S W 1988 A steepest edge active set algorithm for solving sparse linear programming problems *Int. J. Numer. Meth. Eng.* **26** 2671-85
13. Sloan S W 1988 Lower bound limit analysis using finite elements and linear programming *Int. J. Numer. Anal. Met.* **12** 61-77
14. Sloan S W and Kleeman P W 1995 Upper bound limit analysis using discontinuous velocity fields *Comput. Method. Appl. M.* **127** 293-314
15. Ukritchon B, Whittle A J and Sloan S W 1998 Undrained limit analyses for combined loading of strip footings on clay *J. Geotech. Geoenviron.* **124** 265-76
16. Ukritchon B, Whittle A J and Sloan S W 2003 Undrained stability of braced excavations in clay *J. Geotech. Geoenviron.* **129** 738-55
17. Yang F, Yang J S, Zhang X M and Zhao L H 2010 Finite element analysis of upper bound solution of shallow-buried tunnel stability in undrained clay *Chin. J. Rock Mech. Eng.* **29** 3952-59
18. Huang Q W, Huang M S and Wang G H 2007 Calculation of bearing capacity of strip footing using lower bound limit method *Chin. J. Geo. Eng.* **29** 572-9
19. Huang Q W and Jia C Q 2007 3D analysis of ultimate bearing capacity by numerical limit analysis *Chin. J. Geo. Eng.* **29** 1809-14
20. Li Z and Wang J X 2007 Lower bound limit study on plastic limit analysis of rock slope using finite elements based on nonlinear programming *Chin. J. Rock Mech. Eng.* **26** 747-52
21. Salgado R, Lyamin A V, Sloan S W and Yu H S 2004 Two-and three-dimensional bearing capacity of foundations in clay *Geotechnique* **54** 297-306
22. China renewable energy engineering institute 2008 *FD003-2007 Design regulations on subgrade and foundation for wind turbine generator system* (Peking: China Water Power Press)

Applying the AcciMap methodology to investigate the tragic Mirsharai road accident in Bangladesh

Omar Faruqe Hamim[1,*], *Md. Shamsul* Hoque[1], *Rich C.* McIlroy[2], *Katherine L.* Plant[2], and *Neville A.* Stanton[2]

[1]Department of Civil Engineering, Bangladesh University of Engineering and Technology, Dhaka, Bangladesh
[2]Transportation Research Group, University of Southampton, Southampton, UK

Abstract. A serious road accident happened on Monday, 11 July, 2011 afternoon in Mirsharai sub-district of Bangladesh's Chittagong district, that killed at least 44 schoolboys injured many more when the truck in which they were traveling skidded off a highway and plunged into a canal. In this study, the AcciMap methodology is applied, an extension to Rasmussen's Risk Management Framework. This sociotechnical framework provides an insight into the interactions between different levels of a system, and supports the development of recommendations that could mitigate or prevent the outcomes of such an incident in the future. In the case of Mirsharai accident, the police and social media put the blame on the truck driver alone; however, this study, based on popular media reports, accident investigation reports and local interviews revealed mismanagement and an apparent disregard for safety across different levels of the socio-technical system. By integrating the different levels of the system, from international committees to end users, the road accident scenario of Bangladesh can be improved through developing a systematic approach to accident analysis.

1 Introduction

The investigation of major accidents in safety-critical systems can explore a variety of contributing factors, both from within involved organizations and from dysfunctional interactions between them in a broader perspective [1]. The sociotechnical systems perspective posits that accidents are not due to the act of an isolated individual or a front-line operator, but due to highly interactive and collective processes that involve decision-makers and policy-makers at all levels of society [2]. Given the current focus in accident analysis on the end user and immediate environment, there is a significant need for accident investigation frameworks that integrate the analysis of contributing factors from different parts of a sociotechnical system, and the interactions between them [1].

Sociotechnical methods include those such as the systems-theoretic accident model and processes (STAMP) by Leveson [3, 4], Reason's model of organizational accidents [5], and

*Corresponding author: hamim@ce.buet.ac.bd

Rasmussen's AcciMap approach [6]. In this paper, an extension of Rasmussen's AcciMap approach [7] will be used to analyze a road accident that took place in Mirsharai, Bangladesh.

The accident at Mirsharai killed at least 44 students as a truck carrying the students drove into a ditch while the driver was reported to be talking over the phone. Police said a truck carrying 70 school students fell into a watery ditch when its driver lost control over the steering at about 1:30 pm. Relatives and locals recovered 17 bodies from the truck sunk in a nearly 6 foot deep canal.

Eyewitnesses, survivors, and locals told The Daily Star (a daily newspaper) that reckless driving was the cause of the crash. It was reported that Mofiz, a helper of the truck owned by Haji Rice Mill, was in a hurry as he was supposed to go to Feni to bring a rice consignment for the mill. "The man did not listen to repeated warnings from us and kept talking on the phone," a survivor was quoted as saying. Some other local residents alleged that the truck was overloaded and was driven by a helper, not the driver, adding that the driver escaped after the accident. Investigators found that Mofiz did not possess a valid driver's license at the time of the accident and was speaking on his mobile phone while driving. A court sentenced Mofiz to five years in jail and a fine of Tk. 20,000 (~£180) on December 8, 2011.

The aim of this paper is to apply the AcciMap methodology to identify the higher system factors that contributed to the accident and provide some recommendations based on the sociotechnical analyses. The AcciMap approach is based on Rasmussen's risk management framework, a six-levelled, hierarchical representation of a system that not only considers the activities of actors acting in each level but more importantly, the interactions between them, which eventually take the form of decisions propagating downward and information propagating upward [8, 9, 10].

The methodology has been used in a number of domains, including chemical processing [11, 12], public health [13], transportation [8, 14], aviation [9], anti-terrorism [15, 16], gas production [17], and most recently, oil and gas drilling [18]. For the purpose of this paper, an extension to Rasmussen's risk management framework was used; this included two additional upper levels, namely International Committee and National Committee (proposed by Parnell et al. [7]).

Often "a quite normal variation in somebody's behavior" can cause an accident [6]. In reality there are several factors that lead to an accident, so even if a particular factor is eradicated, it is likely another factor will cause the accident. One crucial point of consideration in this regard is to observe the actions of workers or the errors that triggered an accident in a broader sociotechnical context since those actions are impacted by the decisions and activities of other actors in all the stated levels of the system.

2 AcciMap framework for Mirsharai road accident

A commonly performed first step in conducting an AcciMap analysis is to identify the actors involved in the system through the development of an Actor Map. This is a graphical representation of those actors, organised by system level. The Actor Map defines all actors or organizations involved in a system. It is a method to model actors and interactions at all levels of a system. It provides a hierarchical description of a sociotechnical system. It originates from Rasmussen's Risk Management Framework [6] and can be applied to any complex domain where safety management is a concern.

The Actor Map for the Bangladesh road transport system was indeed developed, and is presented in McIlroy et al. [19]. From this, the AcciMap for the Mirsharai road accident was developed (Figure 1). Analysis started from the bottom and propagated to the top of the system. The developed AcciMap framework in Figure 1 illustrates the factors contributing to the Mirsharai road accident and the interrelations between them. The inter-connections between the elements of the AcciMap show where one factor leads to another factor.

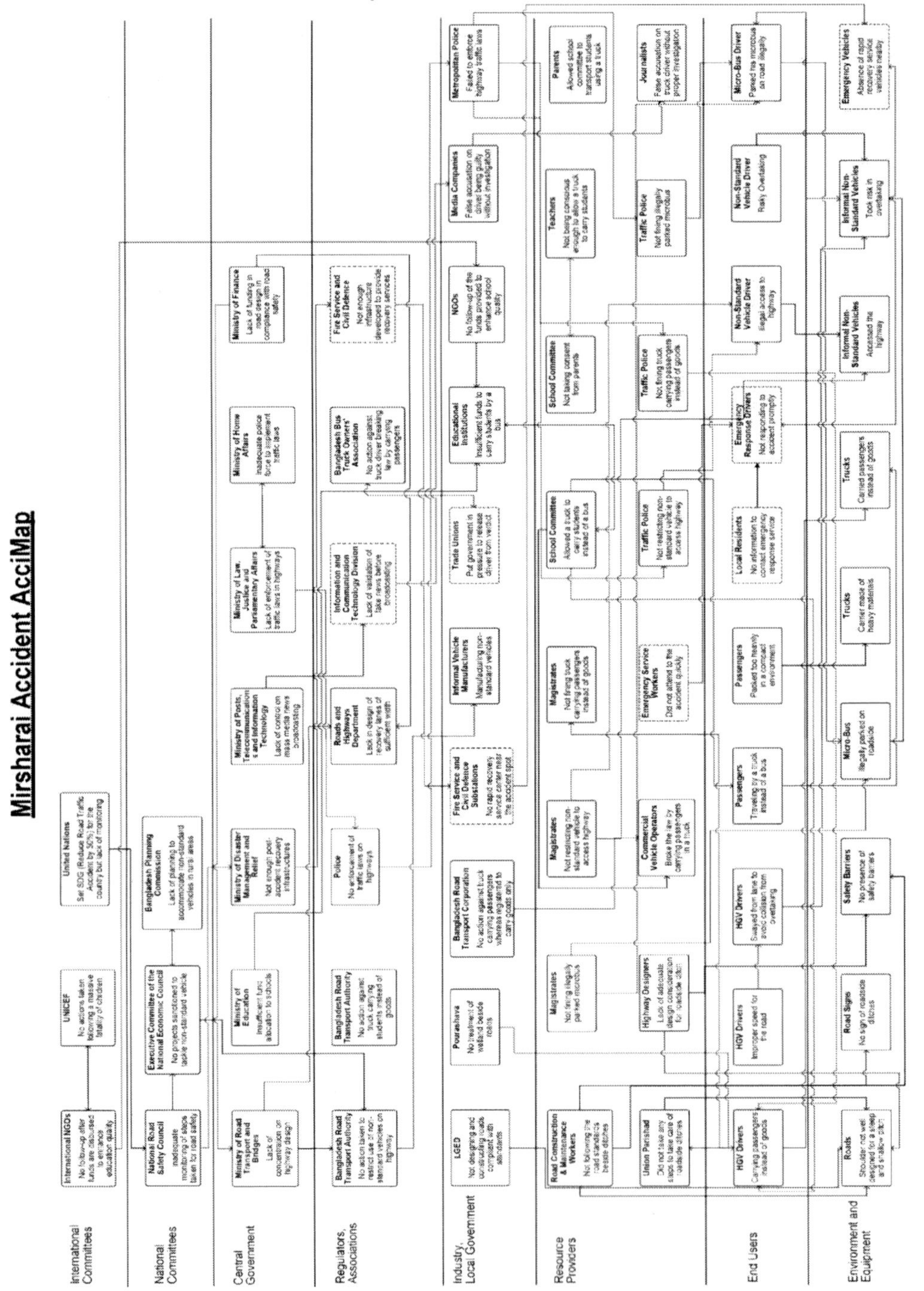

Fig. 1. AcciMap representing the various interrelated factors contributing to Mirsharai accident.

The AcciMap highlights a number of factors; truck carrying students instead of goods was moving at a high speed whilst an informal non-standard vehicle (locally called "Nasimon") was overtaking an illegally parked microbus on a two-lane two-way road. The truck tried to move towards the shoulder to avoid collision with the informal non-standard vehicle. There was a shallow wetland beside the road and the condition of the shoulder was poor. As a result of this, and the absence of road side safety barriers, the truck exited the carriageway and entered the shallow ditch. Due to the rigid body of the truck carrier and the shallow depth of the ditch, the students were trapped inside that carrier. Also, in the absence of nearby emergency response service, the local residents could not rescue all of the students trapped inside the truck carrier in the ditch. All of these factors lead to the death of 44 school going students.

By analyzing the sequence of events in this accident, it is evident that there are a number of higher level contributory factors. For example, the school committee allowed the students to travel to the football tournament by a truck that was not meant for human transportation; however, the school committee had insufficient funds to provide other transportation. The parents did not object to this decision. The truck that was given a route permit only to carry goods in its carrier was carrying passengers in a cramped condition, thereby violating the law; however, police and administration rarely stop vehicles illegally carrying passengers. The police also did not take any steps against the illegally parked microbus or the informal, non-standard vehicle that was illegally travelling on a regional road. Considering the contribution of the "Nasimon" (informal non-standard vehicle) in the rural economy in Bangladesh, the Planning Commission could take steps to provide feeder roads for these vehicles in rural areas in conjunction with Ministry of Road, Transport and Bridges and Local Government Department.

The local media laid all the blame on the driver for harsh driving while talking over the phone. The police arrested the driver and sentenced him to jail. No attempt was made to investigate the whole scenario and find the causal factors higher up in the sociotechnical system. As such, false allegation without proper investigation lead to blaming only the driver ignoring other factors. The flow of activities shown in the AcciMap (Figure 1) shows how one factor triggers several other factors leading to the occurrence of a disastrous accident, indicating that blame is shared across system levels, not place only on the driver.

3 Recommendations

A very important characteristic of the AcciMap approach is placing and matching the events and conditions that finally released the accident [1]. This characteristic avoids the unfair blame of the frontline operators, since it provides a broader perspective and background on where those events and conditions come from and what the sources of actors' decisions are [1]. The AcciMap template (8 levels) can be used to define recommendations at each level for different actors. In this section, for every event or decision that lead to the accident, a recommendation is defined (Table 1). These recommendations can be useful for legislating a systematic framework for developing a safer road system in Bangladesh.

Table 1. Recommendations corresponding to specific level of the system based on accident events.

Level	Accident Event	Recommendation
International Committees	Inadequate follow-up by the international NGOs after funds are disbursed to enhance education quality	Legislate that all funds invested by the international NGOs must be monitored to ensure proper allocation
	Sustainable Development Goal (Reduce Road Traffic Accident by 50%) was set for the country but lack of monitoring failed the	Legislate that proper monitoring system must be present for all targets set by international organizations

Level	Accident Event	Recommendation
	achievement of the target	
National Committees	National Road Safety Council took insufficient monitoring steps for ensuring road safety	Develop a framework for monitoring the road safety programs
	Executive Committee of the National Economic Council sanctioned insufficient projects to tackle non-standard vehicle intrusion into roads	Develop a rehabilitation program to relocate the non-standard vehicles from the highways and stop their intrusion
	Bangladesh Planning Commission has not portrayed any plans to accommodate non-standard vehicles in rural areas	Non-standard vehicles have a major contribution to the flourishing economy of rural people, so there should be planning for feeder roads in rural areas to ensure the use of non-standard vehicles
	Ministry of Education allocate insufficient funds to rural schools	Adequately resource essential school transport (e.g. shared bus for use in a multi-school district)
	Ministry of Home Affairs has limited police force to enforce traffic laws	Police force must be enriched with more skilled traffic policemen for better implementation of traffic laws
	BRTA has taken minimal action against the truck carrying students instead of goods	BRTA must cancel the driving license of drivers disobeying laws and getting involved in accidents
	Police do not enforce traffic laws properly in highways	Develop a monitoring cell which will look up the activities of traffic police
Industry and Local Government	LGED (Local Government Engineering Department) roads are inadequately designed and constructed in compliance with standards in accordance with the traffic volume	Proper design methods and technology must be developed by research and strictly enforce monitoring system in construction stage
	Pourashava does not treat wetlands beside roads	Proper treatment must be provided to roads adjacent to wetlands and ditches
	Informal non-standard vehicles are being manufactured locally	Rules for use of informal non-standard vehicles only in rural areas must be made by BRTA
	Media made false accusation on driver being guilty without any investigation	Provision should be made to enforce strict punishment for false allegation by media without investigations
Resource Providers	Magistrates did not fine illegally parked microbus, truck carrying passengers instead of goods and non-standard vehicle that accessed the highway	Develop a framework to facilitate routine monitoring of road activities by magistrates to fine any illegal activities made by the road users
	School committee allowed a truck to carry students instead of a bus without taking consent from parents	Legislate that no trucks or any vehicle designed to carry goods can carry passengers for any purpose
	Teachers and parents allowed the students to be carried by a	Create social awareness about the hazardous effects of transporting

Level	Accident Event	Recommendation
	truck	people via trucks
End Users	Truck driver was carrying passengers instead of goods at a high speed with respect to the class of the road	Strict enforcement must be followed for punishing drivers carrying passengers in a vehicle designed to carry goods and high speed driving
	Non-standard vehicle driver illegally accessed the highway and got involved in risky overtaking	Provide proper training to non-standard vehicle drivers to make them literate about the traffic rules and regulations
	Micro-bus driver illegally parked his microbus on road	Incorporate provisions of emergency vehicle stop bays in highways
Equipment and Environment	Shoulders are not designed for steep and shallow ditch beside roads	Design roads with proper care taken for roadside ditches
	Road traffic signs were not present for road side ditches or speed limit	Provide adequate road signs and ensure its proper maintenance
	Safety barriers were not present	To reduce fatalities from road accidents, safety barriers must be provided

4 Conclusion

This paper describes the application of AcciMap methodology for investigating the Mirsharai road accident, identified as one of the most tragic accidents that has happened in the road traffic accident history of Bangladesh. For any road accident, a proper and detailed accident investigation accompanied with a systematic, sociotechnical framework of analysis that takes into account higher system factors, provides benefit over traditional investigation approaches. From the AcciMap analysis presented in this article, the root causes of the accident were identified that go beyond blaming only the driver. The possible links between the causes that incurred the accident were also duly recognized. The responsible factors at every level of the sociotechnical system are marked and corresponding recommendations have been proposed. These recommendations will be vital for any future road safety intervention in Bangladesh, and to future accident investigation activities.

Acknowledgments

This research was commissioned by the National Institute for Health Research using Official Development Assistance (ODA) funding. The views expressed are those of the authors and not necessarily those of the NHS, the NIHR, or the Department of Health and Social Care.

References

1. Lee, S., Moh, Y. B., Tabibzadeh, M., Meshkati, N., 2017. Applying the AcciMap methodology to investigate the tragic Sewol Ferry accident in South Korea. *Applied Ergonomics,* 59, 517-525.
2. Le Coze, J.E., 2015. Reflecting on Jens Rasmussen's legacy. a strong program for a hard problem. Saf. Sci. 71, 123-141. Part B.

3. Leveson, N., 2004. A new accident model for engineering safer systems. Saf. Sci. 42 (4), 237-270.
4. Leveson, N., 2011. Engineering a Safer World: Systems Thinking Applied to Safety. The MIT Press.
5. Reason, J., 1997. Managing the Risks of Organizational Accidents. Ashgate Publishing Limited.
6. Rasmussen, J., 1997. Risk management in a dynamic society: a modeling problem. Saf. Sci. 27 (2), 183-213.
7. Parnell, K.J., Stanton, N.A., Plant, K.L., 2017. What's the law got to do with it? Legislation regarding in-vehicle technology use and its impact on driver distraction. *Accident Analysis and Prevention,* 100, 1-14.
8. Rasmussen, J., Svedung, I., 2000. Proactive Risk Management in a Dynamic Society, first ed. Raddningsverket, Risk and Environmental Department. Swedish Rescue Services Agency, Karlstad, Sweden.
9. Branford, K., 2011. Seeing the big picture of mishaps: applying the AcciMap approach to analyze system accidents. Aviat. Psychol. Appl. Hum. factors 1 (1), 31-33.
10. Salmon, P.M., Cornelissen, M., Trotter, M.J., 2012. Systems-based accident analysis methods: a comparison of Accimap, HFACS, and STAMP. Saf. Sci. 50, 1158-1170.
11. CSB Final Investigation Report, 2015. Chevron Richmond Refinery Pipe Rupture and Fire: Chevron Refinery#4 Credit Unit. US Chemical Safety and Hazard Investigation Board. Report No. 2012-03-I-CA.
12. CSB Report, 2014. Catastrophic Rupture of Heat Exchanger: Tesoro Anacortes Refinery. US Chemical Safety and Hazard Investigation Board. Report No. 2010-08-I-WA.
13. Vicente, K.J., Christoffersen, K., 2006. The walkerton E. coli outbreak: a test of Rasmussen's framework for risk management in a dynamic society. Theor. Issues Ergonomics Sci. 7, 93-112.
14. Underwood, P., Waterson, P., 2014. Systems thinking, the Swiss cheese model and accident analysis: a comparative systemic analysis of the grayrigg train derailment using the atsb, AcciMap and STAMP models. Accid. Analysis Prev. 68, 75-94.
15. Jenkins, D.P., Salmon, P.M., Stanton, N.A.,Walker, G.H., 2010. A systemic approach to accident analysis: a Case Study of the Stockwell Shooting. Ergonomics 53, 1-17.
16. Waterson, P.E., Jenkins, D.P., 2011. Lessons learnt from using AcciMaps and the risk management framework to analyse large-scale systemic failures. In: Anderson, M. (Ed.), Contemporary Ergonomics and Human Factors 2011: Proceedings of the International Conference on Ergonomics & Human Factors 2011, Stoke Rochford, Lincolnshire. CRC Press, p. 6. April 12-14, 2011.
17. Hopkins, A., 2000. Lessons from Longford: the ESSO Gas Plant Explosion. CCH, Sydney, Australia.
18. Tabibzadeh, M., Meshkati, N., 2015, April 27e30. Applying the AcciMap Methodology to Investigate a Major Accident in Offshore Drilling: a Systematic Risk Management Framework for Oil and Gas Industry. SPE-174020-MS. SPE Western Regional Meeting. Society of Petroleum Engineers, Garden Grove, CA.
19. McIlroy, R.C., Plant, K.A., Hoque, M.S., Jianping, W., Kokwaro, G.O., Vũ, & N.H, Stanton, N.A., 2019. Who is responsible for global road safety? A cross-cultural comparison of Actor Maps. Accident Analysis and Prevention, 122, 8-18.

The spatial value of local houses around Islamic Boarding School in Mlangi Moslem settlements, Yogyakarta, Indonesia

Nensi Golda Yuli*

Department of Architecture, Universitas Islam Indonesia, Yogyakarta, Indonesia

Abstract. Many houses around Mlangi Muslim settlement refer to the historical aspect of settlement in its spatial layout as the Moslem settlements. Mlangi well known as Kampung Santri - village with many of Islamic Boarding Schools (Pondok Pesantren in Indonesia term) where students live in there- since 1920an. This paper will describe how the spatial value of local houses around Islamic Boarding Schools in Mlangi Moslem settlements. The naturalistic-phenomenology method was the research method for data collection. Data got from subsequent interview after the appearance theme from one research object to the next. Any argument that correlated with the spatial value grouped as the same theme and analyzed as the substantive concept about the spatial value in the houses. The room layout from each home drawn one by one as the existing condition and the value of the room got from the comments of the owner. Cross-checking information technic from the members in the house helped the research to mention many spatial values in-house and then conclude as the room concept and its correlation to the stigma as the Kampung Santri and Moslem Settlements. It also compares the similarity of the indoor room layout and zone in Pondok Pesantren. From existing condition and interviews found that all the houses have the similar function and room layout with the Pondok Pesantren around it. 6 observed houses and 15 interviewed owners stated that they want to make their house as the places and spaces for teaching holy Al Qur'an same like in Pondok Pesantren. That is like a dream come true if a house can hold any recitation and teach holy Al Qur'an and Islamic Knowledge to the neighbor. From this condition, researchers conclude that there is close relation about the significance building in one settlement to their neighbor building, not only in the activity in each house but also in room layout and zone. Religious value and people understanding and dream about the ideal space even get impact from it.

1 Introduction

The spatial value transforms from the spatial meaning while a home entails a physical space that influences the emotional and mental needs of individuals [12] and is organized according to a particular order of social principles [9]. Thus, spatial relation must be

* Corresponding author: nensi@uii.ac.id

considered in the design of living spaces, especially in houses where individuals perform many of their daily activities [2].

The spatial value in Moslem settlements can draw how the people concept about their house related to the arrangements which stamped as Moslem settlements. As one of the oldest Moslem settlements in Yogyakarta, Mlangi village contributed in the spread of Islam in the west site of Yogyakarta Special Region. Nowadays, many pesantrens (Islamic Boarding School) build around its grand mosque, Pathok Negoro Mlangi Mosque. It was the first idea from 1st King of Yogyakarta to make the settlement at west site Yogyakarta as the center of Islam spread in Yogyakarta. As the old city planning, another place of Yogyakarta, in north, east and south also built A Pathok Negoro Mosque to develop Islam from the corners of the city. Recently, only in Mlangi village the indigenous atmosphere of Moslem settlements still sustain which didn't have any significant changes from its first existence.

The research was conduct to explore how the houses in the village have similarity in room layout with exist Pesantren int there. Room layout indicated spatial value from house owner and its family who dedicated their house for the same activities like in Pesantren. An only rich person or people has a broad relation and networking to other pesantren outside village has the ability to build pesantren in there. The other has same dream and vision about their houses, but the economic condition makes they only try to set the house like pesantren. This condition explored by this research to seek how far the houses implemented the owner value about spatial related to pesantren around it.

2 Method

The naturalistic-phenomenology method used as data gathering technics. Six houses which explored by interview to the occupants observed by phenomenology features in the site. The information from users that stated the room function in their houses plotted in floorplan drawing. The room layout plot based on the statement of the user, not the researcher interpretation. Floorplan picture interpretation and crosschecking to the occupants for the spatial value behind that compared to the general room layout in Mlangi pesantren. It draws how the people's houses around pesantren in spatial value. The location of the observed house in this research as in figure 1:

Fig. 1. The location of observed houses.

3 Theoretical review

The spatial value has the close relationship to the spatial quality that identified as view, privacy, lighting, spatiality, spatial arrangements, the transition between public and private spaces, and perceived, built, and human densities [1]. In its conceptual meaning, a house bears such notions as tranquility, peace of mind, spiritual and metal as well as local safety that full fill these needs [7]. From sociology and psychology literature, values are typically conceptualized as fundamental ideas and enduring principles that inform peoples' judgments [3];[10]; [13]. The value could be understood as the value of people; assigned values is how places, species and ecosystems are valued by people [5]. Assigned values may be related to the underlying environmental concerns and worldviews that people hold [11]; [15]. Assigned values could be a better predictor of spatial behavior than held values, together with beliefs and norms, influence assigned value, and therefore are only indirectly linked to environmental behavior [14]. Relational values reflect that people could have strong bonds to places that influence their preferences for uses in a location as well as their attitudes towards management actions [4].

The most critical challenge in eown territory is the incompatibility of the physical and space of the residential complex with the cultural and behavioral properties of the residents [8]. Social relation between people in a time and place is spread dialectic semantic of culture that is explained in a relationship with the human environment and place [6].

The concept and function of space created have close ties to understand the way of life of the community. Lifestyle factors influence the shape and arrangement of area in the house [16].

4 Research findings

From existing condition, six houses observed by its layout, function of the room and spatial value from occupants in-depth interview using phenomenology method. The research findings is in table 1:

Table 1. Summary of The Houses Exploration in Research Area.

Number of Explored House	Floor Plan	Dominant Function Area	Hidden reason from occupant about the large zone allocation for living room
House 1	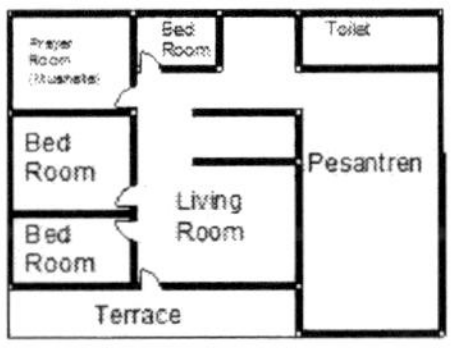	The Living room use for Al Qur'an Recitation and There are *Pesantren* as Educational Facilites for communities in the House	Most of the house zone dedicated for Pesantren activities, like the occupant vision to build pesantren after his convection business bankrupted
House 2	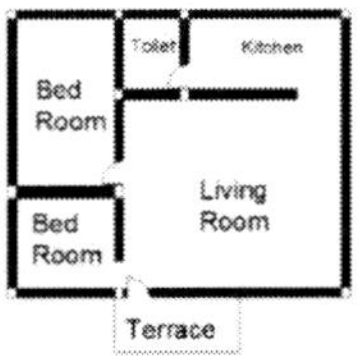	Living Room for Multiactivities i.e: Recitation for Neighbour and minibusiness place activities	All activities in the house only for business and recitation, although the users only teach kids around the house in the evening for holy qur'an recitation

House 3	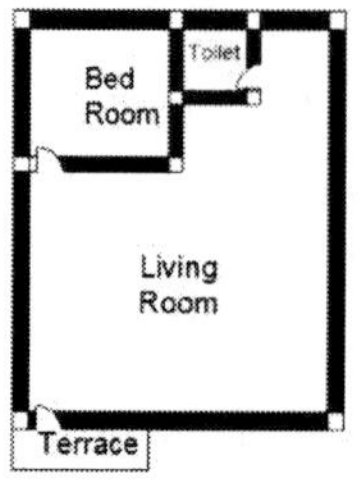	Living room for Holy Qur'an Recitation	House as the pesantren and place for alqur'an learning for neighbor
House 4	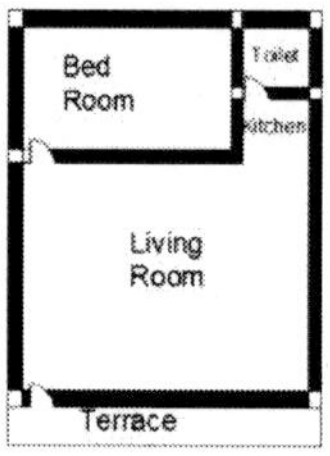	Living Room for Holy Qur'an Recitation	Good house which can accomodate islamic thought activities
House 5	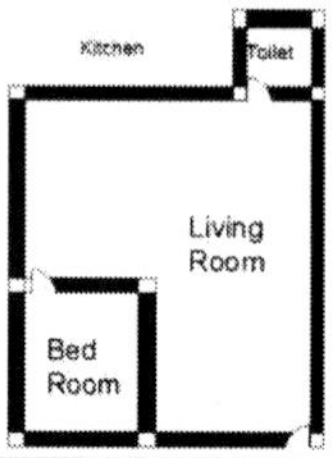	Living room for guest and Holy Qur'an Recitation	House as the religious space to support community activites in Al Qur'an recital
House 6	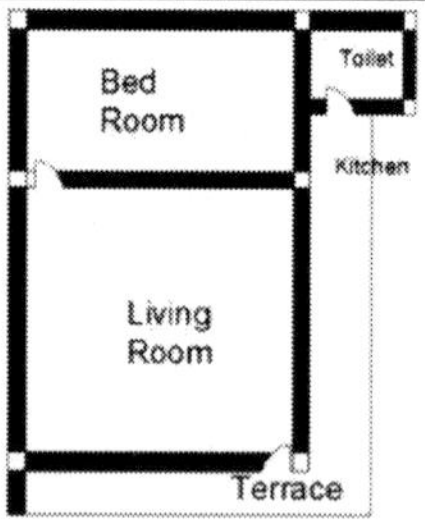	Living room for guest and Al Qur'an recitation	House like pesantren is the best place to family

House 1 is a house with *pesantren* (Islamic boarding school) in it. Students of *pesantren* live in there. There are three families in that house, the house owner, the *pesantren* owner (first son of house owner) and the youngest daughter of the family. They live together with their role for the community. The house owner hs significant role for cities, as the head of Mosque organizer (*Takmir Masjid*). The son is the head of Pesantren (Kyai), and the daughter has the small shop near the main road that sell Moslem wear. In the house, they also have mushalla, little place to pray together. All the room in the houses dedicated for religious activities, such as received the guest from other *pesantren*, hold the Al Qur'an recitation, religious meeting with neighborhood and also to entertain business relation for the daughter. All the activity, from entertaining guests, meeting with a neighbor, hold the recitation including teaching small group for Islamic teaching do in the living room. Their understanding for space in the house is a space for multifunction activities to support *pesantren* activities from teaching and learning, Al Qur'an recitation and give the services for guest who want to know so far about Islamic knowledge.

For house 2, the main activity is convection business. House separates into the living room, bedroom, kitchen, and toilet. The house owner also holds recitation in the evening for around children. They have a dream to be their house as the Islamic teaching place. In their mind, a good home is a house who has dominant religious activity in spatial layout owner set living room for multifunctional room, both for business and recitation. But in the same time, the recitation or other religious activities have higher priority than business activity.

In-house 3,4,5 and 6, all types are similar in spatial layout. Majority of spaces is for the living room, only small space for bedroom and service (toilet). These houses often use for communities meeting exclude pesantren activities. The people around Mlangi settlements have two different activities, pesantren activities, and non pesantren activities. Both support each other. The spatial use for both activities also help each other. But in some activities, there are recitation group for non-*pesantren*, where held in people houses at a regular schedule, from one house to another house. So, the people's house in Mlangi Muslim Settlements should accommodate regular activities by large living room in that house or they should provide a place in front of their house using tend to held regular recitation and meeting. These types similar to pesantren's indoor layout like in figure 2. In figure 2, all the room dedicated to eduational activities i.e teaching Islamic knowledge, reciting Holy Al Qur'an and also study about the history of muslim world.

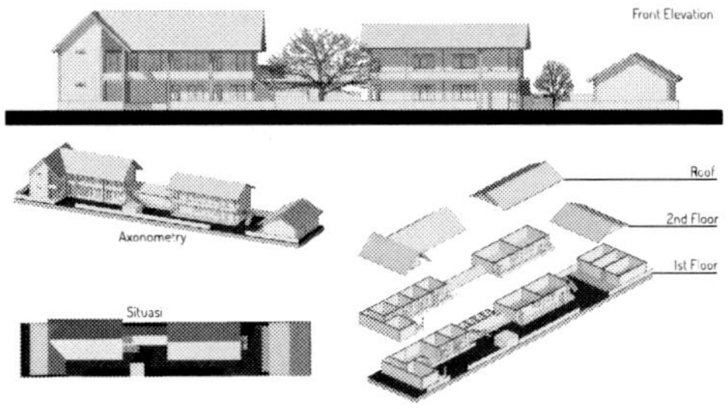

Fig. 2. One Pesantren Type in Mlangi.

5 Discussion

In Mlangi village, almost activities using the living room area. From business, entertaining guests, Al Qur'an recitation and small teaching about Islamic knowledge occur in the living room. Different from research in 2015 by Wahab in Melaka residents found that in Malay house, people using veranda area for reciting Al Qur'an as a public area. From this research, we know there is open space in the house to accommodate religious activity like in Mlangi. Also in Malay, private user Al Qur'an recitation held in private room in the main house. This area is the center and isolated area and can only access by permission. It is just for family. Beside the public space in house scale, a house in Malay divide the house for different user in use. But in Mlangi, people feel they are coming from one big family in the settlement, so there are no private or public for space allocation. In history, Mlangi ancestor is the person who develops the settlement to be a Moslem settlement, Kyai Nur Iman, a family member of Yogyakarta Sultanate. And all the houses around the mosque had close relation to each other and claimed as the blue blood generation from Yogyakarta Sultanate.

If we compare to the research in Iranian house by Alitajer 2016 indicated that privacy is not respected in modern houses than traditional houses. Alitajer stated today the privacy in the house as the rare thing to develop. Both current and classic house, connectivity and integration between room are minimal. From this research findings also confirmed that in some country spatial value which can indicated from the privacy, actually not find both in traditional and modern houses. The houses need private space for the occupants but if the owner live around the his/her central family, like in Mlangi, there are no privacy in the house. All room in house is the public place for anyone who come. There is some space that could be intervention by non family member. Most of traditional house in Malay use *serambi* or veranda as a space to held prayer activities [17].

6 Conclusion

From six houses, all users commented that the house related to *pesantren* in room layout. Their spatial values implemented how they set room allocation. The manifestation of the value entirely draws for living room allocation portion in the house. Spatial value also brings in their mind

From this condition, research conclude that there is close relation about the significance building in one settlements to their neighbor building, not only in the activity in each house but also in room lay out and zone. Religious value and people understanding and dream about the ideal space even get impact from it.

Acknowledgement

The Fundamental Research ostensibly supports this research for Excellency University Grant from Ministry of Research, Technology and Higher Education of The Republic of Indonesia (PDUPT RISTEK DIKTI) 2019 entitled The Moderation of Pondok Pesantren Building Design Standard based on Vernacularity and the Development of Its Activities for Intelligent, Sustainable and Disaster Responsive Masterplan in Mlangi Village, Nogotirto, Sleman, Yogyakarta.

References

1. Acre, F., Wyckmans, A. 2014, Spatial quality determinants for building renovation: a methodological approach to the development of spatial quality assessment. J. Sustainable Technol. Urban Dev. SUSB. http://dx.doi.org/10/1080/2093761X.2014.923793
2. Alitajer, Saeid and Ghazaleh Molavi Nojoumi. 2016. Privacy at home: Analysis of behavioral patterns in the spatial configuration of traditional and modern houses in the city of Hamedan based on the notion of space syntax. Frontiers of Architectural Research (2016) 5, 341–352
3. Brown, T. C. (1984). The concept of value in resource allocation. Land Economics, 60(3), 231–246. https://doi.org/10.2307/3146184.
4. Chan, K. M. A., Balvanera, P., Benessaiah, K., Chapman, M., D.az, S., G.mez-Baggethun, E., et al. (2016). Opinion: Why protect nature? Rethinking values and the environment. Proceedings of the National Academy of Sciences, 113(6), 1462–1465. https:// doi.org/10.1073/pnas.1525002113.
5. Ives, C. D., & Kendal, D. (2014). The role of social values in the management of ecological systems. Journal of Environmental Management, 144, 67–72. https://doi.org/10.1016/j.jenvman.2014.05.013.
6. Jencks, Charles (1996),What is post-modernism, Michigan, the University of Michigan Press
7. Kooshali, Atefeh Deljooy, Reza parvisi, Amir Reza Karimi Azeri, S. Bagher Hosseni. 2015. A Comparative Study on the Effect of Nature on Satisfaction with Residence at

Detached Houses (Single Unit) and Residential Building Complex (Apartment). Social and Behavioural Sciences 201 (2015) 243-254: Elsevier 2015

8. Mashayekhi, Mohammad; Mohammadjavad Mahdavinejad, and Soghra Niroumand. 2012. Designing Family Friendly Architecture Environments. Elsevier: 2012
9. Reis, A. 2003. Original and converted social housing: spatial configuration and residents' altitude: Proceeding of the Paper presented at the 4th International Space Syntax Symposium London.
10. Rokeach, M. (1973). The nature of human values [by] milton Rokeach. New York: Free Press.
11. Rossi, S. D., Byrne, J. A., Pickering, C. M., & Reser, J. (2015). 'Seeing red' in national parks: How visitors' values affect perceptions and park experiences. Geoforum, 66(Suppl. C), 41–52. https://doi.org/10.1016/j.geoforum.2015.09.009.
12. Saruwono, Masran. 2012. Shouting in Silence: expression of self in private homes a dwelling, house, or home ? . Procedia – Social and Behaviour Sciences 42 (2012) 34-41
13. Schwartz, S. H. (1994). Are there universal aspects in the structure and contents of human values, Journal of Social Issues, 50(4), 19–45. https://doi.org/10.1111/j.1540-4560. 1994.tb01196.x.
14. Seymour, E., Curtis, A., Pannell, D., Allan, C., & Roberts, A. (2010). Understanding the role of assigned values in natural resource management. Australasian Journal of Environmental Management, 17(3), 142–153. https://doi.org/10.1080/14486563. 2010.9725261
15. Van Riper, C. J., & Kyle, G. T. 2014. Capturing multiple values of ecosystem services shaped by environmental worldviews: A spatial analysis. Journal of Environmental Management, 145(Suppl. C), 374–384. https://doi.org/10.1016/j.jenvman.2014.06.014.
16. Wahab, Mohamad Hanif Abdul et al. 2015. Malay Furniture: Design, Function and Meaning. *, Procedding on ASEAN-Turkey ASLI (Annual Serial Landmark International) Conference on Quality of Life 2014, ABRA International Conference on Quality of Life, AQoL2014, 26-28 December 2014, Istanbul, Turkey
17. Wan Teh, W.H., Nasir, A.H., 2011. The Traditional Malay House (Fourth ed.). Penerbit Fajar Bakti, Shah Alam, Malaysia.

HHT based analysis on seismic response recordings for a base-isolated building

Fei Wang*, *Xiandong* Kang, *Ting* Yan, and *Ying* Liu

Beijing Earthquake Agency, Beijing 100080, China

Abstract. Hilbert-Huang transform (HHT) is proposed to process the seismic response recordings in an 8-story frame-shear wall base-isolated building. Empirical Mode Decomposition (EMD) method is first applied to identify the time variant characteristics and the data series can be decomposed into several components. Hilbert transform is well-behaved in identifying the frequency components. The first 5 intrinsic mode functions (IMFs) are decomposed with their different frequencies. The analytical function is reconstructed and compared with the original signal. They are extremely consistent in amplitude and phase. Based on the IMFs obtained, frequencies of the original signal are inferred at 5 Hz and 1.6 Hz. The higher frequency is regarded as the vibration excited by surface waves. 1.6 Hz is suggested as the dominant frequency of the building. Analysis indicates that HHT is accurate in extracting the dynamic characteristics of structural systems.

1 Introduction

Structural vibration signals are commonly analyzed to extract the dynamic characteristic parameters including natural frequency, damping ratio and mode shapes [1]. Structural health states are identified based on the changes of structural dynamic characteristics. By comparing the characteristics of the undamaged and original structures, the damages are detected. Various methods including Fast Fourier transform (FFT) and wavelet transform have been proposed for natural characteristics identification in structural health monitoring and diagnosis. Fast Fourier transform is an analysis method in frequency domain and suitable for analyzing stable time series in linear process. However, it has obvious deficiency in nonlinear process due to low resolution and external noise in fitting original time series. Wavelet transform realizes analysis in time domain by transforming the signals with scalable and translatable wavelets [2,3]. Wavelet analysis is essentially a Fast Fourier transform with adjustable windows and signals in time windows are required to be stationary. Therefore, it has the similar limitations with Fast Fourier transform [4,5].

The method of Hilbert-Huang transform (HHT) has been developed to analyze signals in the frequency-time domain. It is consisted of Empirical Mode Decomposition (EMD) and Hilbert transform. Thereinto, EMD is the critical part for data analysis. HHT method has already been proved more precise than the wavelet analysis[6,7]. Furthermore, HHT method is applicable to process non-stationary time series. In this study, HHT is introduced

* Corresponding author: wangfei@bjseis.gov.cn

to analyze the structural seismic response recordings in an 8-story base-isolated building and the inherent dynamic characteristics are identified.

2 Basic principle of HHT

The method of EMD is crucial to HHT and it was recently proposed to decompose data series $x(t)$ into intrinsic mode functions (IMFs). An IMF has two basic requirements. The number of extreme points and the number of zero crossings for the entire data must be equal or must not be more than one if they differ. The mean of the upper and lower envelopes is computed to be zero at any extreme point. The IMF components can be determined by repeated application of sifting process, an iterative procedure. Han and Li repeated the process for HHT and applied to analyse the data in shake table test [8].
For data series $x(t)$, by introducing h_1 it can be sifted firstly as

$$h_1 = x(t) - m_1. \quad (1)$$

Here m_1 stands for Instantaneous mean of the upper and lower envelopes. Usually h_1 does not meet the basic requirements. Therefore, the sifting process can be repeated k times and h_{1k} is obtained meeting the IMF requirements. The first IMF component C_1 is computed out with the component of highest frequency,

$$r_1 = x(t) - C_1. \quad (2)$$

Analysis is carried out with r_1 instead of $x(t)$ and the second IMF component is obtained. The signal can be finally decomposed into n IMFs and the mean trend of the signal or a constant as $r_n(t)$, indicated in

$$x(t) = \sum_{j=1}^{n} C_j(t) + r_n(t). \quad (3)$$

$\tilde{x}(t)$ is the Hilbert transform of $x(t)$,

$$\tilde{x}(t) = HT[x(t)] = \frac{1}{\pi} P \int_{-\infty}^{+\infty} \frac{x(\tau)}{t-\tau} d\tau. \quad (4)$$

Here P stands for Cauchy principal value. Analytical signal of $x(t)$, $Y(t)$ is calculated as,

$$Y(t) = x(t) + i\tilde{x}(t) = A(t)e^{i\theta(t)}. \quad (5)$$

$A(t)$ is denominated as the instantaneous amplitude, $\theta(t)$ instantaneous phase angle and $\omega(t)$ instantaneous frequency. They can be computed out as

$$A(t) = \left[x^2(t) + \tilde{x}^2(t)\right]^{1/2}, \quad (6)$$

$$\theta(t) = \arctan\left(\frac{\tilde{x}(t)}{x(t)}\right), \quad (7)$$

and

$$\omega(t) = {d\theta(t)}/{dt}. \quad (8)$$

Now $Y(t)$ can be transformed into

$$Y(t) = \sum_{j=1}^{n} A_j(t)\, e^{i\int \omega_j(t)\, dt} = \sum_{j=1}^{n} A_j(t,\omega_j)\, e^{i\int \omega_j(t)\, dt}. \quad (9)$$

$A_j(t,\omega_j)$ = amplitude of the jth IMF at time t with frequency ω_j. Thus, the amplitude of the frequency-time decomposition of $x(t)$, denoted by $A(t,\omega;x)$ is given by

$$A(t,\omega;x) = \sum_{j=1}^{n} A_j(t,\omega_j). \quad (10)$$

$A(t,\omega;x)$ is referred to as the Hilbert-Huang spectrum of $x(t)$. The marginal spectrum, denoted by $h(\omega,x)$, is obtained as

$$h(\omega,x) = \int_0^{\infty} A(t,\omega;x)dt, \quad (11)$$

which is the decomposition of $x(t)$ in the frequency domain. Consequently, each IMF can be calculated using the Hilbert transform to obtain the first n different frequency components at any time. HHT is applied in processing the seismic response recordings in a base-isolated building and the dominant frequencies are inferred.

3 Building profile

The base-isolated building is frame-shear wall structure with 8 stories and 2 basements. The total height is 33.0 meters and the plan view is shown in figure 1. An intense seismic response array was instrumented in the building with 33 sensors on basement 2, 1st floor, 3rd floor, 6th floor and the roof. Five 3-channel sensors are installed on the geometrical center and six 3-channel sensors are on the two ends of the building. A 3-channel seismometer was deployed in a 100-meter borehole. Sensor layout is demonstrated in figure 2.

4 Seismic response recordings

After the array was instrumented, several earthquakes were recorded including the earthquake at the border of Baodi District, Tianjin and Yutian County, Hebei in China. The earthquake occurred in June 18th, 2012 with magnitude 4.0. The epicenter distance was approximately180 kilometers and the amplitudes of the recordings are small. The seismic response recordings of the building were corrected and shown in figure 3. The peak acceleration is 0.20 cm/s^2 in North-South direction and 0.21 cm/s^2 in East-West direction.

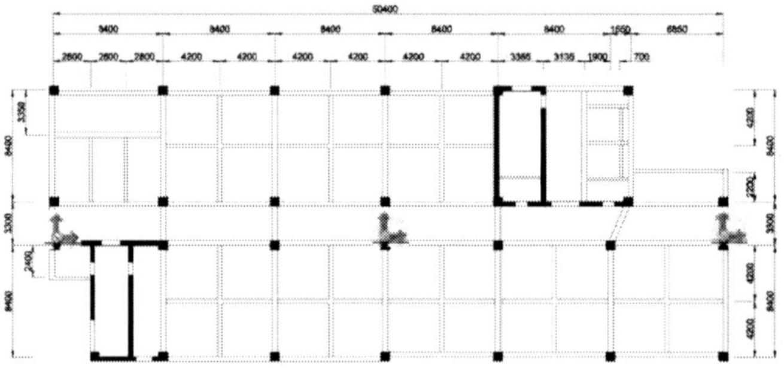

Fig. 1. structure plan view. Red arrows represent the sensor and the monitoring direction.

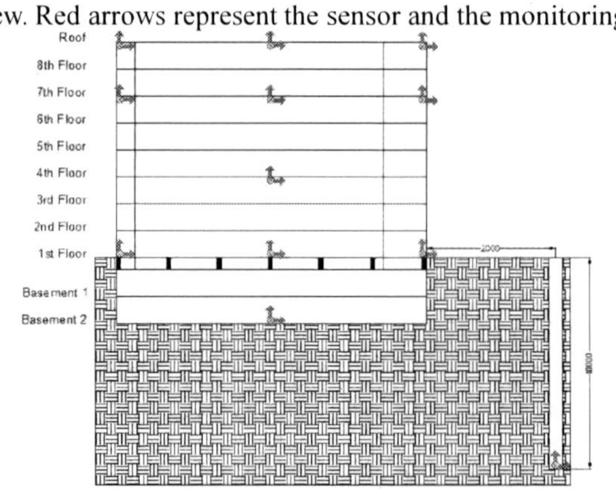

Fig. 2. layout of 36 sensors in the building. Red arrows represent the sensor and the monitoring direction.

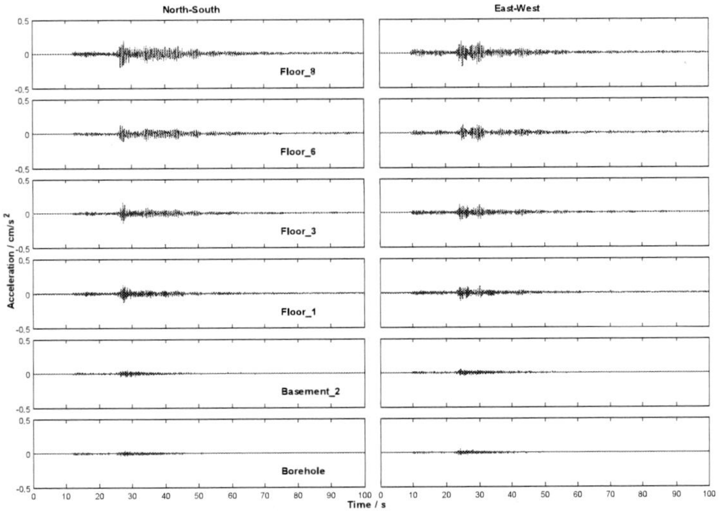

Fig. 3. time-histories of seismic response recordings in North-South and East-West directions.

5 HHT analysis

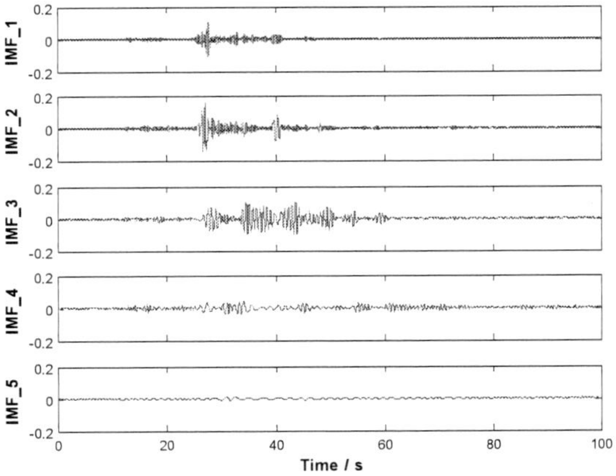

Fig. 4. IMFs obtained in EMD analysis defined from IMF_1 to IMF_5 with the unit as cm/s^2.

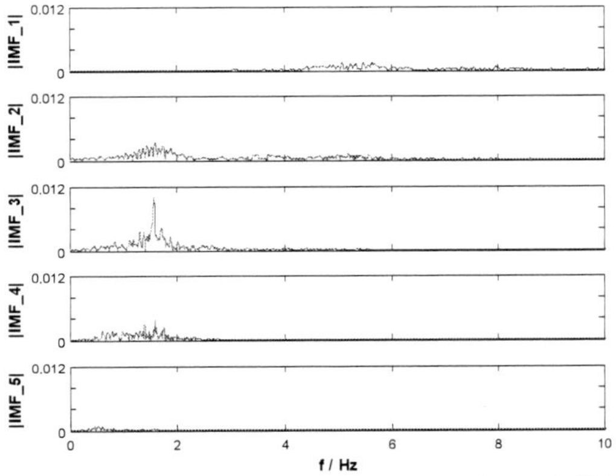

Fig. 5. Hilbert transform of the IMFs and the frequency contents for each IMF with the unit as amplification factor.

The time-history in North-South direction on the 8th floor is selected for HHT analysis. EMD is applied to the recording and the first 5 IMFs are decomposed. They are nominated from IMF_1 to IMF_5. Figure4 demonstrates the IMFs and IMF_5 becomes almost vanished. The first 4 empirical mode components of the recording are dominated at a specific frequency throughout the time history. After EMD analysis, Hilbert transform is used to each IMF and the frequency components are obtained for all IMFs, as indicated in figure 5. Herein, IMF_1 contains higher component and the frequency decreases as the number of IMF increases. Frequency component in IMF_5 is greatly weakened.

To verify the effectiveness in component decomposition, the first 5 IMFs are summated and the analytical function is reconstructed. It is compared with the original signal. The plot in figure 6 suggests the differences between the analytical function and the original signal. The main vibration from 25s to 50s are extracted and demonstrated. Tiny differences indicate that 5 cyclic analyses for IMFs are adequate to obtain the majority frequency components. Therefore, all the frequencies in different IMF are added together and the frequency component of the analytical function is determined. The frequency contents of the first 3 IMFs represent the frequency components in the recorded seismic response. As shown in figure 7, the dominant frequency of the building under this earthquake is implied as 1.6 Hz. The dominant frequency was analysed with the value of 1.62 Hz using other analytical method. The difference is 1.3%. They are considered to be consistent. There is another frequency between 4 Hz and 6 Hz. It is regarded as the vibration excited by the surface wave. As shown in figure 5, it is the frequency of the IMF_1 and it is visible, but very weak.

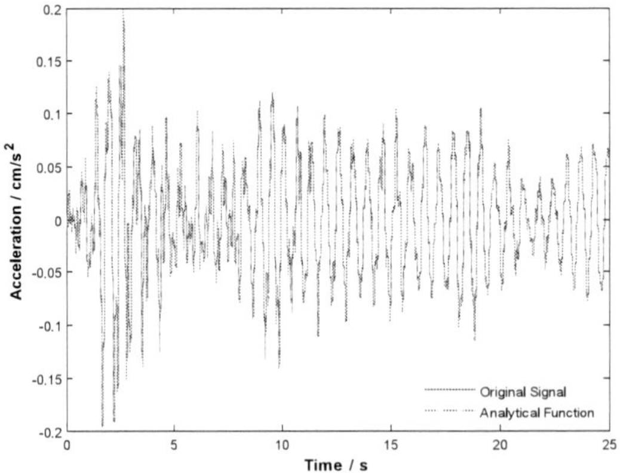

Fig. 6. Comparing between the seismic response recording on the 8th floor and its analytical function.

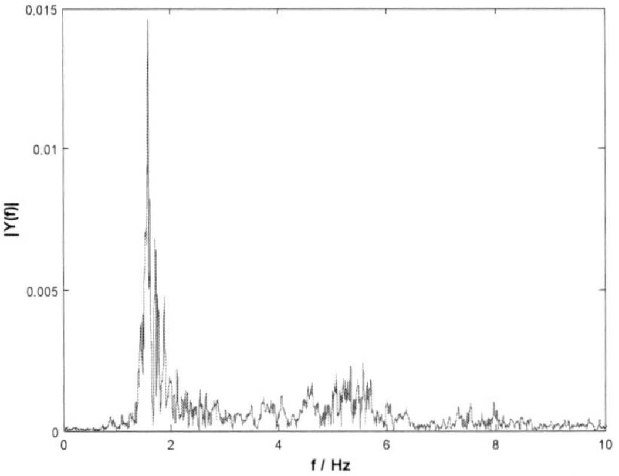

Fig. 7. The frequency content of the analytical function and the dominant frequency is 1.6Hz.

6 Conclusion

Hilbert transform can be used to identify the time variant characteristics of the structural system. The data series have to be decomposed into its components to admit a well-behaved Hilbert transform. EMD method is first implemented for component decomposition. The EMD method and Hilbert transform construct HHT, which is developed to analyse signals in the frequency-time domain. HHT is introduced to process the seismic response recordings in an 8-story frame-shear wall base-isolated building. EMD method is applied to decompose the frequency content of the signal on the roof and 5 IMFs are inferred from higher frequencies to small ones. The components with their frequencies are indicated after Hilbert transform. The origin signal and its analytical function are compared and no obvious differences are detected. The HHT analysis can achieve satisfactory accuracy in reconstructing the origin signal and inferring the frequency contents.

Acknowledgement

We would like to thank the anonymous reviewers for thorough reviews, which helped to improve the technical quality and presentation of this article. Special thanks are extended to Beijing Earthquake Bureau for providing the waveform data. We also thank Erol Kalkan for sharing his computer codes, which we modified significantly for this study. This study is financially supported by Project of Science for Earthquake Resilience, China Earthquake Administration, Granted No. XH17001.

References

1. Huan S L, Mcinnis B C, and Denman E D 1983 Analysis of the random decrement method Int. J. Syst. Sci. 14 (4) pp 417-423
2. Abdulkareem M, BakharyN, Vafaei M 2015 Application of Wavelet Transform to Damage Detection in Plates using Response-only Measurements In 16th Asia Pacific Vibration Conference pp 24-26
3. Vafaei M, Alih S C, Rahman A B A, Adnan A B 2015 A wavelet-based technique for damage quantification via mode shape decomposition Structure and Infrastructure Engineering 11(7) pp 869-883
4. J. N. Yang, Y. Lei, S. Lin, and N. Huang 2004 Hilbert-Huang Based Approach for Structural Damage Detection JOURNAL OF ENGINEERING MECHANICS 130 pp 85-95
5. Ridong Du , Yongbo Yuan and Miao Chen 2012 Empirical Mode Decomposition Application for Structural Seismic Responses Applied Mechanics and Materials and Materials256-259, pp 2096-2101
6. Huang N E et al. 1998 The empirical mode decomposition and hilbert spectrum for nonlinear and nonstationary time series analysis Proc. R. Soc. London Ser A 454 pp 903-995
7. Huang N E, Shen Z and Long S R 1999 A new view of nonlinear water waves The Hilbert spectrum Annu Rev Fluid Mech. 31 pp 417-457
8. Jianping Han and Dawen Li 2010 Modal parameter identification based Hilbert-Huang Transform and natural excitation technique ENGINEERING MECHANICS 27 (8) pp 54-59

Facility management and building information modeling during operation and maintenance

Michal FALTEJSEK*, and *Blanka* CHUDIKOVA

VŠB - Technical University of Ostrava, Faculty of Civil Engineering, Ludvíka Podéště 1875/17, Ostrava-Poruba, 708 33, Czech Republic

Abstract. BIM is a concept and also a method currently applied throughout the world in the modern building construction. The impact and benefits of the BIM method are clear across all stages of the life cycle of buildings (project, realization, operation, demolition). The longest and most expensive phase of buildings life is the operation and usage part. Many common aspects and linkage of processes between the BIM method and the facility management, i.e. a modern asset management method, can provide greater efficiency for these activities, help reduce the cost of required building operations and maintenance and save time for individual activities. The BIM can be understood as an information management of buildings. Many common elements can create a synergy of processes related to the management of buildings. The article discusses these aspects and emphasizes the listing of the basic data and processes that are necessary for mutual interconnectedness.

1 Introduction

Operation and maintenance of buildings are the longest and most expensive parts of the life cycle of buildings. They are parts of many processes that must work efficiently, clearly and sustainably in the long run. The life of buildings is in the order of decades, and throughout the period it is necessary to maintain a standard for the users of these buildings. Properly set up facility management processes using efficient and effective methods can help extend the useful life of buildings and delay their degradation. One of these methods that can be applied is BIM method, i.e. Building Information Modeling / building management.

BIM is a concept that brings a new look at the life cycle of buildings. BIM method can be applied in two ways. The first is the implementation of BIM method already in the construction project, through implementation and then in operation and maintenance. This can only be applied to new constructions. For buildings that are already built, we need to implement BIM afterwards. BIM implementation process for existing buildings may vary, depending on the available documentation, information, software used, data formats, the level of detail, or the expected benefit. f[1]

The procedures and methods for implementing BIM in these two cases are different. For already constructed buildings, the data is available at the very beginning of BIM implementation process, as well as the construction, equipment and overall construction.

* Corresponding author: michal.faltejsek@vsb.cz

Most of the companies that has launched BIM is still focused on implementing this concept in designing and constructing new buildings. [2]

To optimize the facility management processes it is necessary to provide a detailed overview of structures and elements, and the general state of the buildings already built. This guarantees BIM and, if properly managed, guarantees long-term benefits in many forms. Individual activities of a facility management can be simplified by BIM to achieve greater work efficiency, reduce time for individual activities and cut down costs. [3]

BIM is also about cooperation. Co-operation and correctly set procedures and processes already at the implementation stage of BIM can be crucial for the following work and maintenance of BIM. Just as designers and developers of buildings should communicate with the end user and facility manager, i.e. to take into account the following stages of the building life cycle, the future work with BIM model (including maintenance and updates) should also be made provision for in the creation of an additional BIM model and its implementation into the facility management. The implementation of BIM should then be built on that. [4]

2 Detailed passportization of property

The real asset overview and its technical and economic features are key to the proper functioning of facility management. This requires high-quality passportization of the existing building. A passportization itself is a process consisting of several sub-parts. First of all, it is necessary to gather the available documents, sort them out and check their up-to-dateness. The next step is to fill in the missing information. Mostly by a personal survey of the building or by obtaining it from external sources.

It is also useful to take a photo of individual facilities, especially the technical equipment, and enclose it to an element in the CAFM system as an attachment. This ensures checking and a better overview of the equipment, its location in the space, attachment(s), components, etc.

Personal research is time-consuming and requires access to individual spaces. In rooms, it may be difficult to locate or physically access the searched for elements. They may be under lock and key or covered. In many cases they might not be in places indicated by available data. All this information has to be recorded and then re-edited into the correct structure and assigned to the corresponding areas (if it involves checking a facility).

2.1 Quality records and data sources

Quality records of managed assets' data are a basic prerequisite for the efficiency and cost-effectiveness of downstream activities. Quality level of records is a very important element in any property management. The records should be uniform, comprehensive, clear, of high-quality and standardized. In the service life cycle of buildings, we assume a long-term records of assets. That's why their accuracy and up-to-dateness is necessary. [5]

The quality level of the records serves as a prevention of error-making by decisions that ultimately may have negative impacts due to lack of information and associated poor evaluation. This creates additional cost items associated with removing or correcting errors.

Data sources should be in the first place verified and up-to-date. For example, the sources can include:

1. Electronic drawing documentation
2. Digitization of paper drawing documentation
3. Physical data collection on site
4. Suppliers of design services
5. Construction contractors

6. Suppliers of technical equipment and technologies
7. Service suppliers
8. General Central Registers (Land Registry, Territorial Identification Registry of Addresses, Registers of Economic Bodies, Roads Register) [6]

3 Information model of building in facility management

The interconnectedness of facility management and BIM methods is primarily in data. The BIM model forms a basic graphical and informative basis for managed asset. Facility management and support software (CAFM systems) provide detailed inventory of assets and enable planning of processes and operation and maintenance activities. The interrelation of these activities is therefore indisputable and lies above all in data that allow better visualization in a BIM model environment, refurbishment or modernization planning, their status verification, provide accessibility, etc. [3]

Digital background information created in 2D CAD software can be a good basis for BIM creation. This is evidenced by the claim that the introduction of BIM for existing buildings (with 2D only documentation) is possible with the use of this background information and there is no need to create a completely new model and without background material.

The 3D model itself does not yet form BIM, but only a suitable background. At present, "Information Building Management" is increasingly mentioned instead of "Information Modeling of Buildings", which shifts the importance to the information and subsequent work with it rather than to the 3D model itself. However this does not mean we should do BIM without a 3D model. Its use is wide and the creation of such a model in accordance with the BIM concept is therefore desirable for facility management.

3.1 Additional BIM building model creation

Original drawings of a building usually in 2D serve as a basis for creating a 3D model (figure 1). This drawing will be a ground plan of the future model and will guarantee the retention of the ground plan dimensions, equipment layout, etc.

In the Revit software it is very easy to create such a background from a DWG drawing. For individual floors a corresponding drawing is always attached of the modelled floor and aligned with the preceding floor so that the individual floors link to each other when drawn in 3D (that is why it was necessary to modify individual 2D drawings).

The model is drawn on the basis of a defined LoD and the available software options. For example, structures and rough construction can be drawn in a very detailed and realistic way. For specific types of devices and other equipment, there may be a problem with the model. Atypical shapes and many kinds of equipment (such as chairs, tables, cabinets, but also technical equipment) are often not available in libraries. Also, in cases of existing buildings the equipment is mostly obsolete and not produced anymore, therefore not available. In such cases, the placeholder model of a universal element can be applied and distinguished only by the information / description. Another option may be to create your own model. However, it is very labor-intensive and the efficiency / time ratio shows that it is usually not worth it. [1]

Fig. 1. 3D graphic model of the MSDK building (author, Autodesk Revit software).

3.2 Adding basic information to individual elements

What constitutes BIM is the information. It is important, therefore, that individual elements of the model should be enriched by the information or be properly interconnected with such information (figure 2). Some information may be inserted directly into the model, but if there is much of information to the individual elements it is more appropriate to feed it into the CAFM system and link it to the graphical model. [2]

Therefore, only basic but important information is included in the model. It may be information about material, colour, type of equipment, or price. Naturally there is also information on the dimensions and location of the equipment.

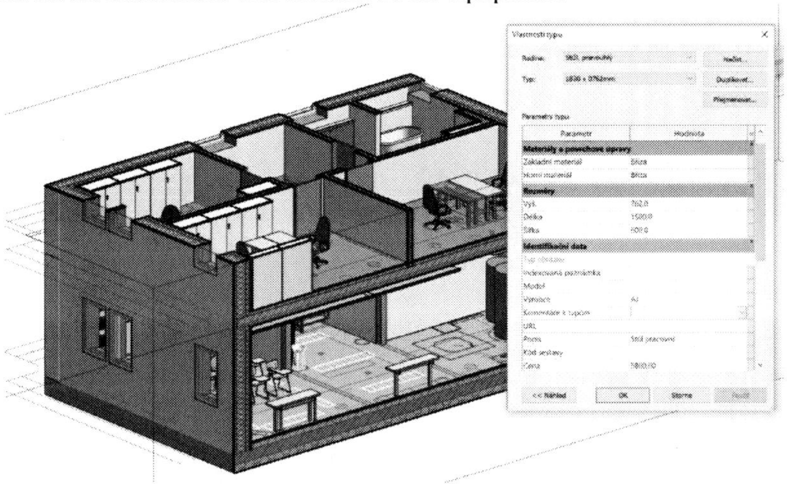

Fig.2. Information in the 3D model of construction for individual elements (Author, Autodesk Revit software).

4 CAFM systems and BIM

Computer Aided Facility Management are systems that should make facility management's asset management work easier. These systems are mainly used to manage the FM support processes, their clarity of records and any planning and evaluation. CAFM systems are continually improving to meet more modern and newly incoming customer requirements. It could be said that due to the development of systems around the world, we should only designate true CAFM system only those systems that feature a statistical, dynamic and output data area. These are, above all, systems that cover all known areas of support processes.

The CAFM systems' deployment aims in particular to:

- reduce operating costs,
- improve the quality of services provided and improve the quality of the environment,
- optimize the relationship between worker, environment and processes,
- extend the life of the monitored constructions and objects,
- introduce standards, rules and working processes in the given field and in the system of embedded business logic,
- introduce and distribute internal costs and their assignment to departments, divisions, activities, projects, etc.,
- manage and maintain documentation, moves, benchmarking, inventory and inspection,
- prepare for unexpected events and accidents, processes required by legislation (audits, inspections, ...), sustainable development. [6]

4.1 Connecting to a CAFM system

The 3D BIM model itself is not sufficient enough and needs to be linked to the CAFM system. This allows for the mutual localization of the elements: from the model to the CAFM system and vice versa, the exchange of information again in both directions, or the possibility of colouring the search areas. CAFM (Computer Aided Facility Management) is a facility management system with a wealth of different asset management functions (reactive and proactive maintenance, planning of activities, inventory of small and investment assets, migration planning, room reservation, key management, OSH and FS, waste management etc.).

5 Linking BIM and Facility Management Processes

The first step is to implement existing FM processes to the BIM model and the CAFM system. Thus, we create typical operations and maintenance activities in relation to the areas and equipment we know and that are already being routinely used.

The main interaction consists in visualizing the data into the graphical realistic form of the building. With the help of virtual reality, this effect can be further enhanced. Facility management can save the facility managers a lot of time in the case of large complexes and complicated buildings when they need not go through the building personally to see the individual facilities, the layout of the equipment in the rooms, the accessibility of the individual facilities, etc., as a clear and realistic visualization in BIM model with a link to all important data in the CAFM system is for such tasks sufficient. Also, all changes (reconstruction, upgrading, transfers, moving, etc.) can be planned through the BIM model thus creating options for subsequent final decisions.

Basic facility management activities that can be applied:

- reactive maintenance - ad hoc troubleshooting and repairs
- proactive maintenance - planned inspections and checks
- management of lease relations
- energy, media and consumption meter management

- key management
- waste management
- OSH and FS
- cleaning of spaces and elements
- management of held events
- room reservations, room occupancy (offices by internal / external staff)
- document management system (document administration) [6]

There are many activities falling under facility management. Generally speaking, their task is to ensure that the subject and its staff are carefree, so they do not have to worry about dealing with the above-mentioned activities in any way.

In the CAFM system, you need to define the processes to support the respective FM activity (listed above, but also others that are not mentioned here but are associated with asset management, operation, and maintenance). The linking of activities and their records in the CAFM system will ensure greater clarity, traceability and access to follow-up information (by whom and when the activity is performed, for what purpose, which documents are attached to the activity, responsible persons or entities contact particulars, further information on the managed element, etc.).

5.1 Optimization of Facility Management Processes

The optimization involves primarily analysis and evaluation of results / comparison with the original state. For this purpose it is advisable to pre-determine the quantities that we will track over time and compare. On this basis, it is possible to define the differences and clearly define the gaps in the individual processes. You can then optimize these based on your experience, options and data.

Optimization is a long-term and repetitive process. It is necessary to have the process set for some time and to monitor and record it in detail. After performing optimization (change), it is necessary to again analyze and track newly established procedures, which are then compared with the original ones and their contribution evaluated.

In the long run the overview makes it much easier to evaluate and analyze facility management processes, track back their workflows, and then optimize them accordingly. [6]

6 Conclusion

On the whole, facility management and use of the BIM model offers clarity: if we have a large complex and are looking for a specific facility, room or equipment, we can easily visualize this element or person through the CAFM system and display it in a 3D model. Realistic graphical depiction will help to avoid lengthy search and visualization during team work will help without having to physically visit a particular part or equipment.

Other uses may be in the colouring of areas: e.g. planning for cleaning or moving, reconstruction and upgrading (e.g. colouring of all areas where the floor covering is tiled or where a non-load-bearing partition of 100 mm thickness is used, etc.).

Using a BIM model interconnected with CAFM system during modernization and reconstructions we can plan the work activities, the applicable bill of quantities and find out where the technical building installation systems are concealed in walls, what the properties of the current structure are, what materials have been used, and much more.

An important element for the operation and collaboration of facility management and BIM method is the aspect of the data update. BIM itself, if it remains inactive, lacks sense in the long term. This problem is very common in new constructions, when a BIM model was created and the data passed to the implementation stage. Based on the BIM concept, the construction is subsequently implemented, but most of the time the BIM process ends there and does not continue into the operation and maintenance phase. Which is a problem that is still very much underestimated. Yet, the operating phase of the construction is the longest and most costly part of the life cycle during which the wide scope of the BIM model and data use could be utmost beneficial.

Acknowledgement

This work was supported from the funds of the Students Grant Competition of the VSB – Technical University of Ostrava. Project registration number is SP SP2018/113.

References

1. BRUMANA R, DELLA TORRE S, PREVITALI M, BARAZZETTI L, CANTINI L, ORENI D and BANFI F. Generative HBIM modelling to embody complexity (LOD, LOG, LOA, LOI): surveying, preservation, site intervention—the Basilica di Collemaggio (L'Aquila). *Applied Geomatics.* 2018, 10(4), 545-567, DOI: 10.1007/s12518-018-0233-3. ISSN 1866-9298. Available from: http://link.springer.com/10.1007/s12518-018-0233-3
2. SMITH, DANA K and TARDIF M. Building information modeling: a strategic implementation guide for architects, engineers, constructors, and real estate asset managers. *Hoboken, NJ: Wiley*, 2009, ISBN 978-0-470-25003-7.
3. Černý M a kol. BIM příručka, 1. vydání, *Praha*, 2013, ISBN 978-80-260-5297-5, s. 80.
4. SANCHEZ A X, HAMPSON K D and VAUX S. Delivering Value with BIM, A whole-oflife approach. *London a New York: Routledge*, 2016. ISBN: 978-1-138-11899-7.
5. EASTMAN Ch, TEICHOLZ P, SACKS R and LISTON K. BIM Handbook, A Guide to Building Information Modeling for Owners, Managers, Designers, Engineers, and Contractors. New Jersey: *John Wiley & Sons, Inc*, 2008. ISBN: 978-0-470-18528-5.
6. Kuda F, Beránková E and Soukup P.. Facility management v kostce pro profesionály i laiky, *nakladatelství FORM Solution*, 2012, první vydání, ISBN 978-80905257-0-2.

Design practice of interactive imaging art in the new media art——Taking “Ink-wash Tai Chi” as an example

Lu Geng[1,*], and *Lei* Cao[2]

[1]College of Sino-Korea Multimedia Design, Shanghai University of Engineering Sciences, Shanghai, China
[2]College of Electronic and Electrical Engineering,Shanghai University of Engineering Sciences, Shanghai, China

Abstract. In order to be accepted by the extensive public, the traditional culture requires digital media interactive methods featuring "diversified sensual experiences" and "interactivity" to adapt better to the development tendency of cultural communication under the background of new media. The design practice, with Tai Chi culture as the content and using motion data captured by infrared sensor Kinect combined with digital art creation platform *Processing* as the technical measures to implement interactive imaging. The work will give interactive experience to the experiencer and promote the spread of Tai Chi culture.

1 Interactive imaging art overview

The interactive imaging art, as a new art form emerged recently in the world, belongs to the subject of new media art. The new media refer to the media based on digital technology and featuring interactive communication and creative forms[1]. The interactive imaging art is the best manifestation of the interactivity of new media art. It grew out of but is different from the video art and computer graphic art[2].This new art form requires the strong support of information technology of the time. The interactive imaging art depends on various sensors to capture the spatial or moving data, which will then be collected and put into computer for calculating and processing. In this way, uncertain imaging feedbacks with random beauty are gained. This interactive form breaks the previous rigid and non-interactive art form. The participation of people and environment can change many characteristics and even the meanings of a work. McLuhan said in the book of understanding media that "Media is an extension of people," and people is the world's most primitive information media[3]. The interactive imaging can encourage the emotional exchanges between people and narrow the distance between people and the art.

In the *Video Place*, an interactive imaging device work by Myron Krueger in the 70s, the first system allowing users to interact with virtual objects was created. The device mixed the imaging pictures of the users with other images through a computer. This is significant as the initial appearance of interactive imaging. This work created a new art

* Corresponding author: genglu_2005@163.com

form based on the "creator - computer system - experiencer" interaction. Taking the images of the audiences as an important part of the work, this interaction mode will become the basic model for many of the interactive media works in the future[4].

2 Interactive form in interactive imaging art

As an interactive method, interactive image art mainly uses interactive hardware and programming languages a human-machine interface to perform "dialogue" with the experiencer. There are three commonly used methods in interactive images: touch interaction, gesture and action interaction, and voice interaction. Gesture and action interaction are easy-to-understand languages that are consistent with popular cognitive habits, enabling natural interactions between humans and computers.

With the device interactive work named *Quantum Space*, the movements of experiencers who entered the space are identified by the Xtion depth camera and computer system to project one or more digital figure images on the digital wall in time. These digital figures are momentary, now colorful swirls, now geometrical neon, and now particle system, parting and gathering in different shapes. The device generates shapes and colors based on particle physical modelling and then casts digital figures in various postures(see figure 1). As the designer Sodazot explained: "In the quantum space, you are divided into light quantum and empowered with the force to communicate with the universe".

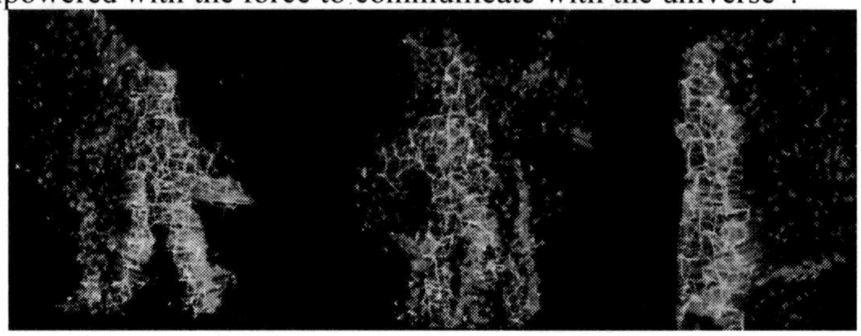

Fig. 1. Imaging of a Quantum Space Device Interactive Work.

Established in 2001, TeamLab, a technology art institution from Japan, created many immersion works featuring mixed lights and shadows and fabulous beauty. For example, the theme work" Flower Forest, Lost and Immersed" builds a fancy world for the audience. The exhibition hall is all dark when empty. If someone steps into the hall, flowers will bloom on his or her body and spread to the feet, and then link with the flowers on the next person; the flowers bloom and wither automatically, in cycles without an end; when the visitor is sitting or standing still, flowers are in extraordinary bloom; when visitors are leaving, flowers begin to wither and disappear. Then the hall returns to darkness(see figure 2). The TeamLab team created this work to break the boundaries in modern people's minds, those between themselves and the world and between the world and others, enlightening the future living space of digital art.

Fig. 2. An interactive Imaging Work from TeamLab,the "Flower Forest, Lost and Immersed".

3 Design practice of the interactive imaging work named "Ink Painting Tai Chi"

3.1 Work contents

The Tai Chi culture, as a representative of Chinese traditional culture, is very famous at home and abroad. The Tai Chi exercise is also the most popular Chinese traditional fitness project in the world. Chinese Tai Chi, applying to be one of the world intangible cultural heritages, requires urgent improvement in exposure and influence on the world cultural stage.

Driven by the social background of cross-border integration and multicomponent integration, the traditional arts are all trying to become diversified, in order to keep a foothold and make improvement. There are practices of interaction + dance, interaction+ drama, and interaction+ music, integrating technology into the traditional arts and realizing stronger participation and interactivity. The Ink Painting Tai Chi is a design practice work trying to combine interaction with Tai Chi. Through the picture effects like ink painting, it combines the movements of Tai Chi exercise with twenty-four forms with the Tai Chi theories and shows them by means of interactive imaging art, reflecting the ideas of Tai Chi in four interactive aspects, namely the opening, developing, changing and concluding.

Opening: the start of the Tai Chi exercise with twenty-four forms requires concentration, as calm as the mountains. The work simulates the ink painting mountains with Processing. The opening movements in Tai Chi, together with breathes, will control the ups and downs of the mountains. Breathing in for ups and breathing out for downs. There is a sun rising behind the mountains.

Developing: smooth connection, natural linkage, and movements in line with will. The filmy figure of the experiencer shows up in the work. When the experiencer is moving from left to right slowly, the postures can be seen in the screen. The moving traces of hands stay for a while like the smoky particles. The "Qi" of Tai Chi is represented by particles, displaying the relaxed and full movements in Tai Chi.

Changing: variegated, motion-stillness integrated, and virtual-real transformable. There are two interactions in this part. Firstly, when the experiencer is moving, each movement is recorded and, along with the time, the records are made into the static shadows of a series of different movements. When the experiencer is still, the shadows begin to move. Secondly, the pictures show the outlines of the twenty-four forms in addition to the shadow of the experiencer. If the experiencer's movements match with the form outlines, it will turn to the next form.

Conclusion: the ending should be with integrated harmony, and you will know about Yin and Yang if you can not only open but also conclude. The figure of the experiencer is no longer concrete. It can interact with the numerous particles in the pictures, which do not disappear or rest and will gather into a shape of Tai Chi in the end.

3.2 Technical structure

The Tai Chi movements and body images of the experiencer are sensed in time by the Kinect depth camera. The data collected will be sent to a computer for the processing by SimpleOpenNI, an open human-computer language interactive framework. The computer can identify the body location and the joint positions in the 3D space through the original depth data. According to the feedbacks of different movement signal sources, image changes preset by the Processing are casted synchronously. The real-time interactions between the experiencer's body and the images are realized(see figure 3).

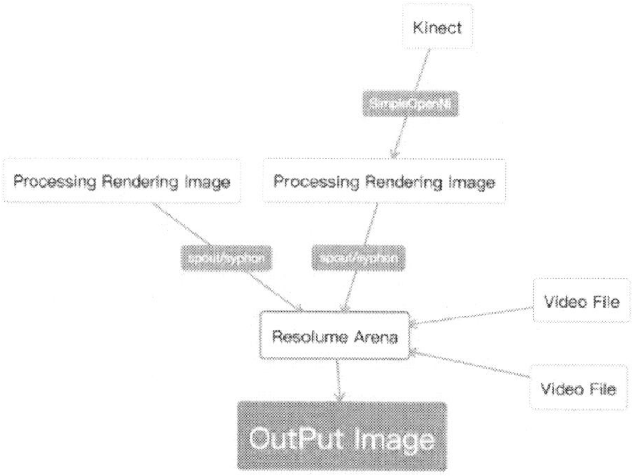

Fig. 3. Technical Structure of the Work.

The User mechanism of the SimpleOpenNI enables the interactions between more local users. Once a new user is identified, the response function onNewUser(), is motivated to register the user ID and start the Tracking Skeleton. The user outlines traced by the device is stored in the form of binary image matrix, which can be used through userMap() function in the form of arrays.

3.3. Key techniques implementation approaches and results show

3.3.1 Ink painting mountains simulation by processing

Realize the overlaps and fluctuations of mountains with a self-defined class, which consist of: image variables to record a transverse mountain, coordinate of the mountains in the picture, two-dimensional variables to generate the Perlin noise of the mountains, and the dynamic moving distances of mountains. A transverse mountain is completed through the drawing function and the Perlin noise in Processing. The program records and shows the mountain pictures generated along with the ups and downs of the experiencer's hands, creating mountain images of different heights and distances[5] (see figure 4).

Fig. 4. Ink Painting Mountains.

3.3.2 Haze effects simulation by processing

Haze effects are realized by identifying two classes. One for haze particle effects realization, setting the initial moving speed and directions of particles with the random numbers produced by random Gaussian(). The other for particle swarm management, controlling the lifespan of particles with a member variable of Lifespan, regulating the running of each particle and removing those ended particles(see figure 5).

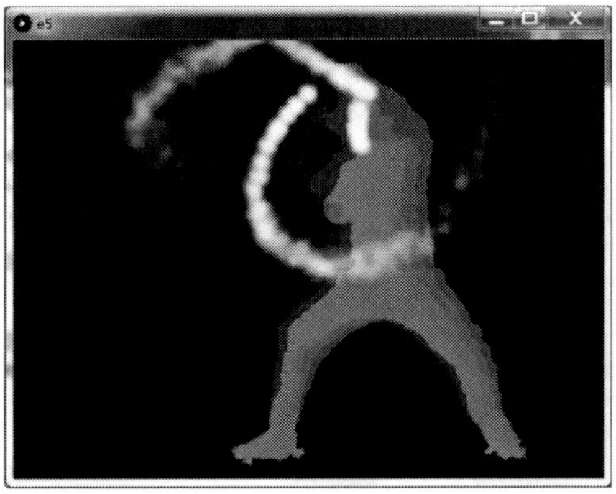

Fig. 5. Haze Particles.

3.3.3 Realization of multi-linear record and display through processing

The recording process is divided into four stages: In the reset stage, all relevant parameters are reset and the second stage will come in the next frame. In the second stage of recording, the experiencer's body moving process is recorded and the frame will be displayed if it is the integral multiple of the space, so that the experiencer images will stay for a long time. The third is the pause stage. Since estimation of the duration of experiencer's each movement is key to the implementation of recording, a threshold value is set in order to stop the recording when the experiencer is holding a posture. The recording will stop when the holding time surpasses the threshold value and continue when the user moves again. If the pause time reaches the threshold value, the display will begin. The last is the multi-linear display stage. The breaking points are displayed at the same time and it will return to the first stage until the first breaking point has been displayed(see figure 6).

Fig. 6. Multi-linear Recording.

3.3.4 Particle Tai Chi images realized through processing

In order to realize the production of Tai Chi images with particles, a picture of Tai Chi in the format of png or jpg should be prepared in advance. During the reviewing of all pixels of the picture, the pixel is identified as the coordinate of a particle if its lightness is lower than the threshold value. And the identified coordinates are used for the recovery of Tai Chi images. The number of particles depends on their intervals. Before the Tai Chi images are produced, interactions between the particles and figures are achieved by using repulsive particles(see figure7 and 8).

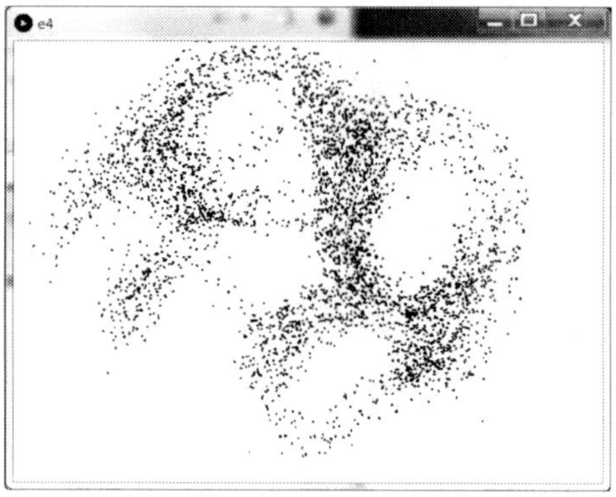

Fig. 7. Interactions between the Particles and Figures.

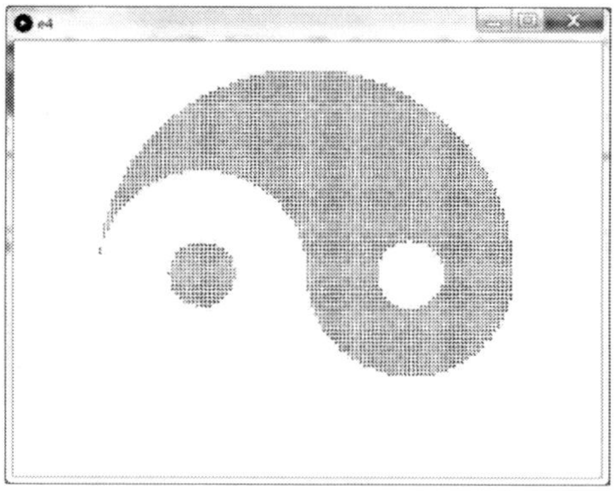

Fig. 8. Particles Aggregate into Tai Chi Images.

4 Conclusion

Interactive imaging is one of the expressions of new media arts. It has strong connectivity and interaction. Through the different connections, integration, interaction and transformation between the experiencer and the work, the image is finally presented. The imaging at this time is a work created by the creator and the experiencer. It is not only the image itself with the meaning of artistic creation, but also an extension of people's actual experience. Through the design practice of this article, it is also proved that in the context of cultural communication design in the digital context, it is necessary to integrate technology and art reasonably, establish emotions through interaction and exchange, enhance the sense of cultural identity, and create a dialogue-style cultural experience.

References

1. Baidu Encyclopedia 2017 *New Media* http://baike.baidu.com/view/339017.htm
2. Hua Yang 2009 *New Media Art Interactive Imaging Installation Art* (China: Shandong Fine Arts Publishing House)
3. Marshall McLuhan 2011 *Understanding Media: The Extensions of Man* (Canada: Yilin Press)
4. Ran Zhang 2013 *Real-time Interactive Image Rendering Based on Processing Platform* (China: Beijing University of Technology)
5. Daniel Shiffman 2017 *Learning Processing* (China: China Machine Press)

Deep convolutional neural networks for cardiovascular vulnerable plaque detection

Lincan Li[1,*], *Tong* Jia[2], *Tianqi* Meng[2], and *Yizhe* Liu[2]

[1]College of Mechanical Engineering and Automation, Northeastern University, Shenyang, China
[2]Institute of Machine Learning and Statistics, College of Information Science and Engineering, Northeastern University, Shenyang, China

Abstract. In this paper, an accurate two-stage deep learning method is proposed to detect vulnerable plaques in ultrasonic images of cardiovascular. Firstly, a Fully Convonutional Neural Network (FCN) named U-Net is used to segment the original Intravascular Optical Coherence Tomography (IVOCT) cardiovascular images. We experiment on different threshold values to find the best threshold for removing noise and background in the original images. Secondly, a modified Faster R-CNN is adopted to do precise detection. The modified Faster R-CNN utilize six-scale anchors (12^2,16^2,32^2,64^2,128^2,256^2) instead of the conventional one scale or three scale approaches. First, we present three problems in cardiovascular vulnerable plaque diagnosis, then we demonstrate how our method solve these problems. The proposed method in this paper apply deep convolutional neural networks to the whole diagnostic procedure. Test results show the Recall rate, Precision rate, IoU (Intersection-over-Union) rate and Total score are 0.94, 0.885, 0.913 and 0.913 respectively, higher than the 1st team of CCCV2017 Cardiovascular OCT Vulnerable Plaque Detection Challenge. AP of the designed Faster R-CNN is 83.4%, higher than conventional approaches which use one-scale or three-scale anchors. These results demonstrate the superior performance of our proposed method and the power of deep learning approaches in diagnose cardiovascular vulnerable plaques.

1 Introduction

Vulnerable plaques are unstable plaques which prone to form thrombus. They closely related to the occurrence of acute cardiovascular events such as thrombosis, acute coronary syndrome, coronary heart disease and even sudden death. About 19 million patients die each year due to acute coronary syndrome (ACS) and sudden cardiac death, patients who suffer from myocardial infarction caused by the rupture of unstable coronary atherosclerotic plaques account for 70% of the total number [1, 2]. Therefore, accurate detection of cardiovascular vulnerable plaques is significant in preventing cardiovascular diseases.

Developed in recent years, Optical Coherence Tomography (OCT) is a new coronary artery imaging technology both at home and abroad. OCT not only has a resolution rate of 10μm, 10 times the resolution rate of IVUS, but also has the ability to identify

* Corresponding author: lincanli@outlook.com

organizations (e.g. the main components and characteristics of plaque can be showed in OCT images [3]), hence Intravascular Optical Coherence Tomography (IVOCT) has become preferred method in clinical detection of vulnerable cardiovascular plaques.

In recent years, with the fast development of deep learning and computation power, computer aided diagnosis has been applied to some medical fields, making medical image diagnosis more effective [4]. However, they yield some distinctive drawbacks when it comes to cardiovascular vulnerable plaques detection:

1. Cardiovascular vulnerable plaques adhere to the inner wall of blood vessels, which are more difficult to detect than common object detection targets [5].
2. IVOCT cardiovascular images belong to ultrasonic image. Owing to the imaging process of ultrasonic image, image contents and the interference of instrument catheters, IVOCT cardiovascular images have a lot of speckle noise [6].
3. Medical images require manual annotation by professional physicians, which is more difficult than common object annotation. Besides, the number of available images is limited.

In order to solve these drawbacks mentioned above, we propose a novel cardiovascular vulnerable plaque detection method using deep convolutional neural networks to detect vulnerable plaques. In the proposed method, we first use a Encoder-Decoder structure neural network named U-Net to segment the IVOCT cardiovascular image dataset, removing background and instrument catheters. Then, state-of-the-art object detection framework Faster R-CNN is applied to detect vulnerable plaques. We experiment different threshold values on U-Net and choose the threshold which has best segmentation performance. We also design six-scale anchors for cardiovascular vulnerable plaques detection, so it takes account of a wider range of plaques' size than conventional methods which use one-scale or three-scale anchors.

We evaluate our method on CCCV2017 Cardiovascular OCT Vulnerable Plaque Detection Challenge dataset. The Recall rate, Precision rate and IoU (Intersection-over-Union) rate are 0.94, 0.885 and 0.913 respectively, higher than the 1st team of CCCV2017 cardiovascular OCT vulnerable plaque detection challenge. AP of the designed Faster R-CNN is 83.4%, higher than both conventional approaches and the 1st team of the challenge. Thus the method presented in this paper has great value for clinical application.

2 Methodology

In this section, we first give a description of data preprocessing and data augmentation, then describe the two-stage deep learning method presented for cardiovascular vulnerable plaques detection. In stage one, we use U-Net to do segmentation, removing background and instrument catheters in the original images. In stage two, we apply Faster R-CNN to do precise detection. The framework of our proposed two-stage deep learning algorithm is illustrated in figure 1.

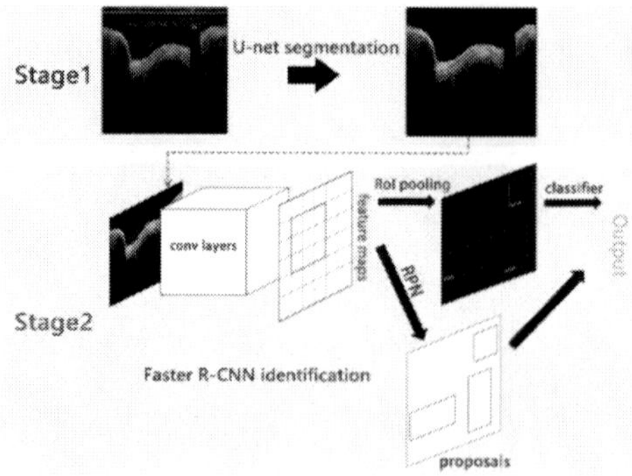

Fig. 1. The framework of our proposed detection algorithm.

2.1 Data preprocessing and data augmentation

The dataset used in this paper is provided by the CCCV2017 cardiovascular OCT vulnerable plaque identification challenge official[1]. The dataset consists of 2000 images, with 1000 positive samples and 1000 negative samples, the size of each image is 720*352 pixels. Ground Truth of the dataset are manually labeled by pathologists. We rotate, translate, shear and zoom the images in a proper range to enlarge the dataset using Keras [7].

2.2 Formatting cardiovascular image segmentation based on U-Net

Image segmentation is a crucial step in medical image diagnosis. It can remove unrelated background, instrument catheters and other irrelevant information from the original images, only keeping body organs or tissues which will be detected. Image segmentation is often the first step of computer aided diagnosis algorithm, preparing for subsequent accurate detection [6].

We mentioned in introduction that there are three problems in cardiovascular vulnerable plaque detection, in order to solve these problems, we compare several approaches and adopt U-Net to do segmentation. Proposed by Olaf Ronneberger et al in 2015, U-Net is a Fully Convolutional Neural Network based on encoder-decoder architecture [8]. U-Net won the first place in the ISBI challenge for segmentation of neuronal structures in electron microscopic stacks in 2015 and has been popular since then. Though U-Net has a relatively simple structure, it's able to have an excellent performance even on small datasets [8]. Besides, U-Net has been proved to have better performance on medical image segmentation than any other currently existing convolutional neural networks [9, 10].

In this paper, we use U-Net to segment IVOCT cardiovascular images. The insight of U-Net is to classify every pixel in a given image. Suppose the segmentation task require pixels to be classified into N class, when the label of original image is made, each pixel of label should belong to one of the N classes. In the presented method, we classify pixels into two classes. One is human cardiovascular area, the other is unrelated background and instrument catheter area. All labels used in this paper were made by pathologists, they sketched the contours of human cardiovascular area and filled unrelated area into black. Figure 2 gives an example of IVOCT image and it's label. Set p to represent an arbitrary pixel in an image, we normalize p to get p' using the equation: $p = p'/255$. Set threshold range as : $0 < threshold < 1$. If $p' > threshold$, we set the label as 1, otherwise the label is 0. Notice that threshold value is directly related to the quality of segmentation results, we make several experiments adjusting threshold values to find the best value. The curve of threshold and validation loss is shown in figure 3. From figure 3. we can observe when $threshold = 0.2$ the presented method attain the minimum loss on validation set. We further illustrate the results of U-Net segmentation using different threshold in figure 4.

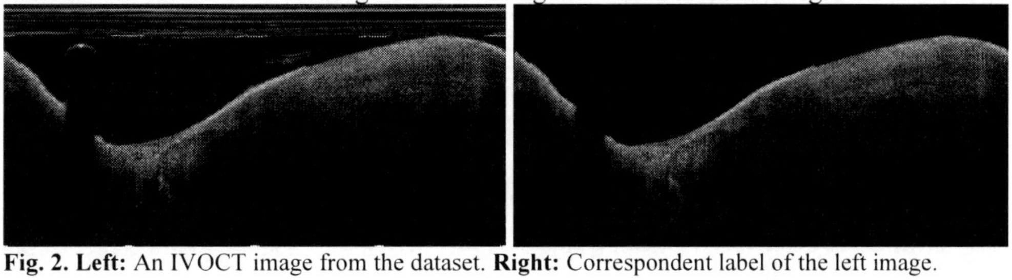

Fig. 2. Left: An IVOCT image from the dataset. **Right:** Correspondent label of the left image.

[1] http://ccf-cccv.org/2017/competition3/

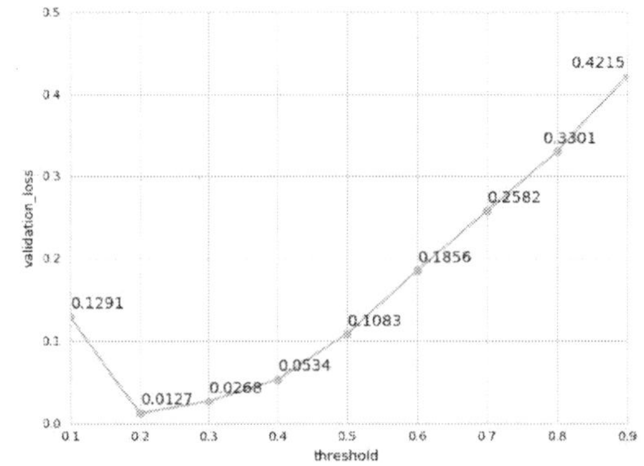

Fig. 3. Curve of threshold and validation loss. We experiment on different threshold values and found validation loss is minimum when threshold = 0.2.

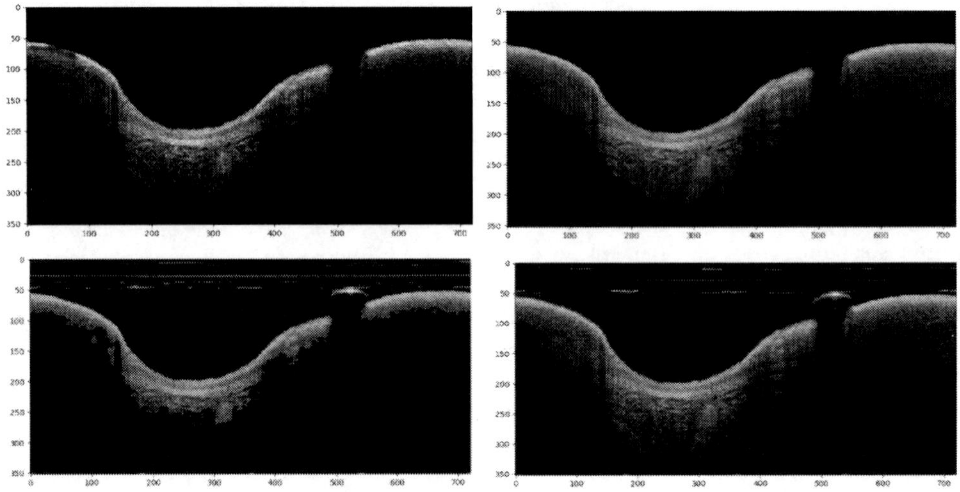

Fig. 4. Comparison of different threshold results. **Upper Left:** U-Net segmentation result when threshold = 0.2. It's the best result and we utilize it in our proposed method. **Upper Right:** Label image manually labeled by pathologists. **Lower Left**: U-Net segmentation result when threshold = 0.1. **Lower Right**: U-Net segmentation result when threshold = 0.6.

2.3 Precise cardiovascular plaque detection using faster R-CNN

In this section, we present how we utilize state-of-the-art detection algorithm Faster R-CNN to detect vulnerable plaques in IVOCT cardiovascular images which have been operated by U-Net. In detection algorithm, vulnerable plaque regions are marked with rectangle boxes, coordinates of the four corners can be obtained from the label. Since vulnerable plaques vary greatly in size, we utilize six-scale anchors (12^2,16^2,32^2,64^2,128^2,256^2) and compare our six-scale approach with the conventional one-scale and three-scale approaches.

Faster R-CNN algorithm consists of two modules:

1. Region Proposal Network.
2. Fast R-CNN [11] detection network.

First, we utilize RPN (Region Proposal Network) to extract detection regions. RPN (Region Proposal Network) can share full-image convolutional features with the detection network [12], making region proposal almost spending no time [12]. The RPN network first extracts features, then generates candidate regions. In the presented method of this paper, we utilize six scale (12^2,16^2,32^2,64^2,128^2,256^2) anchors and three aspect (1:2,1:1,2:1) ratios, then classify boxes and refine boxes location.

Second, in order to train the RPN network, we set class labels for each candidate area $\{\text{object},\ \text{non object}\}$. The label settings for RPN network are described as follows:

Object: candidate regions which have overlap ratio larger than x with arbitrary Ground Truth bounding boxes $(IoU > x)$.

Non Object: candidate regions which have overlap ratio equal or smaller than $(1-x)$ with all Ground Truth bounding boxes.

Since RPN network extracts a large number of proposed boxes and there are lots of overlapping regions between these proposed boxes, we use NMS (Non-Maximum Suppression, NMS) to merge and delete proposed boxes. To be specific, all candidate boxes are sorted by score and box with the highest score is selected, then we delete boxes which have overlap ratios larger than $x\ (IoU > x)$. We then sort proposed regions and take top N proposed regions for detection.

Third, After RPN network got the proposed regions, Fast R-CNN is used to realize the final detection and identification. We train RPN network and Fast R-CNN to let them share the convolutional layers of ResNet101 [13].

3 Experimental results

The goal of CCCV2017 Cardiovascular Vulnerable Plaque Detection Challenge is to detect cardiovascular vulnerable plaques in IVOCT images precisely and instantly. Performance of detection algorithms are evaluated by 3 indicators, they are Precision rate, Recall rate and IoU (Intersection-over-Union) rate. We evaluate our presented two-stage deep learning method on the official dataset of CCCV2017 Cardiovascular Vulnerable Plaque Detection Challenge. Let TP be the positive class which the algorithm also predicts as positive class, FP be the negative class which the algorithm predicts as positive class, FN be the positive class which the algorithm predicts as negative class, S_g demotes the area of Ground Truth bounding box, S_p denotes the area of predicted bounding box. The computational formula of the three indicators are shown below:

Precision rate: $P = TP/(TP+FP)$

Recall rate: $R = TP/(TP+FN)$

IoU rate: $IoU = (S_g \cap S_p)/(S_g \cup S_p)$

Total score: S=0.5*(2*P*R/(P+R)) + 0.5*IoU

We compare our detection algorithm with other submitted algorithms, the result is illustrated in table 1 and table 2. Table 1 shows the performance comparison of top3 team and us, we can see our method has the highest Precision rate and IoU rate, thus we get the highest total score. Table 2 compares AP value of our method and top3 team.

Table 1. Performance comparison with the top3 teams of CCCV2017cardiovascular vulnerable plaque detection challenge.

Team	Precision rate	Recall rate	IoU rate	Total score
1st team	0.825	0.932	0.883	0.879
2nd team	0.865	0.838	0.869	0.860
3rd team	0.868	0.713	0.831	0.807
Our	0.943	0.885	0.913	0.913

Table 2. AP comparison with top3 teams of CCCV2017 cardiovascular vulnerable plaque detection challenge.

Team	AP(%)
1st team	76.3
2nd team	72.5
3rd team	61.9
Our	83.4

To better demonstrate the performance of our presented method, we draw ROC curve of our method as it shown in figure 5.

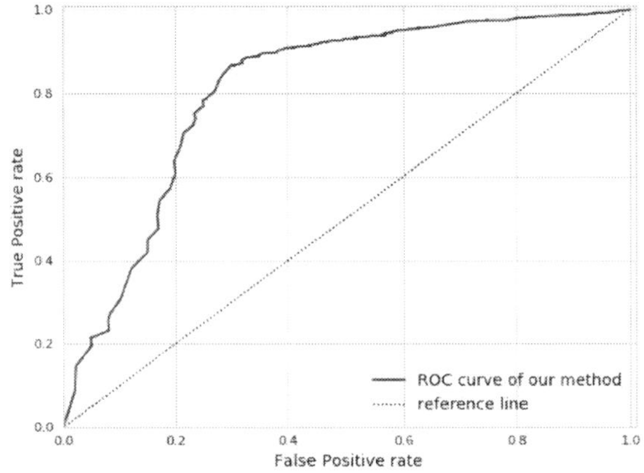

Fig. 5. ROC curve of our presented method for detecting cardiovascular vulnerable plaques.

Also, we evaluate our six-scale anchors approach to prove it has better performance than the conventional one-scale and three-scale approaches. The results are shown in table 3. A sample detection result of our proposed six-scale Faster R-CNN is shown in figure 6 (right).

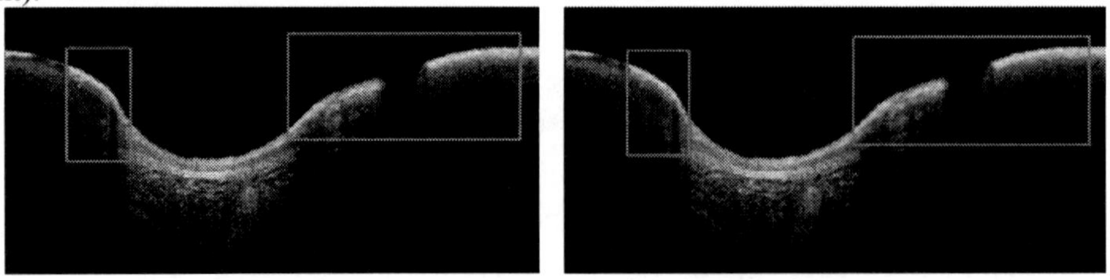

Fig. 6. Ground Truth image (left) and Faster R-CNN predicted result (right).

Table 3. AP of Faster R-CNN when using different scales and ratios.

Settings	Anchor scales	Aspect ratios	AP(%)
1 scale, 1 ratio	128^2	1:1	79.6
1 scale, 1 ratio	256	1:1	78.3
1 scales, 3 ratios	128^2	{1:1, 1:2, 2:1}	81.2
1 scale, 3 ratios	256^2	{1:1, 1:2, 2:1}	80.2
3 scales, 1 ratio	$\{128^2, 256^2, 512^2\}$	1:1	81.7
3 scales, 3 ratios	$\{128^2, 256^2, 512^2\}$	{1:1, 1:2, 2:1}	82.3
6 scales, 3 ratios(our method)	$\mathbf{\{12^2,16^2,32^2,64^2,128^2,256^2\}}$	**{1:1, 1:2, 2:1}**	**83.4**

4 Conclusion

In this paper, we propose a novel cardiovascular vulnerable plaque detection method based on deep convolution neural networks. In the first stage we experiment on different thresholds to attain the best segmentation result, in the second stage we design an improved Faster R-CNN which has six-scale anchors to do precise detection. Experimental results on the CCCV2017 Cardiovascular Vulnerable Plaque Detection Challenge demonstrate the proposed method has superior performance than the 1st team of the challenge. The Recall rate, Precision rate, IoU rate and Total score of our method are 0.94, 0.885, 0.913 and 0.913 respectively, AP of the designed Faster R-CNN is 83.4%, higher than conventional one-scale and three-scale approaches. We believe our proposed method would be a very powerful tool for clinical cardiovascular vulnerable plaques diagnosis.

Acknowledgments

Research is supported by the National Natural Science Foundation of China under Grant U1613214 and in part supported by the Fundamental Research Funds for the Central Universities of China under Grant N170402008, by the Fund for the Shenyang City Science and Technology Plan under Grant F16-2051-04, by the 111 Project under Grant B16009. We thank all the CCCV2017 Cardiovascular OCT Vulnerable Plaque Detection Challenge organizers with special acknowledgments to Dr. Jianan Li, who comes from Xian Institute of Optics and Fine Mechanics, Chinese Academy of Sciences.

References

1. Fleg L, Stone W and Fayad A 2012 Detection of high-risk atherosclerotic plaque: report of the NHLBI Working Group on current status and future directions. *JACC Cardiovasc Imaging* **5(9)** 941-955.
2. Ambrose A and Srikanth S 2010 Vulnerable plaques and patients:improving prediction of future coronary events. *Am J Med* **123(1)** 10-16.
3. Takashi K, Takashi A and Junya S 2013 OCT compared with IVUS in a coronary lesion assessment: The OPUSCLASS Study. *JACC Cardiovascular Imaging* **6(10)** 1095-1104.
4. Wang S, Luo J and Ling H 2013 Computer-aided diagnosis data platform by using medical imaging. *Chin. J. Biomed. Eng* **32(1)** 105.
5. Valgimigli M, Agostoni P and Serruvs PW 2007 Acute coronary syndromes: an emphasis shift from treatment to prevention; and the enduring challenge of vulnerable plaque detection in the cardiac catheterization laboratory. *Cardiovasic Med (Hagerstown)* **8(4)** 221-229.
6. Chen X 2014 *Research on the Application of Sparse Representation Theory in Medical Image Processing and Analysis* (Wuhan: Huazhong University of Science and Technology Press) p 4-10.
7. Keras Documentation/Image Preprocessing. https://keras.io/preprocessing/image/. last accessed 2018/08/14.
8. Ronneberger O, Fischer P and Brox T 2015 *Int. Conf. on Medical image Computing and Computer-Assisted Intervention (Munich)* pp.234-241.

9. Li X, Chen H, Qi X, Dou Q, Fu C and Heng P 2018 H-DenseUNet: hybrid densely connected UNet for liver and tumor segmentation from CT volumes. *IEEE Trans Med Imaging* **PP(99)**: 1-1.
10. Dong H, Yang G, Liu F, Mo Y and Guo Y 2017 Automatic brain tumor detection and segmentation using U-Net based fully convolutional networks. *Conf. on Medical Image Understanding and Analysis.*
11. Girshick R 2015 Fast R-CNN. *IEEE Int. Conf. on Computer Vision.*
12. Ren S, He K, Girshick R and Sun J 2015 Faster R-CNN: towards real-time object detection with region proposal networks. *Conf. on Neural Information Processing Systems (Montreal)* p 91–99.
13. He K, Zhang X, Ren S and Sun J 2015 Deep residual learning for image recognition. *arXiv.*1512.03385.

Multi-layer attention for person re-identification

Yuele Zhang, *Jie* Guo, *Zheng* Huang, *Weidong* Qiu*, and *Hexiaohui* Fan

School of Cyber Security, Shanghai Jiao Tong University, Shanghai, 200240, China

Abstract. Person re-identification has been a significant application in the field of video surveillance analysis, yet it remains a challenging work to recognize the person of interest across disjoint cameras of different viewpoints. The factors affecting the identification results include the variation in background, different illumination conditions and the changes of human body poses. Existing person re-identification methods mainly focus on the feature extraction of the whole frame and metric learning functions. However, most of those algorithms treat different areas without distinction. It is worth emphasizing that different local regions make different contributions to image representaion, which exactly conforms to the attention mechanism. In this paper, we introduce a novel attention network which explores spatial attention in a convolutional neural network. Our algorithm learns the visual attention in multi-layer feature maps. The proposed model not only pays attention to the spatial probabilities of local regions, but also takes the features in different levels into consideration. We evaluate this multi-layer spatial attention model on three benchmark person re-identification datasets: Market-1501, CUHK03, and DukeMTMC-reID. The experiment results validate the advances of our adopted network by comparing with state-of-the-art baselines.

1 Introduction

Recently, person re-identification (Re-ID) task, as an indispensable part of video behaviour analysis field, has received widespread attention. The task of person re-identification aims at recognizing the same people from multiple different cameras with non-overlapping views. Person re-identification problem has broad potential application prospects in many occasions, especially the security systems. It remains a challenging job since the same person changes a lot under different shooting conditions. As can be seen from the image sets in figure 1, the variation in pose, illumination and background has a great impact on the recognition results.

When dealing with person re-identification task, given an image captured by Camera A (probe image), it is compared with all images which come from Camera B (gallery images). The results are ranked according to the degree of similarity between the probe image and gallery ones. In order to achieve good performance, two steps are essentially important: (i) extract features that better describe the images; (ii) find a proper similarity measurement. Many attempts have been made to improve the above two steps. Some algorithms [1, 2, 3, 4] focus on the feature extraction schemes, including the representation of images in color and

* Corresponding author: wendypenny@sjtu.edu.cn

texture, and the fusion of different levels of features. Regarding to the similarity learning, several metric learning methods [5, 2, 6] have been proposed to learn a feature space in which the calculated distance of the feature vectors belong to the same person are smaller than those belong to different pedestrians.

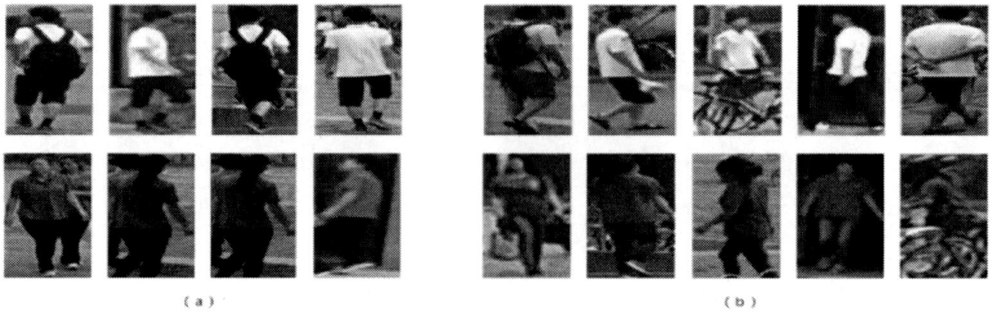

Fig. 1. Sample image pairs from the Market-1501 Re-ID dataset. The images in (a) belong to the same pedestrian from different cameras. The images in (b) are different persons that look similar to the person in (a). It is obvious that the variation in human pose, the change of viewpoints and background, and the low photo quality make it hard to judge whether two similar images point to the same person.

With the success of deep learning models, many methods [7, 8, 9, 10, 11, 12, 13, 14] rely on deep neural networks to tackle the feature extraction and metric learning steps. The deep network architectures achieve more representative feature expressions than hand-crafted features. The most commonly employed convolutional neural network (CNN) models can be categorized into two types: (i) classification networks [8, 12, 13], which are originally used in the image classification task and object detection task; (ii) siamese networks [7, 10, 11, 14], which take a pair of images or triplet images as the network inputs. However, few algorithms take the human attention mechanism into consideration. With the assumption that when processing images, human visual system tends to selectively focus on the important regions instead of treating all regions equally, attention mechanism dynamically allows different levels of features to gain different weights. The adoption of attention model has been proved effective in various tasks such as image captioning [15, 16, 17], machine translation [18, 19], and question answering [20, 21].

In this paper, we adopt a novel algorithm which explores visual attention models in a CNN network. Our contributions are: (1) We propose a spatial attention-based convolutional neural network for person re-identification task. Existing models generally ignore the selection of attentive regions. The formulated attention of spatial regions conforms to the natural visual attention mechanism, resulting in a better expression of the whole image. (2) We explore the attention mechanism in multi-layer feature maps. The multiple layers of weighted attention features are fused. Since CNN networks are multi-layer, the proposed attention network takes full advantage of the CNN architecture. (3) The attention-based deep model completes the person re-identification procedure in an end-to-end manner. We validate the effectiveness of our framework by comparing the results with the state-of-the-art algorithms on three Re-ID benchmark datasets: Market-1501 [22], DukeMTMC-reID [23] and CUHK03 [9].

The rest of this paper is organized as follows. Section 2 briefly reviews the related work of Re-ID task. Section 3 describes the framework of our attention-based approach in details. Section 4 shows the experiment results on three public benchmark datasets and gives the corresponding analysis. Section 5 concludes our work.

2 Related Work

For an image-based person re-identification system, feature extraction and distance learning are typically considered as the two main components. Various algorithms have been proposed to solve the Re-ID problem. Some of them complete the above two steps separately, others utilize the deep learning-based method to achieve a jointly-learning procedure.

In order to present a better representation of given images, the basic principle is to find features unrelated to illumination, pose, background and viewpoints. Many existing algorithms have extracted more discriminative and viewpoint-invariant features. To get more information about the mean attributes of pixels, Matsukawa [3] exploit a hierarchical Gaussian distribution descriptor (GoG), modelling both means and covariances. In [24], An et al. conduct the matchings by projecting the original features into a reference subspace instead of matching directly. Experiments validate the effectiveness of the reference descriptors (RDs) generated with the correlations of the reference sets. To enhance the Re-ID performance, [1, 4] utilize different fusing schemes to combine the features obtained at different levels, including pixel-based low-level features (e.g. SIFT (scale-invariant feature transform) [25]), mid-level features (e.g. BoW (Bag of Words) [22]), and high-level features (e.g. features jointly learned in CNN [14]).

In addition to methods concerned with image representation, many approaches aim at finding a proper similarity measurement which makes the distance metric discrimitive. In [26], a null space is learned in which the in-class distance is minimized and between-class gap is maximized. Yu et al. [27] propose an unsupervised approach to deal with the limitation of lacking labelled sample images. The algorithm learns an asymmetric metric with projections on each viewpoint respectively. Considering the cross-view distortion of features, Chen et al. [28] formulate a Camera coRrelation Aware Feature augmenTation (CRAFT) framework, which automatically calculates the cross-view camera correlation. With learned features projected into an adaptive subspace, the CRAFT framework obtains view-specific features for Person Re-ID. [29, 30, 31] take the advantage of deep learning architecture and optimize the task of feature selection, similarity learning and ranking jointly.

Most of the Re-ID methods are proposed under the assumption that the probe and gallery images are well-aligned. However, due to the different viewpoints and the change of pedestrian poses, misalignment is a key issue to be considered. To solve this problem, visual attention schemes and saliency-based techniques are adopted. Some studies pay attention to saliency learning. Zhao et al. [32] apply pedestrian saliency distribution learning and estimate the scores based on the matching of constrained patches. The whole learning and matching procedure is unified into a RankSVM framework. In [33], a weighted integration scheme is adopted, combing human salience information with SDALF (Symmetry-Driven Accumulation of Local Features) [34]. With the rotation invariant attributes, the experimental performance is improved. Attention mechanism can also be seen as a concern for salient regions. With the recent success of exploiting attention modules in other fields, a few attention-based deep learning networks have been adopted to tackle the problem of misalignment in Re-ID. Gradient-based attention mechanism is exploited in [35]. In [36], both the CNN-extracted features and color histograms are fed into the recurrent attention architecture to perform a coarse-to-fine selection process. Liu et al. [37] formulate a Comparative Attention Network (CAN) architecture which takes image triplets as the training input. The CAN algorithm compares different local regions repeatedly instead of taking just one glimpse of the whole image. The global features learned from CNNs are delivered to a LSTM-based attention module to obtain the visual attentive features. The model in [38] also utilizes LSTM units to generate the spatial attetion encoding. In [39], a Harmonious Attention CNN (HA-CNN) model is proposed. HA-CNN adopts multi-branch scheme to process both local level and global level attention.

For each branch, soft attention and hard attention are combined to fully explore the complementary information.

Different from the attention-related algorithms mentioned above, our proposed attention network not only explores spatial attention, but also fuses the attention weighted feature maps learned at different layers for a richer representation. Consequently, the adopted method outperforms the state-of-the-art approaches.

3 Network Architecture

3.1 Overview

We adopt the attention-based model for person Re-ID task. The overall framework is illustrated in figure 2. We formulate an end-to-end Convolutional Neural Network to learn attentive features. Through exploiting spatial attention on multiple layers, our proposed network is able to make the original CNN-based feature maps adaptive to the more discriminative local regions and attributes. The spatial attention aims at encoding "where" to pay attention to. The feature integration scheme results in a more robust feature representation.

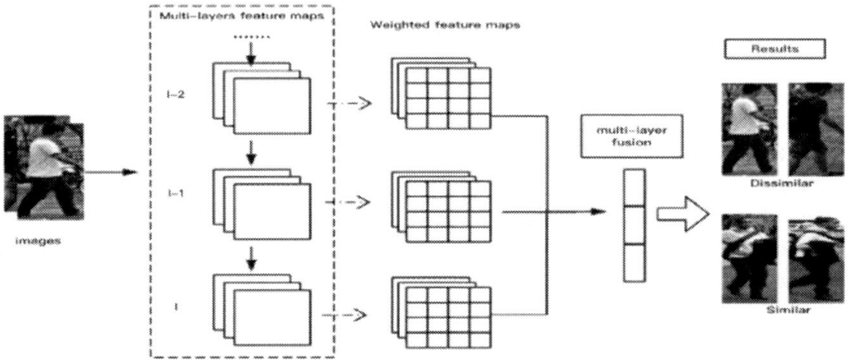

Fig. 2. The overview of the proposed multi-layer attention network architecture. For the feature map $\boldsymbol{V}^l$ on the l-th convolution layer, we learn a spatial attention weighted feature map $\boldsymbol{A}^l$ which is of the same size as $\boldsymbol{V}^l$. The attention modules are applied to multiple layers. The multi-layer attentive features are finally fused.

Given $\boldsymbol{n}$ distinct pedestrians, $\boldsymbol{m}$ training images captured from cameras of different viewpoints, the task is to learn a model that better describes all the pedestrians in spite of the variation of poses and changes. Our proposed model learns the attention-related features on the basis of a CNN base network. This basic CNN structure can be replaced by any CNN model, e.g. VGG-19, ResNet-101. The key components are the attention learning module and multi-layer fusion module. In additional to the final layer, one or more mid-layers of the CNN base network are exploited to construct the multiple attention weighted features.

For attention selection mechanism, we formulate an attention module which learns spatial attention. At the $\boldsymbol{l}$-th layer of the base CNN network, the feature map is denoted as $\boldsymbol{V}^l$. The $\boldsymbol{l}$ -th layer attention weights $\boldsymbol{\gamma}^l$ are learned from $\boldsymbol{V}^l$. The functions are formulated as:

$$V^l = CNN(V^{l-1}) \tag{1}$$

$$\gamma^l = \phi(V^l) \tag{2}$$

$$A^l = \eta(V^l, \gamma^l) \tag{3}$$

where ϕ refers to the attention learning function, which will be elaborated in the next attention sections. η is a linear weighting calculation which combines both original feature representation V^l and learned weights γ^l. A^l is the weighted map, considered as the output of an attention module.

Note that the obtained output map $A^l \in R^{h\times w\times c}$ is of the same size as the CNN original input map $V^l \in R^{h\times w\times c}$, where h, w and c denote the pixels of the feature map in height dimension, width dimension and channel dimension. The spatial attention maps are generated from the representation map V^l.

3.2 Spatial Attention

Spatial attention can be intuitively explained and conforms to the visual processing mechanism naturally. Instead of treating all the local regions equally, the spatial attention focuses on the selected regions and helps to enhance the representation. Inspired by the region enhancement characteristic, we adopt the spatial attention module in our formulated network to tackle the misalignment problem for person Re-ID.

Given the original generated feature map $V^l \in R^{h\times w\times c}$, the spatial attention module aims at calculating a probability map that indicates the attention probabilities of all local regions. The spatial attention weighted feature can be formulated as:

$$f^S = \sum_i^h \sum_j^w f_{i,j}^S = \sum_i^h \sum_j^w \alpha_{i,j} f_{i,j}^V \; (0 \le \alpha_{i,j} \le 1) \qquad (4)$$

where $f_{i,j}^V$ represents the feature at the location of (i, j) in feature map V^l, and $\alpha_{i,j}$ represents the corresponding attention probability at the same location. At (i, j), the learned spatial attended feature is denoted as $f_{i,j}^S$. In the equation, f^S refers to the final spatial attention representation.

3.3 Multi-layer Fusion

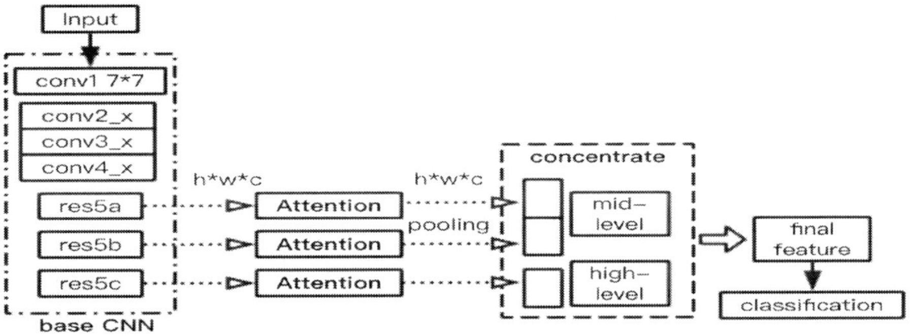

Fig. 3. The adopted CNN network structure. ResNet-50 is adopted as the base CNN network for Re-ID task. For the last three conv-layers ***res5a***, ***res5b*** and ***res5c***, the attention module is applied to obtain attention-based feature maps. After pooling methods, the mid-level features and high-level features are concentrated for final classification.

We choose ResNet-50 [40] as our CNN base network for person Re-ID task. Figure 3 shows the network structure of our proposed method. For ResNet-50 network, ***res5c*** represents the last convolutional layer, which contains the image representation in high-level. To take full advantage of the mid-level semantic information, we also select layers ***res5a*** and ***res5b*** as a supplement for better description. The feature maps of the above layers are denoted as V^{res5a}, V^{res5b} and V^{res5c}. Instead of fusing the multiple original feature vectors, we concentrate the weighted feature maps after the attention modules, denoted as A^{5a}, A^{5b} and A^{5c}.

By adopting the appropriate pooling scheme, the feature vectors are constructed from multiple attention-weighted layers. For the maps A^{5a}, A^{5b}, A^{5c} that is corresponding to the selected layers, we leverage the global average pooling layer to generate the vectors f^{5a}, f^{5b} and f^{5c}. The concentration of mid-level feature vectors f^{5a} and f^{5b} are connected by a fully connected layer, which helps to reduce the feature dimension. After that, the generated mid-level vector f^{mid} is fused with f^{5c} to obtain the final feature vector for classification.

4 Experiments

In this section, we evaluate the performance of our proposed attention-based multi-layer integrated method. We validate the effectiveness of the multi-layer fusion strategy. The assessment is conducted on three benchmark datasets comparing with the state-of-the-art baseline algorithms.

4.1 Dataset

We perform the evaluation experiments on three public Re-ID benchmark datasets, i.e. Market-1501 [22], DukeMTMC-reID [23] and CUHK03 [9].

4.1.1 Market-1501 Dataset

The Market-1501 Dataset [22] contains 1,501 pedestrians captured from 6 non-overlapping cameras, with 5 of high resolution and 1 of low resolution. There are altogether 32,668 annotated bounding boxes, using DPM (Deformable Part Model) as the detector for pedestrian. In the experiment, we set the training and test sets as provided, including 750 persons in training set and 751 in testing set.

4.1.2 DukeMTMC-reID Dataset

The DukeMTMC-reID Dataset [23] involves 1,404 identities and 36,411 bounding box images. All the dataset images are cropped from video frames recorded by 8 different high-resolution cameras. The whole set is partitioned into two parts, with 702 individuals forming the training set, and 702 identities forming the testing set.

4.1.3 CUHK03 Dataset

The CUHK03 Dataset [9] consists of 13,164 images from 1,360 individuals captured by 6 cameras. Part of the images are manually cropped and the rest are detected by DPM. There are 767 individuals in training set and 700 identities in the test set.

4.2 Experimental Setup

In our formulated network, we exploit the widely-adopted CNN model: ResNet-50 [40] as the basic CNN network. The method is implemented using the deep learning framework Pytorch. All the pedestrian images in the datasets are resized to 160×64. The human body ratio is thus kept to remain undistorted. For optimization step, we choose Adam as the optimizer. The initial learning rate and the decay factor is set to 0.0005 and 0.95 respectively. We set the batch size as 32, and the maximum training epoch as 100. For performance evaluation, we use the standard CMC (Cumulative Matching Characteristic), which represents the relationship between the correct identification rate and rank numbers, and mAP (Mean Average Precision), which measures the precision of classification. Our whole attention-based framework is completed in an end-to-end procedure.

4.3 Evaluation of Multi-layer Attention

In this section, we investigate the improvement effect of our multi-layer attention fusion mechanism. We conduct the experiments by reducing the number of layers followed by the attention module. For our adopted ResNet-50 network, 1-st layer, 2-nd layer, 3-rd layer refer to ***res5a***, ***res5b*** and ***res5c*** respectively. Table 1 shows the experiment results. In the table, ***1-layer*** refers to the method of applying the attention module to ***res5c*** layer only. ***2-layer*** represents the fusion of learned attention features on ***res5b*** and ***res5c***. And ***3-layer*** denote our adopted approach.

Table 1. Evaluation on the multi-layer fusion scheme (Market-1501 Dataset).

	Rank1	Rank5	Rank10	Rank20	mAP
1-layer	83.1	91.6	93.9	95.3	66.3
2-layer	86.6	93.3	95.1	96.4	69.5
3-layer	**87.1**	**94.8**	**96.6**	**97.8**	**70.1**

As can be observed from table 1, compared to modulating the attention network directly to the last convolutional layer, the 2-layer fusion method gains 3.5% for Rank-1, 2.3% for Rank-5, 2.8% for Rank-10, 0.9% for Rank-20 and 3.2% for mAP on Market-1501. The accuracy of 3-layer fusion method has a minor increase of 0.5% for Rank-1 accuracy. The experiments indicate that the multi-layer attention mechanism has a positive effect for better image representation.

4.4 Comparison with State-of-the-arts

4.4.1 Evaluation on Market-1501

We compare the results of our method with recent state-of-the-arts on Market-1501 dataset, as shown in table 2. The algorithms consist of feature extraction methods, i.e. CRAFT [28], GLAD [41], Zhao et al. [42], metric learning methods, i.e. SCSP [43], DNS [26], and deep network-based methods, i.e. CAN [37], PIE+Kissme [44], PDC [45]. As can be observed from table 2, our method outperforms the 2-nd best algorithm PDC by 3.0% in Rank-1 accuracy and 6.7% in mAP. Among all the compared approaches, CAN and PDC use attention mechanism for person Re-ID task in their algorithms as well. Statistics indicate that our attention-based deep network outperforms other approaches using attention information. It is worth noting that PIE also uses ResNet-50 as the base network architecture, with part-aligned representation based on the pose estimator. Our proposed method surpasses PIE by 8.4% for Rank-1 accuracy. These statistics show the advantage of our multi-layer attention model over the state-of-the-arts baseline.

Table 2. Evaluation on the Market-1501 Dataset.

Method	Rank1	Rank5	Rank10	Rank20	mAP
SCSP [43]	51.9	72.0	79.0	-	26.4
CAN [37]	60.3	-	-	-	35.9
DNS [26]	61.0	-	-	-	35.7
CRAFT [28]	68.7	-	-	-	42.3
GLAD [41]	61.4	76.8	81.6	85.9	34.0
PIE [44]	78.7	90.3	93.6	95.7	53.9
PIE+Kissme [44]	79.3	90.7	94.4	96.5	56.0
Zhao et al. [42]	81.0	92.0	94.7	-	63.4
PDC [45]	84.1	92.7	94.9	96.8	63.4
Ours	**87.1**	**94.8**	**96.6**	**97.8**	**70.1**

4.4.2 Evaluation on DukeMTMC-reID

Table 3 indicates the superiority of our multi-layer attention integration method compared to the state-of-the-arts on DukeMTMC-reID dataset. Compared to Market-1501 dataset, DukeMTMC-reID dataset has more complex scenes and more changes in background. Except for LOMO+XQDA [2], all the other compared approaches, i.e. SVDNet [46], APR [47] and PAN [48], formulate the framework on the basis of the deep learning network. On DukeMTMC-reID dataset, we achieve Rank-1 accuracy at 78.8%, and mAP at 60.0%, surpassing the 2-nd method SVDNet (ResNet-50) by 0.1% and 3.2%, respectively. This suggests that the attention mechanism plays a significant role in feature representation and our proposed network adapts to the complex scenes well.

Table 3. Evaluation on the DukeMTMC-reID Dataset.

Method	Rank1	Rank5	Rank10	mAP
LOMO+XQDA [2]	51.9	72.0	79.0	17.0
SVDNet (CaffeNet) [46]	60.3	-	-	45.8
APR [47]	61.0	-	-	51.9
PAN+Re-rank [48]	61.4	76.8	81.6	56.7
SVDNet (ResNet-50) [46]	78.7	90.3	93.6	56.8
Ours	**78.8**	**90.6**	**93.9**	**60.0**

4.4.3 Evaluation on CUHK03

Table 4 shows the CMC Rank accuracy and mAP of several recent proposed algorithms reported on CUHK03 dataset. We conduct the evaluation on both manually cropped bounding boxes and automatically detected images. Compared to the manually labelled images, the detected ones have more misalignment problems, presenting a more difficult task. In our evaluation, the compared counterparts include HACNN [39], LOMO+XQDA [2], LOMO+XQDA+re-rank [49], Ahmed et al. [7] and IDE+Re-rank [49]. HACNN proposes a harmonious network which combines soft attention in pixel and hard attention in region, exploiting attention mechanism in deep learning network. Our attention method achieves the performance of Rank-1 = 59.2%, mAP = 57.9% for manually labelled images, and Rank-1 = 56.3%, mAP = 54.8% for detected images. It shows the superiority of the adopted attention learning approach over other attention driven mechanism.

Table 4. Evaluation on the CUHK03 Dataset.

	Manual		**Detected**	
Method	Rank1	mAP	Rank1	mAP
HACNN [39]	44.4	41.0	41.7	38.6
LOMO+XQDA [2]	49.7	56.4	44.6	51.5
LOMO+XQDA+Re-rank [49]	50.0	56.8	45.9	52.6
Ahmed et al. [7]	54.7	-	45.0	-
IDE+Re-rank [49]	57.2	**63.2**	54.2	**60.5**
Ours	**59.2**	57.9	**56.3**	54.8

Despite the disparities among these three Re-ID baseline datasets, i.e. scenes, camera views, and image processing methods, we show that our proposed approach improves the performance for Re-ID task. The experiment results demonstrate the excellence of our attention-based framework.

5 Conclusion

In this paper, we present a novel attention-based method for person re-identification task. The whole framework is formulated in an end-to-end procedure. Our network benefits from both attention mechanism and multi-layer integration mechanism. For attention selection, we explore spatial attention on different local regions. The introduction of attention module takes full advantage of the complementary attention information. For multi-layer fusion module, the concentration of mid-level features and high-level features performs a better image representation. The detailed analysis of feature fusion scheme on multiple layers is provided. We conduct the performance evaluation on three baseline Re-ID datasets, i.e. Market-1501, DukeMTMC-reID and CUHK03. Experiments validate the effectiveness of our proposed model when compared to the state-of-the-art approaches.

Acknowledgments

This work is supported by the National Key Research and Development Program of China under Grant 2017YFB1002401.

References

1. Aske R Lejbølle, Nasrollahi K and Thomas B Moeslund 2018 Enhancing person re-identification by late fusion of low-, mid- and high-level features *Iet Biometrics* **7(2)** 125-135
2. Liao S, Hu Y, Zhu X and Li S Z 2015 Person re-identification by Local Maximal Occurrence representation and metric learning *IEEE Conference on Computer Vision and Pattern Recognition* **Vol.8** 2197-2206
3. Matsukawa T, Okabe T, Suzuki E and Sato Y 2016 Hierarchical Gaussian Descriptor for Person Re-identification *Computer Vision and Pattern Recognition* 1363-1372
4. Tan F, Liu K and Zhao X 2017 Person Re-Identification Based on Multi-Level and Multi-Feature Fusion *International Conference on Smart City and Systems Engineering* 184-187
5. Jose C and Fleuret F 2016 Scalable Metric Learning via Weighted Approximate Rank Component Analysis *European Conference on Computer Vision* 875-890
6. Yang Y, Liao S, Lei Z and Li S Z 2016 Large scale similarity learning using similar pairs for person verification *Thirtieth AAAI Conference on Artificial Intelligence* 3655-3661
7. Ahmed E, Jones M and Marks T K 2015 An improved deep learning architecture for person re-identification *Computer Vision and Pattern Recognition* 3908-3916
8. Krizhevsky A, Sutskever I and Hinton G E 2012 ImageNet classification with deep convolutional neural networks *International Conference on Neural Information* Processing *Systems* **Vol.60** 1097-1105
9. Li W, Zhao R, Xiao T, and Wang X 2014 DeepReID: Deep Filter Pairing Neural Network for Person Re-identification *IEEE Conference on Computer Vision and Pattern Recognition* 152-159
10. Varior R R, Haloi M and Wang G 2016 Gated Siamese Convolutional Neural Network Architecture for Human Re-identification *European Conference on Computer Vision* 791-808

11. Varior R R, Shuai B, Lu J, Xu D and Wang G 2016 A Siamese Long Short-Term Memory Architecture for Human Re-identification *European Conference on Computer Vision* 135-153
12. Yu Q, Chang X, Song Y Z, Xiang T and Hospedales T M 2017 The devil is in the middle: exploiting mid-level representations for cross-domain instance matching *ArXiv e-prints* 1711.08106
13. Wu S, Chen Y C, Li X, Wu A C, You J J and Zheng W S 2016 An enhanced deep feature representation for person re-identification *Applications of Computer Vision* 1-8
14. Yi D, Lei Z, Liao S and Li S Z 2014 Deep Metric Learning for Person Re-identification *International Conference on Pattern Recognition* 34-39
15. Chen L, Zhang H, Xiao J, Nie L, Shao, J and Liu W 2017 Sca-cnn: spatial and channel-wise attention in convolutional networks for image captioning *Computer Vision and Pattern Recognition* 6298-6306
16. Xu K, Ba J, Kiros R, Cho K, Courville A and Salakhutdinov R 2015 Show, attend and tell: neural image caption generation with visual attention *Computer Science* 2048-2057
17. You Q, Jin H, Wang Z, Fang C and Luo J 2016 Image Captioning with Semantic Attention *Computer Vision and Pattern Recognition* 4651-4659
18. Choi H, Cho K and Bengio Y 2018 Fine-grained attention mechanism for neural machine translation *Neurocomputing* 284
19. Luong M T, Pham H and Manning C D 2015 Effective approaches to attention-based neural machine translation *Computer Science*
20. Chen K, Wang J, Chen L C, Gao H, Xu W and Nevatia R 2015 Abc-cnn: an attention based convolutional neural network for visual question answering *Computer Science*
21. Lioutas V, Passalis N and Tefas A 2018 Visual Question Answering using Explicit Visual Attention *IEEE International Symposium on Circuits and Systems* 1-5
22. Zheng L, Shen L, Tian L, Wang S, Wang J and Tian Q 2015 Scalable Person Re-identification: A Benchmark *IEEE International Conference on Computer Vision* 1116-1124
23. Ristani E, Solera F, Zou R, Cucchiara R and Tomasi C 2016 Performance Measures and a Data Set for Multi-target, Multi-camera Tracking *European Conference on Computer Vision* 17-35
24. An L, Kafai M, Yang S and Bhanu B 2016 Person reidentification with reference descriptor *IEEE Transactions on Circuits & Systems for Video Technology* **26(4)** 776-787
25. Zhao R, Ouyang W and Wang X 2014 Learning Mid-level Filters for Person Re-identification *IEEE Conference on Computer Vision and Pattern Recognition* 144-151
26. Zhang L, Xiang T and Gong S 2016 Learning a Discriminative Null Space for Person Re-identification *IEEE Conference on Computer Vision and Pattern Recognition* 1239-1248
27. Yu H X, Wu A and Zheng W S 2017 Cross-View Asymmetric Metric Learning for Unsupervised Person Re-Identification *IEEE International Conference on Computer Vision* 994-1002
28. Chen Y C, Zhu X, Zheng W S and Lai J H 2018 Person re-identification by camera correlation aware feature augmentation *IEEE Transactions on Pattern Analysis & Machine Intelligence* **40(2)** 392-408.

29. Chen W, Chen X, Zhang J and Huang K 2017 A multi-task deep network for person re-identification *AAAI Conference on Artificial Intelligence*
30. Li W, Zhu X and Gong S 2017 Person re-identification by deep joint learning of multi-loss classification *International Joint Conference on Artificial Intelligence* 2194-2200.
31. Wang F, Zuo W, Lin L, Zhang D and Zhang L 2016 Joint Learning of Single-Image and Cross-Image Representations for Person Re-identification *IEEE Conference on Computer Vision and Pattern Recognition* 1288-1296
32. Zhao R, Oyang W and Wang X 2014 Person re-identification by saliency learning *IEEE Transactions on Pattern Analysis & Machine Intelligence* **39(2)** 356-370
33. Yang M, Wan W, Hou L and Zhang Y 2016 Person re-identification using human salience based on multi-feature fusion *International Conference on Smart and Sustainable City and Big Data* 5
34. Bazzani L, Cristani M and Murino V 2014 Sdalf: modeling human appearance with symmetry-driven accumulation of local features *Person Re-Identification* **63(4)** 43-69
35. Rahimpour A, Liu L, Taalimi A, Song Y and Qi H 2017 Person re-identification using visual attention *IEEE SigPort* http://sigport.org/2046
36. Zhuang Z, Ai H, Shang C and Xiao L 2017 Person re-identification with coarse-to-fine visual attention *IEEE International Conference on Image Processing* 1097-1101
37. Liu, H., Feng, J., Qi, M., Jiang, J and Yan S 2017 End-to-end comparative attention networks for person re-identification *IEEE Transactions on Image Processing* **26(7)** 3492-3506
38. Wu L, Wang Y, Li X and Gao J 2018 Deep attention-based spatially recursive networks for fine-grained visual recognition *IEEE Transactions on Cybernetics* **PP(99)** 1-12
39. Li W, Zhu X and Gong S 2018 Harmonious attention network for person re-identification *IEEE Conference on Computer Vision and Pattern Recognition* 2285-2294
40. Yang Y, Liao S, Lei Z and Li S Z 2016 Large scale similarity learning using similar pairs for person verification *Thirtieth AAAI Conference on Artificial Intelligence* 3655-3661
41. Wei L, Zhang S, Yao H, Gao W and Tian Q 2017 GLAD: Global-Local-Alignment Descriptor for Pedestrian Retrieval *ACM Multimedia*
42. Zhao L, Li X, Zhuang Y and Wang J 2017 Deeply-Learned Part-Aligned Representations for Person Re-identification *IEEE International Conference on Computer Vision* 3239-3248
43. Chen D, Yuan Z, Chen B and Zheng N 2016 Similarity Learning with Spatial Constraints for Person Re-identification *IEEE Conference on Computer Vision and Pattern Recognition* 1268-1277
44. Zheng L, Huang Y, Lu H and Yang Y 2017 Pose invariant embedding for deep person re-identification *ArXiv e-prints* 1701.07732
45. Su C, Li J, Zhang S, Xing J, Gao W and Tian Q 2017 Pose-driven deep convolutional model for person re-identification *IEEE International Conference on Computer Vision* 3980-3989
46. Sun Y, Zheng L, Deng W and Wang S 2017 SVDNet for Pedestrian Retrieval *IEEE International Conference on Computer Vision* 3820-3828
47. Lin Y, Zheng L, Zheng Z, Wu Y and Yang Y 2017 Improving person re-identification by attribute and identity learning *ArXiv e-prints* 1703.07220

48. Zheng Z, Zheng L and Yang Y 2017 Pedestrian alignment network for large-scale person re-identification *ArXiv e-prints* 1707.00408
49. Zhong Z, Zheng L, Cao D and Li S 2017 Re-ranking Person Re-identification with k-Reciprocal Encoding *IEEE Conference on Computer Vision and Pattern Recognition* 3652-3661

Image tampering detection using genetic algorithm

Ritu Agarwal*, and *Mallika* Pant

Department of Information Technology, Delhi Technological University, New Delhi - 110042, India

Abstract. As digital images become an indispensable source of information, the authentication of digital images has become crucial. Various techniques of forgery have come into existence, intrusive, and non-intrusive. Image forgery detection hence is becoming more challenging by the day, due to the unwavering advances in image processing. Therefore, image forensics is at the forefront of security applications aiming at restoring trust and acceptance in digital media by exposing counterfeiting methods. The proposed work compares between various feature selection algorithms for the detection of image forgery in tampered images. Several features are extracted from normal and spliced images using spatial grey level dependence method and many more. Support vector machine and Twin SVM has been used for classification. A very difficult problem in classification techniques is to pick features to distinguish between classes. Furthermore, The feature optimization problem is addressed using a genetic algorithm (GA) as a search method. At last, classical sequential methods and floating search algorithm are compared against the genetic approach in terms of the best recognition rate achieved and the optimal number of features.

1 Introduction

In today's era of digitalization and computerization, information is mostly conveyed through digital images and videos. Fields like weather forecasting, forensic investigation, journalism have raised the need for digital images. Due to the increase in the availability of a wide range of image-editing software and advancement in processing techniques, image forgery is on the rise and thereby effecting diverse areas of life, such as politics, cybercrime investigations, medical diagnoses based on images, politics, businesses etc. Crimes, such as defaming of websites, personalities etc., are on the rise and make use of such tools to do the act. Different type of forging techniques has come up of which, Image Splicing is performed by cutting and merging more images and is common among all. Copy move forgery is another forgery technique, detection of which by human visual system is very difficult. In this type of forgery, a part of an image is replicated in the same image in another region that depicts as an authenticated image.

Although, Human visual system has difficulty in perceiving spliced images and copy move forged images. New and efficient methods to detect forgery in images are proposed

* Corresponding author: ritu.jeea@gmail.com

by researchers. The available techniques for tampering detection can be divided into broadly into active approaches [1][2] and passive approaches [3][4]. The former needs addition of traces of authenticity to be able to detect later for any manipulations, for example a digital watermark where we feed some information prior to image sharing and the detection can be done based on modifications to that information, whereas the passive approaches use a blind approach, perform the same task without any prior information. They make use of the distinct properties of original images and traces of changes made to them to perform authentication. The proposed method uses a blind approach to find out whether the given image is altered or not.

2 Digital image forgery detection methods

2.1 Active approach

Active forgery detection approaches [1][2] use a digital watermark or a signature embedded in the actual image to prove or reject the authenticity of the image. This approach holds a strong limitation that the watermark which is embedded must be performed either by a person authorized to process the image or by the acquisition device.

2.2 Passive approach

Passive/Blind approaches [3][4] use the fact that statistical changes occur in images or camera fingerprints during the process of creation/modification, to detect the forgery. Unlike active methods, blind approaches do not require any data regarding authenticity. Blind forensics can be classified into six categories [5] i.e. pixel-based, camera based, format-based, geometric-based, physics-based and source camera identification-based.

3 Related work

Image splicing detection detects if a given image is a composite one created by combining or separating different portions of two or more images. Copy move forgery detects if an image contains a copied portion of the same image. In a series of paper [6] [7] higher order moment spectra, bi-coherence are treated as features in order to detect images that are spliced. When tested, bi-coherence was found sensitive to quadratic phase coupling. The accuracy of 72% was found when these features are performed on a dataset. A combination of Hilbert Huang Transform and wavelet decomposition [8] is proposed in which Hilbert Huang Transform is used to examine disturbance in linearity caused by image splicing and a statistical model using wavelet decomposition is proposed for calculating moments of the characteristic function. Detection accuracy was 80.15% was found when these two features were used collectively. Hsu and Chang [9] used camera response function and geometry invariants to detect splicing in a semi-automatic manner. In order to automate their work, image segmentation is incorporated [10]. The measure of blurriness [11] and local sharpness [12] was taken as an advantage to check if image under detection is blurred/smoothed or not. A scheme to detect a tampered region in JPEG images was proposed in [13]. In [14] a scheme used for steganalysis was used to check image splicing. He et al. used Run- Length based scheme to detect forgery [15]. Statistical moments obtained from wavelet characteristic function and 2D phase congruency [16] were used to pull out spliced ones from original ones. On detection of these features on Columbia Image Splicing Evaluation Dataset [17] using SVM classifier, an accuracy of 82.32% was attained. Shi et al [18] checked two different types of statistical features proposing a natural image model on Columbia Image Splicing Dataset attaining detection accuracy of 91.87%. In [19] presented an algorithm for tampering detection using SVD. A small window of size B x B is slid over the input image to separate the image into overlapping blocks. SVD is applied to these separated blocks to obtain feature vectors, sort them and store it in the matrix. k-d tree is constructed using the feature vectors and it is searched for similar blocks. The matched blocks satisfying a threshold t will be labeled as suspected regions and these suspected regions are merged together to determine the tampered region. With the above discussion, we can conclude that there are several methods for locating the altered region.

4 Proposed approach

The approach starts by extracting features from given input images. To achieve a high recognition rate, a subset of best features need to be selected. Exhaustive search to evaluate best feature subset is not feasible as it increases required computational effort. The genetic approach, being a heuristic search provide a feasible approach for such an optimization problem and therefore used for feature selection in the proposed approach as shown in figure 1. Classical sequential methods are compared with genetic algorithm approach in context with recognition rate and cardinality of feature subset chosen.

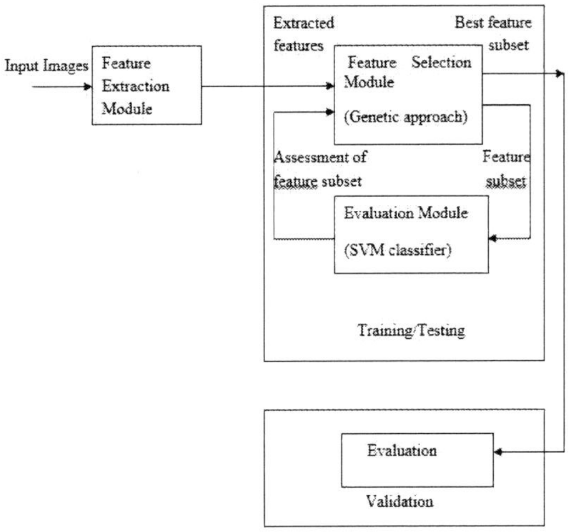

Fig. 1. The architecture of detection algorithm.

4.1 Feature Extraction

A set of ten features are pulled out from authentic as well as spliced images. They are:

4.1.1 Local binary pattern

It is an efficient texture feature which works by labeling the pixels of an image and then thresholding each of the neighbors. Finally, the result is displayed as a binary number.

4.1.2 Entropy

$$ENT = -\sum_{i=1}^{N}\sum_{j=1}^{N} C(i,j)\log(C(i,j)) \quad (1)$$

where C denotes the co-occurrence matrix and N is the number of levels that are gray in the quantized image.

4.1.3 Contrast

$$CON = \sum_{n=0}^{N-1} n^2 \left\{\sum_{i=1}^{N} \sum_{j=1}^{N} \{C(i,j)\}\right\} \quad (2)$$

4.1.4 Histogram of oriented gradients (HOG)

These descriptors use the fact that the occurrences of gradient orientation help to localize the parts in the image. Unlike SIFT, it performs on dense grid structure of cells spaced uniformly. It performs local contrast normalization on blocks that are overlapped.

4.1.5 Inverse difference moment

$$IDM = \sum_{i=1}^{N} \sum_{J=1}^{N} \frac{1}{1+(i-j)^2} C(i,j) \quad (3)$$

4.1.6 Speeded up robust features

It uses the multi-resolution pyramid technique to extract features in images which are transformed into coordinates.

4.1.7 Inertia

$$INER = \sum_{i=1}^{N} \sum_{j=1}^{N} (i-j)^2 * C(i,j) \quad (4)$$

4.1.8 Cluster shade

$$CS = \sum_{i=1}^{N} \sum_{j=1}^{N} \left(i + j - u_x - u_y\right)^3 * C(i,j) \quad (5)$$

4.1.9 Cluster prominence

$$CP = \sum_{i=1}^{N} \sum_{j=1}^{N} \left(i + j - u_x - u_y\right)^4 * C(i,j) \quad (6)$$

4.1.10 Angular second moment

$$ASM = \sum_{i=1}^{N} \sum_{j=1}^{N} \{C(i,j)\}^2 \quad (7)$$

4.2 Feature Selection

Feature selection is done to use a smaller feature set while maintaining the accuracy of classification.

4.2.1 Deterministic search method

Feature selection algorithms are divided into the following categories, which are Exponential, randomized and sequential algorithms. The algorithms of sequential search use hill-climbing strategy to select or reject features. Sequential forward selection (SFS) starts by evaluating all feature subsets which contains one feature and select the feature with the best performance. This subset is then combined with the feature that gives the best performance for larger size subsets. The process repeats until no improvement is done by extending the current subset. Sequential backward selection (SBS) starts with a set which contains all features and repeatedly removes a feature whose rejection gives a maximal improvement in performance. Both these methods are suboptimal methods that create nesting of features in a straight forward manner. Due to these previous additions and removal of features cannot be corrected. To overcome this problem, these algorithms are modified and are termed as floating sequential algorithms. Sequential floating forward selection (SFFS) works by applying backward steps after each forward step until the corresponding subsets are better than previous ones.

4.2.2 Genetic algorithm approach

In order to select the best set of features, a genetic algorithm which is based on global search method is used. Based on Darwin principle, the algorithm states that the initial population of individuals have a high probability of survival. Here each feature subset is encoded in form of a binary string which is called a chromosome. Each bit is associated with a feature in the binary vector. If the ith bit of this vector is 1 then it means that i^{th} feature is allowed to participate in classification. In case the bit is a 0, then the feature is not selected. The fitness value is assigned based on classification performance. Here genetic approach has been used to select the best features by tournament approach. Diagnostic accuracy, sensitivity, and specificity measure are considered for evaluation. Also, the number of features that are not selected (zeros in a chromosome) with a coefficient is added.

Fitness function:

$$fitness = Accuracy + Sensitivity + Specificity + 0.05 * Number \quad (8)$$

In order to maintain diversity in solution, replacement technique is used and the whole population is replaced with the new generation. Parents having a high selection rate are selected. For simplicity, single point crossover is adopted here. The feature subset among all the generations with the highest classification rate is regarded as the optimum.

4.2.3 Classification

The proposed approach improves the usefulness of machine learning techniques that are used for classification. The main objective as a whole, is to improve the accuracy of classification and reduce the size of feature subsets using a genetic algorithm approach. To save time, the population size and the number of generations used by genetic algorithm are relatively small. To achieve better results, it is possible to allow more iterations or larger population size. SVM and Twin SVM [20] used as classifiers. SVM is a non-probabilistic supervised learning method which develops a model which assigns a class label based on patterns.

Twin SVM is a binary SVM classifier that solves two related SVM type problems and obtains two non-parallel planes. Both problems are smaller than in a standard SVM. The Twin SVM classifier is obtained by solving a pair of quadratic programming problems. For each class, it finds two hyperplanes and then classifies point. The first term in the objective function is the sum of squared distances from the hyperplane to points of one class. Therefore, minimizing it tends to keep the hyperplane close to points of one class. The second term of the objective function minimizes the sum of error variables. In short, Twin SVMs consists of a pair of quadratic programming problems. In each QPP, the objective function corresponds to a particular class and the constraints are determined by patterns of the other class. Twin SVM classify approximately four times faster than the usual SVM.

5 Results and discussion

The results are evaluated and determined for its accuracy, robustness, and computational complexity with respect to the proposed approach. For evaluation purposes, two datasets are selected. One of them is the most commonly available dataset for splicing detection which is Columbia Image Splicing Detection [20]. It consists of 933 authentic and 912 spliced images with a size of 128 x 128 pixels. The second dataset named MICC – F220 [21] of 220 images is used, 110 images of which are forged images and 110 are originals. The image resolution range varies from 722 x 480 to 800 x 600 pixels. The dataset depicts alterations such as translation, rotation, scaling (symmetric/asymmetric) and/or a combination of these while copying and pasting. The database images are divided into subgroups for implementation issues.

5.1 Performance measures

Performance of an algorithm can be measured in terms of accuracy, sensitivity, and specificity as given in equation 8,9,10. Some of the few terms used in the calculation are:

TN (True negative): - Authentic image is detected as authentic only
FN (False negative): - Forged image is detected as authentic
TP (True positive): - Forged image detected
FP (False positive): - Authentic image is detected as forged

$$\text{Accuracy: } \frac{TP+TN}{TP+TN+FP+FN} \quad (8)$$

$$\text{Sensitivity: } \frac{TP}{TP+FN} \quad (9)$$

$$\text{Specificity: } \frac{TN}{TN+FP} \quad (10)$$

5.1.1 Performance measure on MICC-F220 [21] dataset

Given below some of the result in terms of Accuracy, Sensitivity, and Specificity as mentioned above on MICC-F220 dataset. table 1 and table 2 shows the difference between SVM, and Twin SVM classifier in terms of performance measure. The best chromosomes selected by the genetic algorithm based on the features extracted from the images is given on table 3.

Table 1. Performance parameters on MICC-F220 using SVM as classifier.

Image	Accuracy	Sensitivity	Specificity	Average
Tampered 1	93.33	100	85.71	93.01
Tampered 2	92.84	100	85.03	92.62
Tampered 3	93.33	100	85.71	93.01
Tampered 4	92.56	100	86.20	92.92

Table 2. Performance parameters on MICC-F220 using Twin SVM as a classifier.

Image	Accuracy	Sensitivity	Specificity	Average
Tampered 1	93.33	100	85.71	93.01
Tampered 2	92.84	100	85.03	92.62
Tampered 3	93.33	100	85.71	93.01
Tampered 4	92.56	100	86.20	92.92

Table 3. Best chromosomes selected by Genetic Algorithm for MICC-F220

Accuracy	Image name
93.33	HOG, LBP, CON
93.33	HOG, CON, IDM
92.45	LBP, SURF, CON
93.33	LBP, CS,INER
92.48	SURF,ENT,CS
93.02	HOG,LBP,CON,CS
92.65	HOG,LBP,CON,CP
91.87	HOG,SURF,INER,LBP
93.01	IDM,SURF,CS,ENT
91.67	IDM,ENT,LBP,CON

5.1.2 Performance measure on Columbia Dataset

Similar to MICC-F220 given below table 4, table 5 and table 6 gives a comparative comparison on Columbia dataset.

Table 4. Performance parameters on Columbia dataset using SVM as classifier.

Image	Accuracy	Sensitivity	Specificity	Average
Tampered 1	88.63	86.90	90.52	88.68
Tampered 2	88.63	86.95	90.47	88.68
Tampered 3	88.67	87.45	88.34	88.15
Tampered 4	88.67	87.56	88.45	88.22

Table 5. Performance parameters on Columbia dataset using Twin SVM as classifier.

Image	Accuracy	Sensitivity	Specificity	Average
Tampered 1	88.63	86.95	90.47	88.68
Tampered 3	86.57	89.47	88.65	88.36
Tampered 5	86.10	85.76	89.56	87.14
Tampered 11	88.63	86.95	90.47	88.68

Table 6. Best chromosomes selected by Genetic Algorithm in 10 iterations for Columbia dataset.

Accuracy	Image name
88.65	HOG, INER, CON
88.67	LBP, HOG, CS
89.56	HOG,INER,LBP
85.64	HOG, CS, CON
86.78	SURF,INER,CON
88.97	SURF,INER,CON,ENT
86.45	LBP,IDM,CON,HOG
85.64	LBP,SURF,CON,CS
85.78	HOG,LBP,CON,CS
86.76	ENT,SURF,IDM,CS

5.1.3 Comparative study

The same set of feature set belonging to MICC F220 and Columbia dataset when applied to classical feature selection algorithms like Sequential floating forward selection (SFFS), sequential backward selection (SBS), sequential forward selection (SFS), and like methods gave the following feature subset with the accuracy as shown below in table 7 and table 8:

Table 7. Performance parameters of sequential feature selection algorithms on MICC F220 dataset using SVM as a classifier.

Feature selection algorithm	Feature set algorithm	Accuracy
SFS	HOG,INER,CS,LBP	93.33
SBS	HOG, INER ,SURF,CS	93.33
SFFS	HOG,INER,LBP,CS	93.33

Table 8. Performance parameters of sequential feature selection algorithms on Columbia dataset using SVM as a classifier.

Feature selection algorithm	Feature set algorithm	Accuracy
SFS	HOG, CON, LBP, INER	88.63
SBS	LBP,IDM,CS, SURF	88.63
SFFS	LBP,CON,CS,HOG	88.63

The execution time of the algorithm using SVM and Twin SVM as classifier can be compared with the following graph as given in figure 2.

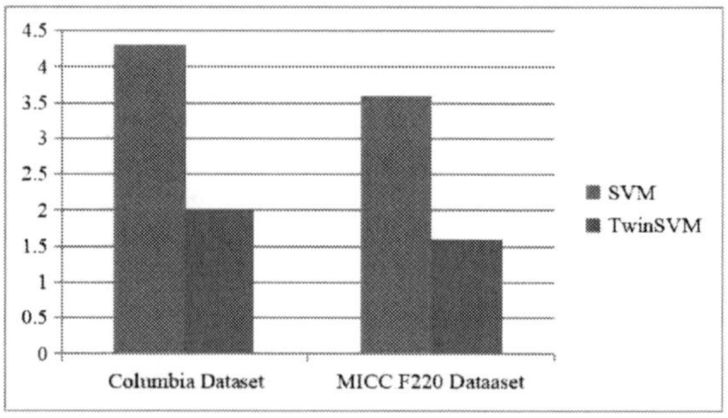

Fig. 2. The execution time of Twin SVM and SVM classifier.

6 Conclusion

This paper describes a technique to detect alteration in images with competing for accuracy and significantly low computational complexity and robust to various kinds of manipulations such as compression, rotation, scaling, etc. as can be seen from the database selection. As we know that different features of an image are used to detect different types of forgery accurately and efficiently. So a system is designed to collect the best of all features and classify them to detect forgery. To achieve this, set of ten features are selected and from them, the best of features are selected using genetic feature selection algorithm, which can further reduce the feature dimensions to get more optimal results. The reduced feature set is passed to SVM and Twin SVM for classification to get the final class results and the performance parameters are accuracy, sensitivity, and specificity. We obtain comparative results with a significant low computational complexity

References

1. C Rey and J-L Dugelay, "A survey of watermarking algorithms for image authentication.," *EURASIP Journal on Applied Signal Processing*, pp. 613-621, 2002.
2. V M Potdar, S Han, and E Chang, "A survey of digital image watermarking," in *3rd IEEE International Conference on Industrial Informatics*, Perth,WesternAustralia, 2005, pp. 709-716.
3. Granty Regina Elwin J, Aditya T S, and Madhu Shankar S, "Survey on Passive Methods of Image Tampering Detection," in *Proceedings of the International Conference on Communication and Computational Intelligence*, 2010, pp. 431-436.
4. B Mahdian and S Saic, "A bibliography on blind methods for identifying image forgery," in *Signal Processing:Image Communication*, 2010, pp. 389-399.
5. M Ali Qureshi and M Deriche, "A Review on Copy Move Image Forgery Detection Techniques," in *IEEE*, 2014.
6. T T Ng and S F Chang, "Blind Detection of Digital Photomontage using Higher," ColumbiaUniversity, 2004.
7. T T Ng and S F Chang, "A model for image splicing," in *IEEE International Conference on Image Processing*, Singapore, 2004, pp. 1169–1172.
8. D Fu, Y Q Shi, and W Su, "Detection of image splicing based on Hilbert–Huang transform and moments of characteristic functions with wavelet decomposition," *Lecture Notes in Computer*, vol. 4283, pp. 177–187, 2006.
9. Y F Hsu and F S Chang, "Detecting image splicing using geometry invariants and camera characteristics consistency," in *IEEE ICME*, Toronto,Canada, 2006, pp. 549–552.
10. F Y Hsu and S F Chang, "Image splicing detection using camera response function consistency and automatic segmentation," in *IEEE ICME*, Beijing, China, 2007, pp. 28–31.
11. Q Zheng, W Sun, and W Lu, "Digital spliced image forensics based on edge blur measurement," in *IEEE International Conference on Information Theory and Information Security*, Beijing,China, 2010, pp. 399–402.
12. Y Sutcu, B Coskun, H T Sencar, and N Memon, "Tamper detection based on regularity of wavelet transform coefficients," in *IEEE International Conference on Image Processing*, SanAntonio,TX,USA, 2007, pp. 397–400.
13. Z Lin, J He, X Tang, and C K Tang, "Fast automatic and fine-grained tampered JPEG image detection via DCT coefficient analysis," in *PatternRecognition*, 2009, pp. 2492–2501.
14. J Dong, W Wang, T Tan, and YQ Shi, "Run-length and edge statistics based approach for image splicing detection," *Lecture Notes in Computer Science*, pp. 76–87, 2009.
15. Z He, W Sun, W Lu, and H Lu, "Digital image splicing detection based on approximate run length," in *Pattern Recognition Letters*, 2011, pp. 1591–1597.
16. W Chen, Y Q Sh, and W Su, "Image splicing detection using 2-D phase congruency and statistical moments of characteristic function," in *Imaging Security, Steganography,andWatermarking of Multimedia Contents*, 2007.
17. Ng and S F Chang, "A DataSet of Authentic and Spliced Image Blocks," Columbia University, Technical Report 2004.
18. Y Q Shi, C Chen, and W Chen, "A natural image model approach to splicing detection," in *MM & Sec*, Dallas,TX,USA, 2007, pp. 51–62.
19. S. Prasad and B. Ramkumar, "Passive copy-move forgery detection using SIFT, HOG and SURF features," 2016 *IEEE International Conference on Recent Trends in*

Electronics, Information & Communication Technology (RTEICT), Bangalore, 2016, pp. 706-710.

20. Jayadeva , R Khemchandani, and Suresh Chandra, "Twin Support Vector Machines for Pattern Classification," *IEEE TRANSACTIONS ON PATTERN ANALYSIS AND MACHINE INTELLIGENCE,* vol. 29, pp. 905-910, MAY 2007.
21. Amerini, Irene, Lamberto Balian, Roberto Caldelli, Alberto Del Bimbo, and Giuseppe Serra. "A sift-based forensic method for copy-move attack detection and transformation recovery", *IEEE Transactions on Information Forensics and Security,* vol. 6, no. 3 ,pp. 1099-1110, 2011.

A novel method for SIFT features matching based on feature dimension matching degree

Yao Yang[1,*], *Jinkang* Wei[1], *Ximing* Zhan[1], and *Xikui* Miao[2]

[1]School of Astronautics Northwestern Polytechnical University Xi'an ,shannxi, China
[2]Key Laboratory of Electro-Optical Countermeasures Test & Evaluation Technology, LuoYang, China

Abstract. We proposes a method for fast matching SIFT feature points based on SIFT feature descriptor vector element matching. First, we discretize each dimensional feature element into an array address based on a fixed threshold value and store the corresponding feature point labels in an address. If the same dimensional feature element of the descriptor vector has the same discrete value, their feature point labels may fall into the same address. Secondly, we search the mapping address of the feature descriptor vector element to obtain the matching state of the corresponding dimensions of the feature descriptor vector, thus obtaining the number of dimensions matching between feature points and feature dimension matching degree. Then we use the feature dimension matching degree to obtain the suspect matching feature points. Finally we use the Euclidean distance to eliminate the mismatching feature points to obtain accurate matching feature point pairs. The method is essentially a high-dimensional feature vector matching method based on local feature vector element matching. Experimental results show that the new algorithm can guarantee the number of matching SIFT feature points and their matching accuracy and that its running time is similar to that of HKMT, RKDT and LSH algorithms

1 Introduction

Image feature matching is essential and crucial for computer vision, and can be employed in image retrieval[1], image stitching[2], visual reality[3], and scene reconstruction[4-6]. Firstly, local features are extracted from each image in these applications, and then the local descriptor with invariance is computed. By applying the nearest neighbor search of features, the most similar feature can be found to achieve the image matching. SIFT[7] is known as the most representative feature extraction and matching method currently, and it had been improved by many researchers[8-10]. The feature-points extracted from SIFT descriptors are extremely robust, it is adapt to illumination change and deformation. Thus, it is still the most popular local descriptors to solve the image matching problems. However, the SIFT descriptors is high-dimension, namely 128. It is difficult to find a way to deal with the computational complexity during the matching process. The proposed method is a high-dimensional feature vector matching method based on the partial matching of feature vector.

* Corresponding author: yangyao61@163.com

The experimental results show that the performance of our proposed method is similar to that of HKMT, RKDT and LSH.

2 Related work

SIFT operators are considered to be highly robust, but their descriptors have 128 dimensions. The existing algorithms use the exhaustive method as the search matching algorithm. The exhaustive search algorithm can ensure that the correct matching point pairs are searched out, however, its time cost is high; therefore many approximately nearest neighbor search algorithm comes into being. These algorithms can not only ensure the search precision to a certain degree but also save much time compared with the exhaustive search algorithm. Generally speaking, these algorithms can be divided into two types: those based on tree indexing method and those based on hashing indexing method. The KD tree search method[11] is a typical search matching algorithm based on trees but may fail when its number of dimensions is more than 15[12] or may even be inferior to the linear exhaustive search algorithm. The best-bin-first (BBF) algorithm [12] is the extension of the KD tree and breaks through the restriction of the number of dimensions and can search well high-dimensional features. On this basis, Hartley et al proposed the RKD-tree search algorithm[13] and improved the traditional BBF search algorithm. With GPU, Hu et al achieved the parallel processing of the RKD-tree[14], thus greatly enhancing the matching speed. But experimental data show that[15] its correct matching rate is less than that of the traditional SIFT image feature point matching algorithm. The IDistance[16] is another type of tree-based indexing algorithm, uses the K-means clustering method to analyze data distribution and then indexes the B+- tree. This reduces the search scope to a certain degree and increases the search speed. However, because the algorithm takes some clustering time to establish its index structure, it is not effective for the applications based on image feature matching. Different from the indexing method based on tree structure, the locality sensitive hashing (LSH) algorithm[17] is another type of approximate nearest neighbor search matching algorithm based on the hashing indexing. The method is: first, the data points are projected to the Hamming space and the indexes are established with a set of hash functions that satisfy certain constraint conditions, thus obtaining the nearest neighbor of an enquiry point with a rather big probability. To solve the problem that the algorithms based on LSH have a rather big space cost, scholars improved many LSH algorithms[18-19], thus ensuring the accuracy rate and search speed and optimizing the space cost. Ref. [20] applied the LSH algorithm to retrieving SIFT descriptors, thus reducing search time and obtaining satisfactory results.

Starting from the matching of each dimensional element of a feature point, this paper relies on the dimensional matching degree of a feature point pair to screen the suspicious matching point pair and then purifies it with the Euclidean distance, thereby obtaining the ultimate matching pair. Compared with the Euclidean distance exhaustive algorithm, the KD tree algorithm and the LSH algorithm, through discretizing the feature address mapping, our algorithm can directly match each dimensional element, obtain the dimensional matching degree among the feature points, screen out a large number of unmatched points with the dimensional matching degree, reduce the search scope, guarantee the accuracy rate and achieve fast search.

3 The feature dimension matching algorithm(FDMA)

In matching two images, the 128-dimensional SIFT feature descriptor point sets $\{P_i\}, i = 1, 2, \cdots n$ and $\{Q_j\}, j = 1, 2, \cdots m$ of the two images are obtained with the SIFT algorithm. The feature descriptor point set $\{P_i\}$ contain n number of descriptor points; the feature descriptor point set $\{Q_j\}$ contain m number of descriptors. We employ $\{P_i\}$ as the benchmark set and $\{Q_j\}$ as the target set to match the feature points of the two images. The feature point matching method proposed by Lowe uses the exhaustive search method, with which the Euclidean distance of each SIFT feature point in $\{P_i\}$ and each SIFT feature point in $\{Q_j\}$ is calculated. The feature point pair that satisfies the Euclidean distance threshold requirements and is purified with the secondary nearest neighbor is the matching pair. When $\{Q_j\}$ is a matching pair which satisfies the Euclidian distance threshold requirements, the SIFT feature point P_i,Q_j satisfies the following relationship:

$$\sqrt{\sum_{k=1}^{k=128}(P_i(k)-Q_j(k))^2} \le \tau \qquad (1)$$

where τ donates the set threshold value; P_i(k) is the k-dimensional element of the *i* number of descriptors in the $\{P_i\}$ benchmark set; Q_j(k) is the k-dimensional element of the *j* number of descriptors in the $\{Q_j\}$ target set. Then for all the descriptors, the k-dimensional element has the following function:

$$\max\left|P_i(k)-Q_j(k)\right| = \sigma \le \tau \qquad (2)$$

Using $\min(P_i(k))$ as the starting point, we discretize each-dimensional element according to the scale σ:

$$N = \left\lfloor \frac{P_i(k)-\min(P_i(k))}{\sigma} \right\rfloor, N \in Z \qquad (3)$$

The k-dimensional feature discretized coordinate axis is formed and shown in Fig. 1.

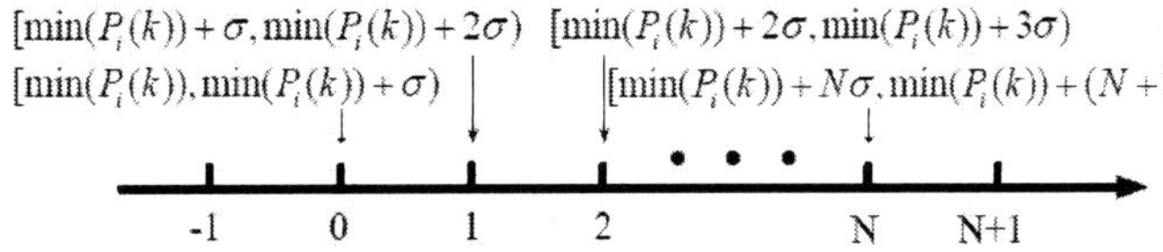

Fig. 1. The k-dimensional feature discretized coordinates.

According to Equation. (2), the $\langle P_i, Q_j \rangle$ matching pair meets the following condition.

If

$$P_i(k) \in [\min(P_i(k)) + N\sigma, \min(P_i(k)) + (N+1)\sigma] \qquad (4)$$

Then

$$Q_j(k) \in [\min(P_i(k)) + (N-1)\sigma, \min(P_i(k)) + (N+2)\sigma] \qquad (5)$$

In other words, on the k-dimensional feature discretized coordinates axis, the mapping of $P_i(k)$ and $Q_j(k)$ is certainly adjacent or equal:

If $P_i(k) \to N$

Then $Q_j(k) \to N-1$ or $Q_j(k) \to N$ or $Q_j(k) \to N+1$

If extended to other 128-dimensional feature discretized coordinates axis, the $\langle P_i, Q_j \rangle$ matching pair is shown in Fig.2.

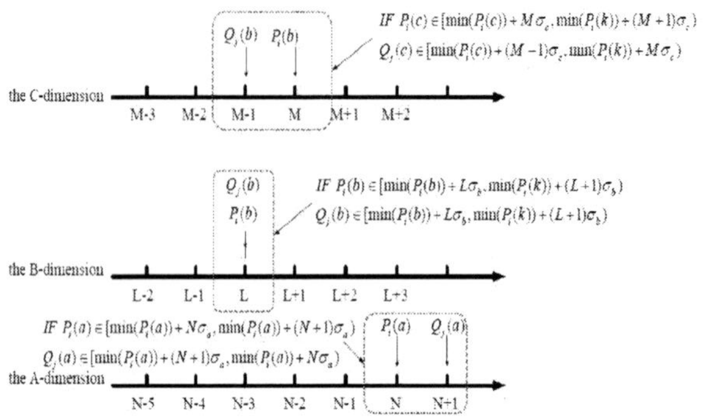

Fig. 2. The distribution of the i number of feature points in the benchmark set and the j number of feature points in the target set along each-dimensional feature discretised coordinate axis.

The above analysis draws the following conclusion: For the 128-dimensional SIFT feature descriptor point sets $\{P_i\}, i = 1, 2, \cdots n$ and $\{Q_j\}, j = 1, 2, \cdots m$, if the discrete feature obtained after every dimensional feature of one pair of feature points P_i and Q_j is discretized with a given threshold σ is adjacent or equal, then $\langle P_i, Q_j \rangle$ is the matching pair whose every dimensional distance is smaller than σ and whose Euclidean distance is smaller than 128 σ.

Therefore, after using the threshold σ to discretize the feature coordinate axis, each-dimensional feature matching descriptor automatically obtains its adjacent or equal discrete values. In actual application, we transform into unit cell array address the discrete coordinate of the each-dimensional element of feature descriptor in $\{P_i\}$ and label the corresponding feature points stored and mapped according to the array address, thus transforming the each-dimensional feature descriptor in $\{P_i\}$ into their array address and obtaining the unit cell array T_P, which contains $(N+3) \times n$ number of cells. $T_P(i, j)$ is the (i, j) number of cells in T_P; the elements in the cell denote the labels of all the feature points in the i -th discrete intervals that correspond to the feature of this dimension which the j-dimensional feature satisfies. If no feature points are satisfied, then the element is Ø label. For example, the unit cell array element $T_P(N, j) = \{a, b, c, \cdots\}$ denotes that all the j-dimensional features of the feature points $a, b, c, \cdots$ in $\{P_i\}$ fall within the N-th discrete interval at the discrete coordinates of the j-dimensional feature. The diagram for creating the j-dimensional unit cell array is shown in Fig. 3:

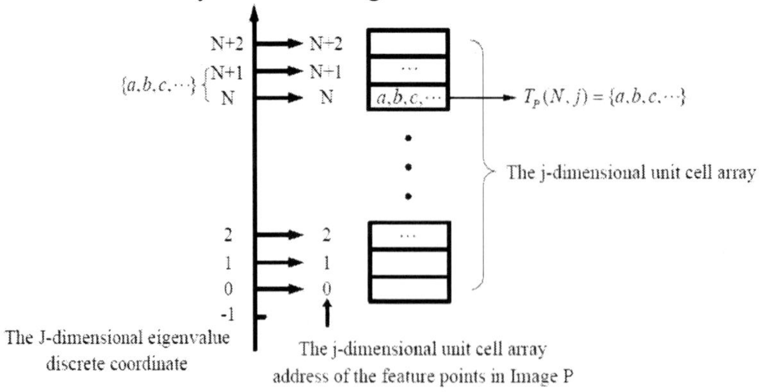

Fig. 3. The diagram for creating the j-dimensional unit cell array.

We can calculate the elements $Q_j(k)$ of each dimension through the feature-points of $\{Q_j\}$ by Eq. (6) to obtain the corresponding cell array address,

$$\begin{aligned} &if \quad n_{Q_j(k)} \le 0, n_{Q_j(k)} = 0; \\ &if \quad n_{Q_j(k)} \ge M+3, n_{Q_j(k)} = M+3; \\ &else \\ &n_{Q_j(k)} = \frac{T_2(n,k) - \min(T_1(i,k))}{\sigma_k} + 2 \end{aligned} \tag{6}$$

With the direct query of $T_P(ad,k), T_P(ad+1,k), T_P(ad-1,k)$ elements, the labels of the extracted correspondent feature points are the feature points that match P in the K-dimension and Q that has the N-th feature-points in the K-dimension.

The feature dimension matching is achieved by the ratio between number of matching dimensions and total number of dimensions using the feature points in $\{P_i\}$ and $\{Q_j\}$, $D = I/L, 0 \le I \le L$,where I represents the number of matching dimensions; L represents the total number of feature dimensions; D represents the feature dimension matching degree. Because the different dimensional values of the feature points with a certain invariance response differently in different variations, we assume that the matching feature points may not necessarily satisfy the requirements for all-dimensional matching; namely their feature dimension matching should be D<1. Therefore, the feature dimension matching degree threshold must be introduced into practical applications. After calculating the feature dimension matching degree between P_i and Q_j , it should be compared with the given feature dimension matching threshold. If it is larger than the threshold, it is assumed to be mutually matching. The feature dimension matching threshold that has SIFT features generally assumes to be around 0.7; the larger the value, the fewer the matching points. The smaller the value, the more matching errors.

The above operations can achieve the suspect matching point pair sets between all the feature points in P and all the feature points in Q , thus establish the suspect feature point matching matrix $T_{match} \in \mathbb{N}^{2\times c}$, where the j column of the matrix denotes the alternative feature point matching pair in P and Q .

Table 1. Algorithm process and steps.

1. Input two images P and Q
2. Using SIFT algorithm to calculate the SIFT feature description sets of two images: benchmark set $\{P_i\}$ and target set $\{Q_j\}$
3. Use $\{P_i\}$ to construct the 128-dimensional feature discrete coordinates
4. Transform the feature discrete coordinate of $\{P_i\}$ into unit cell array and use the feature point label inside P as the array element value of the corresponding address
5. Use $\{Q_j\}$ to calculate the unit cell array address that corresponds to the each-dimensional value of feature points in Q
6. Extract the feature point labels in the array according to the address in 5,Obtain the dimension matching degree of each point, and remove the feature points that do not satisfy the dimension matching degree threshold.
7. Purification by Euclidean distance to eliminate mismatches and achieve one-for-one matching
8. Delete the farthest Delete-counts pairs of feature point
9. Output the matching point pairs

However, the analysis and experiments show that the matching point pairs do not have one-for-one correspondence. One point in P may correspond to several points in Q or several points in P may correspond to one point in Q. To solve this problem, we calculate the Euclidian distance of the repeatedly matching feature point pair and use the nearest point as the matching point, ultimately obtaining the one-for-one correspondent feature point pair, whose number is c'.

Finally, to purify the experimental results on the algorithm, we introduce the ultimate screening threshold $DeleteCounts = floor(c'/beta)$ and use $beta$ as the screening threshold to calculate the Euclidian distance in the same way, rank it and delete the $DeleteCounts$ feature point pair which is the farthest away from the Euclidean distance and the least matching. Algorithm process and steps are as shown in Table1.

4 Experimental results and analysis

We use the Visual Studio 2013 to implement the novel feature point matching method proposed in this paper and calculate the matching time and feature points. The configuration of the computer is Intel (R) Core (TM) i5-3210M CPU @ 2.50Hz, 8GB memory.

We first perform the scale variation, rotation variation, noise variation, illumination variation and deformation variation of the Lena images to analyze the matching accuracy and speed, and then use the standard image dataset [21] for evaluation.

The FDAM algorithm uses the following parameters: $N = 20$, $\alpha = 0.7$, $beta = 3$, where N is the number of effective discrete segments; α is the feature dimension matching degree threshold; $beta$ is the screening threshold. The threshold value of the exhaustive algorithm for comparison is 0.2.

In order to evaluate the effectiveness of the proposed FDAM algorithm, we compare its performances with those of the three ANN algorithms: the hierarchical k-means tree (HKMT) algorithm, the random KD tree (RKDT) algorithm and the local sensitive hashing (LSH) algorithm and the exhaustive algorithm. We use the algorithms in the FLANN library[22] in the OpenCV3.0 to implement the three algorithms.

4.1 Evaluation of artificial transformation test

This paper carries out the artificial transformation of the Lena images (512*512, the number of SIFT feature points is 1998),.In this case, we can know the exact position of the feature points, so as to better analyze the matching results. The experimental results are shown in Fig 4. In Fig. 4(a), salt and pepper noise is added to the Lena image, and the number of SIFT feature points is 1931. In Fig. 4(b), the ratio of areas of the two matched images after scale variation is 1:0.25, and the number of SIFT feature points is 1192. In Fig 4(c), the illumination variation of the Lena image is conducted, and the number of SIFT feature points is 2256. In Fig. 4(d), the Lena image is rotated 45 degrees clockwise, and the number of SIFT feature points is 2927. In Fig. 4(e), the Lena image is deformed, and its upper edge varies into the original 3/4; the number of SIFT feature points is 2105. Fig 4 shows the matching results of HKMT, RKDT, LSH, FDMA and Exhaust algorithm from left to right.

The running time of the algorithm and its logarithm pairs of matching feature points are obtained, as given in Tables 2 and 3, where the running time only includes that of feature point matching but does not include that of feature point detection.

Table 2. Match point number comparison result.

	HKDT	RKDT	LSH	Exhaust	FDMA
Noise	129	122	152	204	121
Scale	175	172	143	137	171
Illumination	153	143	228	376	139
Rotational	484	467	610	548	463
Deformation	69	51	226	359	49

Table 3. Match time comparison table (unit s).

	HKDT	RKDT	LSH	Exhaust	FDMA
Noise	2.5	1.72	2.39	4.51	0.84
Scale	0.601	1.38	1.4	1.28	0.26
Illumination	1.767	1.71	2.63	5.241	0.96
Rotational	1.611	1.76	2.57	7.271	1.26
Deformation	1.581	1.67	2.73	4.99	0.92

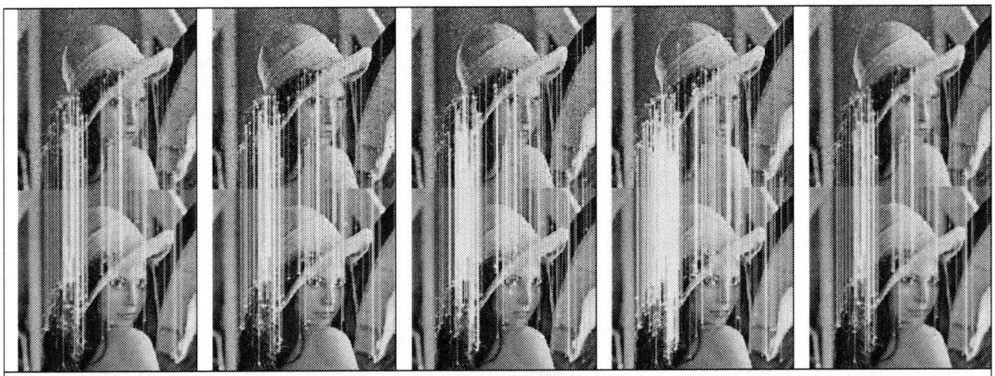

(a) Noise Variation

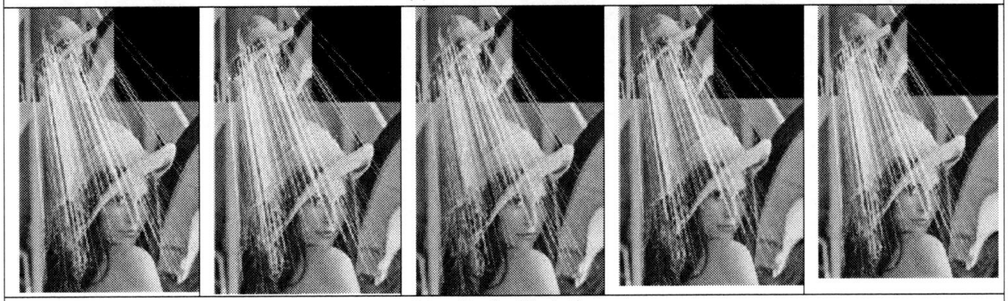

(b) Scale Variation

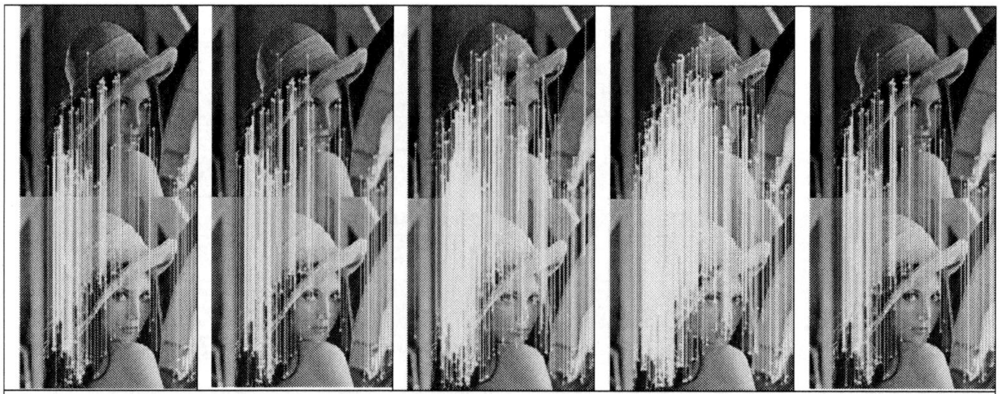

(c) Illumination Variation

(d) Rotational Variation

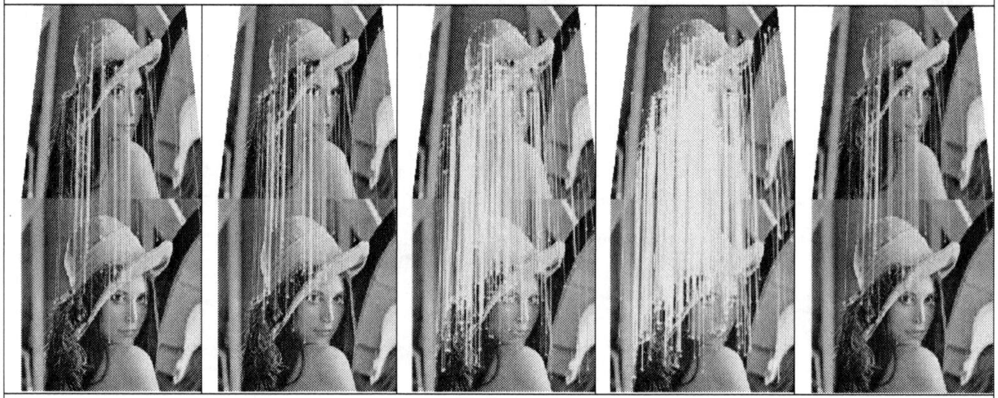

(e) Deformation

Fig. 4. Matching effect after transformation of Lena image.

4.2 Evaluation of the benchmark

To further evaluate the search performance of the FDAM algorithm, we use data sets as ours test data set.to compare the matching accuracy and matching speed. The test data set uses six different scene images, with a total of 164 images, and 809471 SIFT feature points. For each set of images, the first image is used as a query image, and its feature is matched with that of another image (reference image) in the image set respectively. We record the time consumption and the number of matched feature used by the Exhaustive, HKMT, RKDT, LSH and our FDAM algorithms. Table 4 gives the average search time spent on the

16 sets of images by the above-mentioned algorithms. Table 5 shows the average number of points in the 16 sets of images matched by the algorithms.

Compared with the other conventional algorithms, our FDAM algorithm performs better in terms of matching speed. Even when some image sets have a small number of matching points, the FDAM algorithm has a much higher number of matching points and faster matching speed than other algorithms. However, the FDAM algorithm is relatively slower when there are a large number of matching points because it spends some time on ranking during final screening.

Table 4. the average search time of the algorithm on 16 sets of images(unit s).

DataSet		HKMT	RKDT	LSH	FDMA	Exhaust
Rotation	VanGogh	2.645	3.258	4.525	2.232	14.162
	Marseil	5.449	11.236	25.117	20.848	138.182
	Monet	2.795	3.726	7.164	3.155	18.427
Zoom	Bip	4.106	6.064	10.140	5.687	34.478
	Belledonnes	5.200	5.757	7.030	4.057	30.125
	Crolles	3.719	5.082	6.979	4.422	29.802
Rotation&zoom	UBC	3.282	5.370	9.096	4.879	41.851
	Resid	3.333	5.592	13.299	6.926	37.917
	Inria	4.176	6.910	15.442	7.945	48.641
Viewpoint	Graffiti2	3.787	6.203	13.149	8.773	46.727
	Graffiti6	2.693	4.087	11.277	6.033	31.385
	Graffiti7	2.490	3.798	9.821	5.189	26.723
Illumination	MOVI	0.109	0.280	0.377	0.055	0.172
	graph	0.479	0.839	1.332	0.354	1.189
	fruit	2.561	2.144	2.884	1.213	6.647
Blur	bikes	7.468	8.076	12.620	6.592	40.040

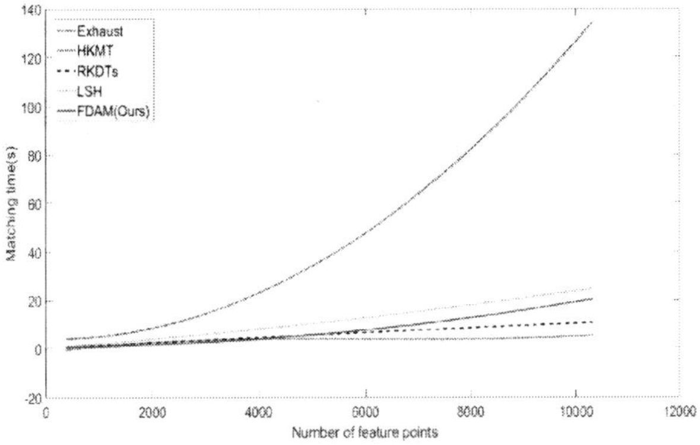

Fig. 5. The curve of matching time between RKDT, HKMT, LSH,Exhaust and FDAM with feature points.

Fig. 5 shows the curves of variation of matching time with the number of feature points among the RKDT, HKMT, LSH and FDAM algorisms.

The Table 4 and Table 5 shows that with suitable parameters, when images have variations in zoom, illumination, rotation, blur and Viewpoint, the FDAM algorithm does not have much difference in number of matching points from other algorithms and may not affect the robustness of the existing algorithms. Even in the case of viewpoint variation, the number of feature matching points searched is much large than that of other algorithms. This is because the distance function selected is different. Other algorithms use the ratio between nearest neighbor search and secondary nearest neighbor search as the main reference quantity. In doing initial selection, the FDAM algorithm uses the feature dimension matching degree as main reference quantity and allows some dimensions to have a rather big difference. The Euclidean distance is only taken as a relative quantity to do final purification and mainly used for screening one-for-many and many-for-one cases; the FDAM algorithm also screens out a certain proportion of matching pairs that sequence to the rear in the suspect matching pairs screened out through distance sequencing after initial selection is done. This is in agreement with the actual conditions of feature points after image variations.

Table 5. The average matching points in 16 sets of images.

DataSet		HKMT	RKDT	LSH	FDAM	Exhaust
Rotation	Van Gogh	555	520	679	635	510
	Marseil	2172	2111	2725	2253	2026
	Monet	400	385	570	485	379
Zoom	Bip	135	118	176	117	256
	Belledonnes	79	67	96	162	68
	Crolles	256	234	318	458	217
Rotation&zoom	UBC	762	729	862	690	701
	Resid	125	121	267	602	114
	Inria	81	67	292	541	64
Viewpoint	Graffiti2	6	5	56	229	4
	Graffiti6	60	49	283	671	43
	Graffiti7	44	40	199	481	33
Illumination	MOVI	45	44	59	85	44
	Graph	295	292	300	270	297
	Fruits	38	40	14	40	41
Blur	Bikes	185	176	299	260	173

5 Conclusion

This paper proposes a new SIFT feature point matching algorithm based on feature dimension matching (FDAM). The descriptor vector element is discretized into a cell array address, and the nearest neighbor search is conducted near the mapping address of the descriptor vector element. The matching speed increases linearly with the increase of the number of feature points. The proposed FDAM algorithm has matching accuracy and speed similar to, or even better than, those of the conventional RKDT, HKMT and LSH algorithms. This is particularly true when an image has viewpoint variation because the FDAM algorithm uses the feature dimension matching degree.

Acknowledgment

This work is supported by the Open Project Fund for the Key Laboratory of Photoelectric Confrontation Testing and Evaluation Technology (GKCP2016003).

References

1. Li L I, Lin F, Jun W U, et al. Exploiting global and local features for image retrieval[J]. Journal of Central South University, 2018, 25(2):259-276.

2. Luo Q M, Shi L. Review on image stitching methods[J]. Transducer & Microsystem Technologies, 2017..
3. Bi J, XU J, Xin Q, Shang D. An Augmented Reality Method Based on SURF and Geographic Grid Model [J]// JISUANJI YU XIANDAIHUA, 2017(6):47-53.
4. Jianfang X U, Liu Z, Han Z, et al. Application of Point Cloud Registration in 3D Reconstruction of Catenary Parts Based on SIFT and LBP[J]. Journal of the China Railway Society, 2017, 39(10):76-81..
5. Huang D, Zhang G, Ardabilian M, et al. 3D Face recognition using distinctiveness enhanced facial representations and local feature hybrid matching[C]// IEEE International Conference on Biometrics: Theory Applications & Systems. IEEE, 2016:1-7.
6. Lin C, Lin J, Chen Q, et al. A 3D Reconstruction Method for The Common View Field in New Hybrid Vision System[J]. Machine Design & Research, 2017.
7. Lowe, D.: Distinctive Image Features from Scale-Invariant Keypoints. IJCV 60 (2004) 91-110
8. Ke, Y., Sukthankar, R.: PCA-SIFT: A more distinctive representation for local image descriptors. In: CVPR (2). (2004) 506–513
9. Bay, H., Tuytelaars, T., Gool, L.: SURF: Speeded Up Robust Features. In: ECCV (2006) 404 – 417
10. Milkolajczyk, K., Schmid, C.,: A performance evaluation of local descriptors. IEEE Transactions on Pattern Analysis & Machine Intelligence (2005) 1615-1630
11. Benltley, L.: K-D trees for semi-dynamic point sets. In: Symposium on Computational Geometry (1990) 187–197
12. Nene, S., Nayar S.: A simple algorithm for nearest neighbor seareh in high dimensions. IEEE Transactions 0n Pattern Analysis and Machine Intelligence 19(9), 989–1003(1997)
13. Silpaanan, C., Hartley, R.: Optimised KD-trees for fast image descriptor matching. In: CVPR. (2008) 1–8
14. Hu L, Nooshabadi S, Ahmadi M. Parallel randomized KD-tree forest on GPU cluster for image descriptor matching[C]// IEEE International Symposium on Circuits and Systems. IEEE, 2016:582-585.
15. Muja, M., Lowe, D.: Fast approximate nearest neighbors with automatic algorithm configuration. In: ICCV (1). (2009) 331-340
16. Jagadish, H., Ooi, B., Tan, K., Yu, C., Zhang, R.: Adaptive B+-tree based indexing method for nearest neighbor search. Acm Transaction on Database Systems 30(2), 364-397(2005)
17. Yu, C., Ooi, B., Tan, K., Jagadish, H.: Indexing the distance： An efficient method to KNN processing. Vldb (2001)
18. Kulis, B., Grauman, K., Kernelized,: Locality-sensitive hashing for scalable image search. In: ICCV(2009) 2130-2137
19. Chakrabarti A, Satuluri V, Srivathsan A, et al. A Bayesian Perspective on Locality Sensitive Hashing with Extensions for Kernel Methods[J]. Acm Transactions on Knowledge Discovery from Data, 2015, 10(2):1-32.
20. Yang,K,Chen ,L.,Liu, Y． ,Fast Image Feature Matching Based on Extended Cascade Original Locality Sensitive Hashing． Computer Engineering，2016，42(8)：211-219
21. Image database:http://lear.inrialpes.fr/people/Mikolajczyk/Database/index.ht ml
22. FLANN:https://docs.opencv.org/3.0-beta/modules/flann/doc/flann.html

MATEC Web of Conferences **277**, 02028 (2019) https://doi.org/10.1051/matecconf/201927702028
JCMME 2018

Detect-and-describe: Joint learning framework for detection and description of objects

Adeel Zafar[1,*], and *Umar* Khalid[2]

[1]School of Software, Shanghai Jiao Tong University, Shanghai, China
[2]School of Electrical Engineering, Shanghai Jiao Tong University, Shanghai, China

Abstract. Traditional object detection answers two questions; "what" (*what the object is?*) and "where" (*where the object is?*). "what" part of the object detection can be fine grained further i-e. "what type", "what shape" and "what material" etc. This results in shifting of object detection task to object description paradigm. Describing object provides additional detail that enables us to understand the characteristics and attributes of the object ("*plastic* boat" not just boat, "*glass* bottle" not just bottle). This additional information can implicitly be used to gain insight about unseen objects (e.g. unknown object is "*metallic*", "*has wheels*"), which is not possible in traditional object detection. In this paper, we present a new approach to simultaneously detect objects and infer their attributes, we call it Detect-and-Describe (*DaD*) framework. *DaD* is a deep learning-based approach that extends object detection to object attribute prediction as well. We train our model on aPascal train set and evaluate our approach on aPascal test set. We achieve 97.0% in Area Under the Receiver Operating Characteristic Curve (AUC) for object attributes prediction on aPascal test set. We also show qualitative results for object attribute prediction on unseen objects, which demonstrate the effectiveness of our approach for describing unknown objects.

1 Introduction

Detecting objects in images has been one of the prime focus of researchers in the field of computer vision [1, 2, 3, 4]. Detection is the traditional naming task in which an object is identified by the name and localized by a bounding box in the image. Describing an object goes beyond the traditional naming task and identify few attributes and properties of the object [5, 6]. Detection when combined with description opens a door for whole lot of new applications in computer vision. For example, we can infer few properties and attributes of the object (e.g. "*round*" bottle), we can say what is unusual about the object (e.g. "*plastic*" car), even if an object is unrecognizable we can still say something about it (e.g. the unknown object "*has legs*", "*has head*"), and something about the object's environment can also be said (e.g. train "has leaf" means train is in trees).

In this paper, we propose an end-to-end deep learning framework for object detection and description which can detect objects and infer their attributes simultaneously, we call it

* Corresponding author: adeelz92@gmail.com

Detect-and-Describe (*DaD*) framework. As described in Fig. 1, image is passed through a deep convolution neural network (CNN) [7], which outputs object's bounding box and its label along with attributes of that object. Joint end-to end learning has multiple advantages over distinct learning. Firstly, simultaneous detection and attribute inference provide additional information about the identified object. This allows the user to perform appropriate actions based on the additional information gained from the object's description. For example, consider an automated driving car detects a traffic light signal, if it receives related knowledge about the state of the traffic signal ("*Red*" traffic signal), it can perform appropriate actions or take decisions accordingly (i-e "*Stop*"). Secondly, when an object is detected, a bounding box around the object is returned. Bounding box usually includes some area from the surroundings of the object as well. Simultaneously inferring attributes, along with object detection can provide some information about the surroundings as well. For example, in Fig. 1, the attribute "*has leaf*" provide information that the detected object "train" is surrounded by trees.

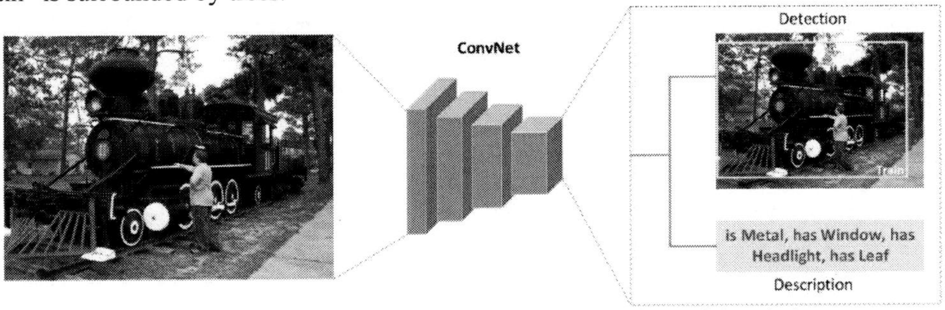

Fig. 1. Overview of Detect-and-Describe framework for object detection and description.

Different object categories share attributes (e.g. "Cat" and "Dog" both have "*skin*", "*leg*" and "*tail*") and same object categories may also have different attributes (e.g. "*Wooden*" boat, "*Plastic*" boat). Therefore, attribute prediction must be able to generalize accurately both within categories and across categories. In order to overcome this challenge, we learn attribute prediction separately from the object detection. It means, we learn separate convolutional features for attribute predictions and object detection. Detection uses its own sub-network to learn deep convolutional features specifically for detection task and description uses its own sub-network to learn features specifically for attributes prediction as shown in Fig. 2.

There are two types of object detectors in the deep learning literature, one-stage object detectors [4, 8, 9] and two-stage object detectors [10, 11, 12]. One-Stage object detectors simultaneously propose object proposals and then detect objects while two-stage object detectors propose object proposals in the first stage and then detect objects in the second stage. For real-time object detection, one stage object detectors demonstrated high speed but suffered from low accuracy and precision. Recently, [4] showed that one stage object detectors suffered from accuracy issues because of high density of background ("No object") proposals and low density of foreground ("object") proposals in large number of candidates bounding boxes or anchors. They introduced dynamically scaled cross-entropy loss called *Focal loss* to overcome the problem of high loss due to large density of background anchors. Since we also predict object attributes based on anchors therefore we use the *Focal loss* from [4] as the minimization objective for attributes learning. We propose Detect-and-Describe (*DaD*) framework (Fig. 2), which extends the one-stage object detector from [4] for simultaneous object detection and description.

To demonstrate the effectiveness of the proposed *DaD* framework, we perform experiments on aPascal attributes dataset [5]. We train our model on aPascal train split and report performance on aPascal test split. We achieve 97% in Area under the Receiver

Operating Characteristic Curve (AUC) for object detection on aPascal test set. We also show qualitative results of attribute predictions on unknown/unseen object categories (these categories are different from our training categories) by randomly selecting images from Google Image Search to demonstrate the generalization ability of our approach.

2 Related Work

Attributes detection in objects goes almost as back as object detection in the history of computer vision. As attributes are merged within the object itself, researchers have been learning the role of attributes detection in object detection e.g. [13]. In feature integration theory of attention [14], Treisman and Gelade theorized that human brain recognizes objects in two stages. At first stage, human brain perceives information about the basic features of the visual area in focus and then in second stage, focused visual attention is used to combine the individual features of an object to recognize the object. It states that object attributes play an important role in understanding the nature of the object. Since the inception of feature integration theory of attention, researchers in computer vision have adopted attributes recognition for object categorization as well as object description.

Ferrari and Zisserman [15] presented a probabilistic generative model to learn simple color and pattern attributes in image segments in a weakly supervised setting using Google image search. Later, [5] presented a new dataset for attributes based on Pascal VOC 2008 [16] object detection dataset and presented a classification model that learned to predict attributes using HOG features [17]. They also learned to categorize objects based on the annotated semantic features and learned discriminative features. The most relevant work is [18] which learned to predict object location via iterative procedure in a weakly supervised manner, where images are only labelled with categories but not object locations. They use initial model to predict saliency score for the object and use visual cues from attributes model to improve the model for object detection (*location and class prediction*), which is their prime goal. In contrast to their work, our goal is to improve the performance of attributes prediction while retaining object detection capability using end-to-end deep learning model. In *DaD* framework, we build object and attributes subnetworks on top of feature pyramid network [12] to detect objects and their attributes in the anchor-boxes/bounding-boxes.

3 Detect-and-Describe Framework

In one stage object detectors, multiple anchor boxes of different sizes and aspect ratios are proposed on each spatial location of some deep CNN layer of the backbone network. Anchors are assigned positive (*"Object"*) or negative (*"No object"* or *"Background"*) labels based on IoU value with the ground truth bounding box. Final anchors for the output are chosen based on top objectness score and filtering best anchors based on non-maximum suppression. The class label and bounding box offsets are computed for these anchors. In *DaD* framework, in addition to object's label and bounding box offsets, the log probability for multi-label attributes is also computed for final anchors. The attribute set for the detected object is the attributes probability score for each final anchor box, in which score of 0 depicts *attribute not present* and score of 1 depicts *attribute is present*. As most of the anchors are assigned negative label due to low IoU with the ground truth bounding box, they produce a negative learning signal which results in high loss while training. [4] presented focal loss to overcome this problem.

3.1 Focal Loss

The standard cross entropy (CE) loss for binary classification is defined as

$$CE(p,y) = \begin{cases} -\log(p) & if\ y = 1 \\ -\log(1-p) & otherwise \end{cases} \quad (1)$$

where $y \in \{\pm 1\}$ specifies ground truth label and $p \in [0, 1]$ is the model's estimated probability for the class with label $y \in 1$.
The $CE(p,y)$ can be rewritten as $CE(p_t) =-\log(p_t)$ where p_t is defined as:

$$p_t = \begin{cases} p & if\ y = 1 \\ 1-p & otherwise \end{cases} \quad (2)$$

Then the *focal loss* is defined as

$$FL(p_t) =-\alpha(1-p_t)^{\gamma}\log(p_t) \quad (3)$$

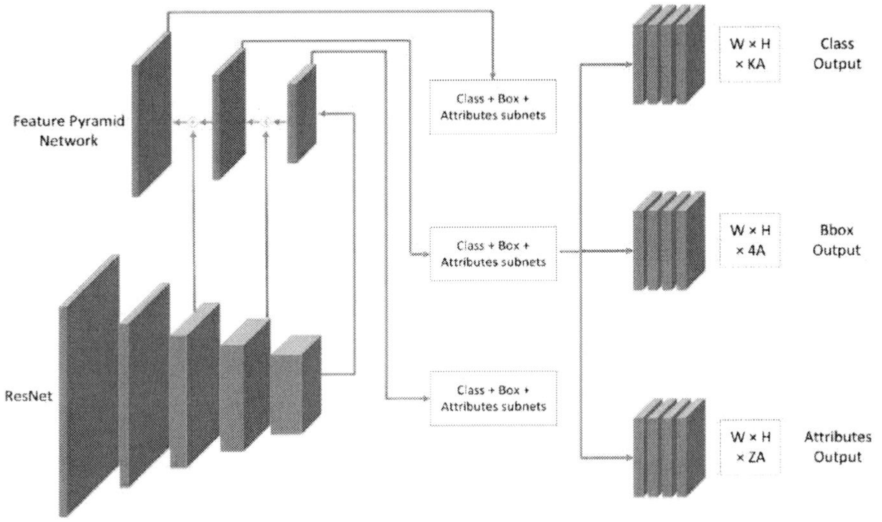

Fig. 2. In *DaD* framework, image is first passed through Resnet [19] which consists of 5 convolution blocks, a feature pyramid network [12] is built on top of the final convolution layers of each block (Conv3, Conv4 and Conv5 layers) of ResNet. Each layer in feature pyramid is connected to their respective bounding box, class, and attributes subnets, where $W = width$, $H = height$ of the last convolution layer of subnets and K = number of classes, 4 = Bbox offsets, A = number of anchors and Z = number of attributes.

where $\alpha \in [0, 1]$ is the weighting factor to address class imbalance (high ratio of negative anchors w.r.t positive anchors) and $\gamma \geq 0$ is the tuneable focusing parameter that can decrease or increase the weight of easily classifiable examples. As large number of anchors in one stage detectors are backgrounds, they are easy to classify, so when background anchor is correctly classified then $p_t \to 1$ and $(1-p_t)$ is small which down-weights the loss for background classes. This way negative signal due to background anchors is eliminated.

3.2 Model Architecture

We extend one stage object detector called RetinaNet detector from [4] for attributes recognition. RetinaNet consists of backbone network and two task specific sub-networks.

Backbone network is the convolutional neural network (ResNet [19] in our case) and two subnetworks are class label predictor and bounding box predictor. A feature pyramid network (FPN) [12] is built on top of the backbone network and these sub-networks are connected to each layer of the FPN. In addition to the class sub network and bounding box sub-network, we add an attribute sub-network, as shown in Fig. 2.

3.2.1 Attribute subnetwork

The design of the attribute sub-network is exactly the same as class subnet and box subnet. A fully convolutional network (FCN) is built on top of each layer of FPN, it takes C input channels from the given layer of the pyramid and convolve the input four times using 3×3 kernel and $C = 256$ channels each time. Each convolution layer in this FCN is followed by ReLU activations. The sigmoided final layer outputs ZA channels expressing probabilities of Z attributes for A anchors on each spatial location of the input. The translation-invariant anchors of multiple sizes $(32^2\ to\ 512^2)$ and aspect ratios $(1{:}2, 1{:}1, 2{:}1)$ are used to detect objects of different sizes and shapes. Our attribute predictor subnetwork uses the same anchors to predict attributes for the object.

4 Experiments

In this section, we report the details of several experiments to evaluate the performance of *DaD* framework on the attribute prediction task using the aPascal dataset [5]. We first perform training on aPascal train split and choose the optimal parameters. In the end, we perform testing on aPascal test to report within category generalization and randomly chosen images from Google to report across category generalization.

4.1 Dataset

The aPascal dataset [5] is built on top of Pascal VOC 2008 object classification and detection dataset [16]. The dataset consists of 2113 images in train split with 6340 object instances and 2227 images in test split with 6355 object instances. aPascal dataset is annotated with 64 different types of sematic attributes for each object instance in both splits. These attributes contain shape, parts and material attributes. We choose images which are closely associated to aPascal classes but different object categories from Google to report across category generalization. We choose images of donkey, monkey, goat, wolf, jetski, zebra, centaur, mug, statue, building, bag and carriage. The classes in test images are closely related to classes in aPascal, e.g. "wolf" in test images is related to "dog", "statue" is related to "person" etc in aPascal, therefore attributes should generalize well to test images as well.

4.2 Experimental setup

We use aPascal train split for training and use 25% randomly chosen subset of aPascal test for validation. We use Adam optimizer [20] with $\beta 1 = 0.9$ and $\beta 2 = 0.99$. The base learning rate is set to 1×10^{-5} which decays by a factor of 0.5 if training loss is not reduced after 2 epochs. We use ResNet-152 as backbone network and use batch size of 1 image because of high memory consumption of the model. Image is resized to minimum side of 800 and maximum side of 1333 to facilitate detection of small objects. We terminate training at 50 epochs and choose the best model for testing based on validation accuracy. We report results on aPascal test. All experiments are implemented in Keras [21].

Table 1. AUC score for different types of backbone networks in feature pyramid network.

Backbone Network	AUC
ResNet-50	92.71%
ResNet-101	94.64%
VGG-Net	93.11%
ResNet-152	**97%**

5 Results

5.1 Performance on aPascal

As mentioned earlier, within category generalization can be checked on dataset where distribution of instances is from the same set of classes for train and test splits. Therefore, to test within category generalization, we train on aPascal train split and report performance on aPascal test split. We use area under ROC curve (AUC) as performance metric for attribute predictions because it is class invariant. We achieve 97% in AUC for attributes prediction. Table 1 shows different ablations of experiments on different backbone networks.

5.2 Detecting and describing objects

Fig. 3(a). This figure shows detection and attributes for randomly chosen images for some of the classes in aPascal. 8 randomly chosen attributes for each of the image are shown. Attributes starting from *"No"* means attribute is not present, otherwise attribute is present.

Fig. 3(b). This figure shows attribute predictions on images of unknown objects.

We now show some qualitative results on real images from aPascal test to verify the detection and description ability of *DaD* framework in Fig. 3(a). Some interesting and unusual results can be seen in the images, e.g. in the "cow" image (1st row, 4th column), the attribute *"No Eye"* is predicted, because eyes are covered by hair. In the same image, attribute *"No Leather"* is predicted, because this cow's skin is hairy and leather surface is invisible. Similarly, in the "sheep" image (1st row, 3rd column), *"No Mouth"* is predicted because sheep's mouth is too small and almost invisible. This shows the capability of *DaD* framework to recognize what is unusual about the object while retaining the detection ability.

5.3 Describing unknown objects

We show 12 different images for each of the test classes mentioned in Section 4.1 in Fig. 3(b). Model is able to generalize attributes learned from other categories to new categories fairly well. For example, in "goat" image (2nd row, 3rd column) model predicts *"Horn"* attribute for goat, which during training is learned from "cow" category. Similarly, in "carriage" image (1st row, 3rd column), model is able to predict the presence of *"wheel"*, although these wheels are wooden, while during training is learned from "cars" which have metal and rubber wheels. New classes which do not have any close association with the training classes, do not generalize well, e.g. in "building" image (1st row, 1st column), system predicts attributes for the known objects, but these known objects are in vicinity of the building therefore it predicts some attributes for the building such as *"Glass"*.

6 Conclusion

In this paper, we present Detect-and-Describe (*DaD*) framework to simultaneously detect objects and recognize their attributes. Compared with existing approaches to predict attributes, the *DaD* framework achieves significant performance improvement both within object categories and across categories. The framework can recognize attributes of the unknown objects fairly well due to high generalization ability of our approach. *DaD* framework is an extension of one stage object detector called "RetinaNet", in which we add an additional attribute sub-network to predict attributes, along with class label and bounding box. We find out that attributes learning can be transferred fairly well to new objects without having prior knowledge about them, but generalization is better for those objects which have same symmetry, material, body parts and texture to the learned objects which is an understandable limitation.

References

1. Kaiming He, Xiangyu Zhang, Shaoqing Ren, and Jian Sun. Spatial pyramid pooling in deep convolutional networks for visual recognition. In ECCV, 2014.
2. Ross Girshick, Jeff Donahue, Trevor Darrel, and Jitendra Malik. Region-based convolutional networks for accurate object detection and segmentation. IEEE transactions on pattern analysis and machine intelligence, 2015.
3. Shaoqing Ren, Kaiming He, Ross Girshick, and Jian Sun. Faster r-cnn: Towards real-time object detection with region proposal networks. In NIPS, 2015.
4. Tsung-Yi Lin, Priya Goyal, Ross Girshick, Kaiming He, and Piotr Dollar. Focal Loss for dense object detection. *arXiv preprint arXiv:1708.02002*, 2017.
5. Ali Farhadi, Ian Endres, Derek Hoiem, and Devid Forsyth. Describing objects by their attributes. In Computer Vision and Pattern Recognition, 2009.

6. Christoph H Lampert, Hannes Nickisch, and Stefan Harmeling. Attribute-based classification for zero-shot visual object categorization. IEEE Transactions on Pattern Analysis and Machine Intelligence, 2014.
7. Yann LeCun, Bernhard E Boser, John S Denker, Donnie Henderson, Richard E Howard, Wayne E Hubbard, and Lawrence D Jackel. Handwritten digit recognition with a back-propagation network. In Advances in neural information processing systems, pages 396-404, 1990.
8. Joseph Redmon, Santosh Divvala, Ross Girshick, and Ali Farhadi. You only look once: Unified, real-time object detection. In CVPR, 2016.
9. Wei Liu, Dragomir Anguelov, Dumitru Erhan, Christian Szegedy, Scott Reed, Cheng-Yang
Fu, and Alexander C Berg. Ssd: Single shot multibox detector. In ECCV, 2016.
10. R. Girshick. Fast R-CNN. In ICCV, 2015.
11. S. Ren, K. He, R. Girshick, and J. Sun. Faster R-CNN: Towards real-time object detection with region proposal networks. In NIPS, 2015
12. T.-Y. Lin, P. Dollar, R. Girshick, K. He, B. Hariharan, and S. Belongie. Feature pyramid networks for object detection. In CVPR, 2017.
13. Nancy Kanwisher and Jon Driver. Objects, attributes, and visual attention: Which, what, and where. *Current Directions in Psychological Science*, 1992.
14. Anne M Treisman and Garry Gelade. A feature-integration theory of attention. *Cognitive psychology*, 12(1):97–136, 1980.
15. Vittorio Ferrari and Andrew Zisserman. Learning visual attributes. In NIPS, 2008.
16. M. Everingham, L. Van Gool, C. K. I. Williams, J. Winn, and A. Zisserman. The PASCAL Visual Object Classes Challenge 2008 (VOC2008) Results. http://www.pascalnetwork.org/challenges/VOC/voc2008/workshop/index.html, 2008.
17. Navneet Dalal and Bill Triggs. Histograms of oriented gradients for human detection. In *Computer Vision and Pattern Recognition, 2005. CVPR 2005. IEEE Computer Society Conference on*. IEEE, 2005.
18. Gang Wang and David Forsyth. Joint learning of visual attributes, object classes and visual saliency. In *Computer Vision, 2009 IEEE 12th International Conference on*. IEEE, 2009.
19. Kaiming He, Xiangyu Zhang, Shaoqing Ren, and Jian Sun. Deep residual learning for image recognition. In *Proceedings of the IEEE conference on computer vision and pattern recognition*, 2016.
20. Diederik P Kingma and Jimmy Ba. Adam: A method for stochastic optimization. *arXiv preprint arXiv:1412.6980*, 2014.
21. Fran¸cois Chollet et al. Keras. https://keras.io, 2015.

A survey on joint object detection and pose estimation using monocular vision

Aniruddha V Patil[1] , and *Pankaj* Rabha[2,*]

[1]IIIT-Hyderabad, Gachibowli, Hyderabad, Telangana, India 500032
[2]Intel, Bellandur, Bangalore, Karnataka, India 560103

Abstract. In this survey we present a complete landscape of joint object detection and pose estimation methods that use monocular vision. Descriptions of traditional approaches that involve descriptors or models and various estimation methods have been provided. These descriptors or models include chordiograms, shape-aware deformable parts model, bag of boundaries, distance transform templates, natural 3D markers and facet features whereas the estimation methods include iterative clustering estimation, probabilistic networks and iterative genetic matching. Hybrid approaches that use handcrafted feature extraction followed by estimation by deep learning methods have been outlined. We have investigated and compared, wherever possible, pure deep learning based approaches (single stage and multi stage) for this problem. Comprehensive details of the various accuracy measures and metrics have been illustrated. For the purpose of giving a clear overview, the characteristics of relevant datasets are discussed. The trends that prevailed from the infancy of this problem until now have also been highlighted.

1 Introduction

Object detection is the process of finding instances of real-world objects in images or videos. Object detection algorithms typically use extracted features and learning algorithms to recognize instances of an object category.

The position and orientation of an object is called the pose of the object. To determine the pose of an object in an image is called pose estimation. A general task in the various applications of pose estimation is to determine the position and orientation of each of the object in the scene with regard to some coordinate system. There have been many works on object detection and pose estimation separately. This work aims to cover all works that treat these two problems simultaneously.

Monocular vision, by definition is vision with one eye/camera. This type of vision greatly limits the perception of depth. For all of joint object detection and pose estimation, the goal remains the same, given an input image, provide what objects exist in the input image and their pose. The methods used vary based on the level of sophistication of the input. In this paper, we discuss methods where the input is a single image (RGB or Grayscale).

* Corresponding author: pankaj.rabha@intel.com

The various applications of this problem include augmented reality [1], autonomous driving [2], robotic manipulation [3,4] and inventory management [5]. Applications like augmented reality and autonomous driving need real-time approaches whereas this constraint can be relaxed a little when it comes to robotic manipulation or inventory management in situations where taking time does not pose a threat.

2 Input types

Advancement in technology has enabled us to capture input data at various sophistication levels.

2.1 Monocular (Colored/Grayscale)

One can say that this is the most common form of input. A colored monocular image is easily available from any camera. Medical applications that interpret signals as images commonly use grayscale images.

2.2 RGB-D

These are colored images along with depth data. The apparatus required for these images has recently become more economically viable in the form of the Microsoft Xbox Kinect. Other popular sensors are the Asus Xtion PRO LIVE and The Leap.

2.3 LiDAR

Light Detection and Ranging is an optical remote-sensing technique that uses laser light to densely sample the surroundings, producing accurate measurements of distance to each point the laser hits. The major hardware components of the LiDAR system include a vehicle, laser scanner system, Global Positioning System (GPS) and an Inertial Navigation System (INS).

2.4 RADAR

Radio Detection and Ranging is a radio wave based detection system that helps determine the location and velocity of objects. There exist three kinds of RADAR - Short Range (SRR), Medium Range (MRR) and Long Range (LRR). In the context of autonomous driving, these RADAR systems perform different tasks (SRR for object/pedestrian detection and emergency braking, MRR for lane change assistance and cross traffic alert and LRR for Autonomous Cruise Control (ACC) and collision warning).

3 Formulation of object detection and pose estimation

3.1 Object detection

Object detection has two main components – localization and classification of the object of interest. Localization can be formulated as estimating a vector (x,y,h,w), where x, y, h and w are respectively the x-coordinate, y-coordinate, height and width of the object. These are mostly specified in the camera coordinate system. Classification can be formulated as estimating a single number c that corresponds to the class number of the object of interest. Combining these two, object detection can be formulated as estimating a vector (c,x,y,h,w).

3.2 Pose estimation

There are two ways in which we can formulate the pose estimation problem. We can formulate it as a classification problem where the all the possible configurations of the object are quantized into various classes. The pose of the object is then classified into the appropriate class.

Assume that the set of all possible configurations of the object is $\boldsymbol{D}$ then, on dividing this set into k dimensions such that all possible configurations in the ith dimension can be depicted as $\boldsymbol{D}^i$ where $i \in \{1..k\}$. Further division of $\boldsymbol{D}^i$ into n subsets gives us $\boldsymbol{D}_j^i$ where $j \in \{1..n\}$. Thus, the problem becomes to correctly place the pose of the object into one of the $\boldsymbol{D}_j^i$ classes.

In the formulation of pose estimation as a regression problem, we regress an initial pose to best fit the pose of the object. When we assume the pose of the object to be k-dimensional, the regression-based approach should estimate a k-dimensional vector as the pose of the object. Now we look at different types of approaches for joint object detection and pose estimation. They can be categorized into vision based, hybrid and deep learning based approaches.

4 Vision based

The vision based approaches heavily use domain knowledge and involve modeling the input space in such a way that the expected output can be learnt based on the model. The mathematical formulation, information contained in the selected features, determining what features to use and using the right learning technique for efficiency are key to approach this problem with this method.

The different contributions of works that fall under this approach can be broadly categorized into descriptors/models and estimation methods.

4.1 Descriptors/Models

4.1.1 Chordiograms

Are shape representations based on geometric relationships of object boundary edges ensuring that foreground elements are prioritized. A chordiogram is a k dimensional histogram of all chord features on the boundary of a segmented object. A chord is a pair of points (p,q) on the boundary points. The chord feature d_{pq} is given as

$$d_{pq} = \left(l_{pq},\ \varphi_{pq},\ \theta_p -\ \varphi_{pq},\ \theta_q -\ \varphi_{pq}\right) \quad (1)$$

where l_{pq} is the chord length, φ_{pq} is the chord orientation and θ_p and θ_q are normals of the object boundary at p and q. They take into account the chord vector lengths, their orientations and normals given the silhouette of the object of interest and capture the geometric relationships like relative location and normals between pairs of boundary edges as a high-level descriptor [6].

4.1.2 Shape aware deformable parts model

A deformable parts model (DPM) is a star shaped conditional random field (CRF), with a root part for the whole object and several parts connected to the root. Each model part has an anchor point which is relative to the root part. These parts are allowed to move around their anchor point with a penalty. For simultaneous object detection and pose estimation from only model silhouette information, a shape aware modification of the DPM can be made. Every coarse pose of the object maps to a component of the learned S-DPM. By grouping nearby viewpoints, a certain number of discrete poses are arrived at. Silhouettes of a coarse pose cluster are used in the training of a S-DPM component as positives. The silhouettes of other poses, objects and random background edges are used as negatives. The coarse pose of the object is determined by the component with the maximum score at each image location [6].

4.1.3 Bag of boundaries (BOB)

After extracting contours of the image containing the object of interest, the bounding boxes of various views of the object are resized to their average size. A template is created in which each pixel counts the number of contours it is a part of. This gives us a midlevel representation of the image as a bag of boundaries for shape matching [7]. A BOB operates on information apart from shape context. It operates on image edge information by detecting and aggregating boundaries in the neighbourhood of the BOB. The best matching subset of BOBs to the given shape templates, such that maximal coverage of the image contours estimated to be boundaries, are found in the case of object detection.

4.1.4 Distance transform templates

Templates based on the distance transform have proved to be effective in detecting and estimating the pose of texture less planar objects. In [8], an edge map is created with the Canny edge detector [9], followed by applying the distance transform on this edge map. The distance transform template is then created by computing the level curve at a given value. The level curve is essential as it makes sure that small gaps do not cause the object of interest to be detected as two distinct objects.

4.1.5 Natural 3D markers

Natural 3D markers [10] are a set of landmarks on the object of interest. They are chosen based on their visibility from different views and that are robust to illumination, partial occlusions, clutter, etc. They are extracted under multiple viewpoints. In the case of the N3Ms being clustered in a single region, the occlusion of that region would cause this method to fail. Thus, the points are taken such that they are evenly spread across the entire object of interest.

4.1.6 Facet features

Given the image contours, polygons are approximated to reduce the number of vertices. Prior knowledge of the facets (depicted in figure 1) of the object of interest is then used for detection and pose estimation [1].

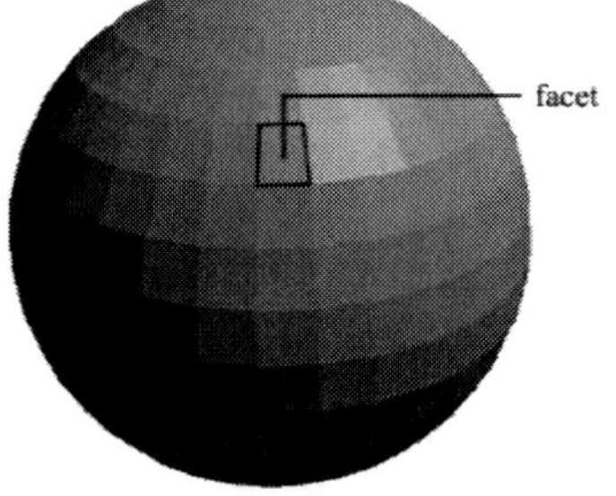

Fig. 1. Depiction of a facet.

4.2 Estimation methods

4.2.1 Expectation Maximization (EM) and Iterative Clustering Estimation (ICE)

EM finds the maximum likelihood parameters in statistical models. These methods are conceptually similar (ICE is the manifestation of EM that considers local features that map to an object instance and evaluates pose hypotheses). The expected membership of each data point to one of the distributions is given by the expectation step and the parameters for each distribution are given by the maximization step. Similarly, the expected membership of each local feature to an object instance is given by the clustering step and the best object pose corresponding to the feature memberships is given by the estimation step [4,7].

4.2.2 Probabilistic Network

This tree-structured network of probabilistic classifiers uses boosting for quick rejection of negatives while the positives traverse multiple branches. The fundamental total probability law is leveraged to compute the probability of being an object, given the pose parameters [11].

4.2.3 Iterative genetic matching

This method uses a genetic algorithm with a given population, crossover rate and mutation rate for the purpose of matching. The matching rate is the metric used for determining the fitness of an individual. This method is proposed as an alternative to the EM and ICE methods [12].

5 Hybrid approaches

In the case of hybrid approaches, the handcrafted features, representations or intermediate results of some other technique work with a neural network. Again, the features are as important as the architecture of the network to make this approach accurate and efficient.

Geometric constraints applied to a predicted 2D bounding box followed by regression by a CNN to give a 3D bounding box is shown in [2]. Though hybrid methods try to achieve best of both worlds, extending the network to also extract features has proved to be much more effective.

6 Pure deep learning based approaches

Pure deep learning based approaches use end-to-end networks to map the inputs directly to the expected outputs without any intermediate representations or extracted features. Though computationally expensive, due to recent breakthroughs in technology, accurate real-time approaches are feasible on high-end GPUs. Recently, CNNs have been outperforming DPM based methods for the task at hand. We can further classify these approaches into single stage or multistage methods.

A network is said to be multistage if it has stages performing distinct tasks linearly. A single stage network has no intermediates and provides the output directly from the input.

6.1 Single stage

SSD-6D [13] proposes an extension of the SSD [14] detection framework for the task of 3D detection and 3D rotation estimation. The 3D rotation space is decomposed into discrete viewpoints and in-plane rotations. The rotation estimation is then treated as a classification problem. Appropriate sampling for the rotation space is required for obtaining good results. This approach requires an offline stage to estimate 3D translation. This stage is used to precompute bounding boxes w.r.t. all possible sample rotations.

PoseCNN [15] jointly carries out segmentation and pose estimation of the segmented objects. It relies on a fully convolutional network to localize objects which might make it harder to deal with multiple instances of the same object.

Deep-6DPose [16] proposes adding a branch to the prominent Mask-RCNN [17] for the purpose of pose estimation such that no further pose refinement is necessary. Decoupling of pose parameters into translation and rotation so that the rotation can be regressed via a Lie algebra representation has been shown.

In [18], the authors attempt to combine detection and pose estimation at the same level. The scores for the presence of an object category, the offset for its location and the approximate pose are all estimated on a regular grid of locations in the image.

6.2 Multi stage

In BB8 [19], the idea of a cascade of multiple CNNs is proposed. In the first stage, a segmentation network is applied to the input image to localize objects. The next stage involves another CNN to predict 2D projections of the corners of the 3D bounding boxes around objects. Pose in 6D is then estimated for the correspondences between the projected 2D coordinates and the 3D ground control points of bounding box corners.

7 Accuracy measures

The accuracy measures are specific to the proposed methods in the earlier works of pose estimation. Intersection over Union (IoU) and Mean Average Precision (mAP) has been the standard for reporting object detection accuracies. For pose estimation as a classification problem, mAVP is used, whereas for regression, 2D Projections [20], 5cm5° [21] and ADD (or 6D pose) [22] are used.

7.1 Object detection

7.1.1 IoU

IoU gives a measure of overlap between the detected and the ground truth bounding boxes. It has become the standard for reporting detection accuracies. Generally, 0.5 and 0.9 are the thresholds used for considering a detection as successful.

7.1.2 mAP

Average precision (AP) is the average of maximum precision values across eleven recall values (from 0 to 1.0) with a step of 0.1. Then the mean average precision is the mean of APs across all classes.

$$AP = \frac{1}{11}(AP_r(0) + AP_r(0.1) + AP_r(0.2) + \dots + AP_r(1.0)) \tag{2}$$

7.2 Pose estimation

7.2.1 mAVP

AVP is an extension of the AP metric that includes the extra constraint that the pose classification is also correct, in addition to the IoU being greater than 0.5. AVP is evaluated at different levels according to how many total pose bins exist. It is generally evaluated at 4, 8, 12 and 24 bins. The average AVP across all the objects is the mAVP of the pipeline.

7.2.2 2D projections

A pose is considered correct if the average of the 2D distances between the projections of the object's vertices from the estimated pose and the ground truth pose is less than 5 pixels.

7.2.3 5cm5°

The 5cm5° metric states that the predicted pose is said to be correct if it is within 5cm of translational error and 5° of rotational error of the ground truth.

The ADD metric states that an estimated pose is accepted if the average distance between the ground truth pose and the estimated pose is smaller than 10% of the diameter of the object.

$$ADD = \frac{1}{m} \sum_{x \in \mathcal{M}} \left\| (\boldsymbol{R}\boldsymbol{x} + \boldsymbol{T}) - (\boldsymbol{R}\widetilde{\boldsymbol{x}} + \widetilde{\boldsymbol{T}}) \right\| \tag{3}$$

where $\boldsymbol{T}$, $\boldsymbol{R}$, $\widetilde{\boldsymbol{T}}$ and $\widetilde{\boldsymbol{R}}$ are the respective ground truth translation, ground truth rotation, estimated translation and estimated rotation of the 3D model $\mathcal{M}$.

8 Datasets

Many approaches have proposed their own datasets for this task, however over time, LineMOD [22] and the Tejani *et al.* dataset [23] have become popular. The T-LESS dataset [24] has come up recently with a large number of instances per class using various imaging tools. More specific datasets like OCCLUSION [25] are trained on for making the pipelines robust to occlusion.

8.1 LineMOD

It has over 18,000 images with 15 different objects and labels for ground truth pose. It also has the 3D models saved as a point clouds.

8.2 Tejani et al.

Has around 5,000 images with 6 object categories. There are an average of 870 instances per category of indoor setting. It also has non-centred objects and cluttered backgrounds. Labels are provided for orientation.

8.3 Pascal3D+

The Pascal3D+ dataset [26] has 12 object categories with around 3,000 instances per category of both indoor and outdoor setting. It also has non-centred objects and backgrounds that are cluttered. Labels are provided for occlusion level and orientation.

8.4 T-LESS

This dataset has around 38,000 training and 10,000 test images from each sensor of 30 different textureless objects. The training images are of the objects against a black background and the testing images are real scenes ranging from simple to the ones with high clutter and occlusion. The sensors used are Primesense CARMINE 1.09 (a structured-light RGB-D sensor), Microsoft Kinect v2 (a time-of-flight RGB-D sensor), and Canon IXUS 950 IS (a high-resolution RGB camera). They provide a manually created CAD model and a semi-automatically reconstructed one.

Table 1. Characteristics of the discussed datasets

Dataset	Images	Classes	Instances	Scene Type
LineMOD [22]	~18,000	15	~1,200	Indoor
Tejani et al. [23]	~5,000	6	~870	Indoor
Pascal3D+ [26]	~36,000	12	~3,000	Indoor and outdoor
T-LESS [24]	~48,000	30	~1,600	Indoor

Table 2. The reported results of the discussed methods on the discussed datasets

Dataset	Method	Measure	Value of Measure
LineMOD [22]	SSD-6D [13]	IoU (0.5)	99.4
		ADD	76.3
	Deep6DPose [16]	IoU (0.5)	99.7
		2D projection	99.3
		5cm5°	68.5
	BB8 [19]	2D projection	89.3
		ADD	62.7
		5cm5°	69
Tejani et al. [23]	Deep6DPose [16]	IoU (0.5)	99.6
		2D projection	99.3
		ADD	62
		5cm5°	64.5
	SSD-6D [13]	IoU (0.5)	96.3
		2D projection	98.8
Pascal3D+ [26]	Share300 [18]	mAVP (24)	27.7
T-LESS [24]	BB8 [19]	ADD	54

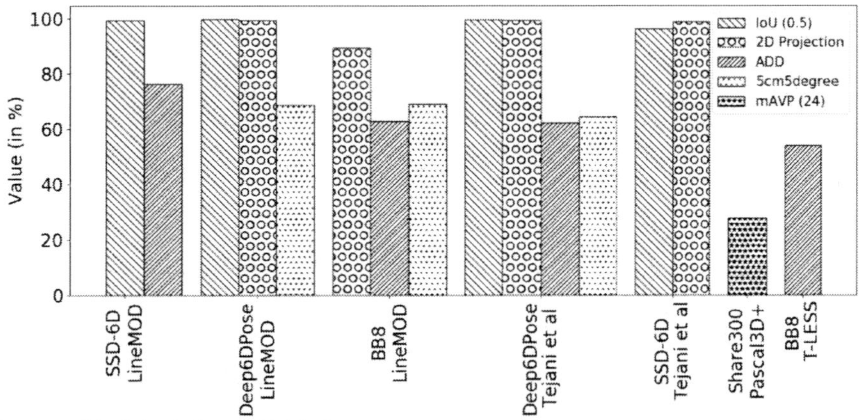

Fig. 2. A graphical depiction of Table 2.

9 Trends

9.1 Accumulation of data

There was a scarcity of data when the relatively old works were published, hence they all proposed their own dataset and accuracy measures with respect to datasets in their experiments. Hence, there was no normalized method to report accuracies. Over time, the works have consistently performed experiments on the LineMOD and Tejani *et al.* [23] datasets. The accuracy measures of 5cm5° and ADD have become the standards for pose estimation, whereas IoU and mAP are used for detection.

9.2 Evolution of approaches

The earlier works showcased their accuracy and the feasibility of their representations/methods using metrics specific to the dataset used. On the other hand, in the recent times, feasibility has become a bare minimum, in addition to which, methods must prove that their accuracies and speeds are also feasible using standard metrics on standard datasets compared to existing methods.

10 Problems

10.1 Occlusion

An object is said to be occluded if its contour cannot be seen completely as it is obstructed by another object. When the properties of models are learnt or modelled considering the object as a whole, there may be severe drops in their accuracies due to occlusion. To overcome this, DPMs have come up that allow predictions based on the visible parts of the object. As mentioned earlier, since CNNs implicitly model the object parts unlike the explicit modelling as in the case of DPMs, CNNs tend to perform better than DPM based models for occluded objects also. The use of higher potentials for detection and pose estimation along with segmentation is described in [20]. It is based on HOG templates for objects and occlusion that do not model a specific object that causes occlusion. Hence they are termed as discriminatively learnt HOG templates.

10.2 Illumination

Varying types of illumination that cause varying pixel intensities may cause issues with the extracted features and learnt models. One way to get over this problem is normalization during learning and inference. However, in the case that only the object in the scene is poorly illuminated, normalization will not help. One can borrow from the idea of [10] to use evenly distributed keypoints across the object that are invariant to illumination for the purpose of its detection and pose estimation.

10.3 Clutter

The objects of interest must be detected regardless of the background. Learning the properties of the object against a fixed background has been shown to cause a drop in performance during inference. Thus, the randomness that clutter introduces in the background is necessary for robustness. Ideally, the training data should have different levels of clutter from zero to heavy for good accuracies at all the possible levels of clutter. Methods proposed in [15] are robust to cluttered scenes.

10.4 Scene diversity

Scene diversity has quite an impact on the performance. If the learning process is carried out only in a fixed environment, there is less variation in the scene properties, which causes the inference in another environment to suffer in terms of accuracy. Training must be carried out in all the possible environments in which inference would be carried out. The use of PASCAL3D+ [26] that has both indoor and outdoor environments is shown in [18] to make the system more robust to scene diversity.

11 Conclusion

Through this survey, we have provided an overview of the developments in joint object detection and pose estimation approaches that use monocular vision. We outlined descriptors/models and estimation methods used in traditional approaches. A brief illustration of hybrid approaches was made. Single stage and multi stage pure deep learning based pipelines were discussed. Characteristics of various datasets were discussed and the standard accuracy measures were illustrated. Wherever possible, statistics of the datasets and results on them were tabulated as shown in table 1, table 2 and figure 2. The trends from the infancy of this problem up until now were also explained.

References

1. Shahrokni A, Vacchetti L, Lepetit V and Fua P 2002 Polyhedral object detection and pose estimation for augmented reality applications *Proc. of Computer Animation* (Lausanne: Geneva) pp 65–9
2. Mousavian A, Anguelov D, Flynn J and Košecká J 2017 3d bounding box estimation using deep learning and geometry *IEEE Conf. on Computer Vision and Pattern Recognition* (Fairfax: Honolulu) pp 5632–40
3. Collet A, Berenson D, Srinivasa SS and Ferguson D 2009 Object recognition and full pose registration from a single image for robotic manipulation *IEEE Int. Conf. on Robotics and Automation* (Pittsburgh: Kobe) pp 48–55
4. Collet A, Martinez M and Srinivasa SS 2010 The moped framework: object recognition and pose estimation for manipulation *IEEE Int. Conf. on Robotics and Automation* (Pittsburgh: Anchorage) pp 1284–306
5. Lim JJ, Pirsiavash H and Torralba A 2013 Parsing ikea objects: fine pose estimation *IEEE Int. Conf. on Computer Vision* (Cambridge: Sydney) pp 2992–9
6. Zhu M, Derpanis KG, Yang Y, Brahmbhatt S, Zhang M, Phillips C, Lecce M and Daniilidis K 2014 Single image 3d object detection and pose estimation for grasping *IEEE Int. Conf. on Robotics and Automation* (Philadelphia: Hong Kong) pp 3936–43
7. Payet N and Todorovic S 2011 From contours to 3d object detection and pose estimation *IEEE Int. Conf. on Computer Vision* (Corvallis: Barcelona) pp 983–90
8. Holzer S, Hinterstoisser S, Ilic S and Navab N 2009 Distance transform templates for object detection and pose estimation *IEEE Conf. on Computer Vision and Pattern Recognition* (Garching: Miami) pp 1177–84
9. Canny J 1986 A computational approach to edge detection *IEEE Trans. on Pattern Analysis and Machine Intelligence* **6** 679–98
10. Hinterstoisser S, Benhimane S and Navab N 2007 N3m: natural 3d markers for real-time object detection and pose estimation (Garching: Rio de Janerio) *IEEE Int. Conf. on Computer Vision* pp 1–7

11. Zhang J, Zhou SK, McMillan L and Comaniciu D 2007 Joint real-time object detection and pose estimation using probabilistic boosting network *IEEE Conf. on Computer Vision and Pattern Recognition* (Chapel Hill: Minneapolis) pp 1–8
12. Kayanuma M and Hagiwara M 1999 A new method to detect object and estimate the position and the orientation from an image using a 3-d model having feature points *IEEE Int. Conf. on Systems, Man, and Cybernetics* (Yokohama: Tokyo) pp 931–6
13. Kehl W, Manhardt F, Tombari F, Ilic S and Navab N 2017 Ssd-6d: making rgb-based 3d detection and 6d pose estimation great again *IEEE Int. Conf. on Computer Vision* (Munich: Venice) pp 1530–8
14. Liu W, Anguelov D, Erhan D, Szegedy C, Reed S, Fu CY and Berg AC 2016 Ssd: single shot multibox detector *European Conf. on Computer Vision* (Chapel Hill: Amsterdam) pp 21–37
15. Xiang Y, Schmidt T, Narayanan V and Fox D 2017 arXiv:1711.00199
16. Thanh-Toan D, Ming C, Trung P and Ian R, 2018 arXiv:1802.10367
17. Kaiming H, Gerogia G, Piotr D and Ross G 2017 Mask r-cnn *IEEE Int. Conf. on Computer Vision* (Menlo Park: Venice) pp 2980-8
18. Poirson P, Ammirato P, Fu CY, Liu W, Kosecka J and Berg AC 2016 Fast single shot detection and pose estimation *4th Int. Conf. on 3D Vision* (Chapel Hill: Stanford) pp 676–84
19. Rad M and Lepetit V 2017 Bb8: a scalable, accurate, robust to partial occlusion method for predicting the 3d poses of challenging objects without using depth *IEEE Int. Conf. on Computer Vision* (Graz: Venice) pp 3848–56
20. Brachmann E, Michel F, Krull A, Ying Yang M and Gumhold S 2016 Uncertainty-driven 6d pose estimation of objects and scenes from a single rgb image *IEEE Conf. on Computer Vision and Pattern Recognition* (Dresden: Las Vegas) pp 3364–72
21. Shotton J, Glocker B, Zach C, Izadi S, Criminisi A and Fitzgibbon A 2013 Scene coordinate regression forests for camera relocalization in rgb-d images *IEEE Conf. on Computer Vision and Pattern Recognition* (Cambridge: Portland) pp 2930–7
22. Hinterstoisser S, Lepetit V, Ilic S, Holzer S, Bradski G, Konolige K and Navab N 2012 Model based training, detection and pose estimation of texture-less 3d objects in heavily cluttered scenes *Asian Conf. on Computer Vision* (Munich: Daejeon) pp 548–62
23. Tejani A, Tang D, Kouskouridas R and Kim TK 2014 Latent-class hough forests for 3d object detection and pose estimation *European Conf. on Computer Vision* (London: Zurich) pp 462–77
24. Hodan T, Haluza P, Obdržálek Š, Matas J, Lourakis M and Zabulis X 2017 T-less: An rgb-d dataset for 6d pose estimation of texture-less objects *IEEE Winter Conf. on Applications of Computer Vision* (Prague: Santa Rosa) pp 880–8
25. Brachmann E, Krull A, Michel F, Gumhold S, Shotton J and Rother C 2014 Learning 6d object pose estimation using 3d object coordinates *European Conf. on Computer Vision* (Dresden: Zurich) pp 536-551
26. Xiang Y, Mottaghi R and Savarese S 2014 Beyond pascal: a benchmark for 3d object detection in the wild *IEEE Winter Conf. on Applications of Computer Vision* (Ann Arbor: Steamboat Springs) pp 75–82

MATEC Web of Conferences **277**, 02030 (2019) https://doi.org/10.1051/matecconf/201927702030
JCMME 2018

Handwritten capital letter recognition based on OpenCV

Yuncong Lu*

Zhengzhou University, Zhengzhou , Henan Province, China

Abstract. Handwriting capitalization recognition is a function of distinguishing handwritten capital letters by means of machine or computer intelligence, which is classified into the field of optical character recognition. Given that capital letters are widely used around the world, identification and analysis are often used as the main components of some control systems. Therefore, the research on handwritten capital letter recognition is also very practical and has important practical significance. The key part of the research contained in this paper is the image preprocessing and the optimal selection of feature vectors, and finally completes the design of handwritten digit recognition system. In this paper, the Fourier and Bayesian commonly used are compared, and eventually the Fourier transform feature is applied to the system classification identification. After completing the test on the relevant experimental data, the results show that the handwritten capital recognition system established in this paper has a high recognition accuracy for handwritten capital letters after repeated training.

1 Introduction

1.1 Subject background and significance

Due to the prevalence of information technology in daily life, the application of computer networks is constantly evolving. In the actual production life, there are handwriting functions on mobile phones and computers, which involves the identification of handwritten capital letters.

Optical Character Recognition Technology (OCR) (optical character recognition) has the intelligent function of using computer to send information on paper in a short period, using relevant software systems to convert information into text application technology, and can this text is adjusted[1]. Handwriting recognition has always been a hot research direction, and data collection, processing and screening are all dependent on this technology.

1.2 Research status

* Corresponding author: 1138862948@qq.com

Dausheck International patented optical recognition technology in the early 19th century. For nearly a century, we have seen the results of today's research. OCR has become the most promising research project in all kinds of identification categories in the 21st century. One. The technology has the advantages of image processing, computer graphics and artificial intelligence, and then enters the electronics field and is promoted. Through years of continuous improvement, relatively mature OCR products have flooded into people's lives.

1.3 The main research content and arrangement of the paper

The environment designed in this paper is VC platform, which recognizes the handwritten English letters connected to the machine. The functions of all systems include mouse movement track reading, mouse movement track data discrete processing, track isometric feature data reading, and track external maximum quadrilateral. Obtain, de-dimensionalize the data (so that the feature data is independent of its size), obtain five key harmonic data of the Fourier transform , and adopt several techniques such as random tree type identification and estimation.

2 System development technology

2.1 Visual2010

Microsoft Visual Studio 2010 is the best version of the VS system product IDE enhancement, developed by Microsoft Corporation, is the only IDE that supports the .NET Framework 4.0 development path. VS 2010 is a multi-view development that includes code editors, compilers/interpreters, debugging tools, and installation package building tools. Currently, Windows code is basically developed by it, and its relationship with the .NET Framework is shown in Fig 1.

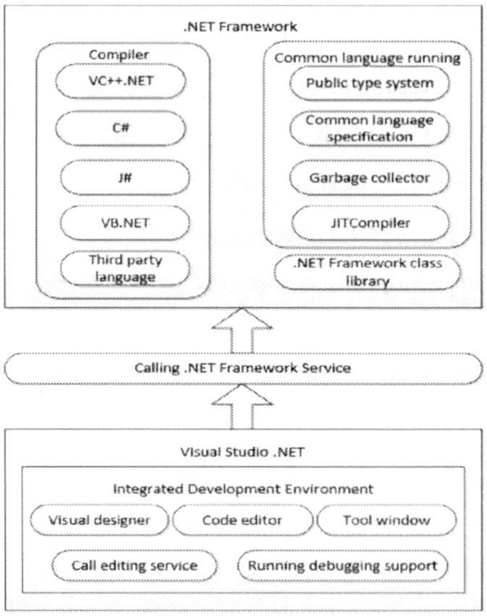

Fig. 1. Relationship between the .NET Framework and VS

2.2 MFC Introduction

MFC, the Microsoft Foundation Class (Microsoft Foundation Class), both output the VCL application structure with Microsoft's Visual C++ development platform. This system library has released a list of development libraries that can be widely used.

MFC is actually derived from the structure and startup part of Microsoft's C++ platforming software, MFC is attached to a development tool library in a supporting software, such as MFC's VC++ key location (note C++ and VC++) The difference: C + + belongs to a programming language, which helps humans understand the development of software. VC + + belongs to the compiler or compiler + IDE source code editor[2].

In the underlying operations, the use of MFC is very simple, in the combination with the Windows API or the driver can use the original program of the API function, and the driver API and data are both based on the C language, allowing VC R&D engineers make it easy to use Windows API.

3 Demand analysis

3.1 Feasibility study

The workbench of the billing machine should use a PC with P4 or higher. In terms of personnel, it is required to be able to operate software development tools and perform tasks associated with the system.

Programming language and development tools:

- Programming language:C++
- Development tools:Microsoft Visual C++2010、etc.

With Windows and VC++ development, the price is relatively cheap and can be developed at low cost.If there is a problem, people can solve it by means of information related to computer hardware software.There are many control elements inside Microsoft Visual C++2010, which can write a large number of programs, which can shorten the development process time.With the above points, we can conclude that the research of the data set conversion device can be realized.

3.2 Development design ideas

- This software can adapt to a wide range of computer configuration environments and is inexpensive.
- The system should match the goals of the data conversion to meet its final needs.
- To achieve standards that are easy to develop, easy to expand, and easy to maintain.

3.3 Functional requirements analysis

The template matching method identifies the handwritten uppercase English letters, which is determined according to the design steps of the system. This design is completed on the VC development platform, and the handwritten English letters connected to the machine are recognized. The functions of all the systems include mouse movement track reading, mouse movement track data discrete processing, track isometric feature data reading, and track external connection maximum. Quadrilateral acquisition, data de-dimensionalization (making feature data independent of its size), obtaining five key harmonic data of Fourier transform of feature data, adopting random tree type identification estimation, etc. technical part. Performance requirements:

- Correctness: The system must show the steps of the entire process.
- Accuracy: Recognition has a high degree of accuracy.
- Efficiency: When the input of data is large, the efficiency of processing should be guaranteed.

4 Overall system design

4.1 System overall functional module design

The overall module is shown in Fig 2:

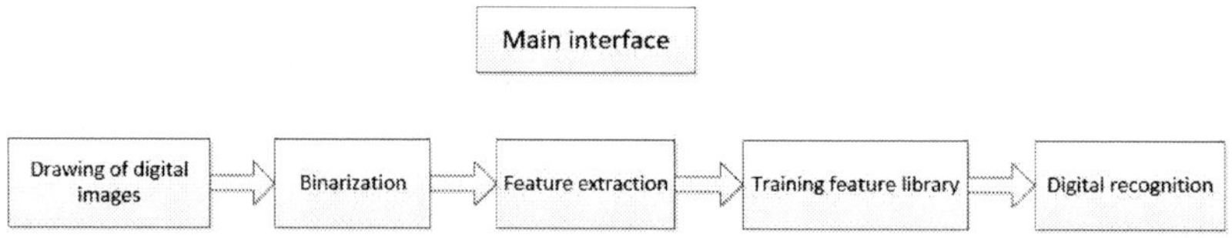

Fig. 2. Overall module

4.2 Interface design

4.2.1 External interface

- User Interface

The user interface is also known as the Windows window. The user shares and learns with the system through menu buttons, text editing and other functions of the web interface.

- Software interface

Development tools are developed using the Microsoft Visual C++ 2010 development environment.

- Hardware interface

Includes previous keyboard, mouse, and monitor interfaces for input and output. An interface that interworks with the printer to help print related resources and lists. The purpose is also the output.

4.2.2 Internal interface

The internal interface part borrows function calls, parameter passing, and return values to interact with data between different parts. The data structure design section below will describe the structure of each data in detail. After the data is encapsulated by the data structure, it is passed to different modules in the form of the return value or parameter transfer, and this is sent as a connection part[3].

4.3 Code design

4.3.1 Principle of code design

- Uncertainty: Each entity or feature is rendered by a specific code.
- Standardization: Because each information module unit can basically be consistent with global practices and national unit interests.

- Versatility: Use the original code and facilitate the gradual promotion of the program.
- Simplicity: Do not affect the coding and scalable coding position, as far as possible with easy to understand features.

4.3.2 System code design

The function of the code: making it convenient and effective to save, detect and find data and input information to the information system; the efficiency and accuracy of computer execution have been greatly increased due to the existence of the program; the comprehensive and unified information has been enriched.

The code design needs to be concerned:the code should be designed according to the user's thinking order. The existence of the code is unique. Save the appropriate expansion location to meet the development requirements. Need to make the composition of the code have a certain realistic effect. It is harder that other characters should not be used to avoid unclear occurrences. For relatively long code, try to separate.

5 System implementation

5.1 Identification system basic principle

5.1.1 Character trajectory depiction

The completion steps of the handwritten letter drawing command are: handwritten letters are completed by the encapsulation class CDC in MFC, and the canvas is created by the CStatic control in Visual C++. MouseDown moves the mouse down, MouseMove moves the mouse, and MouseUp moves the mouse up. These are messages that respond to mouse events. They are all displayed in the dialog window. If the program saves a specific coordinate of a point, it means this time. The mouse is clicked on the CStatic control[4]. The mouse down event marks the start of the handwritten letter drawing, and the left mouse button indicates the completion.

5.1.2 Image preprocessing

To indicate the characteristics of handwritten letters, the image must be preprocessed. This part of the design roughly includes: image binarization operation. Setting the gray value of the processed image to 0 or 255 makes the black and white hue of the image very prominent. We call this the binarization process. In the system, if a letter is written in a fixed-size control, this control exists in the VC visual editing interface, then we receive a series of coordinates with space upper left position as the origin. Set a two-dimensional data column similar to this control element in the memory (that is, treat the pixel as a unit, produce a picture with a shape and size similar to the area), and take out the specific coordinates in the handwriting area. Because the previous operation has made the handwritten area coordinates the same as the array of data areas in the memory image, in order to obtain a binary image of handwritten letters, we need to match the data in the image data area with the coordinate value just obtained. Let it have a gray value of 255 (white), and let other coordinate values in the region corresponding to a gray value of 0 (black)[4].

5.1.3 Creation of the feature library

A feature library should be set in the design process of handwritten capital letter recognition, and extract the feature value of a letter we write. After the program wants to input the letter, use the program to make the written letter correspond to all the feature values. And as a template library record, before the handwritten letters are recognized, the template library must be initialized. It should be noted that the template library should be expanded at the same time as the recognition if it appears in the process of identification. If the error is not recognized or identified, it should be collected into the template library in time to make the inventory of the template more diverse.

5.1.4. Image letter recognition

The identification of letters needs to be based on the characteristics of handwritten letter images. For the recognition of handwritten letters, we use template matching, which is the last step required for the system design.

The detailed steps are: if there is a sample data X to be tested, the analysis obtains the action distance of the X and the sample Xj in the training data (0<j<m, m represents the number of sample points in the training data). To find the existing sample that is most similar to the sample to be tested, we need to calculate the distance between the test sample and all existing samples in the training set by continuous calculation and compare it. Finally, the sample Y to be tested belongs to Type is the closest sample type we have derived[5].

5.2 System function design and implementation

5.2.1 Drawing of uppercase letters

The most primitive window is presented with the system executing commands, as shown in Fig 3. In the place where the image is drawn, use the mouse action to complete the outline of the handwritten letter outline, as shown in Fig 4. Select CPen to create the object pen and assign values to the relevant parameters. For example, pen (PS_SOLID, 3, RGB (0, 0, 255)) can get a variety of drawing tools with different types and sizes. The message response action utilizes the function OnLButtonDown(UINT nFlags, CPoint Point), OnMouseMove(UINT nFlags, CPoint point), OnLButtonUp(UINT nFlags, CPoint point) to complete. By constructing the stack by the function AddPointStack(const CPoint &point), the drawing stage of the handwritten letter trajectory can be saved to the handwritten letters. With the help of the function OnButtonRset(), click the reset button on the system interface to erase the drawn letters after the letter is drawn.

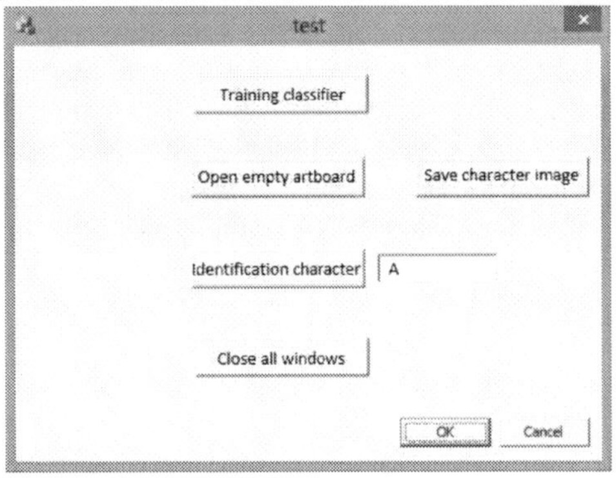

Fig. 3. System initial interface

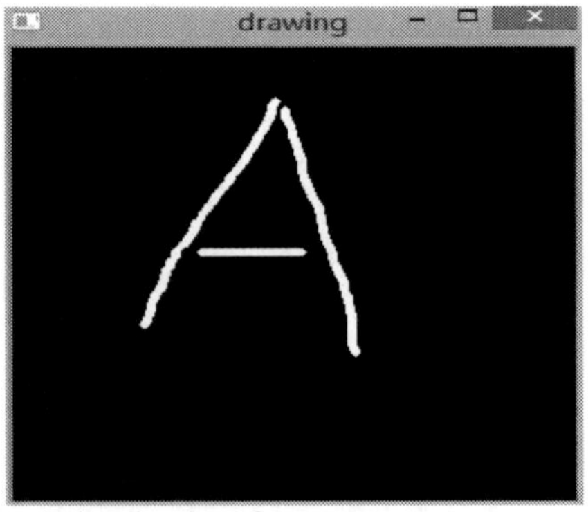

Fig. 4. drawing interface

5.2.2 Capital letter feature extraction

The purpose of the extraction of the letter features is to be able to identify the letters accurately. In this paper, the pixel features of the letters depicted are the same as the above-mentioned conclusions that the number of target pixels is smaller than the total number of pixels in the upper range. There are a total of 64. This paper uses the function GetROI(const cv::Mat drawImage) to complete the function of obtaining feature quantities and uses the SaveImage(const cv::Mat& drawImage) function to implement the storage function.

5.2.3 Establishment of template feature library

The structure of the feature library can be completed by starting the learning classifier button on the system interface. After many learnings, the feature library will be completed and the recognition effect will be more credible. At the same time, the record data is constructed and the learning of the feature data is completed by means of LoadSamples().

5.2.4 Uppercase English letter recognition

In this paper, with the help of PredictSample(const cv:: Mat& drawImage), the eigenvalues of the uppercase letters written are compared with all existing eigenvalues in the template library, and the difference between them is calculated. The minimum distance difference corresponds. The category is the type of letter written, which can be achieved by clicking the recognize letter button on the system window. The final discrimination result will be displayed as shown in Fig 5.

Fig. 5. Identification results

References

1. Zhang Yuxiong, Liu Jianxia: *Development Status and Trends of Data Image Processing Technology*, Friends of Science, No. 3, 2012;
2. Sun Rongchun, Sun Junxi, Song Xue: *DSP-based CMOS Image Acquisition and Processing System*, Semiconductor Optoelectronics, 2011, Issue 6;
3. Huang Zhiwei, Wang Bing, Zhu Wei: *Application Design and Practice of M32F 32-bit Microcontroller*, Beijing University of Aeronautics and Astronautics Press, 2012 edition;
4. Li Deming, Han Jian, Jiang Guoqiang: *Design of Image Acquisition and Display System Based on ov7670*, Journal of Instrument and Instrument, 2010, No. 8;
5. Dong Jia: *Research on Imaging System of CMOS Image Sensor*, Master thesis of Xi'an Institute of Optics and Precision Mechanics, Chinese Academy of Sciences.

Multi-scale foreground extraction on graph cut

Jiayi Liu[1], and *Kun* He[2,*]

[1]School of Software Engineering, Sichuan University, 610065 Chengdu, China
[2]School of Computer Science, Sichuan University, 610065 Chengdu, China

Abstract. In order to improve Grab Cut implementation effect for real images, we propose a novel improvement which extends the Grab Cut in three aspects: 1) a series of edge-preserved components are generated via the TV smoothing model; 2) the number of sub-regions is estimated by histogram shape analysis to remove the negative effects on the unreasonable number of the sub-regions; 3) a segmentation termination condition is constructed by integrating the multi-scale components. The experiment result indicates that this method performs well compared to other methods based on graph cut and is insensitive to sub-regions.

1 Introduction

The foreground, which plays the key role in image analysis and comprehension, is a region with semantic meanings in an image. Existed approaches to foreground extraction are based on low-level or mid-level characteristics of the image, such as edge information [1], appearance [2] and texture features [3]. However, those methods still cannot achieve correct segmentation result because of the ambiguity of those characteristics.

The Grab Cut [4] well extracted foreground with minimal user interaction via combining edge and appearance models. The appearance model is statistically described using the Gaussian Mixture Models (GMMs), in which one problem is that the number of Gaussians has a significant influence on accuracy of the segmentation [5]; the other is that the GMMs is inefficient on the inhomogeneous sub-regions. The implementation effect is low for high-textured images.

To optimize the foreground extraction effect, a novel improvement is proposed. This improvement extends the Grab Cut in three aspects. One is that a series of smoothing components are generated via the TV smoothing model. These components satisfy edge-preserved and region-smoothed. These characteristics are favored to model the inhomogeneous sub-regions via the GMMs. The other is that the number of sub-regions is estimated by histogram shape analysis to remove the negative effects on the unreasonable number of the sub-regions. At last, integrating the multi-scale appearances of an image, a segmentation termination condition is constructed using the IOU of the foreground objects.

In the next section, relevant segmentation algorithms derived from the graph cut will be reviewed. The following section proposes the non-linear smoothing model to optimize the appearance for segmentation, and analyzes the characteristics of smoothing components.

* Corresponding author: hekun@scu.edu.cn

Next, the GMMs is optimized using the histogram shape analysis. Finally, the experimental results and conclusion are given.

2 Previous Work

Users may be interested in different foreground regions for an image. To customize user's demands, foreground extraction approaches usually exploit information of user interaction. However, the former's extraction effect is unsatisfactory for images that the distribution of foreground and background exists a considerable overlap. The latter loses its effectiveness on high textured images.

Foreground extraction methods on graph cut, combining edges and appearance models, successfully extract foreground by the min-cut algorithm [6]. The appearance models can be divided as the given and estimation modals, note that the given modals [7, 8] is intensity distribution of foreground and background. However, this modal highly relies on the known foreground and background by the user's mark, and requires a large amount of the user interaction. The estimation modals [4, 7, 9] require user to define initial boundary box, and an image is divided into the background and foreground. Each region is statistically estimated via the GMMs in the segmentation process.

The fixed number of Gaussian in each GMMs has negative effects on segmentation. To remove the effect of unsuitable number of Gaussian, GMMs is optimized [7] by estimating the optimal number of Gaussians in each GMMs via the CLUSTER algorithm [10]. Since it does not exploit the intensity distribution of an image, it is not effective for an image with inhomogeneous sub-regions. To best model the inhomogeneous sub-regions using GMMs, the Supercut [9] implements the local similarity constraint to optimize GMMs, which guarantees that homogeneous pixels are classified into the same sub-class.

3 Image Multi-scale Analysis

Due to exploiting the given-scale edge and appearance information, foreground extraction methods on graph cut have an unsatisfactory segmentation effect on real images. We use the TV smoothing operator [11] to generate a serious of smoothed components $\boldsymbol{u}^{(i)}$ of an image $\boldsymbol{u}$. The component $\boldsymbol{u}^{(i)}\ i=0,\ 1,\ \cdots$ can be described as $\boldsymbol{u}_0=\boldsymbol{u}^{(0)}\xrightarrow{u_0}\boldsymbol{u}^{(1)}\xrightarrow{u_0}\cdots\xrightarrow{u_0}\boldsymbol{u}^{(k-1)}\xrightarrow{u_0}\boldsymbol{u}^{(k)}\xrightarrow{u_0}\cdots$

The component $\boldsymbol{u}^{(k)}$ contingents on $\boldsymbol{u}^{(k-1)}$, which smears the fine information by the smoothing operators. With the k increases, the component is getting coarser, and the intensity distribution of regions becomes tight cohesion. It is favored to model the foreground and background using GMMs. The other, the involvement of $\boldsymbol{u}_0$ in iterations simultaneously pays equal attention to raw images which contains the intrinsic singularities features, through which the edges are preserved.

4 Foreground Extraction Model

Given the boundary box around the desired object, an image $\boldsymbol{u}$ with N pixels is divided into two parts—the foreground region T_F and a background T_B . Combining edge and appearance model, an energy function is defined whose minimum will identify with a good segmentation.

$$E(\boldsymbol{\alpha},\boldsymbol{\omega},\boldsymbol{u})=R(\boldsymbol{\alpha},\boldsymbol{\omega},\boldsymbol{u})+V(\boldsymbol{u}) \tag{1}$$

Here, $\boldsymbol{\alpha} \in \{0,1\}$ is the segmentation label of each pixel, where 0 and 1 correspond to the mixture region and the background, respectively.

$$V(\boldsymbol{u}) = \sum_{j \in \Lambda_0} \gamma dis(i,j)^{-1} \exp(-\beta(u_i - u_j)^2) \tag{2}$$

The term $V(\bullet)$, denotes edge information, is expressed as a penalty weight which is high with the low gradient and low with the high gradient .To encourage smoothness, the constant γ relaxes the tendency to smoothness in regions of high contrast. In our experiments, γ is fixed to 50. Λ_0 is a set of the 8-neighboring pixels, and factor $dis(\bullet)$ is the Euclidean distance. To ensure the exponential term in (2) could switch appropriately between high and low contrast, the constant is chosen to be:

$$\beta = 0.5\left\langle (u_i - u_j)^2 \right\rangle^{-1} \quad \beta = 0.5\left\langle (u_i - u_j)^2 \right\rangle^{-1} \tag{3}$$

Here $\langle \bullet \rangle$ denotes the expectation over an image sample.

The parameter $\boldsymbol{\omega}$ represents the parameters of the appearance models. The term $R(\bullet)$ evaluates the fitness of the segmentation $\boldsymbol{\alpha}$ to an image $\boldsymbol{u}$, given the models $\boldsymbol{\omega}$.

$$R(\boldsymbol{\alpha},\boldsymbol{\omega},\boldsymbol{u}) = \sum_{i \in T_F} -\log \sum_{m=1}^{K_\alpha} \eta(\boldsymbol{\alpha},m) g(\boldsymbol{\alpha},\boldsymbol{\mu}_m(k),\boldsymbol{\Sigma}_m(k),u_i) \tag{4}$$

Here, $\boldsymbol{\mu}_m(k)$ and $\boldsymbol{\Sigma}_m(k)$ are the mean vector and the covariance matrix of the m^{-th} GMM, $K_\alpha, \alpha = 0,1$ denote the number of Gaussian in the foreground and background, respectively.

Combining edge and region information of the component, the energy function of segmentation be formulated and can be written as follows:

$$\boldsymbol{\alpha}^* = \arg\min_{\boldsymbol{\alpha}} \left\{ \min_{\boldsymbol{\omega}} E(\boldsymbol{\alpha},\boldsymbol{\omega},\boldsymbol{u}) \right\} \tag{5}$$

Minimization is done by using a standard minimum cut [6]

4.1 Clustering via Histogram Shape Analysis

The histogram can describe grey-level distributions of an image and the difference among sub-regions. Each sub-region corresponds to peak in the histogram. And its shape illustrates the fact that the gap between the two peaks probably result from border pixels between objects and background. Thus, the number of objects in an image can be estimated via histogram shape analysis.

If the histogram of an image is multi-modal, that is the image is high-texture or complex, it typically has many local minima and maxima. To weaken the effect of the local minima and maxima on the shape analysis, median filter is used to get the histogram pre-processed to preserve the valley. Let s denotes the grey levels of an image, $h(s):\{0,1,\cdots,255\} \to [0,1]$ denotes the histogram, and $\tilde{h}(s)$ is a smoothed histogram after the median filtering. A series of valleys $\{v_0,\cdots,v_{m-1},v_m,\cdots,v_k\}$ are detected by the signal change of the differential of $\tilde{h}(s)$, with $v_0 = \min\{s\}$ and $v_k = \max\{s\}$. These valleys separate the grey levels of an image into k intervals. The pixels of grey-level in $[v_{m-1}, v_m]$ comprise the m-th subclass, and the area ration of its subclass is defined as:

$$\eta(m) = \int_{v_{m-1}}^{v_m} h(s)ds \tag{7}$$

Finally, image pixels are adaptively clustered into several sub-regions on its intensity distributions via histogram analysis. The median filter is usually short of an efficient decision rule on obtaining good results on different intervals in the histogram. As the sample size of the median filter increasing, the number of valleys falls slightly, which influences the partition of the sub-classes. In this paper, the sample size is determined as 17 by experiments.

4.2 Multi-scale Foreground Extraction

The foreground extraction task is to deduce the foreground object from the image u using the appearance model and edge. However, the appearance model and edge extracted from a single component cannot generate a complete correct segmentation result for the high-texture images. In different smoothing components, the inhomogeneous sub-regions are smoothed and are of compact distribution, leading that the number of Gaussians and parameters for every GMM are not the same. This difference cause that the segmentation results are inconsistency for each component. In order to evaluate segmentation performance, we use the IOU of the foreground object to measure the significant level of segmentation, which is defined as the following:

$$Ls(k) = 1 - \frac{F(T^{(k)}) \cap F(T^{(k-1)})}{F(T^{(k)}) \cup F(T^{(k-1)})} \tag{8}$$

Here, $F(T) = \{i | \alpha_i = 1 \ and \ \alpha_i \in T, i = 1, \cdots, N\}$ denotes the foreground object region; $T^{(k)}$ and $T^{(k-1)}$ represent the segmentation results of the adjacent iteration smoothing component $u^{(k)}$ and $u^{(k-1)}$, respectively. With the number of smoothing increasing, the significant level of segmentation monotonically decreases. According to the changes of significant level with smoothing number, the segmentation termination condition is defined as following:

$$\nabla_k Ls(k) = 0 \tag{9}$$

5 Experimental Results

The experiments are conducted using Visual Studio 2013 on a PC with Intel-Core i5CPU @ 3.40GHz and 4GB of RAM without any particular code optimization.

Aimed at improving segmentation performance, the non-linear smoothing is exploited to optimized appearance of images and using the histogram shape analysis to evaluate the optimal number of Gaussians for GMM construction. An image with several sub-regions (seen in the Fig.1) is used as an example. These sub-regions are smoothed and a serious of edge-preserved components are generated, which firms the segmentation of the foreground from the background. Also, the histogram shape analysis effectively optimizes the estimation of the number of Gaussians.

(a) Original image (b) initial curve (c) the ground truth

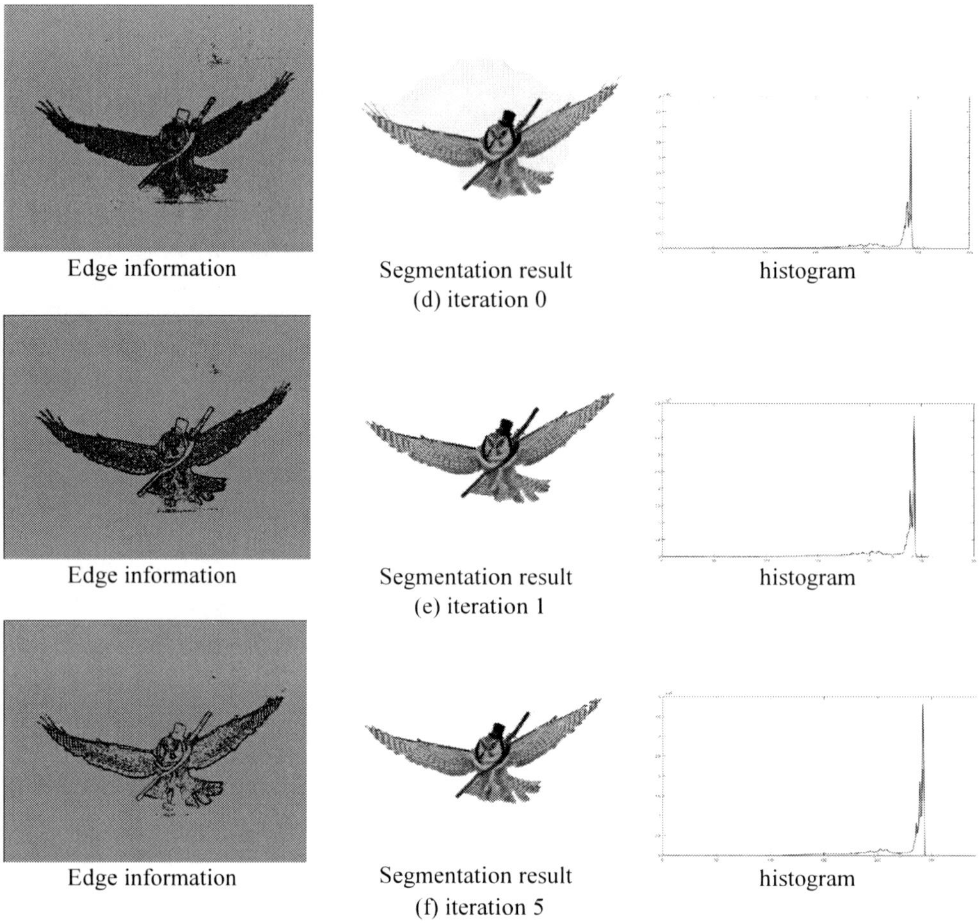

Fig. 1. The segmentation process using this method for the test image. a) the original image, b) the initial curve, c) the ground truth, (d-f) the edge information, the segmentation results, and the histogram (the blue curve denotes the histogram, the red curve is the smoothing histogram with median filter, green spots are valleys) for the component with the different times of iterations: 0,1,5.

To further test segmentation performance of this approach, experiments are carried out to compare with the correlative segmentation methods based on the graph theory, such as the Grab Cut [4] and the Grab Cut with GMM optimization [7]. The results using different methods are shown in the Fig.2. For a simple scene image with strong color contrast between the foreground and the background (i.e. the Fig.2a, b)), the effects of these three algorithms on details of edges are almost the same. And this method and the Grab Cut with GMM optimization have better performance on accurate segmentation of sub-regions than the Grab Cut (show as the red circle in Fig.2a). The reason behind it is a more reasonable estimation of color clusters for the GMM is achieved by using different methods for clustering.

For a complex scene image with multiple sub-regions (i.e. the Fig.2c, d), this method has a better result in dealing with edges comparing to the other two methods, showed as the red circle in Fig.2c. This is because in the course of smoothing, a serious of edge-preserved components are generated, which avoids the losing of edge information while smoothing sub-regions. In this situation, segmentation effect is optimized by enhancing the consistency of sub-regions and preserving edge information at the same time. Integrating the multi-scale appearances of an image and optimizing the GMMs for each region, this method successfully improves segmentation performance for the complex scene images.

(a)Bird (b) Starfish (c) Butterfly (d) Mushroom

Fig. 2. Comparison of this method with the Grab Cut, the Grab Cut with GMM optimization on the simple scene images. Row1 original images and initial curve, Row2 the ground truth, Row3 this method, Row4 the Grab Cut, Row 5 the Grab Cut with GMM optimization.

The IOU metric [12] is adopted to estimate the effectiveness of comparing methods above. It is estimated by the following:

$$\mathrm{IOU} = \frac{F(s) \cap F(g)}{F(s) \cup F(g)} \tag{10}$$

where $F(s)$ and $F(g)$ mean the foreground object of the segmentation and ground truth, respectively. The F-measure is also used for the evaluations, it is computed by the following:

$$\text{F-measure} = \frac{2 \times precision \times recall}{precision + recall} \tag{11}$$

where,

$$\text{precision} = \frac{F(s) \cap F(g)}{F(g)}, recall = \frac{F(s) \cap F(g)}{F(s)} \tag{12}$$

The CPU time and segmentation evaluations are illustrated in Table 1 for images in the Fig. 2. The IOU and F-measure of segmentation using this method is higher than that of the other methods, which shows the superiority of this method. However, this method has higher computational cost in order to achieve better segmentation effects. The surplus CPU-time is mainly spent on the iterative computation of the non-linear smoothing for images. And the number of iteration depends on the inhomogeneous degree of the sub-region.

Table 1. The Comparison Of Cpu Time And Segmentation Evaluations On Figures 2

Segmentation methods	Fig .2a	Fig .2b	Fig.2c	Fig.2d
	465×290	550×430	512×320	320×480
This method				
Precision	0.991	0.995	0.978	0.984
Recall	0.949	0.936	0.947	0.948
F-Measure	0.962	0.966	0.956	0.966
IOU	0.941	0.953	0.949	0.933
CPU-time(s)	7.626	15.123	13.284	9.989
The Grab Cut with the GMM optimization[7]				
Precision	0.987	0.995	0.969	0.982
Recall	0.949	0.935	0.944	0.944
F-Measure	0.960	0.965	0.966	0.970
IOU	0.939	0.951	0.942	0.930
CPU-time(s)	6.495	14.119	12.443	9.125
The Grab Cut [6]				
Precision	0.975	0.994	0.970	0.976
Recall	0.948	0.935	0.943	0.938
F-Measure	0.959	0.964	0.966	0.969
IOU	0.937	0.950	0.937	0.921
CPU-time(s)	5.121	12.859	10.461	8.199

6 Conclusion

This paper proposes two modifications to the Grab Cut object segmentation method to improve segmentation performance without increasing user interaction. For images with inhomogeneous sub-regions, the non-linear smoothing algorithm is shown to optimize image appearances for images with inhomogeneous sub-regions in segmentation. Plus, the histogram shape analysis assists to estimate the suitable number of Gaussians in each GMM to well model the foreground and background. Compared to the original Grab Cut and relative improved methods, the segmentation effect using this method is superior. But it tolerates expensive computations to achieve better segmentation results. To save the computational cost, we will design a method of adaptively labelling the initial curve which adjacent to the boundaries. On the other hands, the fixed sample size of the median filter might mislead the suitable number of Gaussians for each GMM. We plan to construct an algorithm to adaptively determine the sample size of the median filter, which contributes to more accurate clusters' number by the histogram shape.

Acknowledgement

This work was supported by the Sichuan Province Natural Science Foundation of China (2016JZ0014).

References

1. S.Y. Yeo, X. Xie, I. Sazonov, and P. Nithiarasu. "Segmentation of biomedical images using active contour model with robust image feature and shape prior." International Journal for Numerical Methods in Biomedical Engineering, 2014, Vol. 30, No.2, pp. 232-248.
2. P. P. Tunga, and V. Singh. "Extraction and description of tumour region from the brain MRI image using segmentation techniques." IEEE International Conference on Recent Trends in Electronics, Information & Communication Technology IEEE, 2017, pp. 1571-1576.
3. R. Hettiarachchi, and J. F. Peters. "Voronoï region-based adaptive unsupervised color image segmentation." Pattern Recognition, 2017, 65, pp.119-135.
4. C. Rother, V. Kolmogorov, and A. Blake. "GrabCut: interactive foreground extraction using iterated graph cuts." Acm Transactions on Graphics, 2004, Vol.23, No.3, pp. 309-314.
5. Y. Y. Boykov and M. P. Jolly. "Interactive Graph Cuts for Optimal Boundary & Region Segmentation of Objects in N-D Images." Computer Vision, 2001. ICCV 2001. Proceedings. Eighth IEEE International Conference on IEEE, 2001, vol.1, pp.105-112.
6. V. Kolmogorov, and R. Zabih. "What Energy Functions Can Be Minimized via Graph Cuts?" IEEE Transactions on Pattern Analysis & Machine Intelligence, 2004, vol.26, no.2, pp.147-159.
7. D. Chen, B. Chen, G. Mamic, C. Fookes, and S. Sridharan. "Improved Grab Cut segmentation via GMM optimization." Digital Image Computing: Techniques and Applications IEEE Computer Society, 2008, pp. 39-45.
8. M. Tang, L. Gorelick, O. Veksler, Y. Boykov. "Grab Cut in One Cut." IEEE International Conference on Computer Vision IEEE Computer Society, 2013, pp.1769-1776.
9. S. Wu, M. Nakao, and T. Matsuda. "SuperCut: Superpixel based Foreground Extraction with Loose Bounding Boxes in One Cutting." IEEE Signal Processing Letters, 2017, vol. 24, no.12, pp.1803-1807.
10. C. A. Bouman. "Cluster: An unsupervised algorithm for modeling Gaussian mixtures." Technical report, Prude University, July 2005.
11. D. G. Lowe. "Distinctive Image Features from Scale-Invariant Key points." International Journal of Computer Vision, 2004, vol.60, no. pp.91-110.
12. A. Pratondo, CK. Chui, and SH. Ong. "Robust Edge-Stop Functions for Edge-Based Active Contour Models in Medical Image Segmentation." IEEE Signal Processing Letters, 2015, vol. 23, no. 2, pp. 222-226.

MATEC Web of Conferences **277**, 02032 (2019) https://doi.org/10.1051/matecconf/201927702032
JCMME 2018

Virtual ground truth, and pre-selection of 3D interest points for improved repeatability evaluation of 2D detectors

Simon R Lang*, *Martin H* Luerssen, and *David M* Powers

School of Computer Science, Engineering and Mathematics, Flinders University, 1284 South Rd, Clovelly Park SA 5042, Australia

Abstract. In Computer Vision, finding simple features is performed using classifiers called interest point (IP) detectors, which are often utilised to track features as the scene changes. For 2D based classifiers it has been intuitive to measure repeated point reliability using 2D metrics given the difficulty to establish ground truth beyond 2D. The aim is to bridge the gap between 2D classifiers and 3D environments, and improve performance analysis of 2D IP classification on 3D objects. This paper builds on existing work with 3D scanned and artificial models to test conventional 2D feature detectors with the assistance of virtualised 3D scenes. Virtual space depth is leveraged in tests to perform pre-selection of closest repeatable points in both 2D and 3D contexts before repeatability is measured. This more reliable ground truth is used to analyse testing configurations with a singular and 12 model dataset across affine transforms in x, y and z rotation, as well as x, y scaling with 9 well known IP detectors. The virtual scene's ground truth demonstrates that 3D pre-selection eliminates a large portion of false positives that are normally considered repeated in 2D configurations. The results indicate that 3D virtual environments can provide assistance in comparing the performance of conventional detectors when extending their applications to 3D environments, and can result in better classification of features when testing prospective classifiers' performance. A ROC based informedness measure also highlights tradeoffs in 2D/3D performance compared to conventional repeatability measures.

1 Introduction

In Computer Vision (CV), the establishment of ground truth so that new feature classification algorithms can be properly measured is an ongoing topic of research. With 3D scanning, printing, and realistic rendering, there are increasing opportunities for CV to be applied to virtual scenes and a multitude of new approaches are exploiting this newly accesible niche [4]. In the field of 2D CV there are well accepted conventions for measuring interest point/key point based feature detection, the most well known being the work based on research by Schmid and Mikolajczyk [23] [14] that are still regularly used in

* Corresponding author: simon.lang@flinders.edu.au

more recent times [12], and have been used a great deal in other CV research relating to interest point/key point repeatability [17] [24] [16]. It is still challenging however to establish a reliable means of establishing better ground truth of real world environments for the purposes of testing 2D based interest point detectors [15] [12] [5].

Schmid's metric for evaluation of a set of detectors *K*, classifies points between two pixel arrays $\tilde{x}_i$, as either repeated, or not and uses a ratio of true positives and true negatives to measure performance. A threshold based on a radial distance ϵ around each point in the reference scene $\tilde{x}_1$ determines classification. Equations 1, 2 and 3 describe this process, with $\tilde{x}_1$ representing the reference scene as a basis for comparison, and $\tilde{x}_i$ as the scene image I_i is a member of a set of transforms j being compared. A homography H_{1i} of $\tilde{x}_i$ enables threshold distances to be measured with $\tilde{x}_1$, and repeated points to be determined. The default threshold, ϵ=1.5, represents an error rate of 1 pixel distant, also known as the Moore neighborhood, and is considered by Schmid, and researchers in general that apply this metric, to be the optmial tradeoff. Points that don't share the same view area are removed from the validation process as they share no valid repeatable point candidates.

$$r_{K,J(\epsilon)} = \frac{1}{N-1}\sum_{2}^{N} r_{K,I_i(\epsilon)} \tag{1}$$

$$r_{i(\epsilon)} = \frac{|R_i(\epsilon)|}{min(|\tilde{x}_1|,|\tilde{x}_i|)} \tag{2}$$

$$R_{i(\epsilon)} = \{(\tilde{x}_1,\tilde{x}_i)|dist(H_{1i},\tilde{x}_1,\tilde{x}_i) < \epsilon\} \tag{3}$$

1.1 Repeatability in Virtualised Scenes

This paper builds on the work done by Lang et al.[11] [10], where they demonstrated that a virtualised space, whether it be of images, or 3D models, served as a viable testbed for measuring interest point (IP) performance of conventional 2D detectors. Other approaches to IP generation utilise the ground truth of the model directly [4], but classifiers that only utilise 2D data are not designed to utilise extra dimensions. This limitation means that they can be highly optimised for 2D scenes, but not 3D, and subsequently also means their performance can't be properly measured in real-world 3D scenarios. Additionally, the lack of ground truth available for optimisation means that 3D applications for 2D based classifiers are constrained.

2 Methodology

For the purposes of measuring the performance of IP detectors that utilise only 2D images, a rendering context is utilised to maintain consistency between 2D, and 3D. This preserves 2D consistency of detected IP classification, while also allowing for the precision that the world space of the rendering context provides. Unlike a homography H1i of the pixel positions of points within two scenes I_1 and I_i, the virtualised scene uses an inverse affine transform T_{1i}^{-1} , which enables the precise mapping of detected features to each location in world co-ordinates. Standard Schmmid-based repeatability meausures utilise the pixel positions to determine whether a point is repeated or not, however the pre-selection of points represented as floating point coordinates (a world coordinate system for the 3D rendering context). The pre-selection step is described in equations 5 and 6, and replaces the algorithm to determine $R_{i(\epsilon)}$ shown in equation 3, while not interfering with subsequent processing steps shown in equations 1 and 2. Additionally all points now include the z worldspace information as described by equation 4.

To enable 2D/3D pre-selection, D represents the vector dimensions to be utilised when measuring distance, while the function dist determines the distance from the reference point in world space. Pre-selection happens after the removal of points that don't share the same viewport have been removed, but before the points are converted to their pixel positions and s thresholding is applied. By statically pairing the closest point with its corresponding reference point in 3D space before it is measured in 2D, it enables the comparison of 2D and 3D pre-selection with minimal disruption so that later analysis is simplified.

The testing configuration for 3D pre-selection of points follows the methodology done by [11]. It uses a 300x300 image (I_i) which applies 47 transforms (J) of each model in the x and y axis, relative to the viewport as the model is rotated from -50° to +50° in 10° increments (11). The z is rotated from 0° to 180° in 10° increments (19), and the model is scaled in the x,y axis from 1.0, to 4.0 in 0.25 increments (17). This will be applied in two different testing scenarios. The first consisting of a single model, and the other, a dataset of 12 models. Most of the models are 3D scanned, and sourced from commercial, and research sites. The 12 models tested were titled "bowl", "owl", "plaque", "vase", "obelisk", "pot"[1] "marbles"[2], "apple"[3], "Stanford bunny", "happy Buddha", "dragon" and "lucy"[4]. The "Stanford asian dragon" model is tested separately. The bowl, owl, plaque, vase, pot, apple and marbles are textured, and the rest use a generic white mesh.

$$_{i}\widetilde{x} = ({}_{i}x_{1}, {}_{i}x_{2}, {}_{i}x_{3}) \tag{4}$$

$$R_i(\epsilon) = \{(\tilde{x}_1, \tilde{x}_i) \,|\, dist(T_{ij}\tilde{x}_1, \tilde{x}_j)\} \tag{5}$$

$$dist(T_{ij}\widetilde{x}_1, \widetilde{x}_j) = \sqrt{\sum_{j=1}^{D} ({}_{1}\widetilde{x}_j - T_{ij}^{-1}({}_{i}x_j))^2} \tag{6}$$

The IP detectors tested (K) were Harris [6], KLT [8], FAST [22, 2], SIFT [13] and SURF [7] as well as Rohr [21], Foerstner [3], Beaudet [1] and a different implementation of Harris [6], which have been implemented by the Vigra library [9]. The process for using pixel-based interest points in a pre-rendered image I_i, in conjunction with a world co-ordinate space to determine repeatability is summarised in the following steps.

(i) Render scene for each I_{1i} pair with selected affine transform and model

(a) Run detection on 2D image of rendered scenes

(b) Convert 2D interest points to 3D world co-ordinates

(c) Remove 3D points not within model bounding box

(d) Apply inverse affine transform to 3D points where i >1

(e) Remove points that don't share overlapping camera views

(f) Pre-select closest distance of $\widetilde{x}_i$, to each $\widetilde{x}_1$ in world co-ordinates

(ii) Repeat 1 for each model

(iii) Measure repeatability for each model

(a) Convert pre-selected closest 3D point pairs to 2D pixel co-ordinate space

---

[1] http://people.csail.mit.edu/tmertens/textransfer/data/index.html

[2] http://www.sci.utah.edu/~wald/animrep/

[3] http://www.turbosquid.com/3d

[4] http://graphics.standford.edu/data/3Dscanrep/

(b) Calculate repeatability measure of point pairs using corresponding pixel positions to apply ϵ and mark as repeated or not

(iv) Repeat 3 for each model and aggregate results

Two different datasets are being used to measure the effects of generalisation, one of these assess repeatability at a more localised level per transform. The asian dragon model was chosen due to its increased non-homogenous surface, protrusions such as horns, and potential for misclassification of repeated points due to lack of depth due to 2D preselection. This affords an analysis based on the effects of generalisation, as well as seeing the effects of preselection for a single model.

3 Results and Discussion

3.1 Analysis of 2D/3D datasets

When it comes to comparison of the performance of detectors, the first obvious choice is to compare the repeatability at each epsilon threshold. In most cases an ϵ=1.5 is the preferred threshold for discriminating between detectors. Intuitively, it would be expected that interest points that are able to utilise the depth of the scene would result in more reliable and boosted repeatability rates, given that false positives can be avoided, and better candidates chosen. The results in figure 2 and 3 highlight that in most cases the 2D preselection of points provides improved repeatability performance, both across detectors, and across most ϵ thresholds. In many instances more interest points are also detected.

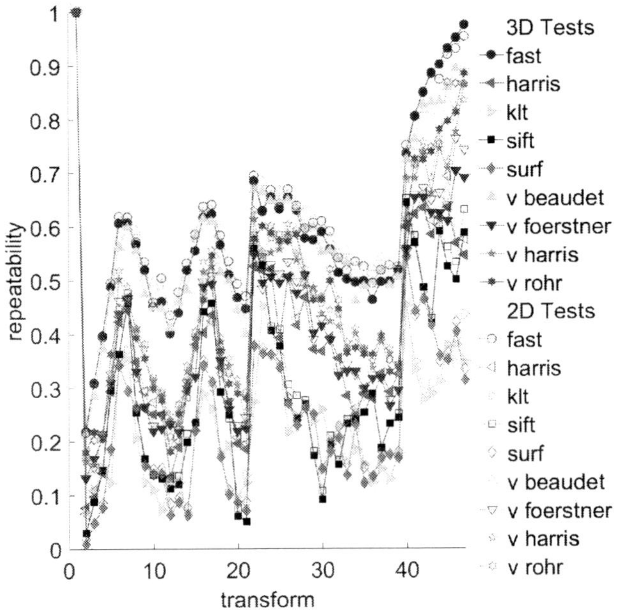

Fig. 1. Repeatability of asian dragon model at the transform level (ϵ = 1.5), with rotation in X at 2-11, Y at 12-22, Z at 23-39, and XY scale at 40-47.

At a superficial level, this could imply that 3D preselection is in fact impacting on performance, and there are indeed a few theoretical corner cases that could justify this. Namely the fact that points could become occluded, and in fact become false positives that are picked up due to them being closer in 2D space compared to other candidate points.

This is difficult to justify, however, as there are only a small number of the 47 transforms that could result in this type of occlusion (namely x, and y rotation of the model), and it also would require a very low number of points in order for more unusual or abnormal point candidates to be preselected. Additionally, when examining the asian dragon model at each transform, which can be seen in figure 1, we can see that repeatability at the scene level shows the same increase for 2D preselection across all scenes. Though it is important to recognise that this is a corner case, the effect (if any) and the criteria necessary to exploit this, requires exceptional circumstances.

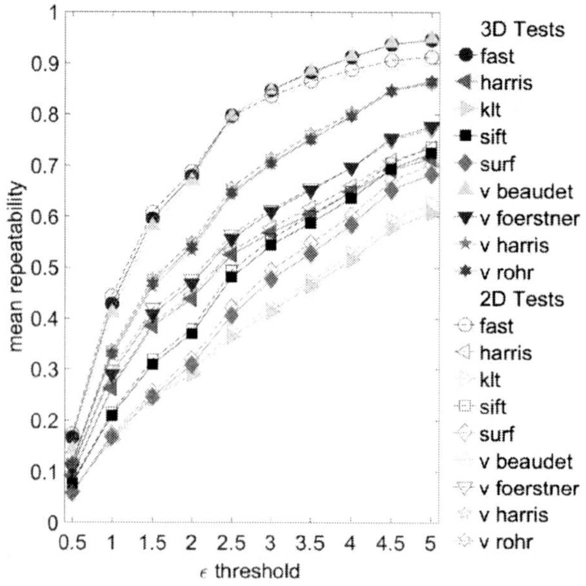

Fig. 2. Repeatability of asian dragon model

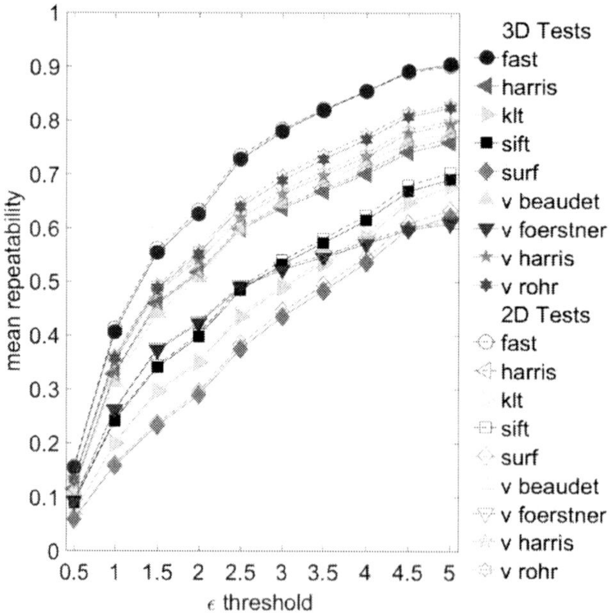

Fig. 3. Repeatability of 12 model dataset

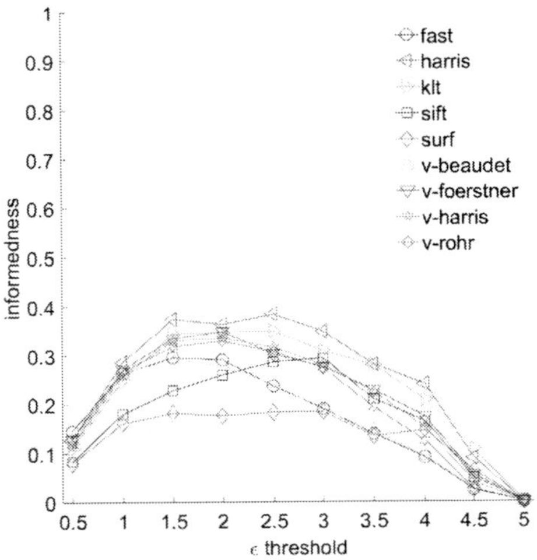

Fig. 4. Informedness of asian dragon model

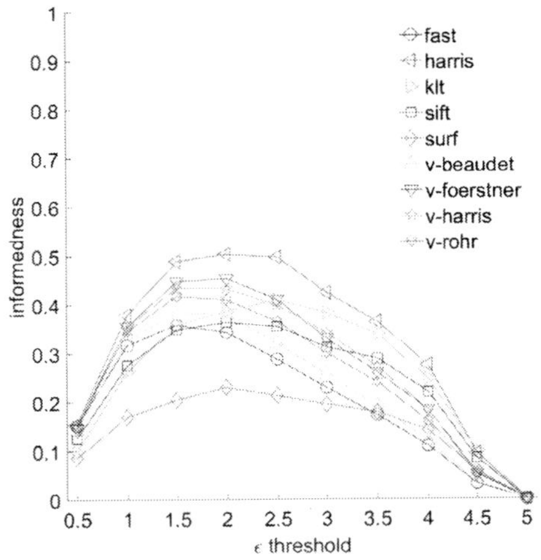

Fig. 5. Informedness of 12 model dataset

To perform a comparison of each dataset that consisted of the singal asian dragon model, and the 12 model dataset, the instances of repeated point pairs between ${}_{i}\tilde{x}_{j}$ and ${}_{i}\tilde{x}_{j}$ for each test were analysed, represented as A and B respectively. To find the *tpr* we intersect D_A and D_B to find true positives common to each testing configuration, and for the *fpr*, intersect and subtract the true positives. This is done at each epsilon which is described in equations 7 and 8. The intersection of repeated points D^3_ϵ, which represents the points that utilised 3D data, and D^2_ϵ which only used 2D data, provides a ratio of the number of true and false positives within each ϵ of a point in the reference image, which can be represented as a ROC graph.

$$tpr(\epsilon) = \frac{|D^2_\epsilon \cap D^3_\epsilon|}{|D^2_R \cap D^3_R|} \quad (7)$$

$$fpr(\epsilon) = \frac{|D^2_\epsilon - D^3_\epsilon|}{|D^2_R - D^3_R|} \quad (8)$$

This provides data sufficient for ROC analysis and calculation of an AUC. However, given the form of analysis that is performed in CV performance, which is to say that it is most common to compare according to the Moore neighborhood (ϵ=1.5), it is difficult to use the data in its current form for comparative analysis. To normalise the *tpr* and *fpr* ratios for better comparison, the informedness at each ϵ threshold can be used, which also provides a performance evaluation that takes into comparison both true positive and false positive detections by each classifier. Informedness is determined by finding the difference between the *tpr* and *fpr* and has been demonstrated as being a reliable metric that can determine to a greater extent the similarity of data sets (compared to randomness) [18] [19] [20]. The informedness of each detector at each ϵ threshold can be seen in figure 4 and 5.

3.2 Informedness Optimisation

Based on the results that are shown in figures 4 and 5, there is a clear divergence in the positions of points when compared to preselection of points that finds closest points in a 2D and 3D environment, even though all other testing conditions are identical. It's clear that, unlike figures 2 and 3 that use only the true positives based on Schmid's approach, there is a substantial mis-classification of points that is not apparent when only true positives are taken into account. This should not be taken as a slight towards true positive repeatability, however, as establishing ground truth is a necessary prerequisite for such an analysis is notoriously hard to reliably or accurately measure in real world environments. It does highlight that there are substantial benefits in adoption of virtualised, or more ideally, 3D scanned real-world objects, so that a more objective ground truth exists that can make these performance analyses possible.

Also of note is the fact that in the case of a singular, as well as more generalised dataset, in figures 4 and 5, the convention of ϵ=1.5, or Moore neighborhood points, is not necessarily indicative of being the most optimal, especially in the case of detections that are not able to preselect points with the assistance of scene depth. In fact, the informedness data suggests that ϵ=2.0 is generally more favorable across the majority of detectors when tested with the 12 model dataset under the current testing conditions. This informedness of 2D detections indicates that 2.0 should be the more preferred threshold when taking into consideration the tradeoffs of true positives, to false positive detections. Not only does it provide a more rigorous examination of 2D performance compared to 3D, but also indicates at which threshold 2D performance is best, which would be ideal for optimisation when it comes to taking classifiers out of the lab and into the real world. These tests demonstrate that the additional metric of informedness, in conjunction with better ground truth testing environments that can effortlessly switch between 2D and 3D, could provide new avenues of performance analysis beyond just concentrating on true positives.

4 Conclusion

This paper explores the topic of IP detectors and their repeatability across multiple scene transformations in virtualised 3D spaces with the assistance of 2D and 3D preselection. Though there is a clear move towards utilising 3D ground truth for classifiers that used 3D ground truth natively, 2D classifiers are not able to leverage this benefit. We have sought to formulate a performance analysis that is able to integrate 3D with the assistance of a vitrualised ground truth that gives a more balanced analysis of performance compared to conventional repeatability. It does so by building on the proof of concept that virtualised 3D spaces can be used for testing 2D based IP classifiers, and expands on this by testing the differences between finding nearest neighbor points via 2D and 3D worldspace co-ordinates, by preselecting best candidates before applying traditional repeatability metrics. Testing configurations consisted of a singular model and a 12 model dataset, to compare the effects of gerneralisation, and 9 conventional 2D detectors were tested across 47 transforms in x, y and z rotation, and x,y scaling. Though conventional 2D based repeatability showed slightly improved performance, more in depth analysis, made possible due to a more reliable ground truth, highlighted that 2D preselection produced considerable false positives compared to those selected using 3D. This was determined via ROC analysis, and was further refined to a singular performance metric using informedness to normalise results at each ϵ threshold. Normalisation via informedness also demonstrated that traditional conventional thresholds like only including the Moore neighborhood points as repeatable (ϵ=1.5) are not necessarily optimal, and other thresholds should be considered in 2D contexts for optimisation of classifiers when applied to 2D scenes, in the absence of 3D data.

5 Future Work

From the results of our work, there is a substantial difference in the repeatability, and by extension reliability of detected points. We have already begun further research building on this, to measure IP detectors in situations of rapidly prototyping and testing classifiers via Genetic Programming. We aim to build on existing research in CV/GP to explore the effects of virtual ground truth with 3D preselection (and without), to determine its effects on classifier design and performance. Another avenue of research being considered is how effective GP based classifier design is, when taken beyond virtualised 3D spaces into real world environments.

Another area that deserves further exploration is developing a means of preventing occluded points from potentially being preselected. It is common in conventional 3D graphics to use back plane culling where the depth of the scene is used to determine if a face is rendered or not for each pixel. Developing a similar process would help to avoid corner cases. This would be an involved process, and would likely require a sophisticated solution at the shader level, but would be a valuable addition to virtual ground truth environments and IP repeatability such as those where more complex transforms are involved and less points appeared in the scene in question. We consider this an important next step in pursuing interest point evaluation for repeatability purposes.

References

1. Beaudet, P.: Rotationally invariant image operators. In: Proc. Intl. Joint Conf. on Pattern Recognition. pp. 579–583 (1978)
2. E. Rosten, T.D.: Machine learning for high-speed corner detection. In: European Conference on Computer Vision (2006)
3. F¨orstner, W.: A feature based correspondence algorithms for image matching. In: Intl. Arch. Photogrammetry and Remote Sensing. vol. 24, pp. 160–166 (1986)
4. Guo, Y., Bennamoun, M., Sohel, F., Lu, M., Wan, J., Kwok, N.M.: A comprehensive performance evaluation of 3d local feature descriptors. International Journal of Computer Vision 116(1), 66–89 (Jan 2016). https://doi.org/10.1007/s11263-015-0824-y
5. H¨ansch, R., Weber, T., Hellwich, O.: Comparison of 3d interest point detectors and descriptors for point cloud fusion. ISPRS Annals of the Photogrammetry, Remote Sensing and Spatial Information Sciences 2(3), 57 (2014)
6. Harris, C., Stephens, M.: A combined corner and edge detector. In: Alvey Vision Conference. p. 147151 (1988)

7. Herbert Bay, Tinne Tuytelaars, L.V.G.: Surf: Speeded up robust features. In: Lecture Notes in Computer Science. vol. 3951, pp. 404–417 (2006)
8. Kanade, C.T.T.: detection and tracking of point features. In: Carnegie Mellon University Technical Report CMU-CS-91-132 (April 1991)
9. K¨athe, U.: Generische Programmierung fr die Bildverarbeitung. Ph.D. thesis, University of Hamburg (2000)
10. Lang, S.R., Luerssen, M.H., Powers, D.M.W.: Proceedings of the 8th International Conference on Computer Recognition Systems CORES 2013, chap. Repeatability Measurements for 2D Interest Point Detectors on 3D Models, pp. 361–370. Springer International Publishing, Heidelberg (2013)
11. Lang, S.R., Luerssen, M.H., Powers, D.M.W.: Automated evaluation of interest point detectors. Int. J. Softw. Innov. 2(1), 86–105 (Jan 2014). https://doi.org/10.4018/ijsi.2014010107
12. Lindeberg, T.: Image matching using generalized scale-space interest points. Journal of Mathematical Imaging and Vision 52(1), 3–36 (May 2015). https://doi.org/10.1007/s10851-014-0541-0
13. Lowe, D.G.: Distinctive image features from scale-invariant keypoints. International Journal of Computer Vision 60, 91–110 (2004)
14. Mikolajczyk, K., Schmid, C.: Scale & affine invariant interest point detectors. International Journal of Computer Vision 60, 63–86 (2004)
15. Moreels, P., Perona, P.: Evaluation of features detectors and descriptors based on 3d objects. Tenth IEEE International Conference on Computer Vision (ICCV'05) Volume 1 1, 800–807 Vol. 1 (2005)
16. Olague, G., Trujillo, L.: Evolutionary computer assisted design of image operators that detect interest points using genetic programming. Image and Vision Computing. Elsevier. 29, 484–498 (2011)
17. Olague, G., Trujillo, L.: Using evolution to learn how to perform interest point detection. Pattern Recognition, International Conference on 1, 211–214 (2006)
18. Powers, D.M.W.: Roc-concert: Roc-based measurement of consistency and certainty. In: 2012 Spring Congress on Engineering and Technology. pp. 1–4 (May 2012). https://doi.org/10.1109/SCET.2012.6342144
19. Powers, D.M.W.:Evaluation evaluation a monte carlo study. CoRR abs/1504.00854 (2015), http://arxiv.org/abs/1504.00854
20. Powers, D.M.W.: Visualization of tradeoff in evaluation: from precision-recall & PN to lift, ROC & BIRD. CoRR abs/1505.00401 (2015), http://arxiv.org/abs/1505.00401
21. Rohr, K.: Modelling and identification of characteristic intensity variations. In: Image and Vision Computing. vol. 10, pp. 66–76 (1992)
22. Rosten, E., Drummond, T.: Machine learning for high-speed corner detection. In: European Conference on Computer Vision. vol. 1, pp. 430–443 (May 2006)
23. Schmid, C., Mohr, R., Bauckhage, C.: Evaluation of interest point detectors. International Journal of Computer Vision 37, 151–172 (2000). https://doi.org/10.1023/A:1008199403446
24. Trujillo, L., Olague, G.: Automated design of image operators that detect interest points. Massachusetts Institute of Technology 16(4), 483–507 (2008)

Animal species classification using machine learning techniques

Fahad Alharbi*, *Abrar* Alharbi, and *Eiji* Kamioka

Graduate School of Engineering and Science, Shibaura Institute of Technology, Tokyo, Japan

Abstract. Animals recognition is one of the research areas in which few effective technologies have been proposed, especially in the predator animals' domain. Predator animals present a great danger to people who are camping or staying in outdoor areas and they are also a menace to livestock. In this paper, a multiple feature detection of predator animals is proposed. This method focuses on the face of the animal, explicitly the eyes and the ears. A database was created by collecting the features of ears and eyes from 10 animals and an experiment was conducted using machine learning techniques such as SVM and MLP to classify them as predators or pets. The evaluation results achieved the classification accuracies of 82% for MLP and 78% for SVM, which justify its effectiveness for the proposed method.

1 Introduction

Animal attacks are a common cause of either human fatalities or injuries. Up to five million people in the US are attacked by animals each year as reported in 2017 [1]. Langley & Morrow [2] estimated that up to two million animal's bites occur each year in the United States. The frequency of animal attacks varies with geographical location. Man-eater is an informal term for a particular animal that preys on humans as a form of hunting habit. For instance, tigers have been recorded that they kill more people than any other of its kind. They have been responsible for more human deaths through direct attack than any other wild animal as reported by Nowak, et al. [3]. Man-eating lions, on the other hand, have been recorded that they actively enter human villages at night as well as during the daytime to hunt for prey. American and Tanzanian scientists report that man-eating incidents in rural areas of Tanzania increased significantly from 1990 to 2005. At least 563 villagers were attacked, and many people were eaten over this period. Warrell [4], stated that injuries caused by animal attacks result in thousands of fatalities worldwide every year. Though it does not seem every country, they keep track of fatalities caused by animals. Most animals, excluding some tigers, do not regularly hunt humans, although some animals will feed on people if they are dead, sick, or unconscious. Attacks on people, livestock, and pets may occur when animals habituate to humans or are in a condition of severe starvation. Attacks are most frequent during night time when animals leave their territory in search of food. No

* Corresponding author: mg17003@shibaura-it.ac.jp

current predator animal detector is found in the literature. In this paper, an animal species recognition method is proposed using image processing techniques, the proposed method is then tested on a dataset containing two species of animals namely pets and predators by using several classifiers. Lastly, the results of the classification are evaluated and discussed regarding accuracy. The outline of this paper is as follows: In section 2, some existing related work will be discussed. In section 3, the background knowledge will aim to provide related information about the proposal. In section 4, the proposal of animal species recognition method will be presented. Next, the effectiveness of the proposed method will be evaluated in section 5. Finally, the conclusion and future work will be stated.

2 Related work

A limited amount of research has been undertaken in the domain of Animal types recognition especially in the predator animal area. One example here [5] is about identifying animal species in a tracking tunnel by capturing ink footprints in a specific area which is not suitable in our case because we wish to detect the animal in real-time. Researchers [6] attempted to detect events like hunting by classifying and tracking of moving object blobs using a neural network. However, in this research, we aim to detect the animal before the hunt not during. Work done here [7] used a background subtraction method for finding the details of the targeted animal from the background, yet it does not classify them. Authors in [8] presented a modified version of Viola and Jones [9] Haar-like detector to be used for detecting and tracking of animals in real-time. They extracted the Haar-like features of a lion and presented a model for monitoring and annotating the presence of an animal in a video. Although, they only used one animal. In this paper, we are proposing a new approach for extracting animal features and using machine learning methods to distinguish between predators and non-predators.

3 Background knowledge

3.1 Feature extraction

Image feature extraction is an essential step for object classification. In order to classify an object in an image, there are various types of features to be extracted, for example, statistical features of raw pixel data, shape and color feature maps and transform coefficient or vectors.

3.2 Classification

Image object classification is one of the famous problems in the area of image processing research. There are numerous methods for solving this problem in an instance, Support Vector Machine (SVM), Decision Tree and Artificial Neural Network (ANN). SVM is among the best known linear methods in image classification and pattern recognition. The main advantage of SVM is the capability to get good generalization with a limited number of training samples. ANN, however, is a nonlinear method that is inspired by the biological neural network of the brain. Some of the advantages of ANN is that it has the ability to learn and model non-linear and complex connections between inputs and outputs, also it is able to generalize, after learning from the primary inputs and their correlation with the outputs, it can derive unperceived connections, thus making the model capable of generalizing and predicting undiscovered data. Decision Tree method uses a tree-like model of decisions. Originally used in data mining research for deriving a scheme to reach a specific goal, its also commonly used in machine learning areas.

4 Predator animals' features

In this paper, a statistical feature extraction method of predator animal face region in outdoor areas is proposed. The region descriptors are used to extract the statistical information of a specific image region of animal faces. We propose that animals' ears and eyes are used as a region for extraction of statistical information because they contain valuable features to classify the type of animal species. Researchers found a correlation between terrestrial species pupil shape and ecological niche, which is to say that, predator animals have different pupil shape than prey animals [8]. In this research we focused on extracting the features of animals eyes and ears, to determine which type of animals it belongs to. The process flow is as follows: First, a dataset was created containing images of ten animals (5 Predators, 5 Pets) The images have large variations in scale, pose and lighting. Samples of the dataset are shown in figure 1.

Fig. 1. Example of the dataset. The top row shows the pet animals and the bottom row shows the predator animals selected for this experiment.

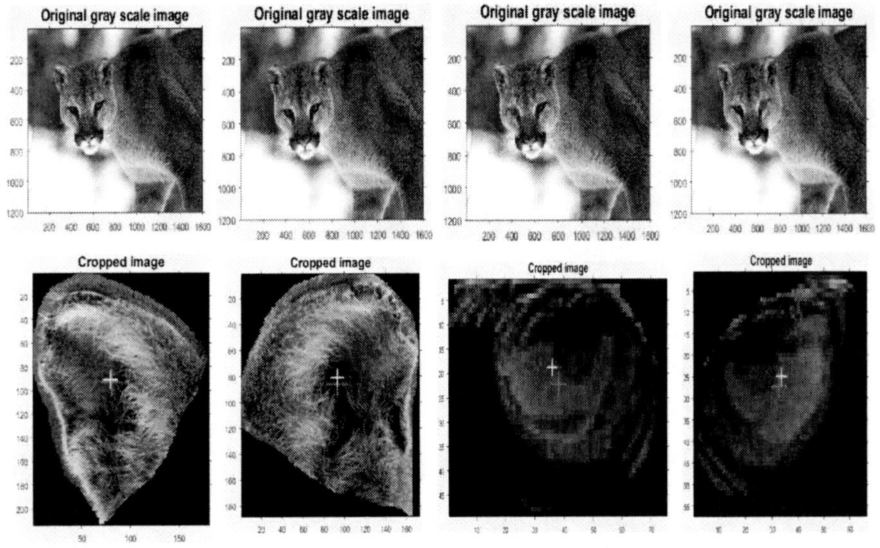

Fig. 2. Shows an example of labelled ROIs of a predator animal.

Second, the ROIs (region of interest) was drawn manually for all animals as demonstrated in figure 2. After that, statistical information such as mean, standard deviation and perimeter were extracted. In the third step, the dataset was split into two. The first was the training set, which contained 150 image information 75 for Predators and 75 for Pets. The second dataset was for testing, and it contained 50 image information 25 for both Predators and Pets. Finally, two classifiers were used for training and testing the dataset by taking the statistical features as in input parameter and outputting two classes pet or predator.

The selected parameters Mean, Standard deviation and Perimeter were chosen as regional descriptors which are an important representation of an object shape and texture. The extracted statistical parameters can then be used as probabilistic descriptor and classifier for objects in an image.

5 Results and Discussion

In this section, the results of the evaluation of the proposed method will be shown. In order to evaluate the results of the classification for the proposed method, the F1 score algorithm was used. In table 1 the results of the SVM and MLP classifiers are computed.

Table 1. Below are the evaluation results of the classifiers for both animal species.

	TP Rate	FP Rate	Precision	Recall	F1 Score	Class
SVM	0.8	0.24	0.769	0.8	0.784	Pet
	0.76	0.2	0.792	0.76	0.776	Predator
MLP	0.88	0.24	0.786	0.88	0.83	Pet
	0.76	0.12	0.864	0.76	0.809	Predator

The F1 score method was utilized as shown in table 1, which considers both the precision P and the recall R of the test to compute the accuracy score. Where P is the number of correct positive results divided by the number of all positive results as given in Equation (1), and R is the number of correct positive results divided by the number of positive results that should have been returned as given in Equation (2). The F1 score shown in Equation (3) is the average of the precision and recall, where an F1 score reaches its best value at 1 and worst at 0.

$$P = \frac{True\,positives}{True\,positives\ + False\,positives}$$

$$R = \frac{True\,positives}{True\,positives\ + False\,negatives}$$

$$F1 = 2 \times \frac{Precision \times Recall}{Precision + Recall}$$

The SVM classifier average accuracy score for both classes is 78%. However, the MLP average score for both classes is higher at 82% which suggests that the neural networks ability to learn and work for non-linear data is a good fit for this proposal. Unlike SVM which is considered a more better fit for liner datasets. Figure 3 indicates that the Receiver Operating Curve (ROC) of the MLP classifier is better in the working area, unlike the SVM classifier.

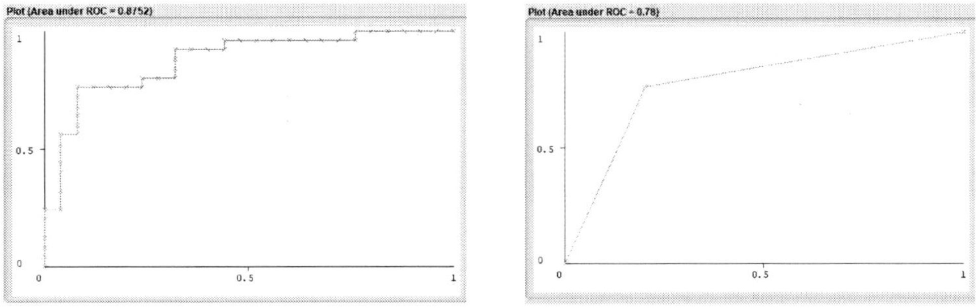

Fig. 3. Shows the ROC curve of the MLP classifier (left) and the SVM classifier (right) where the X-axis represents the false positive rate while the Y-axis represents the true positive rate.

By analysing the ROC curve, we can say it demonstrates several things. First, it shows the trade-off between sensitivity and specificity. Second, the closer the curve to the left-hand border and top border of the ROC space, the higher the accuracy. Comparing our proposed method with the work done here [10] using our dataset. The result of the accuracy was lower at 68% which shows the superiority of our proposal in distinguishing between different species.

6 Conclusion

In conclusion, this research succeeded in developing a multiple feature extraction method for classifying animal species that achieved a good accuracy of 82% for MLP and 78% for SVM.

The proposal used the unique characteristics of animals such as ears and eyes. This work can offer many social significance possibilities, for instance, it can be used to avoid animals in outdoor areas. Users can benefit from this animal species classifier in outdoor activities like camping and hiking. Also, this research can help animal farmers protect their livestock from predator's animal attacks.

For future work, more animals need to add to the dataset and also investigating a way to increase the accuracy by using deep learning

References

1. McBean et al. 2007 *Medical journal of Australia.* **186** pp 38
2. Ricky L and William E 1997 *Wilderness & Environmental Medicine* **8** pp 8-16
3. Nowak et al. 1983 *Johns Hopkins University Press* **4** pp 1088
4. Warrell D A 1993 *Med Aust* **195** pp 11-12
5. Perera, John, Achala and Collins 2015 *Int. Conf. on Sensing Technology (IEEE)* pp 195-199
6. Haering et al. 1999 *Int. Conf. on Multimedia Computing and Systems* pp 905-909
7. Chen and Peijiang 2009 *Int. Symp. on Network and Multimedia* (IEEE) pp 1-4
8. Banks, Martin S., et al. 2015 Science advances 1.7.
9. Viola and Jones 2001 *Int. conf. on Computer Vision and Pattern Recognition* 1 (IEEE) pp 511-518
10. Burghardt, Tilo, and Janko Calic. 2006 Neural Network Applications in Electrical Engineering. NEUREL 8th Seminar on. IEEE.

Action recognition based on 2D skeletons extracted from RGB videos

Sophie Aubry, *Sohaib* Laraba*, *Joëlle* Tilmanne, and *Thierry* Dutoit
University of Mons, TCTS Lab, 31, Boulevard Dolez B-7000 Mons, Belgium

Abstract. In this paper a methodology to recognize actions based on RGB videos is proposed which takes advantages of the recent breakthrough made in deep learning. Following the development of Convolutional Neural Networks (CNNs), research was conducted on the transformation of skeletal motion data into 2D images. In this work, a solution is proposed requiring only the use of RGB videos instead of RGB-D videos. This work is based on multiple works studying the conversion of RGB-D data into 2D images. From a video stream (RGB images), a two-dimension skeleton of 18 joints for each detected body is extracted with a DNN-based human pose estimator called OpenPose. The skeleton data are encoded into Red, Green and Blue channels of images. Different ways of encoding motion data into images were studied. We successfully use state-of-the-art deep neural networks designed for image classification to recognize actions. Based on a study of the related works, we chose to use image classification models: SqueezeNet, AlexNet, DenseNet, ResNet, Inception, VGG and retrained them to perform action recognition. For all the test the NTU RGB+D database is used. The highest accuracy is obtained with ResNet: 83.317% cross-subject and 88.780% cross-view which outperforms most of state-of-the-art results.

1 Introduction

The task of recognition is to name actions and describe them from a sequence of movements. The study of human action is linked to many fields such as computer science, psychology, healthcare. Action recognition can be used in many applications such as video surveillance, human-machine interaction, user interface design, gaming, entertainment, robotics, web-video description, medical diagnosis, sports analysis and many others. Action recognition is a major challenge because of the diversity of movements, the complexity of motion capture and the difficulty of creating a realistic and relevant database. An action, such as drinking, can be done in different ways, depending on the person acting, the context, the environment, the style of the movement and many other parameters. Each action can be done in so many ways that it is difficult to define the features describing the action. Another challenge is how to capture and represent movement. Different representations exist and depending on the final application, many motion capture systems are available. To obtain good results in the recognition task, a large amount of data is necessary and specially to

* Corresponding author: sohaib.laraba@umons.ac.be

train a neural network. All this data is gathered in a database. For action recognition, the creation of a database and manual labelling of all movement sequences one by one takes a lot of time. In addition, the database should be as representative as possible of the reality and conditions under which the recognition algorithm will be used.

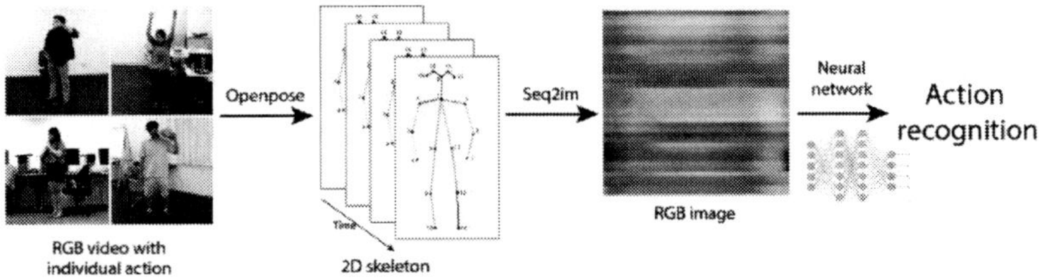

Fig. 1. Methodology used to recognize action: input data are RGB video containing individual actions (from NTU RGB+D dataset), the skeleton is extracted with Openpose then the sequence of motion is converted into a RGB image before going into the neural network which classifies the action.

We present a new methodology to process RGB videos to perform action recognition (Fig. 1). The first step to recognize human activity is to extract the motion from the video. As a result, the essence of the motion is extracted, namely, the skeleton of people. Skeleton extraction is done using OpenPose [1], a DNN-based detection system that extracts a 2D skeleton of 18 joints for each detected body. Second, motion sequences are converted to RGB images. The classification of images and videos has been very successful with the development of deep learning. We exploit these advantages by transforming the skeleton-based action recognition task into an image classification task. The motion parameters are encoded in the three R, G, B channels and an action sequence becomes an RGB image. Finally, image classification neural networks can be retrained to recognize actions.

2 State of the Art

Image and video classification have known a great success in the last few years, and especially with the development of Deep Learning. Some researches tried to exploit these advantages by transforming the skeleton-based action recognition task into an image classification task, and this by transforming 3D skeleton sequences into images. Image classification models can be then used on this data [2]. The motion parameters are encoded into the three R, G, B channels and action sequences become an RGB image. Then the image classification neural network can be retrained to recognize actions.

Laraba et al. [2] encode the (X, Y, Z) position of each joint into R, G, B channels. They obtain with a CNN an accuracy of 74.27% for the cross-subject and 75.74% for the cross-view action recognition using 3D skeleton data recorded by the Kinect from NTU RGB+D dataset [3]. In the same idea, Du et al. [4] obtain an accuracy of 100% with the Berkeley MHAD dataset [5].

In [6], Ding et al. investigated different skeleton features to encode them into RGB images. They test five different features: joint-joint distances, joint-joint vectors, joint-joint orientations, joint-line distances and line-line angles. They encode these features into five RGB images before using them to train five CNN. Then, all the output scores are fused to return a final score of the classification. With this method they achieve an accuracy of 82.31% with the NTU RGB+D dataset [3].

In [7], Ke et al. suggest a new representation of skeleton data. They encode the 3D coordinates (X,Y,Z) into a clip of grey images containing spatio-temporal information. They obtain an accuracy of 84,83% cross-view and 79.57% cross-subject with the NTU RGB+D database [3], 93.57% with the SBU Kinect interaction database [8] and 88.30% with the CMU dataset [9].

Wang et al. [10] propose a joint trajectory map in which 3D skeleton data are encoded. These maps are then classified by a CNN. Their joint trajectory map encodes the 3D coordinate of the joint, the motion direction, the body parts and the magnitude of the motion. They obtain an accuracy of 81.08% cross-view and 76.32% cross-subject with the NTU RGB+D database [3], 94.86% with the MSRC-12 Kinect Gesture Dataset [11] and 96.02% with the G3D DATASET [12].

Li et al. [13] suggest to encode 3D relative coordinates and compute the difference with 3 reference joints (hip centre, right shoulder and left shoulder). The three generated sets of vectors containing relative coordinates are encoded into an RGB image. They train a CNN with these images and get an accuracy of 75.2% in cross-subject and 82.1% in cross-view with the NTU RGB+D database [3].

In [14], Li et al. propose to use skeleton data and divide them into five body parts, the 3D coordinates (X, Y, Z) of joints of each body part is concatenated in a vector and encoded into a RGB image. The RGB images feed the CNN and obtain with the NTU RGB+D dataset [3] an accuracy of 84.6% cross-subject and 90.9% cross-view.

Li et al. in [15] rearrange and select the important skeleton joints automatically. The order of joints during encoding influences the accuracy. The module selects the best order and the most important joints before encoding them into an image and feed the CNN. They obtain an accuracy of 83.2% cross-subject and 89.3% cross-view.

3 Methodology

The proposed methodology is based on RGB videos. Motion information are extracted with OpenPose. These skeleton motion sequences are then converted into images. In this section, OpenPose is presented then the conversion of the motion sequence into RGB image is explained.

3.1 Machine Specification

For this work, all the tests on neural networks for classification algorithms are performed on a computer, equipped with a MSI GeForce_GTX 1080 TI GAMING X 11G GPU and 32 GB (2x16GB DDR4 3000) memory. The computer is also equipped with a Intel Core Skylake-X I7-7800x 3.5Ghz 25Mb LGA 2066 BOX CPU.

3.2 Motion Extraction with OpenPose

OpenPose is able to detect the 2D poses of several people in an image. Detection is done by a bottom-up approach: body parts are detected and then associated with individuals on the image. Cao et al. [1] propose a two-flow CNN architecture: the first branch is dedicated to limb detection and the second to the affinity field of the part that gives the degree of association between body parts to determine which are associated with which body (Fig. 2). This step is done iteratively: each step takes as input the detection confidence cards and the affinity field of the part (the two output branches) of the previous step and the original image. OpenPose extracts the position of 18 joints for each detected body and returns for each joint the coordinates (X, Y) and the confidence score C relative to the detection of each joint.

In this work, we focus on the recognition of individual actions, only the first 49 actions are retained from the NTU RGB+D database. From all RGB videos in the database, the actor's skeleton is extracted with OpenPose. Even if OpenPose does not reach the same performances as 3D motion capture systems, OpenPose requires only RGB camera and extract 2D skeletons from a video. It can be used in indoor and outdoor environments.

OpenPose allows partial skeleton recognition. Some disadvantages of OpenPose are that it returns no information about the depth and it is based on DNN which requires a high-end computer.

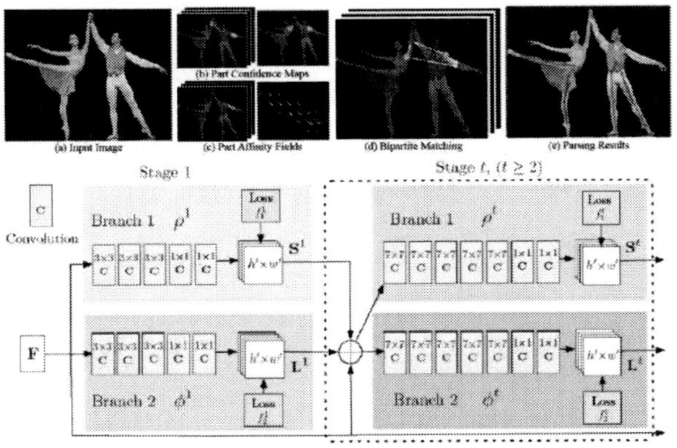

Fig. 2. OpenPose methodology for the 2D pose detection of several people in an image and neural network architecture for detection and association of body parts [1].

On some videos, OpenPose extracts the skeleton of people in the background unrelated to the action recognition task. The extracted data have been cleaned and only the first skeleton of 18 joints remains.

3.3 Data Conversion

Neural networks in the field of image classification have evolved rapidly and many models have emerged. With the large number of models in this field and the good performance of these models, a representation of skeletal motion sequences in images has been proposed by Laraba et al. [2]. The conversion of the motion sequence into an RGB image is done in three steps: extraction of the skeleton data in the matrices (X, Y, Z), normalization between 0 and 255 and mapping with the R, G, B channels. With this transformation, the motion sequence is reduced to an image.

In this work, the motion is recorded through traditional RGB camera. From these videos, the motion is extracted thanks to OpenPose which returns the (X, Y) positions of the joints and the confidence C of the detection of joints for each frame. (X, Y) and C are extracted into three different matrices. For each one, a row corresponds to the position of a joint along the time and a column to the positions of all the joints at one frame. These three matrices are then normalized between 0 and 255 before being mapped into RGB images. With this transformation each sequence of video representing an action is converted into a single RGB image (Fig. 3).

4 Data representation

In this section, an evaluation of the movement parameters influencing the effectiveness of the classification will be conducted. Before feeding them to the neural network, the motion capture data will be transformed into an image. For each joint, three channels are available (RGB). The experimentation consists in finding the best data representation and the best kind of encoding to have the highest accuracy. To realize this experimentation, two models of image classification neural network have been tested: SqueezeNet [16] and DenseNet [17]. SqueezeNet is very small and has very few parameters, which makes it very quick to be trained. DenseNet is a more complex neural network, with a high number of parameters to optimize. Due to its complexity, the training time is higher, but the accuracy is improved compared to SqueezeNet.

All the following tests (summarized in Table 1) have been realized with a deep retraining. We use models pretrained on the ImageNet dataset, and we retrain the whole network. We prefer to use a deep retraining instead of a shallow retraining because our data are very different from the ImageNet dataset.

The best data representation will be then used with more complex neural network architectures. All the data representations have been tested to do cross-subject and cross-view recognition.

4.1 Test 1: Position (X, Y) and Confidence

For the first experimentation, for each motion sequence, (X, Y) position and the confidence score C given by OpenPose for each joint are encoded respectively in the R, G and B channels of the image (RGB) (Fig. 3).

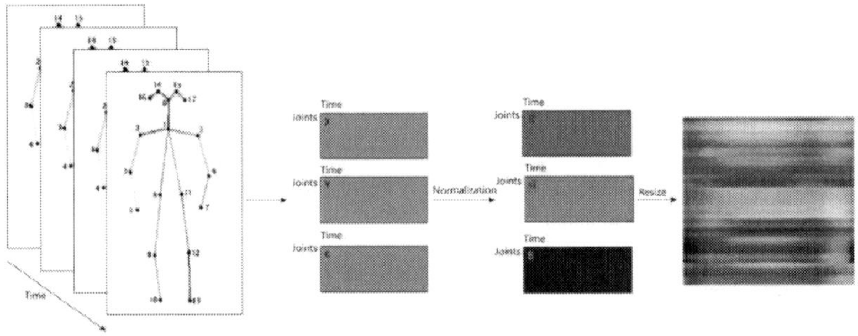

Fig. 3. (X, Y) positions and confidence C of each joint are encoded in R, G and B channels of an image.

The accuracy obtained with SqueezeNet is equal to 74.25% and it is equal to 82.71% with DenseNet (Tables 2 and 3).

4.2 Test 2: Position (X, Y) and mean between (X, Y)

For the second test, we encode the mean value between (X, Y) as input information about the depth. The size of the body in the video gives information about the closeness of the actor. If the actor is far from the camera, he looks small on the image. When he comes closer to the camera, he is larger on the image. Some information about the depth is hence contained into the (X, Y) position of each joint. We propose to encode the (X, Y) positions and the mean between X and Y in the R, G and B channel of an image (Fig. 4).

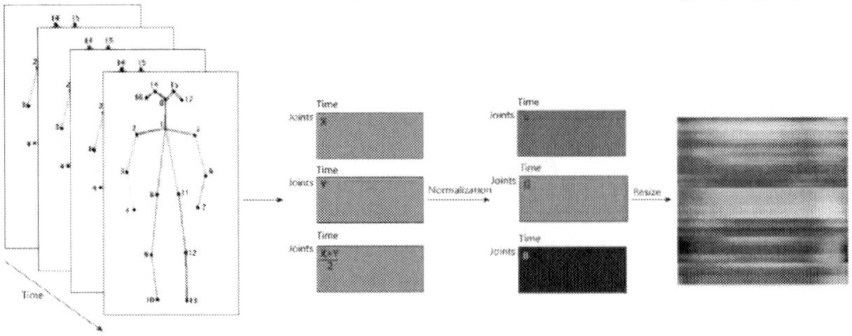

Fig. 4. (X, Y) position and the mean of (X, Y) of each joint are encoded as an image.

The accuracy obtained with SqueezeNet is equal to 71,43% and to 80,33% with DenseNet (tables 2 and 3). These two first tests show that putting the confidence C in the encoding is more relevant than setting the mean value between (X, Y) coordinates.

4.3 Test 3 and Test 4: 14 Joints

For the third and fourth experiments, the two data encoding presented above are tested with only 14 joints instead of 18 (Fig. 5). Some articulation positions of the skeletal extraction performed with OpenPose are missing. In particular, the joints of the eyes and ears are often absent. For some gestures, they are missing in 50% of the frames. The actions we want to recognize do not involve the eyes or ears. The neural network was retrained by keeping only the other 14 nodes.

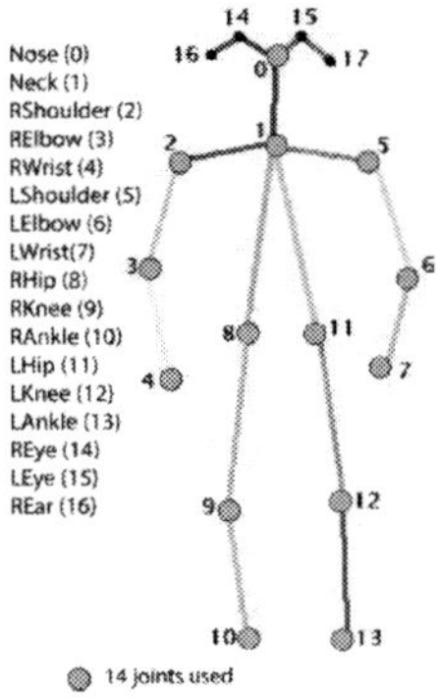

Fig. 5. OpenPose 2D skeleton [18]

The tests with 14 joints are realized one time with the confidence C and a second time with the mean value between (X, Y) coordinates as the third channel B. The results of the tests are in Tables 2 and 3. Using the confidence C, the test 1 and test 3 must be compared together. The removal of the ears and eyes nodes shows an improvement with SqueezeNet [16] and a small regression with DenseNet [17]. Test 2 and test 4, using the mean value between (X, Y) coordinates, show an improvement of the accuracy with the two models when we use 14 joints instead of 18. It seems using 14 joints instead of 18 joints gives better accuracy.

4.4 Test 5 and Test 6: Change Joints Order

For these tests, the (X, Y) positions of the 14 joints are encoded in different orders. The understanding of the motion by the neural network depends on the data representation and the order of the joints in the encoder gives different accuracy results. The first order tried in tests 3 and 4 is the following: 0, 1, 2, 3, 4, 5, 6, 7, 8, 9, 10, 11, 12, 13. The second order tried in tests 5 and 6 is the following: 0, 4, 3, 2, 1, 5, 6, 7, 10, 9, 8, 11, 12, 13 (Fig. 5). This order associates the two arms and the two legs.

When we compare the results of the tests 3 and 4 with tests 5 and 6, we can conclude that the accuracy is influenced by the order of the joints and the order 0, 1, 2, 3, 4, 5, 6, 7, 8, 9, 10, 11, 12, 13 gives a better accuracy (Tables 2 and 3).

4.5 Conclusion: the best data representation

To find the best data representation, the different data representations have been tested with two models: SqueezeNet [16] and DenseNet [17]. These tests show that:

- the more relevant third coordinate is the confidence C returned by OpenPose and it is obvious the (X, Y) coordinates are neccessary.

●the suppression of the most missing joints (eyes and ears) is relevant and improves the accuracy with Squeezenet.

●the order of the joints is important and the best order found is: 0, 1, 2, 3, 4, 5, 6, 7, 8, 9, 10, 11, 12, 13.

Table 1. Parameters used in the tests to evaluate the different data representations.

test	RGB conversion	number of nodes	nodes order
1	(X,Y),c	18	order 1
2	(X,Y),mean	18	order 1
3	(X,Y),c	14	order 2
4	(X,Y),mean	14	order 2
5	(X,Y),c	14	order 3
6	(X,Y),mean	14	order 3

legend
order 1 = (14,15,16,17,0,1,2,3,4,5,6,7,8,9,10,11,12,13)
order 2 = (0,1,2,3,4,5,6,7,8,9,10,11,12,13)
order 3 = (0,4,3,2,1,5,6,7,10,9,8,11,12,13)

Table 2. Results of the tests to evaluate the different data representations with SqueezeNet [16] in cross-subject.

	Memory size	parameters optimized
SqueezeNet	3	747 633

Results	Accuracy (in%)	Training time (min)	Evaluation time (s)
test1	74.253%	22.80	29.2680
test2	71.439%	22.34	29.2983
test3	**74.553%**	22.50	29.9367
test4	73.160%	22.57	29.0531
test5	73.512%	22.83	29.4864
test6	72.487%	22.75	29.2175

As shown in Table 2, with the SqueezeNet model, the training time is approximately the same for all the tests and is between 22.34 and 22.83 minutes. The evaluation time is also approximately the same for all the tests and is between 29.0531 seconds and 29.9367 seconds. The highest accuracy obtained is 74.553%.

The DenseNet model is more complex. Compared to SqueezeNet, DenseNet has a higher number of parameters to optimize and training time and evaluation time are higher. The training time is between 73.16 and 74.63 minutes, which is approximately three times higher than SqueezeNet. The evaluation time is between 55.6769 and 57.2752 seconds which is approximately twice the evaluation time of SqueezeNet. The highest accuracy obtained is 82.718%.

The two models do not agree on the best data representation. According to SqueezeNet, the best data representation is obtained with the parameter configuration of the test 3 and with DenseNet, the best accuracy is obtained with parameters configuration of the test 1. The difference between these two tests is the number of joints: 14 or 18 joints.

In terms of accuracy, the best data representation is the encoding of the (X, Y) position and the confidence C of the 14 (or 18) joints in the following order: (14, 15, 16, 17), 0, 1, 2, 3, 4, 5, 6, 7, 8, 9, 10, 11, 12, 13. This way of encoding is used for the tests in the following section.

Table 3. Results of the tests to evaluate the different data representations with DenseNet [17] in cross-subject.

	Memory size	parameters optimized	
DenseNet	51.1	12 566 065	
Results	Accuracy (in%)	Training time (min)	Evaluation time (s)
test1	**82.718%**	73.39	56.0388
test2	80.338%	74.43	55.6769
test3	82.711%	74.63	57.2752
test4	80.981%	74.56	57.1248
test5	82.688%	73.16	56.1067
test6	80.735%	74.37	57.0115

5 The Image Classification Model for Action Recognition

With the best data representation chosen in the previous section, different models of image classification are compared. Then the importance of the transfer learning approach is demonstrated with different kinds of training with each model: deep, shallow or from scratch. Finally, a comparison of performances is made for the cross-subject and cross-view action recognition tasks.

5.1 Models Comparison

In this section, we compare different models of image classification in terms of accuracy, training time and evaluation time. All these models are tested with two chosen data representations (corresponding to test 1 and 3 in 1): encoding (X, Y) and C in RGB channels for 14 (or 18) nodes in the following order (14, 15, 16, 17), 0, 1, 2, 3, 4, 5, 6, 7, 8, 9, 10, 11, 12, 13. We evaluate the following models: SqueezeNet [16], AlexNet [19], DenseNet [17], Resnet [20, 21], Inception [22, 23, 24], VGG [25].

Three deep retraining of all the selected models have been realized with the two data representations. The best results obtained for each model are summarized in Table 4. The highest accuracy is **83.317%** and is obtained with ResNet152 with 18 nodes. It seems difficult to choose the best data representation between 18 or 14 joints. For some models (SqueezeNet, DenseNet and VGG13), suppressing the more often missing joints can improve the accuracy. With 14 nodes data representation, the highest accuracy is obtained with ResNet152 (**83.317%**). With 18 nodes, the highest accuracy is obtained with DenseNet (**82.651 %**).

The accuracy is not the only criteria to choose a model. Depending on the application, the memory required, or the training and evaluation times can be important. The models requiring the shortest training time are SqueezeNet and AlexNet. SqueezeNet requires the smallest memory size. The slowest model in terms of training time and evaluation time is ResNet152. VGG19 is the model requiring the largest memory size.

5.2 Transfer Learning: Deep, Shallow or from scratch

In this section, we evaluate the importance of the transfer learning. The different models of image classification will be retrained with a deep, shallow or from scratch training [26][27][28]. These models are originally conceived for image classification and we test their ability to perform action recognition by converting the 2D skeleton data into RGB image.

All the models have already been trained for image classification. The following models are tested: SqueezeNet [16], AlexNet [19], Inception [23], DenseNet [17], ResNet [20, 21], VGG [25].

Table 4. Results obtained by a deep retraining using 18 nodes and using 14 nodes parameters configuration in cross-subject.

Model (retrained deep)	Memory size (MB)	Parameters optimized	Accuracy for test 1(%)	Accuracy for test 3 (%)	Training time (min)	evaluation time (s)
SqueezeNet	3	747633	75.788	75.803	22.541	29.263
AlexNet	228.8	57204593	74.545	74.051	22.730	27.619
Inception v3	98.2	24481346	81.985	81.528	87.567	63.738
DenseNet169	51.1	12566065	81.940	**82.651**	74.683	55.322
ResNet34	85.4	21309809	82.591	81.356	40.366	36.053
ResNet152	233.8	58244209	**83.347**	81.693	117.074	74.771
VGG13	516.6	129151601	79.096	78.871	77.244	52.718
VGG19	559.8	139770993	78.497	78.984	109.260	66.915

The two chosen data representations are used to train the neural network models.

Transfer learning can be compared to learning a language. For someone who already knows several languages, it is easier to learn a new one compared to someone who only speaks one language. Thanks to previous knowledge, the learning process is easier. It is the same for neural networks. We start with models designed for image classification. The network is already able to classify images, and with some adaptation of the parameters it is also capable to classify motion sequence images. The three kinds of retraining are tested with different models of image classification neural networks (Table 5). As expected, the accuracy is better when a deep retraining is applied. A training from scratch gives an accuracy between 62.278% and 77.779%. It is better than the accuracy obtained with a shallow retraining which is between 26.607% and 47.025%. The shallow retraining only retrains the last layers. In this case, the difference between the features for the images in the ImageNet dataset and our images representing motion sequences are considerable. It is important to note that the parameters of the simulation are kept the same between the different kinds of retraining, especially the number of epochs which indicates the number of training iterations over the dataset. Training time is smaller for a shallow retraining because only the last layers are retrained. Since the architecture of the models is kept between the different kinds of retraining, the memory required of each model stays the same size. The evaluation time also stays the same for the different kinds of retraining.

Table 5. Results obtained by a training from scratch, deep retraining or shallow retraining and the number of the test for the data representation used (Table 1).

	From Scratch			Deep Retraining			Shallow Retraining		
Models	Accuracy (%)	Training Time (min)	Test	Accuracy (%)	Training time (min)	Test	Accuracy (%)	Training time (min)	Test
SqueezeNet	65.399	23.187	1	75.803	22.541	3	36.389	18.054	3
AlexNet	68.917	22.692	3	74.545	22.647	1	30.147	16.907	3

Inception	75.189	87.291	1	81.985	87.246	1	30.043	40.110	3
DenseNet	77.636	74.458	1	82.651	74.683	3	47.025	34.276	3
ResNet34	77.779	40.195	1	82.591	39.620	1	41.105	22.523	3
ResNet152	72.547	117.145	3	83.317	115.029	1	40.985	46.783	3
VGG13	72.854	77.287	1	79.096	76.006	1	33.366	31.956	3
VGG19	72.337	109.427	1	78.984	109.26	3	26.712	40.226	1

5.3 Cross-subject vs Cross-view

One challenge in action recognition lies in the large variations of action representations. The same action varies depending on the subject or on the view angle [29]. In this section, we compare the capacities of the neural network to treat cross subject or cross view data. The NTU RGB+D database allows two types of action classification evaluation: cross-subject evaluation and cross- view evaluation. These two kinds of evaluation allow to test the robustness of the neural network classification for data with a different view angle and data with different subjects never seen by the neural network before.

All the data representation tested in Table 1 are tried for the cross subject and cross-view simulation on all the models. Table 6 refers the data representation which reach the highest accuracy obtain with the model in cross-subject and cross-view case. Between all the models, the highest accuracy is obtained with ResNet152 in both cross-subject and cross-view configuration. The cross-subject and cross-view do not reach the highest accuracy with the same data representation. It seems more relevant for the cross-subject configuration to encode (X, Y) position and the confidence score C using 14 or 18 nodes with this order (14, 15, 16, 17), 0, 1, 2, 3, 4, 5, 6, 7, 8, 9, 10, 11, 12, 13. At the opposite the higher accuracy for the cross-view configuration are obtained with the encoding of (X, Y) position and the mean value of (X, Y) using 14 nodes in the 2 possible orders: 0, 1, 2, 3, 4, 5, 6, 7, 8, 9, 10, 11, 12, 13 or 0, 4, 3, 2, 1, 5, 6, 7, 10, 9, 8, 11, 12, 13.

Table 6. The highest accuracy obtained for all the models in a cross-subject and cross-view simulation and the data representation used (Table 1).

	Cross-subject		Cross-view	
Models	Accuracy	Test	Accuracy	Test
SqueezeNet1 1	75.803%	3	81.031%	4
AlexNet	74.545%	1	79.157%	5
Inception v3	81.985%	1	87.365%	5
DenseNet169	82.651%	3	88.237%	6
ResNet34	82.591%	1	88.069%	4
ResNet152	**83.317%**	1	**88.780%**	6
VGG13	79.096%	1	85.155%	6
VGG19	78.984%	3	85.885%	6

6 Comparison with the State of the Art

In this section, we compare the results obtained in the state of the art with the same database: NTU RGB+D dataset [3]. This database contains several representations of the same data sequence. We divided the methods in three categories depending on the kind of data used: skeleton (Table 7), RGB video (Table 8) or both (Table 9). Accuracy obtained using skeleton data or RGB+skeleton data is respectively 1.3% and 0.4% higher than the accuracy obtained in our study. They obtain a better accuracy, but they use more complex data containing more information including depth. Nevertheless, for action recognition based on RGB video, we obtain (at the best of our knowledge) the highest accuracy which is 83.317% with ResNet152 in cross- subject. We compare our results with other methods using the NTU RGB+D dataset even if in our case we only use 49 individual actions instead of 60 actions including 11 mutual interactions.

Table 7. Action classification performance for different architectures using skeletal data as input of the neural network

Methods using skeleton data	Date	Accuracy cross-subject	Accuracy cross-view
two streams 3DCNN [30]	August 2015	66,85%	72,58%
LSTM+CNN [31]	July 2017	82,89%	90,10%
5 CNN in parallel [6]	May 2017	82,31%	
conversion into image + CNN [7]	March 2017	79,57%	84,83%
trajectories maps+CNN [10]	December 2016	76,32%	81,08%
conversion into image + CNN [13]	July 2017	75,2%	82,1%
conversion into image +CNN [2]	March 2017	74,27%	75,74%
conversion into image + CNN [14]	April 2017	**84,6%**	**90,9%**
conversion into image +CNN [15]	April 2017	83.2%	89.3%

Table 8. Action classification performance for different architectures using RGB videos as input of the neural network

Methods using RGB videos	Date	Accuracy cross-subject	Accuracy cross-view
body part segmentation +3CNN in parallel l[32]	April 2017	80,8%	
Our method (with ResNet152)	June 2018	data representation test 1 **83,317%**	data representation test 6 **88.780%**
Our method (with DenseNet)	June 2018	data representation test 3 82,651%	data representation test 4 88.237%

Table 9. Action classification performance for different architectures using both skeletal data and RGB videos as input of the neural network

Methods using RGB videos and skeleton data	Date	Accuracy cross-subject	Accuracy cross-view
RNN+CNN [33]	March 2017	**83,74%**	**93,65%**

7 Future Research

Important issues that must still be addressed in future work are scalability of action recognition systems with respect to vocabulary size, recognition in the presence of unknown actions, scenes containing multiple persons and interactions between multiples persons. The increase of number of actions to recognize, involves the increase of the complexity of the recognition and the database size. Only actions contained in the database can be recognized. In presence of an unknown action, the classifier will return a known action class with the highest similarity. In this work, we focus on individual actions, but it should be extended to scenes containing multiples persons and interactions between multiples persons.

8 Conclusion

As artificial intelligence grows, the possibilities and applications multiply in many fields and the action recognition domain is not spared. The recent breakthroughs made in deep learning for the human pose estimation have broadened the potential of the action recognition field by getting rid of the use of dedicated motion capture sensors.

This work demonstrates that the human pose data extracted from traditional RGB videos contain sufficient information to train a classifier capable of recognizing individual actions and obtaining performances similar or superior to the state of the art. These results pave the way for motion capture databases expansion as any video can be used without the use of depth or motion sensors.

OpenPose is a very recent technological leap and to ensure that the data extracted are reasonably accurate, a comparison was carried out between OpenPose, the Kinect v2 camera and the Qualisys system. The Kinect v2 is one of the most popular motion capture systems and has been used in many researches while the Qualisys provides a reliable and precise motion capture data. We concluded that OpenPose offers sufficiently reliable 2D skeleton data to train a machine learning model.

Based on an intensive study of the related work, we chose to use an image classifier based on deep neural networks and the NTU RGB+D database as our first set of data. We converted the motion sequences into RGB images to be able to use existing deep neural network models designed for image classification. We tested several image classifier models: SqueezeNet, AlexNet, DenseNet, Resnet, Inception, VGG. Different data representations have also been tested. The data representation with the encoding of (X, Y) coordinates and the confidence score of 18 joints into RGB channels gave the highest accuracy. Finally, different image classifiers have been tested. DenseNet and ResNet are the ones which gave the highest accuracy. The highest accuracy reached during this study is 83.317% in cross-subject and 88.780% in cross-view evaluation. In the latest state-of-the-art results, higher accuracy was reached with 3D skeleton data or with the fusion of RGB data and skeleton data (+1.3% and +0.4% respectively) due to the use of more complex data containing more information including depth information.

Acknowledgment

This work was partially financed by the institutes Numediart and Infortech of the University of Mons (UMONS). This work is also partially funded by the EU and Wallonia in the framework of the ERDF/DIGISTORM project.

References

1. Z. Cao, T. Simon, S.-E. Wei, and Y. Sheikh, "Realtime multi-person 2d pose estimation using part affinity fields," in *CVPR*, 2017.
2. S. Laraba, M. Brahimi, J. Tilmanne, and T. Dutoit, "3d skeleton-based action recognition by representing motion capture sequences as 2d-rgb images," *Computer Animation and Virtual Worlds*, vol. 28, no. 3-4, 2017.
3. A. Shahroudy, J. Liu, T.-T. Ng, and G. Wang, "Ntu rgb+d: A large scale dataset for 3d human activity analysis," in *The IEEE Conference on Computer Vision and Pattern Recognition (CVPR)*, June 2016.
4. Y. Du, Y. Fu, and L. Wang, "Skeleton based action recognition with convolutional neural network," in *Pattern Recognition (ACPR), 2015 3rd IAPR Asian Conference on*, pp. 579–583, IEEE, 2015.

5. F. Ofli, R. Chaudhry, G. Kurillo, R. Vidal, and R. Bajcsy, "Berkeley mhad: A comprehensive multimodal human action database," in *Applications of Computer Vision (WACV), 2013 IEEE Workshop on*, pp. 53– 60, IEEE, 2013.
6. Z. Ding, P. Wang, P. O. Ogunbona, and W. Li, "Investigation of different skeleton features for cnn-based 3d action recognition," in *Multimedia & Expo Workshops (ICMEW), 2017 IEEE International Conference on*, pp. 617–622, IEEE, 2017.
7. Q. Ke, M. Bennamoun, S. An, F. Sohel, and F. Boussaid, "A new representation of skeleton sequences for 3d action recognition," in *2017 IEEE Conference on Computer Vision and Pattern Recognition (CVPR)*, pp. 4570–4579, IEEE, 2017.
8. K. Yun, J. Honorio, D. Chattopadhyay, T. L. Berg, and D. Samaras, "Two-person interaction detection using body-pose features and multiple instance learning," in *Computer Vision and Pattern Recognition Workshops (CVPRW), 2012 IEEE Computer Society Conference on*, pp. 28–35, IEEE, 2012.
9. "Cmu dataset." http://mocap.cs.cmu.edu/. Acceded on 02-2018.
10. P. Wang, Z. Li, Y. Hou, and W. Li, "Action recognition based on joint trajectory maps using convolutional neural networks," in *Proceedings of the 2016 ACM on Multimedia Conference*, pp. 102–106, ACM, 2016.
11. "Msrc-12 kinect gesture dataset." https://www.microsoft.com/en-us/download/details.aspx?id=52283.Acceded on 02-2018.
12. V. Bloom, D. Makris, and V. Argyriou, "G3d: A gaming action dataset and real time action recognition evaluation framework," in *Computer Vision and Pattern Recognition Workshops (CVPRW), 2012 IEEE Computer Society Conference on*, pp. 7–12, IEEE, 2012.
13. C. Li, S. Sun, X. Min, W. Lin, B. Nie, and X. Zhang, "End-to-end learning of deep convolutional neural network for 3d human action recognition," in *Multimedia & Expo Workshops (ICMEW), 2017 IEEE International Conference on*, pp. 609–612, IEEE, 2017.
14. B. Li, Y. Dai, X. Cheng, H. Chen, Y. Lin, and M. He, "Skeleton based action recognition using translation- scale invariant image mapping and multi-scale deep cnn," in *Multimedia & Expo Workshops (ICMEW), 2017 IEEE International Conference on*, pp. 601–604, IEEE, 2017.
15. C. Li, Q. Zhong, D. Xie, and S. Pu, "Skeleton-based action recognition with convolutional neural networks," in *Multimedia & Expo Workshops (ICMEW), 2017 IEEE International Conference on*, pp. 597–600, IEEE, 2017.
16. [16]F. N. Iandola, S. Han, M. W. Moskewicz, K. Ashraf, W. J. Dally, and K. Keutzer, "Squeezenet: Alexnet-level accuracy with 50x fewer parameters and¡ 0.5 mb model size," *arXiv preprint arXiv: 1602.07360*, 2016.
17. G. Huang, Z. Liu, K. Q. Weinberger, and L. van der Maaten, "Densely connected convolutional networks," in *Proceedings of the IEEE conference on computer vision and pattern recognition*, vol. 1, p. 3, 2017.
18. "Skeleton return by openpose." https://arvrjourney.com/human-pose-estimation-using-openpose- with-tensorflow-part-2-e78ab9104fc8. Acceded on 02-2018.
19. A. Krizhevsky, I. Sutskever, and G. E. Hinton, "Imagenet classification with deep convolutional neural networks," in *Advances in neural information processing systems*, pp. 1097–1105, 2012.

20. K. He, X. Zhang, S. Ren, and J. Sun, "Deep residual learning for image recognition," in *Proceedings of the IEEE conference on computer vision and pattern recognition*, pp. 770–778, 2016.
21. K. He, X. Zhang, S. Ren, and J. Sun, "Identity mappings in deep residual networks," in *European Conference on Computer Vision*, pp. 630–645, Springer, 2016.
22. C. Szegedy, W. Liu, Y. Jia, P. Sermanet, S. Reed, D. Anguelov, D. Erhan, V. Vanhoucke, A. Rabinovich, *et al.*, "Going deeper with convolutions," Cvpr, 2015.
23. C. Szegedy, V. Vanhoucke, S. Ioffe, J. Shlens, and Z. Wojna, "Rethinking the inception architecture for computer vision," in *Proceedings of the IEEE Conference on Computer Vision and Pattern Recognition*, pp. 2818–2826, 2016.
24. C. Szegedy, S. Ioffe, V. Vanhoucke, and A. A. Alemi, "Inception-v4, inception-resnet and the impact of residual connections on learning.," in *AAAI*, vol. 4, p. 12, 2017.
25. K. Simonyan and A. Zisserman, "Very deep convolutional networks for large-scale image recognition," *arXiv preprint arXiv:1409.1556*, 2014.
26. M. Oquab, L. Bottou, I. Laptev, and J. Sivic, "Learning and transferring mid-level image representations using convolutional neural networks," in *Computer Vision and Pattern Recognition (CVPR), 2014 IEEE Conference on*, pp. 1717–1724, IEEE, 2014.
27. H.-C. Shin, H. R. Roth, M. Gao, L. Lu, Z. Xu, I. Nogues, J. Yao, D. Mollura, and R. M. Summers, "Deep convolutional neural networks for computer-aided detection: Cnn architectures, dataset characteristics and transfer learning," *IEEE transactions on medical imaging*, vol. 35, no. 5, pp. 1285–1298, 2016.
28. J. Yosinski, J. Clune, Y. Bengio, and H. Lipson, "How transferable are features in deep neural networks?," in *Advances in neural information processing systems*, pp. 3320–3328, 2014.
29. P. Zhang, C. Lan, J. Xing, W. Zeng, J. Xue, and N. Zheng, "View adaptive neural networks for high performance skeleton-based human action recognition," *arXiv preprint arXiv:1804.07453*, 2018.
30. H. Liu, J. Tu, and M. Liu, "Two-stream 3d convolutional neural network for skeleton-based action recognition," *arXiv preprint arXiv:1705.08106*, 2017.
31. C. Li, P. Wang, S. Wang, Y. Hou, and W. Li, "Skeleton-based action recognition using lstm and cnn," in *Multimedia & Expo Workshops (ICMEW), 2017 IEEE International Conference on*, pp. 585–590, IEEE, 2017.
32. M. Zolfaghari, G. L. Oliveira, N. Sedaghat, and T. Brox, "Chained multi-stream networks exploiting pose, motion, and appearance for action classification and detection," in *Computer Vision (ICCV), 2017 IEEE International Conference on*, pp. 2923–2932, IEEE, 2017.
33. R. Zhao, H. Ali, and P. van der Smagt, "Two-stream rnn/cnn for action recognition in 3d videos," *arXiv preprint arXiv:1703.09783*, 2017.

Deep learning review and discussion of its future development

Zhiying Hao*

University of Electronic Science and Technology of China, No.4, Block 2, North Jianshe Road, Chenghua District, Chengdu, Sichuan, China

Abstract. This paper is a summary of the algorithms for deep learning and a brief discussion of its future development. In the first part, the concept of deep learning and the advantages and disadvantages of deep learning are introduced. The second part demonstrates several algorithms for deep learning. The third part introduces the application areas of deep learning. Then combines the above algorithms and applications to explore the subsequent development of deep learning. The last part makes a summary of the full paper.

1 Introduction

As early as 1952, IBM's Arthur Samuel designed a program for learning checkers. It can build new models by observing the moves of the pieces and use them to improve their playing skills. In 1959, the concept of machine learning was proposed as a field of study that could give a machine a certain skill without the need for deterministic programming. In the process of machine learning development, various machine learning models have been proposed, including deep learning. Due to its complicated structure and the need for a large amount of calculation, the computing cost is very high, so it had not been paid attention to at the beginning. However, with the great improvement in computer performance, the excellent performance of deep learning makes it rose rapidly and has become one of the hottest research areas. In this paper, the main deep learning models will be briefly summarized and the development prospects of deep learning will be analyzed and discussed at the end.

2 Introduction to deep learning

2.1 What is Deep Learning

Deep learning is a branch of machine learning [1]. It is an algorithm that attempts to use the high-level abstraction of data using multiple processing layers consisting of complex structures or multiple nonlinear transforms. In machine learning, deep learning is an algorithm based on characterizing learning data. The concept of deep learning is relative to shallow learning. Shallow machine learning models such as Support Vector Machines and

* Corresponding author: zhiyinghao@std.uestc.edu.cn

Logistic Regression were introduced in the 1990s. These shallow machine learning models have only one layer or no hidden layer nodes, as shown in Fig 1. Deep learning is based on multiple hidden layer nodes. The essence of deep learning is multi-layer neural network. Deep learning uses the input of the previous layer as the output of the next layer to learn highly abstract data features.

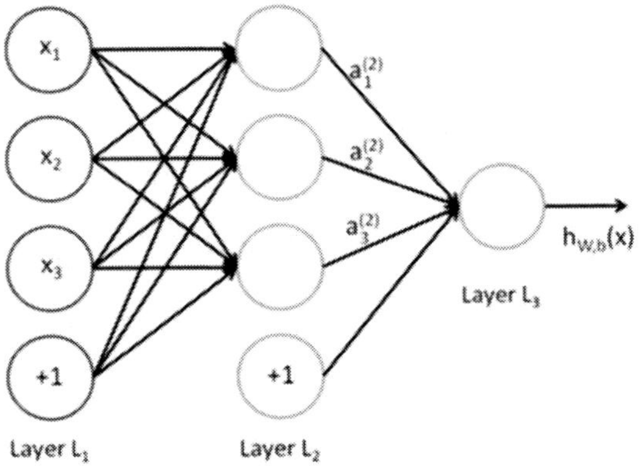

Fig. 1. A single-layer neural network

Like machine learning, deep learning can be categorized into supervised learning, semi-supervised learning, and unsupervised learning. At present, the classical deep learning framework includes Convolutional Neural Networks, Restricted Boltzmann Machines [2], Deep Belief Networks [3], and Generative Adversarial Networks [4]. In the next section, these algorithms will be introduced briefly.

2.2 Advantages and disadvantages of deep learning

Deep learning has shown better performance than traditional neural networks. After a deep neural network is trained and properly adjusted for certain task like image classification, it saves a lot of calculations, and can complete a lot of work in a short time. . Deep learning is also malleable. Usually, for traditional algorithms, if you need to adjust the model, you may need to make copious changes to the code. For the determined network framework used for deep learning, if you need to adjust the model, you only need to adjust the parameters, thus deep learning has great flexibility. The deep learning framework can be continuously improved and then reached the almost perfect state. Deep learning is also more general, it can be modelled based on problems, not limited to a fixed problem.

Deep learning has some shortcomings as well. First of all, its training cost is relatively high. Now, the performance of computer hardware has been improved a lot, and some simple neural networks can be trained on some of the common computing modules. However, the training of some more complex neural networks still requires relatively expensive high-performance computing modules. Although the price of such modules has been greatly reduced compared with the previous ones, the demand for such hardware still makes the training cost of deep learning relatively high. At the same time, not only the economic cost, the training of neural networks requires a large amount of data to be trained to achieve a satisfactory level, but it is often difficult to obtain a sufficient amount of data. Secondly, deep learning can't directly learn knowledge. Although models such as AlphaGo Zero can learn without prior knowledge have emerged, most deep learning frameworks still need to rely on manual feature marking for training. The workload is enormous for marking large-scale datasets, which also increases the training cost of deep learning. Another point is that deep learning lacks sufficient theoretical support. Although deep learning has achieved good results in various application fields, there is still no complete and rigorous theoretical derivation to explain the deep learning model at this stage, which limits the follow-up study and the improvement of deep learning.

3 Main Deep Learning Algorithm Introduction

3.1 Convolutional Neural Network

The convolutional neural network, as seen in Fig 2, is a feedforward neural network whose convolution operation allows its neurons to cover peripheral units within the convolution kernel and has excellent performance in large image processing. A convolutional neural network typically consists of one or more convolutional layers and a fully connected layer, which also includes a pooling layer for integration. Convolutional neural networks give better results in terms of image and speech recognition. It requires fewer parameters to consider than other deep neural networks. The advantages of convolutional neural networks make it one of the most commonly used deep learning models. The basic structure of the convolutional neural network is briefly introduced below.

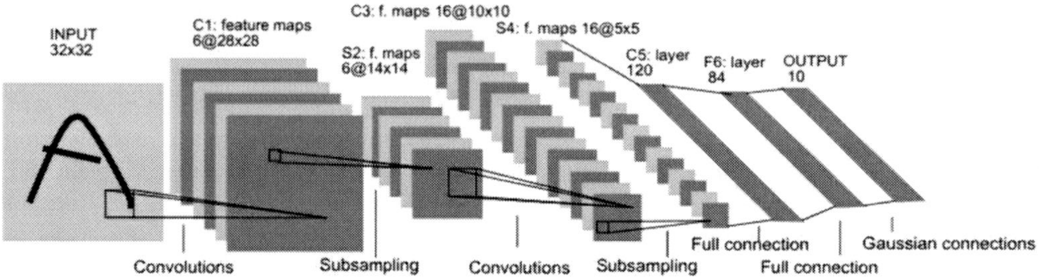

Fig. 2. Convolutional Neural Network, LeNet-5[5]

3.1.1 Convolutional layer.

The convolutional neural network convolves data using multiple convolution kernels in the convolutional layer to generate a plurality of feature maps corresponding to the convolution kernel.

The convolution operation has the following advantages:

1. The weight sharing mechanism on the same feature map reduces the number of parameters;
2. Local connectivity enables convolutional neural networks to take into account the characteristics of adjacent pixels when processing images;
3. There is no object in the image recognition due to the position of the object on the image.

These advantages also make it possible to use a convolutional layer instead of a fully connected layer in some models to speed up the training process.

3.1.2 Pooling layer

After obtaining the features by convolution, we hope to use these features to do the classification. However, the amount of data that is often obtained is very large, and it is prone to over-fitting. Therefore, we aggregate statistics on features at different locations. This aggregation operation is called pooling. In the convolutional neural network, the pooling layer is used for feature filtering after image convolution to improve the operability of the classification.

3.1.3 Fully connected layer

After pooling layer is the fully connected layer, its role is to pull the feature map into a one-dimensional vector. The working mode of the fully connected layer is similar to that of a traditional neural network. The fully connected layer contains parameters in approximately 90% of the convolutional neural network, which allows us to map the neural network forward into a vector of fixed length. We can grant this vector to a particular image class or use it as a feature vector in subsequent processes.

3.2 Deep Belief Network

The deep belief network is a probability generation model. Compared with the neural network which is a traditional discriminative model, the generated model is to establish a joint distribution between observation data and labels, and to evaluate both P (Observation|Label) and P (Label|Observation) while the discriminative model has only evaluated the latter, that is, P (Label|Observation).

The deep confidence network consists of multiple restricted Boltzmann layers, a typical neural network type as shown. These networks are "restricted" to a visible layer and a hidden layer, with connections between the layers, but there are no connections between the cells within the layer. The hidden layer unit is trained to capture the correlation of higher order data represented in the visible layer.

3.3 Restricted Boltzmann Machine

A Restricted Boltzmann Machine is a randomly generated neural network that can learn the probability distribution through the input data set. It is a Boltzmann Machine's problem, but the qualified model must be a bipartite graph. The model contains visible cells corresponding to the input parameters and hidden cells corresponding to the training results. Each edge of the figure must be connected to a visible unit and a hidden unit. In contrast, the Boltzmann machine (unrestricted) contains the edges between hidden cells, making it a recurrent neural network. This limitation of the constrained Boltzmann machine makes it possible to have a more efficient training algorithm than the general Boltzmann machine, especially for the gradient divergence algorithm.

The Boltzmann machine and its model have been successfully applied to tasks such as collaborative filtering, classification, dimensionality reduction, image retrieval, information retrieval, language processing, automatic speech recognition, time series modeling, and information processing. Restricted Boltzmann machines have been used in dimensionality reduction, classification, collaborative filtering, feature learning, and topic modeling. Depending on the task, the restricted Boltzmann machine can be trained using supervised learning or unsupervised learning.

3.4 Generative Adversarial Network

The Generated Adversarial Network was proposed in 2014. The Generative Adversarial Network uses two models, a generative model and a discriminative model. The discriminative model determines whether the given picture is a real picture, and the generative model creates a picture as close to the ground truth as possible. The generated model is designed to generate a picture that can spoof the discriminative model, and the discriminative model distinguishes the picture generated by the generated model from the real picture. The two models are trained at the same time, and the performance of the two models becomes stronger and stronger in the confrontation process between the two models, and will eventually reach a steady state.

The use of generating a network is very versatile, not only for the generation and discrimination of images, but also for other kinds of data.

4 Deep Learning Application

4.1 Image processing

Manually selecting features is a very laborious approach. Its adjustment takes a lot of time. Due to the instability of manual selection, we consider letting the computer automatically learn the features. The automatic learning of the computer can be realized by deep learning.

In image recognition, deep learning utilizes patterns of multi-layer neural networks to pre-process, feature extract, and feature processing images.

Taking the convolutional neural network as an example, the convolutional neural network establishes a multi-layer neural network, and uses the convolutional layer to perform convolution operations to extract feature values, and then performs data processing and training through the pooling layer and the fully connected layer. The detailed process is explained in detail in the Technical Implementation section of 2.2 Neural Network below.

Although neural network image recognition can't reach the accuracy of the human eye at present, the neural network can process a large amount of image data, and the efficiency is much better than manual recognition. Facing huge amount of data that cannot be processed manually, using the neural network method will lead to magnificent improvement.

In addition, deep learning provides an idea for face recognition technology. Face recognition is a biometric recognition technology based on human facial feature information for identification. Face recognition products have been widely used in finance, justice, military, public security, border inspection, government, aerospace, electric power, factories, education, medical care and many enterprises and institutions. And with the further maturity of technology and the improvement of social recognition, face recognition technology will be applied in more fields and has an expectable development prospects. The characteristics of the neural network make it possible to avoid overly complex feature extraction when applied to face recognition, which is beneficial to hardware implementation.

4.2 Audio data processing

Deep learning has a profound impact on speech processing. Almost every solution in the field of speech recognition may contain one or more embedding algorithms based on neural models.

Speech recognition is basically divided into three main parts, namely signal level, noise level and language level. The signal level extracts the speech signal and enhances the signal, or performs appropriate pre-processing, cleaning, and feature extraction. The noise level divides the different features into different sounds. The language level combines the sounds into words and then combines them into sentences.

In the signal level, there are different techniques for extracting and enhancing the speech itself from the signal based on the neural model. At the same time, it is able to replace the classical feature extraction method with a more complex and efficient neural network-based method, which greatly improves the efficiency and accuracy. Noise and language levels also include a variety of different depth learning techniques, and different types of neural model-based architectures are used in both sound level classification and language level classification.

5 Discussion of the future development of deep learning

5.1 Representation Learning

The core of deep learning is the abstraction and understanding of features. Therefore, feature learning plays a very important role in deep learning. Since the essence of deep learning is a multi-layer neural network, some useful information is lost in the process of extracting features and transmitting to the lower layer. However, if too much image features are extracted, it may lead to over-fitting. Therefore, the study of representational learning may be one of the core research issues in deep learning research studying how to accurately extract the required features while avoiding over-fitting. The research progress on this problem will be of great help to the classification and generalization task of neural networks.

5.2 Unsupervised Learning

As mentioned above, training a supervised neural network requires a large amount of labelled data. The workload is very large, adding a lot of extra cost to the training of the neural network. Therefore, if the machine completes the work instead of human, the cost of network training will be greatly reduced. Unsupervised learning can also be used not only for the classification of markers, but also for the Go evaluation program such as AlphaGo Zero. The emergence of AlphaGo Zero proves that in some applications, even without the foundation of human prior knowledge, machines can achieve excellent training results. In this case, the application of unsupervised learning in some areas to automate the learning of the machine from the human knowledge base without the limitations of current state of the art, may contribute to the updating and breakthrough of technology in these fields. Moreover, the current research on unsupervised learning is not intensive now while most people's research focuses on supervised learning. Therefore, unsupervised learning has a rich research potential. In fact, unsupervised learning has recently become one of the hottest research areas. In my opinion, unsupervised learning is also one of the most valuable directions for deep learning in the future.

5.3 Theory Complement

One of the shortcomings of deep learning is that there is no complete theoretical support, which brings a lot of controversy to deep learning and the lack of theory does hinder the development of deep learning. With the increasing attention of deep learning this year, the research on deep learning has become more and more in-depth, and the theory of deep learning is constantly improving.

Nevertheless, the theory behind it is still not enough to rigorously prove the inner principles of deep learning. At present, the research relies on part of the theory combined with the actual test and the experiment-based research method. While the theory can't get further breakthroughs, it only depends on adjusting parameters to improve the models performance, which may easily lead to the bottleneck of the research.

Therefore, it seems that for the future development of deep learning, it is very important to obtain complete theoretical support. In the process of research, the theory of deep learning needs to be continuously improved, and finally reach a level sufficient to explain the structure of the inner principle.

5.4 Perspective of Deep Learning Application

The fourth part of this paper referred to two application areas of deep learning, image recognition and speech processing, which are the two main application areas of deep learning. In addition, recent deep learning has also been used in natural language processing.

Specific applications include, for example, autonomous driving, intelligent dialogue robots such as Siri, image classification, medical image processing, etc. Different deep learning frameworks tend to have slightly different application scenarios, such as convolutional neural networks, which are mainly used for image processing. In the work of medical image processing, for example, the use of convolutional neural networks for brain tumour segmentation has achieved an accuracy of more than 90%. Also, in medical applications, convolutional neural networks can be used to recognize Alzheimer's disease brain image, and more accurate diagnostic results can be obtained combined with manual judgement. Medicine is only a part of deep learning applications. In the application of deep learning, well-trained machines can often calculate some details that are hard to be obtained by human, saving people's workload and improving the quality of results. For example, in order to prevent the traffic light violation at the intersection, the usual way is to take picture by the camera and then manually recognize the license plate to give subsequent punishment. Manually viewing the images and recording them is a very boring job and is not so efficient. If the captured image is recognized by the deep neural network, and the license plate number entry system is automatically extracted, it not only saves manpower, but also improves the efficiency greatly.

And so on, I believe that deep learning applications will be developed in the future in transportation, medical, language, automation, etc., not to mention examples here. Although it seems that it can't completely replace human work now, deep learning and artificial combination can greatly improve the work efficiency.

6 Conclusion

In this paper, we introduced some of the main algorithms and some conjectures for future development of deep learning. Deep learning has already had in-depth research and a wide range of application scenarios, and has been put into practical use in real life with excellent performance. However, there's still a lot to exploit in the area of deep learning and neural network and it has great follow-up research space and considerable application potential.

References

1. Guo Y , Liu Y , Oerlemans A , et al. Deep learning for visual understanding: A review[J]. Neurocomputing, 2016, 187(C):27-48.
2. Chun-Xia Z , Nan-Nan J I , Guan-Wei W . Restricted Boltzmann Machines[J]. Chinese Journal of Engineering Mathematics, 2015.
3. Hinton G E , Osindero S , Teh Y W . A Fast Learning Algorithm for Deep Belief Nets[J]. Neural Computation, 2014, 18(7):1527-1554.
4. Goodfellow I J , Pouget-Abadie J , Mirza M , et al. Generative Adversarial Networks[J]. Advances in Neural Information Processing Systems, 2014, 3:2672-2680.
5. Y. LeCun, L. Bottou, Y. Bengio, and P. Haffner. (1998). "Gradient based learning applied to document recognition". Proceedings of the IEEE, 86(11):2278–2324.

Image quality classification algorithm based on InceptionV3 and SVM

Yu Li[1], and *Lizhuang* Liu[2,*]

[1]School of Mechanical and Electrical Engineering and Automation, Shanghai University, Shanghai, China
[2]Shanghai Advanced Research Institute, Chinese Academy of Sciences, Shanghai, China

Abstract. In this work we investigate the use of deep learning for image quality classification problem. We use a pre-trained Convolutional Neural Network (CNN) for image description, and the Support Vector Machine (SVM) model is trained as an image quality classifier whose inputs are normalized features extracted by the CNN model. We report on different design choices, ranging from the use of various CNN architectures to the use of features extracted from different layers of a CNN model. To cope with the problem of a lack of adequate amounts of distorted picture data, a novel training strategy of multi-scale training, which is selecting a new image size for training after several batches, combined with data augmentation is introduced. The experimental results tested on the actual monitoring video images shows that the proposed model can accurately classify distorted images.

1 Introduction

A large number of surveillance cameras are distributed in the public areas of our daily life. The digital images captured by the camera are transmitted through a wired bandwidth channel, and are susceptible to human and natural factors so as to distort images after a series of processes such as acquisition, compression, processing, transmission and display. The detection of the distortion type of the video image is of great significance to the maintenance of video surveillance system. The current detection methods are mostly based on appropriate handcraft features for specific distortion types. Wang et al proposed a connected-domain pixel ratio algorithm for signal loss anomaly detection and a detection algorithm of Definition of Exception (DOE) [1]. Cheng designed an average gradient detecting algorithm based on modified focus window method to detect blur images and a covariance correlation function based color cast detecting algorithm [2]. Xia et al. presented a quality diagnosis algorithm based on median filtering and frame differential method using the image features of grayscale and gradient value [3].

Convolutional Neural Network is one of the most effective methods to solve image classification problems. Unlike traditional machine learning methods which require handcraft features, CNN can automatically learn the characteristics of data, greatly reducing the dependence of feature selection. Wu et al. introduce a CNN based image classifier

* Corresponding author: liulz@sari.ac.cn

which took small patches segmented from video images as input and introduced the positive and negative sample equalization and an adaptive learning rate to reduce overfitting. Meanwhile, the class of an image is decided using the majority voting rule [4]. In this work we focus on the more complex image quality classification problem in which more distortion types are considered and of which the model is tested directly on monitoring images collected in daily life. The following aspects are studied: 1) Introduce a new image quality classification algorithm combined with convolutional neural network and SVM; 2) Comparing the ability of different deep CNN architectures to extract image quality features; 3) Analyze the features extracted from different processing units in deep convolutional neural networks; 4) Propose a training strategy of multi-scale training to improve the generalization ability.

2 Image quality classification algorithm

The framework of the proposed algorithm for distortion image detection problem is shown in figure 1. We use the pre-trained CNN model on the ImageNet dataset as the base model, and fine-tune it with the surveillance image dataset. The fine-tuned CNN model has two purposes: 1) The model without the final layer is employed as an image describer, from which the features extracted is used as the training data for a SVM model, with Bayesian algorithm to optimize the characteristics. The trained SVM model is used as the image quality classifier; 2) The model with the softmax layer as the final layer is directly trained as the classifier for comparison.

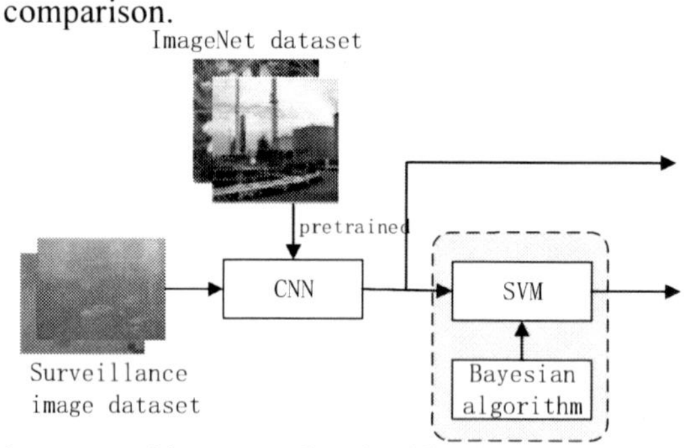

Fig. 1. Flow chart of the proposed image quality classification algorithm.

2.1 Image description using CNN

The generic descriptors extracted from the CNN model are of remarkable generalization capabilities when the model is trained on large scale labeled datasets, and outperforms descriptors of hand crafted, state-of-the-art systems in many image classification tasks. The classic CNN models such as VGGNet [5], GoogLeNet [6], ResNet [7], etc. have shown great classification effects. In this paper, we adopt the InceptionV3 model as the feature extractor, because of its outstanding performance and good generalization ability.

In order to improve the performance of deep CNN model, the common strategy is to increase the depth and width of the network, but this will increase the number of network parameters, which will lead to the over-fitting problem and requires a large amount of computing resources. To solve this problem, the GooLeNet model introduces the Inception architecture, of which the main idea is to approximate the optimal local sparse structure through dense components, so as to maintain the sparseness of the network structure and the high computational performance of the dense matrix. The core of the Inception architecture is three Inception modules, the structure of which is shown in figure 2. The procedures of the Inception modules are as follow: dimensions (or the input channel numbers) of the feature map of the previous layer are reduced by the first 1x1 layers; then, extract the features through the 1x1, 1xn, 3x3 convolutional layers; the last layer is the Filter Concat layer, consisting of an LRN (Local Response Normalization) layer and a DepthConcat layer, wherein the DepthConcat layer fuses the features extracted by convolutional layers. The number of feature channels will increase as the network deepens, enlarging the proportion of convolution kernels in the Inception architecture. If large convolutional kernels such as 3x3,5x5 are selected, the amount of computation will increase significantly. Therefore, the convolutional layers are mostly small convolutional kernels such as 1x1,1xn, the effective receptive fields of which may enlarge to capture global picture characteristics by stacking multiple convolutional layer, meanwhile which will reduce the compute. The Inception architecture also uses an average pooling layer instead of a fully connected layer, and add two auxiliary softmax layers for the forward propagation to avoid vanishing gradient problem.

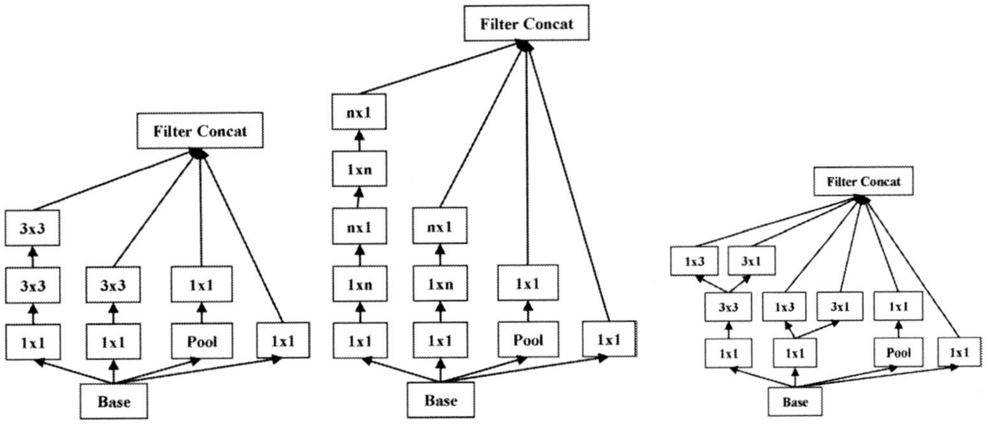

Fig. 2. The Structure of three Inception modules.

The very deep CNN models trained in large datasets offer a powerful classification capability, however, the effect will be limited by a lack of adequate data such as distorted surveillance images [8]. In this paper, the fine-tuning method is introduced to copy with the problem [9]. Fine-tuning has been applied in a variety of different medical imaging modality [10] and image quality evaluation tasks [11], the result of which shows fine-tuning is as effective as training a CNN from scratch while being more robust to the small training data. Compared with directly training a CNN model with small datasets, the fine-tuned model is more robust and is more easy to train. The CNN model we used were pre-trained on the ImageNet natural image dataset. We adapted the CNNs to our problem by replacing the top layers from the last global average pooling layer to the softmax layer with a new layer for the 11 classes in our dataset. The initial CNN weights derived from the ImageNet dataset were fine-tuned through back-propagation so that they can better reflect the modalities in the surveillance image dataset. We used mini-batch stochastic gradient descent to find the optimal weights.

2.2 Multi-scale training

The performance of a deep CNN model generally depends heavily on the size of the available training datasets. However, the collected available surveillance image data is insufficient. Common strategies for attacking this problem are data augmentation technique, which seek to multiply the dataset by rotating, cropping, reflection, and so on. While generating more picture content is simple, ensuring adequate distortion diversity and realism is the harder thing which should be taken into account.

We also resize pictures to generate different size of images not only to augment the dataset but also to make the model robust to running on images of different sizes. Instead of fixing the input image size, we randomly choose a new image dimension size to train the CNN model after several iterations. This regime forces the network to learn to predict well across a variety of input dimensions, making the model capable of predicting at different resolution.

2.3 Support vector machine

Support vector machine is a machine learning method proposed based on the structural risk minimization principle by Vapnik et al [12], which has good generalization ability and is suitable for small sample classification problems. The idea of the SVM is to find the optimal margin separating hyperplane with the maximum classification boundary. In this paper, the SVM is applied to classify the features extracted by the CNN model. In this paper, there are multiple image distortion type need to be detected. The SVM model is adopted with the one-against-one method for the multiple classification problem [13]. The idea of the method is to train a binary classifier between every two kind of image samples. While testing, the label of the test sample is given by these classifiers and the sample is classified into the class with the most votes. Therefore, $k(k-1)/2$ classifiers are constructed where k is the number of classes. For training data from the i th and the j th classes, we solve the following binary classification problem:

$$\begin{aligned} \min_{w^{ij},b^{ij},\xi^{ij}} \quad & \frac{1}{2}\| w^{ij} \|^2 + C\sum_t \xi_t^{ij} \\ s.t. \quad & (w^{ij})^T \phi(x_t) + b^{ij} \geq 1 - \xi_t^{ij},\ if\ y_t = i \\ & (w^{ij})^T \phi(x_t) + b^{ij} \leq -1 + \xi_t^{ij},\ if\ y_t = j \\ & \xi_t^{ij} \geq 0. \end{aligned} \tag{1}$$

where ξ is the slack variable, C is the penalty parameter, and $\phi(x_t)$ is kernel function, which can implicitly map the input feature space to a high-dimensional space, making the features easier to classify without increasing the computational complexity, and effectively avoiding the dimensional disaster. Commonly used kernel functions are as follows: linear kernel, polynomial kernel, radial basis function (RBF) kernel and sigmoid kernel function. Then the decision function is

$$f_n = \mathrm{sgn}((w^{ij})^T \phi(x) + b^{ij}) \tag{2}$$

where f_n is the result of the n th classifier. Input the test sample X_{test} to these classifiers and calculate the votes. Given the numbers of votes of s th class is $\mathrm{D}_s(\mathrm{X}_{test})$, the final classification result is

$$f = \underset{s=1,\dots,k(k-1)/2}{\arg\max}\ \mathrm{D}_s(\mathrm{X}_{test}) \tag{3}$$

We select the RBF function as the kernel function with two hyper-parameters ξ, C. The Bayesian optimization algorithm based on Gaussian process is applied to find the optimal hyper-parameters, where the Gaussian process(GP) is used to model the distribution of the objective function and the Expected Improvement(EI) function is used as the acquisition function to trade off the mean and the variance, the optima are located where the uncertainty in the surrogate model is large and/or where the model prediction is high [14]. Bayesian optimization then select the next query point by maximizing such acquisition functions. By iterating this process, the GP and the hyper-parameters are updated until the optima are found.

3 Experiments

In the experiments, the surveillance image dataset collected in daily life is utilized, consisting of 7 kinds of natural images shown in figure 3, such as color casting, blur, luminance anomaly, brightness anomaly, snow noise, stride noise and normal pictures.

(a) color casting (b) blur (c) signal loss

(d) luminance anomaly (e) snow noise (f) stride noise

Fig. 3. Distorted surveillance images.

The dataset contains 2355 distorted and normal images with different resolutions and is randomly divided into two parts, 80% of which is the training set and 20% as the test set to evaluate the proposed image quality classification model. We adopt three indices as the criteria for evaluating the performance of the surveillance image metrics, which are accuracy, missed diagnosis rate(MDR) and wrong diagnosis rate(WDR). The consistent result means that the classification result of a picture matches its label. If a distorted picture is recognized as a normal image, the classification result is recorded as miss diagnosis. If the type of a distorted picture is misjudged or the normal picture is recognized as a distorted image, the classification result is recorded as misdiagnosis. Given the consistent number a , missed diagnosis number m and wrong diagnosis number w , the accuracy r_a , missed diagnosis rate r_m and wrong diagnosis rate r_w can be expressed as

$$r_a = a / (a + m + w) \tag{4}$$

$$r_m = m / (a + m + w) \tag{5}$$

$$r_w = w / (a + m + w) \tag{6}$$

4 Results

Three CNN architecture of VGG16, Inceptionv3, and ResNet50 are all tested in the experiments, and they are trained in the original dataset and the data-augmented dataset respectively. The performance of these models are shown in table 1. Due to the number of the original dataset is limited, the models trained on this dataset cannot effectively extract the image feature, thus, the accuracy of these models are all worse than that of the model trained in the data-augmented dataset. Specifically, the accuracy of the augVgg16 classifier is improved by 13.24%, the augInceptionV3 classifier improved by 14.18 and the augResNet50 improved by 13%.

Table 1. Test results of CNN classifiers.

model	accuracy (%)	MDR(%)	WDR(%)
Vgg16	74.29	17.92	7.79
InceptionV3	75.26	16.63	8.31
ResNet50	74.29	16.10	9.61
augVgg16	87.53	4.89	7.58
augInceptionV3	89.24	4.40	6.36
augResNet50	87.29	5.87	6.85

The above results also show that the augInceptionV3 classifier has the best performance among all the CNN models, the features extracted by which can best describe the characterization of the surveillance image dataset. Therefore, we select the augInceptionV3 model as the base model of the proposed image quality classifier.

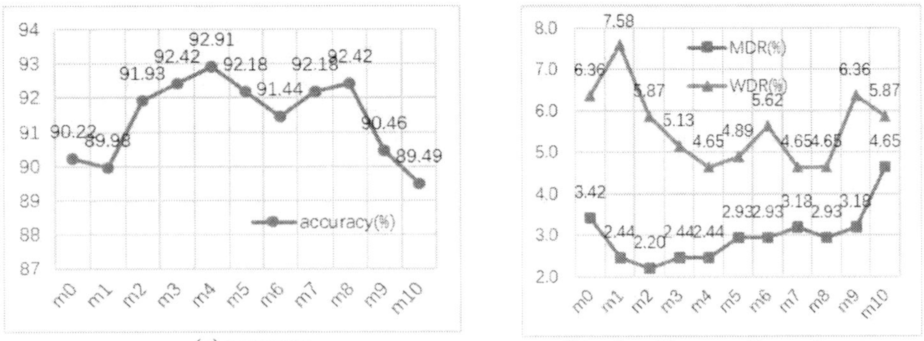

(a) accuracy (b) missed/wrong diagnosis rates

Fig. 4. Results of classifiers with features extracted by different Inception modules.

There are 46 layers in the InceptionV3 model, consisting of 11 Inception modules. In order to analyze the influence of different processing units on the performance of the image quality classifier, we output features extracted by every Inception module to train the SVM model. The classification result of different features is shown in figure 4. The results show that the performance of the higher layer is the worst and the middle layers produce the best accuracy. The accuracy of the 5th module which has the best performance is 3.42% higher than that of the 11st module. This might due to the fact that the lower layers of the CNN model focus on the texture features of an image such as color, edge and corner point, and the higher layers captures higher abstraction of image content which is not sensitive to the image quality. On the other hand, as is shown in the results, the classifier combined the InceptionV3 model with the SVM performs better than the simple CNN model.

5 Conclusion

The results testing on the surveillance images yield a solid evidence that the proposal model based on InceptionV3 and SVM is suitable for the image quality classification task, which has very high classification accuracy, low missed and wrong diagnosis rate. Multiscale training and data augmentation technique can greatly improve the performance of the CNN model. We also analyze the effect of features extracted by different Inception modules, the experimental shows that the middle module especially the 5th module can learn the image quality feature better.

References

1. Wang W H, Liu Y S. Image anomalies analysis and detection algorithm for surveillance video[J]. Journal of Central China Normal University (Natural Sciences), 2015, 49(05): 692-695.
2. Cheng C S. Design and implementation of the real-time detection system for surveillance video quality[D]. Zhejiang University of Technology, 2014.
3. Xia Y J, Sun H. Anomaly detection and quality diagnosis of surveillance video[J]. Computer Applications and Software, 2016, 33(06): 163-167+211.
4. Wu M Y, Chen L, Tian J. Video image distortion detection and classification based on CNN[J]. Application Research of Computers, 2016, 33(09): 2827-2830.
5. Simonyan K, Zisserman A. Very deep convolutional networks for large-scale image recognition[J]. arXiv preprint arXiv:1409.1556, 2014.
6. Szegedy C, Liu W, Jia Y, et al. & RABINOVICH, A. Going deeper with convolutions[C]//Proceedings of the IEEE conference on computer vision and pattern recognition, 2015:1-9.
7. He K, Zhang X, Ren S, et al. Deep residual learning for image recognition[C]//Proceedings of the IEEE conference on computer vision and pattern recognition, 2016: 770-778.
8. Shin H C, Roth H R, Gao M, et al. Deep convolutional neural networks for computer-aided detection: CNN architectures, dataset characteristics and transfer learning[J]. IEEE transactions on medical imaging, 2016, 35(5): 1285-1298.
9. Donahue J, Jia Y, Vinyals O, et al. Decaf: A deep convolutional activation feature for generic visual recognition[C]//International conference on machine learning. 2014: 647-655.
10. Tajbakhsh N, Shin J Y, GURUDU S R, et al. Convolutional neural networks for medical image analysis: Full training or fine tuning?[J]. IEEE transactions on medical imaging, 2016, 35(5): 1299-1312.
11. Gao F, Wang Y, Li P, et al. DeepSim: Deep similarity for image quality assessment[J]. Neurocomputing, 2017, 257: 104-114.
12. Vapnik V. Statistical learning theory[M]. New York: Wiely, 1998.
13. Hsu C W, Lin C J. A comparison of methods for multiclass support vector machines[J]. IEEE Transactions on Neural Networks, 2002, 13(4):1026.
14. Shahriari B, SWERSKY K, Wang Z, et al. Taking the human out of the loop: A review of bayesian optimization[J]. Proceedings of the IEEE, 2016, 104(1): 148-175.

MATEC Web of Conferences **277**, 02037 (2019) https://doi.org/10.1051/matecconf/201927702037
JCMME 2018

Emotions detection scheme using facial skin temperature and heart rate variability

Kahil Mustafa Jamal S*, and *Eiji* Kamioka

Graduate School of Engineering and Science, Shibaura Institute of Technology, Tokyo, Japan

Abstract. Technology nowadays is aiming to provide a better life quality for people, schools and universities are working for the convenient of the students as well as ensuring a high quality of education is attained. Emotions detections system can be a solution for better education results and may also be used as a part of human-computer interaction applications such as robotics, games, and intelligent tutoring system, This study shows potentials method of detecting emotions using mobile computing to recognize and identify emotions (Relax, Fear, Sadness, and Joy) based on facial skin temperature, more specifically 5 spots on the face, Nose, Glabellar line (between the eyes and eyebrows) right\lift cheeks and the chin, in addition to the Heart Rate Variability (HRV). An experiment was conducted with 20 healthy subjects (10 females and 10 males, 20 to 31 years old), Both visual and auditory media were used to induce these emotions in the experiment. By the end of this paper, the output data will be anglicized by an Artificial neural network (ANN) The Multilayer Perceptron (MLP) was selected as a classifier with a result of 88.75 % accuracy. This mechanism proves that human`s emotions can easily identify without physical interaction with the subject and with high reliability with only 0.11 misprediction rate

1 Introduction

In the last decade, [1-12] people have developed serval methods to detect and analyse emotions through different models such as facial expressions, walking gestures, and speech, However, these methods are tending to misjudgement in the emotions detection due to humans restraining form showing their true emotions in public, especially negative emotions and that goes under the social mask theory[1]. Another factor to misclassify these emotions is the using the average of the results as well as normalizing/standardizing methods may lead to high accuracy however in the same time does not take the individual differences in concedes as every person is different from each other. On the other hand, Emotion recognition using physiological signals[2] has shown its reliability of detection different kind of emotions. Take in consideration the Atomic Nerve System (ANS), ANS has a strong connection with the human's emotions[2,7], both sympathetic and parasympathetic outflow which control and regulates the functions of our internal organs (the viscera) such as the heart rate, Heart rate variability, blood pressure and body

* Corresponding author: mg17004@shibaura-it.ac.jp

temperature. From the observation, heart rate variability and facial skin temperature are on the top of the best parameters to detect emotions, facial skin temperature can be easily observed and analysis by using thermal camera, for observing the HRV, an HRV monitor was used to collect the HRV score which was calculated and provided by Ellet.HRV, Essence, human emotions can be easily identified using facial skin temperature and heart rate variability.

2 Related work

Microsoft [4] has a working emotion detecting system. This system is called emotions Application Programming Interface (API), and it can detect anger, contempt, disgust, fear, happiness, neutral, sadness and surprise. The only disadvantage of this system is the use of facial expression, as humans do not show their negative emotions often or have the ability to hide their emotions.

Viviane C.R. Appel, Valdinei L. Belini, Denny H. Jong, Daniel V. Magalhães and Glauco A.P. Caurin [5] proposed a system that uses the thermal camera to detect three emotions (neutral, motivated, overstressed) by taking images of the face, and the detection accuracy was 99.2%. However, the use of facial expressions to classify these emotions is considered a disadvantage as the subject can easily hide his/her emotions.

Valderas MT, Bolea J, Laguna P, Vallverdú M, Bailón R. [6] proposed a new system to detect 4 major emotions (Joy, Fear, Anger, and Sadness) using the heart rate variability (HRV) with varying spectral bands respiratory frequency (RF), On the other hand the use of a database of electrocardiogram (ECG), respiration, blood pressure (BP), skin temperature (ST) and galvanic skin response (GSR) makes this proposal less ubiquitous due to the need to attach several devices to the user to collect the needed data.

3 Proposed method

In this paper, a method to detect human's emotions by using facial skin temperature and HRV is proposed. Figure 1 describes five spots which are investigated to estimate the emotion. Figure 2 shows the HRV results of one subject. The face usually is not covered and its expression definitely does not show human's emotion. Hence, it is a fact that the facial skin temperature has some essential features of the human's emotion. In this research. The temperatures of the five spots are focused on as a senate of the facial skin temperature as well as the HRV. This method will be more suitable for young students Since they usually under much stress and have no time or money to talk to an expert.

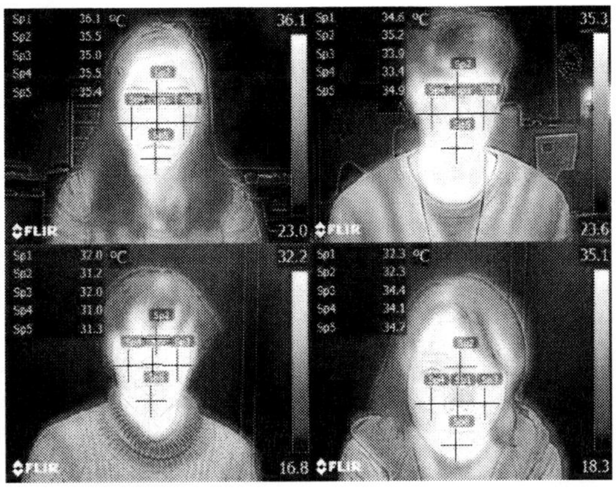

Fig. 1. Considered ROI (five spots) on the face of each subject in thermal images

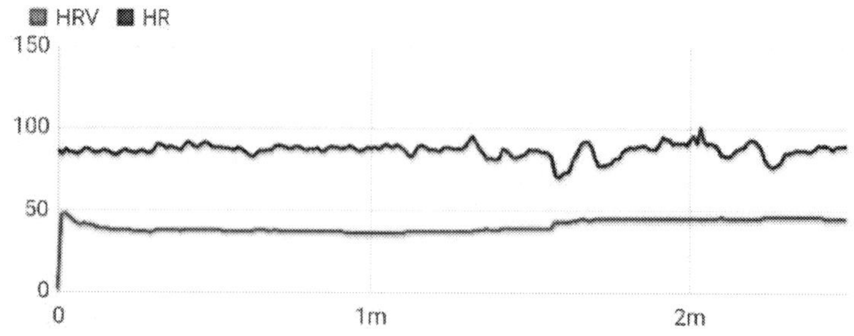

Fig. 2. Shows three minutes of HR and HRV of a subject extracted using (Zoom HRV) monitor

4 Evaluation experiment

4.1 Subjects

An experiment was conducted on 20 healthy volunteer subjects (10 females and 10 males, 20 to 31 years old) without any medical history, the subjects were from 10 different countries (Japan, China, Korea, Saudi Arabia, Indonesia, Senegal, Brazil, Mexico, Malaysia, Thailand) All students at Shibaura Institute of Technology, all subjects agreed on not consuming any caffeine, alcohol, and cigarettes for 6 hours before the experiment.it was confirmed that all subjects have had breakfast and rested well the night before.

4.2 Procedures

The subjects were instructed to accept the emotions and try to be honest in this experiment. The possible risk of fear and sadness stimuli was explained to all the subject, each subject conduct the experiment individually, The subjects were asked to sit in front of a Laptop in a dim light room, facing a thermal camera (FLIR C2) controlled remotely by (Flir tools software) and wearing a HRV monitor (Zoom HRV) connected by Bluetooth to (Huawei Mediapad m5) Using (Elite HRV), (Sony 1000XM2) headphones were used with sound cancellation as seen in Fig.3.the room temperature was sat on 22.5 degrees C. The thermal camera recorded the subject's face as video data. When the subjects became too emotional with fear or sadness stimuli the experiment will be terminated for a while then carried on again after the subject calmed down.

As in previous studies [2], Movie clips method has been confirmed as an effective emotion-inducing tool. All subjects watched the same set of videos in the same orders (Relax, Fear, Joy, Sadness), the content of each clip is described in Table 1. these clips were selected carefully from YouTube according to the number of views, the number of likes as well as the conformations comments of the viewers as an effective emotional stimulus. Each Video triggers one emotion. Each video was 3 minutes long, then the subjects took 3- or 5 minutes break to ensure the mental state of the subject goes back to normal. After the break, the next video will play and repeat until the subject has watched all the videos. Before finishing the experiment, the subjects answered a questionnaire to show how effectively the emotion was induced in each video. Finally, the subjects were rewarded for their contributions.

Fig. 3. Evaluation experiment recording a thermal video of face over watching a video clip

Table 1. Description of video content for each emotion

Emotions	Content of the Videos
Normal state	Beautiful scenery pictures with relaxing music
Fear	Scary ghost and Mystery scenes.
Joy	Videos of try not to laugh challenge.
Sadness	Death and Sick animals.

4.3 Data collection and analysis

After collecting the biological information from the experiment (facial skin temperature and HRV) for all subjects, A database of the max temperature of the five spots on the face and the HRV of each subject was recorded as a baseline for each subject before the experiment started, and then once the video started, the recording starts every 30 seconds for both HRV and the max temperature of the five spots for each emotion, firstly the maximum temperature of the five spots were extracted. Then, that temperature was used as the temperature of the area, each subject provided 28 images for the four emotions; each image has five temperatures of the nose, glabellar line, right/left cheeks and the chin. In total, 560 thermal images were extracted from the thermal's records for all the subjects. on the other hand the HRV monitor was set to start recording as the video start and stopped when the video finished, with the help of (Kubios HRV standard) the HRV Score of each subject was extracted for every 30 seconds for each emotion.

Artificial Neural Network (ANN) was used to estimate the emotion. In the ANN configuration, the five temperatures in each image are used as an input data set with the HRV in a time series database. Then, one of the emotions is output. In the evaluation, cross-validation was performed. Subsequently, all the prepared datasets were divided into training data set and testing data set. First evaluation was made using randomly divided 80% training data set and 20% testing data set. Then, second evaluation was done using randomly divided 50% training data set and 50% testing data set to confirm the validation.

5 Results and discussion

Table 2 shows the result using after processing the data on ANN performing cross-validation, the estimation accuracy of 88.75% was achieved. Table 3 shows the Correct Prediction of each emotion when the cross-validation was performed, which were 95.00% for relax and joy, 90% for fear and 85% for joy and Sadness.

Besides, when using 80% training data set and 20% testing dataset randomly selected from all the observed data set, the total estimation accuracy was 88.75%. Furthermore, when using 50% training data set and 50% testing data set, the total estimation accuracy was 70%.

Since the estimation accuracy using ANN was as the expected, the other algorithms, such as the Decision tree (J48) and Support Vector Machine (SVM), were also applied. The used data sets were the same as ANN. When decision tree J48 algorithm was performed, the lowest estimation accuracy of 62.5% was achieved. On the other hand, when SVM was performed, the total estimation accuracy was 68.75%.

Table 2. Emotion estimation accuracy using ANN

Analyzing Methods	Accuracy
Cross validation (ANN)	88.75%
80% training data set and 20% testing data set	87.5%
50% training data set and 50% testing data set	70%

Table 3. The accuracy of each emotion when ANN with cross-validation was performed

Emotions	Correct Prediction	Accuracy
Relax	19 totals of 20	95.00%
Fear	18 totals of 20	90.00%
Joy	17 totals of 20	85.00%
Sadness	17 totals of 20	85.00%

6 Conclusion and future work

In this paper, a new method was proposed for detecting human's emotions by using facial skin temperature and HRV was proposed and its effectiveness was evaluated using several types of machine learning techniques. As a result, it was revealed that the proposed method with MLP as a classifier achieved the best accuracy of 88.75% in estimating emotions of Normal State (Relax), fear, joy, and sadness.

As future works, more emotions will be including to confirm the proposed method. Also, increasing the number of subjects to investigate the individual difference. Furthermore, the other parameters from the HRV will be conceded such as LF/HF. Also Skin Conductance Response (SCR) will also be discussed as an additional parameter in order to improve the estimation accuracy of detecting emotions.

References

1. Hassani, S., Bafadel, I., Bekhatro, A., Al Blooshi, E., Ahmed, S., & Alahmad, M. (2017, November). Physiological signal-based emotion recognition system. In Engineering Technologies and Applied Sciences (ICETAS), 2017 4th IEEE International Conference on (pp. 1-5). IEEE.
2. Valderas, M. T., Bolea, J., Laguna, P., Vallverdú, M., & Bailón, R. (2015, August). Human emotion recognition using heart rate variability analysis with spectral bands based on respiration. In Engineering in Medicine and Biology Society (EMBC), 2015 37th Annual International Conference of the IEEE (pp. 6134-6137). IEEE.
3. Dobbins, C., Fairclough, S., Lisboa, P., & Navarro, F. F. G. (2018). A Lifelogging Platform Towards Detecting Negative Emotions in Everyday Life using Wearable Devices.
4. Emotion API - Emotion Detector | Microsoft Azure. (2019). Azure.microsoft.com. Retrieved 2 February 2019, from https://azure.microsoft.com/en-us/services/cognitive-services/emotion.
5. Appel, V. C., Belini, V. L., Jong, D. H., Magalhães, D. V., & Caurin, G. A. (2014, August). Classifying emotions in rehabilitation robotics based on facial skin temperature. In Biomedical Robotics and Biomechatronics (2014 5th IEEE RAS & EMBS International Conference on (pp. 276-280). IEEE.
6. Valderas, M. T., Bolea, J., Laguna, P., Vallverdú, M., & Bailón, R. (2015, August). Human emotion recognition using heart rate variability analysis with spectral bands based on respiration. In Engineering in Medicine and Biology Society (EMBC), 2015 37th Annual International Conference of the IEEE (pp. 6134-6137). IEEE.
7. Stoll, A. M. (1964). Techniques and uses of skin temperature measurements. *Annals of theNew York Academy of Sciences*, *121*(1), 49-56.
8. Kulandaidaasan, J., Elara, M. R., & García, E. M. (2012). Comparing Thermography, GSR and Hear Rate During Stimulated Therapeutic Pet Robot Interaction Among Elderly. In *Proceedings of International Conference on Intelligent Unmanned Systems* (**Vol. 8**).
9. Salazar-López, E., Domínguez, E., Ramos, V. J., de la Fuente, J., Meins, A., Iborra, O., ... & Gómez-Milán, E. (2015). The mental and subjective skin: Emotion, empathy, feelings and thermography. *Consciousness and cognition*, *34*, 149-162.
10. Bauer, R. M. (1998). Physiologic measures of emotion. *Journal of clinical neurophysiology*, *15*(5), 388-396..
11. López, R., Poy, R., Pastor, M. C., Segarra, P., & Moltó, J. (2009). Cardiac defense response as a predictor of fear learning. *International Journal of Psychophysiology*, *74*(3), 229-235.
12. Khan, M. M., Ward, R. D., & Ingleby, M. (2009). Classifying pretended and evoked facial expressions of positive and negative affective states using infrared measurement of skin temperature. *ACM Transactions on Applied Perception (TAP)*, *6*(1), 6.

Simulation and experimental study on the air jet pulverization of TiC-containing slag

Dongsheng Wang[1,2,*], *Jianbo* Zhang[1,2], *Qinge* Zhao[2], and *Jiaxu* Huang[1]

[1]Pangang group research institute Co. Ltd., Panzhihua 617000, China
[2]State key laboratory of V and Ti resources comprehensive utilization, Panzhihua 617000, China

Abstract. Process of "high temperature carbonization and low temperature chlorination", which was considered to be one of the most promising technologies, could extracte more than 70% titanium from the furnace slag containing TiO_2 10% - 30%. TiC-containing slag was the product of high temperature carbonization, and to meet the requirement of the following chlorination process, it needs to be crushed and grounded to a suitable size distribution. The air jet pulverization of TiC-containing slag was studied in this paper. On the one hand, By CFD numerical simulation, the distribution regularity of nozzle flow field about the main crushing equipment was obtained with different throat diameters and crushing gas pressures. On the other hand, the influences of nozzle throat diameter, crushing pressure, and collision angles on the particle size of the powder were studied by experiment. With the better parameters, the particle size distribution of the product powder was uniform.

1 Introduction

Vanadium-titanium magnetite is a kind of complex ore, which contains many valuable elements, such as Fe, Ti and V. The found reserve is nearly 48 billion tons in the world, and it is mainly distributed in Russia, South Africa, China and New Zealand. In China, it is mainly located in Sichuan, Hebei and Anhui provinces. In the present utilization way of the vanadium-titanium magnetite, the raw ore is first dealed with a mineral separation process to obtain the iron concentrate, and the obtained tailing is further beneficiated to obtain the ilmenite concentrate. In the ilmenite concentrate, titanium dioxide is concentrated 45%-55% furthermore, the ilmenite is smelted by electric furnace, after which the ilmenite is reduced to iron and the slag contained titanium dioxide 70%-85%. The slag is treated by the chlorination process of molten salt to produce $TiCl_4$, which is raw material of titanium dioxide or sponge titanium. In the iron concentrate, the Fe_2O_3 is gathered to 50%-60%, and by blast furnace smelting, Fe_2O_3 is reduced to iron, at the same time, other materials form blast furnace slag contained TiO_2 10%-30% [1]. At present, this kind of slag is regarded as industrial wastes, and leads to great loss of titanium resources. In recent years, many scholars do lots of work to recovery titanium from this kind of blast furnace slag, such as "making silicon alloy"[2], "enrichment of perovskite and dressing" [3], "separation of titanium carbide after carbon thermal reduction " [4], " acid leaching extraction" [5],

* Corresponding author: 15881250692@163.com

"high-temperature carbonization and low-temperature chlorination" [6] and so on. By high temperature carbonization and low temperature chlorination process, the titanium could recover more than 70% from BF slag, therefore, it is considered to be one of the most possible methods for industrialization. The basic principle of process is that TiO_2 is reduced to TiC by reacting with coke. By this reaction, the BF slag is transformed to slag contained TiC. And the TiC-containing slag could be further reacted with Cl_2 at a low temperature to produce $TiCl_4$. Faninaly, the $TiCl_4$ is used to produce TiO_2. In order to promote the reaction between the TiC-containing slag and Cl_2, the TiC-containing slag needs to be crushed and grinded to a certain size range. The mass percentage of less than 32μm is asked to be less than 20% and the mass percentage of 32μm-150μm is required more than 75%. By traditional ball mill process, the percentage of 32μm-150μm accounted for only 35%. it is far satisfying the raw material demands for low-temperature chlorination. So it is necessary to search new method to fabricate TiC-containing slag powder.

The principle of air jet pulverization is like this: materials are accelerated to a certain speed by high-speed airflow. Then it is crushed, sheared and pulverized by collision between particles and partials or particles and target. Compared with the existing crushing technology, the main merits of air jet pulverization are avoid excessive grinding and the products have a narrow particle size distribution [7]. Airflow crushing equipment is usually divided into 5 types: flat airflow grinding, circulating pipe airflow grinding, jet airflow grinding, target airflow grinding and fluidized bed airflow grinding [8,9]. In this paper, grinding of TiC-containing slag is studied by simulation and experiments. By CFD numerical simulation, the distributed regularity of nozzle flow field is obtained with different throat diameters and gas pressures. Then by experimental study, the effect of process parameters, such as nozzle throat diameter, crushing air pressure and collision angle on the particle size distribution are researched, coupled with rate of product and yield.

2 Materials and methods

2.1 Experimental materials

The water quenched TiC-containing slag was used as experimental feed. The ingredients are as follows: CaO 27.4%, SiO_2 24.69%, TiC 14.14%, Al_2O_3 13.25%, MgO 8.16%, Ti_2O_3 4.46%, C 2.98%, Fe 1.41% and Other 3.51%. The density of water quenching slag is 0.95t/m³, and the particle size distribution is shown in table 1.

Table 1. The size distribution of water quenched slag.

Particle size (μm)	<32	32-150	150-833	833-1397	1397-3000
Mass fraction (%)	1.35	6.79	36.41	33.32	22.13

2.2 Experimental methods

Numerical simulation and experiments were used to study the effect of air jet crushing. The nozzle structure, which is the core component of machine, was analyzed by software of Ansys Fluent 15. For nozzle throat diameter 42 mm and 60 mm, the nozzle flow field and the exit airflow velocity were calculated carefully under different crushing gas pressure 2 kPa, 5 kPa, 10 kPa, 20 kPa and 60 kPa. The 3D model and meshing of pulverized cavity were established as figure 1 and figure 2, respectively. For simulation, a steady state flow model was established. The air was treated as an ideal gas and without considering the influences of the filter wall on the flow field. The airflow field adopted SIMPLE coupling algorithm of pressure and velocity and standard k-ε equation model, followed by the conservation laws of mass, momentum and energy. The governing equations were continuity equation, momentum equation, k turbulent kinetic energy equation and ε dissipation rate equation.

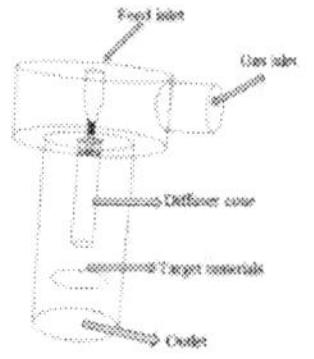

Fig. 1. 3D model of pulverized cavity.

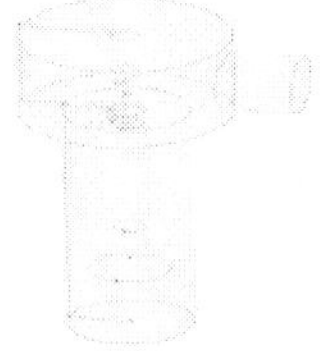

Fig. 2. Meshing by Gambit.

Combined with the characteristics of raw material and the requirements of product size, the airflow grinding machine was selected as the main equipment. Figure 3 shows the experiment flow of air jet pulverization of TiC-containing slag. The particle size distribution of finished slag was analyzed by laboratory standard sieve, and the percentage of less than 32 μm was measured by wet sieving, which refer to the Chinese national standard GB/T 1345-2005.

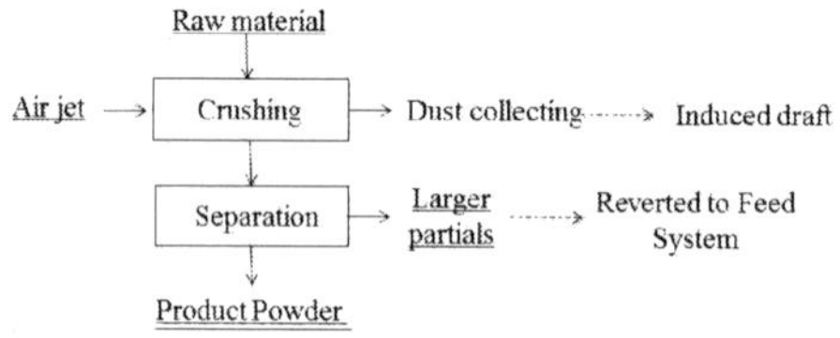

Fig. 3. Flow chart of air jet pulverization process.

3 Results and discussion

3.1 Simulation results analysis

Figure 4 shows the Diagram of simulation results about airflow velocity for nozzle throat dimeter 42mm and 60mm under the crushing air pressure 10 kPa. From the figure 4, it is found that the airflow velocity in the center is about 18m/s smaller than that at the circumference under the condition of nozzle throat diameter 42mm. These results suggested that the airflow velocity isn't very uniform. In contrast, with nozzle throat diameter 60 mm, it seems no big difference between the center and the circumference. Therefore, the materials could be accelerated accordantly. Considering the matching between the air jet flow tube and the nozzle, the nozzle throat diameter 60 mm seems better.

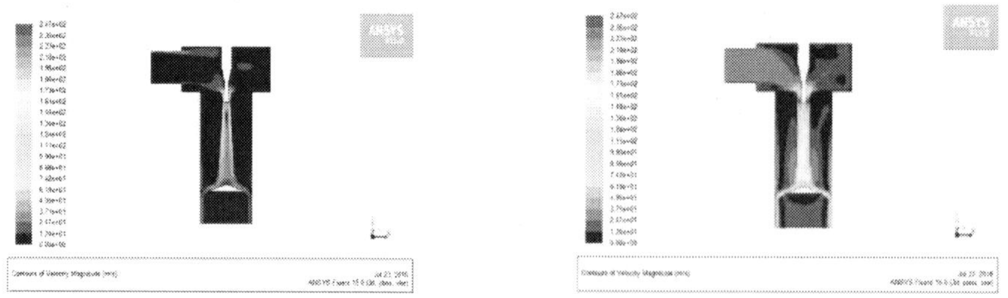

(a) (b)

Fig. 4. Diagram of simulation results about airflow velocity with the crushing air pressure 10 kPa (throat diameter: a-42mm, b-60mm).

The changes of gas mass flow rate with different crushing air pressures are shown in Figure 5. For the same throat diameter, the higher the crushing air pressure, the greater the mass flow. For the same air pressure, the greater the throat diameter, the greater the mass flow rate. Figure 6 shows the change of throat airflow velocity with different crushing pressures. For the same nozzle throat diameter, with the increase of crushing gas pressure, the outlet airflow speed increases, as well as the crushing strength. When the crushing gas pressure increases from 2kPa to 60kPa, the airflow velocity of nozzle throat diameter 42mm increases from 150m/s to 380m/s, and the airflow velocity of nozzle throat diameter 60mm increases from 98m/s to 262m/s. For the same pulverized gas pressure, the greater throat diameter, the lower airflow velocity. The airflow speed of the throat diameter 60 mm is almost half of the number of throat diameter 42 mm.

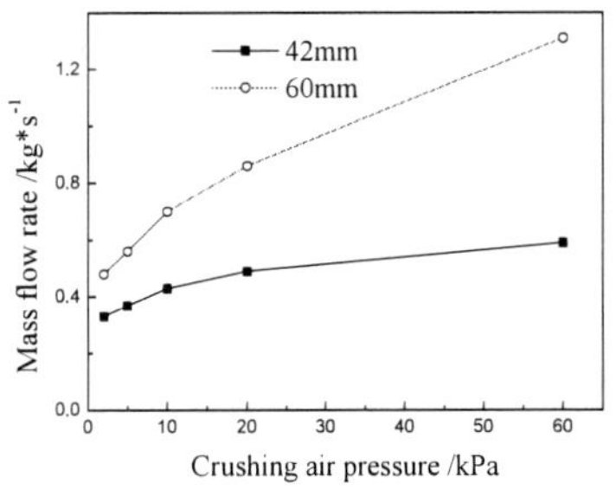

Fig. 5. The change of gas mass flow with different crushing pressures.

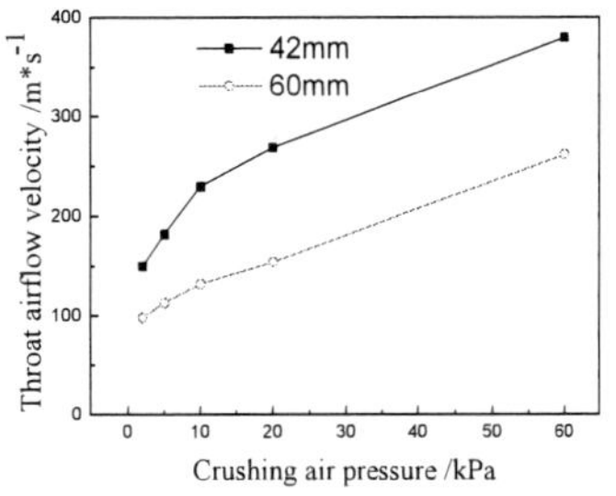

Fig. 6. The change of throat airflow velocity with different crushing pressures.

3.2 Different throat diameters

The particle size distribution with different nozzle throat diameters is shown in figure 7. From the figure 7(a), as the nozzle throat diameter increases from 42 mm to 60 mm, the mass proportion of less than 32μm increases from 15.6% to 17.6%, as well as the proportion of 32μm-150μm adds from 78% to 81.3%. And there is a huge decline of particle size greater than 150μm, which falls from 6.4% to 1.1%. The experimental data is consistent with the simulation results. According to the simulation results, For the throat diameter 60mm and crushing gas pressure 10kPa, the mass flow rate was 0.7 kg/s, which was 0.38 kg/s greater than that of the nozzle throat diameter 42mm. And the throat velocity was 132 m/s, which was 98m/s lower than that of the throat diameter 42mm. In addition, it can be seen from figure 7(b) that with nozzle throat diameter 60mm the mass ratio of product and feed increases about 10% and the output increases about 50%, respectively. Both of them are relatively higher than the numbers with nozzle throat diameter 42mm.

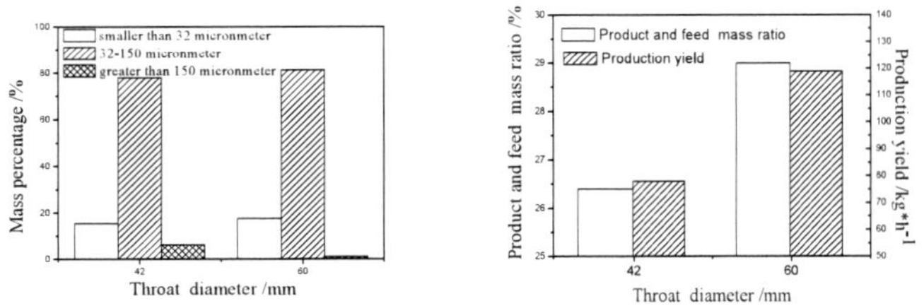

(a) (b)

Fig. 7. The changes of particle size, product rate and yield with different throat diameters.

3.3 Different crushing air pressures

The crushing pressure refers to the surface pressure of dry air entering the inlet of the subsonic nozzle after oil and water removed [10]. Figure 8 shows the influences of different pressures on the particle size, yield and output with throat diameter 42mm. It can be seen from Figure 8(a), with the gas pressure decreases from 60kPa to 10kPa, the particle size proportion of 32μm-150μm increases from 70.8% to 78%, and the proportion of less than 32μm decreases from 28.6% to 15.6%. Under the gas pressure of 10kPa, the mass percentage of particle size 32μm-150μm is the largest and accounts for 78%. As the reduction of crushing pressure, the throat airflow velocity seems to slow within a certain range. According to the results of simulation calculations, along with the crushing gas pressure reduced from 60 kPa to 10 kPa, the airflow speed reduced from 380 m/s to 230 m/s at nozzle exit. As the material speed slowing, the collision energy decreases. The figure 8(b) is the changes of product rate and yield. Along with the crushing gas pressure reducing, the product rate and yield turn to be lower and lower. The mass ratio of product and feed reduces from 37% to 26.4%, and the yield declines from 106 kg/h to 78 kg/h, respectively. By calculation, the crushing capacity under the condition of gas pressure 10 kPa accounts for 60% of that under the condition of gas pressure 60 kPa.

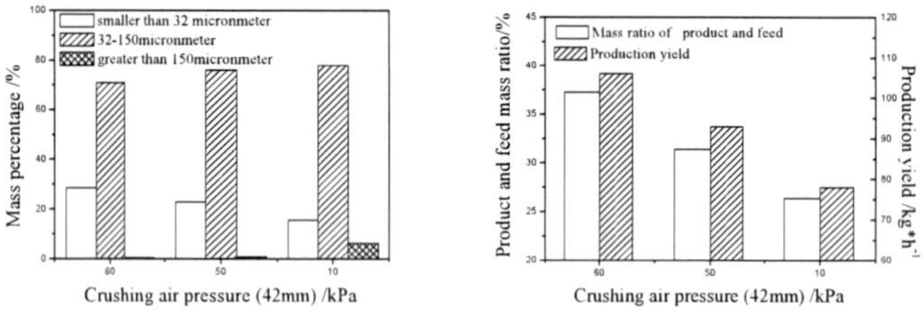

(a) (b)

Fig. 8. The changes of particle size, product rate and yield with different crushing air pressures.

3.4 Different collision angles

Under the condition of nozzle throat diameter 60mm and crushing pressure 10kPa, the impacts of collision Angle is shown in figure 9.

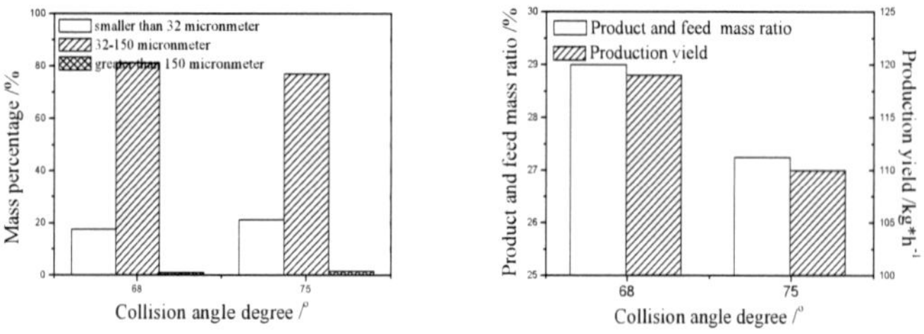

(a) (b)

Fig. 9. The changes of particle size, product rate and yield with different collision angles.

It can be seen from the figure 9(a) with the collision angle increases from 68° to 75°, The particle size of finished slag is obviously finer, and the mass percentage of particle size less than 32μm increases from 17.6% to 21.2% significantly. Figure 9(b) shows that product rate and yield slightly decrease as the collision angle increasing. The mass ratio of product and feed decreases 1.7%, and the yield decreases 9kg/h, respectively. The particle size distribution, product rate and yield are influenced by the shape of the target material. In general, the top Angle of the target material is defined as the target material's angle, and it is usually inelastic and non-vertical collision in the crushing process, in which the part of lost energy is used to break particles. The larger target material's angle, the greater energy loss, and the greater the energy of crushing particles.

3.5 Product size distribution and energy consumption

Under the condition of better process parameters, crushing gas pressure 10kPa, throat diameter 60mm, and collision Angle 68°, the stability of product granularity and equipment energy consumption were evaluated by test which continuous 4 hours. The particles size distribution in products is shown in table 2. The mass proportion of particle sizes less than 32μm, 32~150μm and greater than 150μm was 18.97%, 77.77%, and 3.26%, respectively. And the energy consumption in the test was 70.2kWh/t.

Table 2. The particles size distribution in products.

Sample	<32μm(%)	32~150μm(%)	>150μm(%)
Average	18.97	77.77	3.26

4 Conclusion

1. The data of mass flow rate and airflow velocity under different throat diameters and crushing pressures have been calculated. Compared with nozzle throat diameter 42mm, the nozzle throat diameter 60mm was probable better, and the accelerated characteristic of the materials were more uniform.
2. Nozzle throat diameter and crushing gas pressure were two key factors influencing the particle size distribution and the output. Increasing the nozzle throat diameter or decreasing crushing gas pressure could reduce the airflow velocity with the crushing efficiency reduced.
3. The better process parameters of air jet pulverization of TiC-containing slag were obtained: the nozzle throat diameter 60mm, gas pressure 10kPa, and collision Angle 68°. With the better parameters, The average particle size distribution of TiC-containing slag was as follows: less than 32μm 18.97%, 32～150μm 77.77% and greater than 150μm 3.26%.

References

1. Liu S and Yang S 2007 J. *Light Metal* 7 48-50
2. Li Z, Liu T and Li Z 1996 *J. Chongqing Univ.* **19** 82-86
3. Sui Z, Guo Z and Zhang Li 2006 *J. Mater. and Metall.***5** 93-97
4. Feng C and Zhang J 1997 *J. Miner. Utiliz.* **6** 34
5. Liu X and Sui Z 2002 *J. Chin. Nonferrous Met.* **12** 1281
6. Huang S, Pan J and Zhang R 1994 *Iron Steel V. Ti.* **15** 17
7. Chen H 2007 *PHD dissertation* (Sichuan: Sichuan university) p 1
8. Yang Z 1988 *SuperfineJet Pulverization* (Beijing: chemical industry press) p 263
9. Li F 2000 *Ultrafine Powder Technology* (Beijing: national defense industry press) p 83
10. Zhou X, Chen H and Lu C 2011 *Powder metall. Technol.* **29** 46-49

Effect of diamond particles on the microstructure and composition of pulse plated multilayer Ni-W/diamond composite coatings

Malay Kumar Das[1,*], *Jiaqian* Qin[2], and *Priya* Karn[3,4]

[1]International graduate program of Nano-science and technology, Chulalongkorn University, Bangkok 10330 Thailand
[2]Metallurgy and materials science research institute, Chulalongkorn University, Bangkok 10330, Thailand
[3]Institute of Engineering- Tribhuvan University Purwanchal campus, Dharan, Nepal
[4]PAVIMAM Solutions Education and Research Center of Excellence, India

Abstract. The research encompasses the fabrication of multilayer Ni-W-diamond composite deposits by pulse current electrodeposition. The mean surface diameter of the diamond used for the research was 0.9 microns. The composites were fabricated under various conditions and characterized by means of SEM, EDS and XRD to investigate and analyze their microstructure, grain size and elemental distribution along with variations in morphology under different conditions. The effect of variations in forward and reverse current density along with duty cycle and the combination of various aforesaid conditions on the coatings were studied and analyzed.

1 Introduction

Nanocrystalline Ni-W alloys and composites have a wide arrays of applications such as fabricating the alloys in some sort of barrier/capping layers for ultra large scale integration applications in copper metallization and also potential applications in Microelectromechanical systems and also various allied applications in mold inserts, magnetic heads and relays, bearings, resistors, electrodes accelerating hydrogen evolution from alkaline solutions, environmentally safe substitute for hard chromium plating in the aerospace industry, etc.[1]. Hard-chrome plating has been used as a surface finishing technique in various engineering industries because it has good advantages such as high hardness, corrosion resistance, wear resistance, aesthetic qualities and durability. However, Hard-chrome plating solutions are toxic, carcinogenic and hazardous to human health [2, 3] to substitute chromium; the new alternate surface finishing method must maintain hard-chrome's advantages. One very important and sustainable prospect is to substitute Hard-Chrome plating with metal alloys/composites plating which is certainly better than plating metals with alloys as alloys and composite plating directly on a suitable substrate tends to enhance the abrasion resistance and resistance to surface erosion phenomena while on the

* Corresponding author: malaykumardas90@gmail.com

other hand acts as a highly viable and suitable substitute to hard chrome plating[2]. In various industrial applications (for example boilers) it has been documented that the surface of the equipment and components involved in the running of the plant are highly to prone to mechanical stress, corrosion and damage by wear. This in turn leads to plant closures and subsequent loss in revenue and operations in related industries like mining, mineral processing etc. Hence it becomes imperative to work upon strategies to enhance the surface properties rather than bulk properties [4-6] in a wide scope of engineering materials and components used in industries. It is also an interesting prospect to try and introduce various composites

Ni-W/diamond composite coatings were electrodeposited by means of pulse plating on a 2*2 carbon steel substrate. The operating conditions were modified to obtain a multilayer electrodeposited coating of varying properties, thickness, composition and attributes constituting the aforesaid composites. Diamond composite incorporation into the Nickel-Tungsten alloy matrix could significantly enhance the mechanical properties of the Ni-W base coated matrix in terms of hardness[7], wear resistance, corrosion resistance, abrasion resistance, increased durability, thermal resistance and also the overall roughness of the coatings. However, the major issue that needs to be fixed before embarking on the fabrication of multilayer coatings is to ensure the uniform distribution of diamond particles on the coated matrix. Various contemporary researchers have employed the means of pulse plating to fabricate ni-w and other alloys and composites[8]. Employing suitable pulse parameters is quite crucial to obtain desirable deposits. Fabrication of Multilayer coatings also enhance the versatility of the deposits in terms of varying layers of different surface roughness, friction and wear resistance along with hardness. The influence on the microstructure of a multilayer coating is also profound and very important in terms of enhancement of film structure and quality.

2 Experimental details

Ni-W-diamond composites were fabricated by means of electrodeposition from an ammonia-citrate bath (200 ml) on a 2*2 steel substrate. The bath composition was nickel sulphate 15.8 g/l, sodium tungstate 46.20 g/l, tri-sodium citrate 147 g/l, ammonium chloride 26.7 g/l and sodium bromide 15.8 g/l respectively. The operating temperature was 75 °C the pH, 200 RPM stirring speed and the deposition time was kept constant at 200 RPM and 2 hours, respectively. X-Ray Diffraction technique was employed to analyze crystalline size of the coatings. Brooker D8 advance X-ray diffractometer operated at Cu Kα radiation (nickel filtered) at a rating of 40 kV, 20 mA. The scan rate was 0.02° per step and the measuring time 0.5 second/step. Scherrer's equation[9] was employed for the calculation of the grain size of the electrodeposited crystals. The samples were characterized in terms of morphology and microstructure by JEOL JSM-6400 SEM with EDS capability embedded into it.

2.1 Pulse parameters

Table 1. Pulse parameters for the experiments.

Sample	Diamond Concentration (g/l)	Forward Current Density A/cm^2	Forward Duration (ms)	Reverse Current Density A/cm^2	Reverse Duration (ms)	Duration of Each Layer (minutes)
A	5	0.2	20	0.3	3	-
B	5	0.2	20	0.15	2	-

C	5	0.2	20	0	3	-
D	5	0.2	20	0	0	-
A_ND	0	0.2	20	0.3	3	-
AB_0.1	5	A=0.2 B=0.2	A=20 B=20	A=0.3 B=0.15	A=3 B=3	A=0.1 B=0.1
AC_0.1	5	A=0.2 C=0.2	A=20 C=20	A=0.3 C=0	A=3 C=3	A=0.1 C=0.1
AD_0.1	5	A=0.2 D=0.2	A=20 D=2	A=0.3 D=0	A=3 D=0	A=0.1 C=0.1
AC_1	5	A=0.2 C=0.2	A=20 C=20	A=0.3 C=0	A=3 C=3	A=1 C=1
AC_2	5	A=0.2 C=0.2	A=20 C=20	A=0.3 C=0	A=3 C=3	A=2 C=2

3 Results

3.1 Grain size

Upon analyzing the grain size, a progressive reduction in grain size is observed from A to D. For the Multilayer AD has the optimal size of 7 nm whereas grain size is subsequently enlarged in case of AB and AC with each alternating layer deposited for 0.1 minutes respectively. The incorporation of diamond composites into the alloy matrix/grains increases the grain size from 18 to 23 nm. Therefore, the incorporation of diamond particles into the coating matrix is subsequently established. It is clear that the diamond particles have been embedded into the coating matrix by means of a chemical bonding with the base alloy matrix and not merely by means of physical adhesion or mechanical cohesion. This is a relevant result to further study the means to enhance the level of incorporation of diamond particles and the distribution of diamond particles in the coatings.

Table 2. Grain size of the coatings.

Sample	Grain size(nm)
A_ND	18
A	23
B	19
C	4
D	5
AB_0.1	16
AC_0.1	8
AD_0.1	7

3.2 SEM and EDS analysis

Comparative analysis for figure 1 and 2 reveals the morphological differences between the samples with and without diamond addition[10]. This reveals the importance and the indispensability of diamond composites addition in the Nickel-Tungsten. There is a marked difference in the microstructure for samples with and without diamond addition. Figure 1 depicts a uniform film structure and distribution of diamond particles. Moreover, there are no cracks and pit formations in the samples containing diamond composites encapsulated into the matrix. Unlike the former the samples without diamond have a lot of quasi circular formations for almost all magnifications. There seems to be a lot of pit formations in all of the images without diamond. There is a clear evidence of numerous cracks and fissures on the microstructure for fig. 2 (b). The coating is discontinuous with evidence of internal strain which causes cracks and lack of continuity and uniformity in the film thereby severely affecting the microstructure and in turn the properties associated with them. The coatings with diamond reveal uniform crystalline formations also with no sharp grain boundaries unlike the samples without diamond which has highly discontinuous formations on the surface with no uniformity in terms of deposition and film quality. The comparison between the morphologies of Figure 1 and 2 clearly established the need for use of diamond composites in Ni-W alloy matrix in order to enhance the properties of the coatings. This is a very relevant and substantial result in terms of our justifications for the use of the diamond particles as composites in the coating matrix.

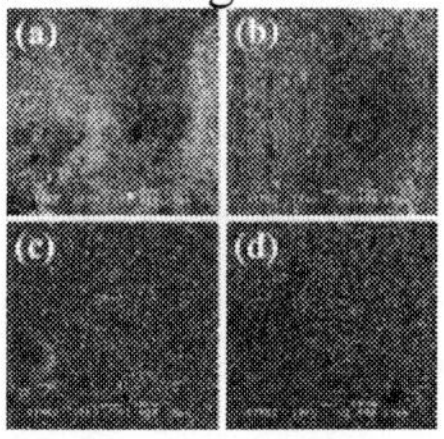

Fig. 1. SEM Images of (a) A monolayer layer (b) B monolayer (c) C monolayer (d) D monolayer.

Figure 2 depicts the SEM images for the Multilayer coating AB_0.1, AC_0.1 and AD_0.1. The image 3 (b) exhibits stark differences in terms of distribution of diamond particles, cracks and irregularity in crystal structure. The pairs AB_0.1 and AD_0.1 are similar and have a uniform crystal structure. The important results that is evident from Figure 3 is that the while fabrication of multilayer AC layer is not desirable in terms of morphology and microstructure. Therefore, either AB or AD should be used a template layer for fabrication of multilayer Ni-W-Diamond.

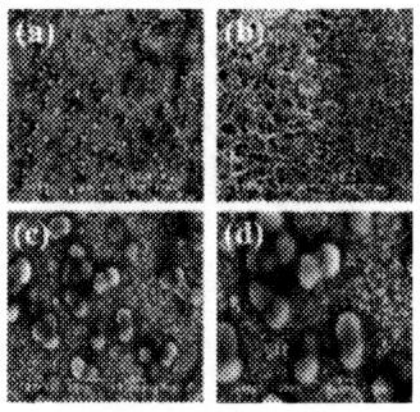

Fig. 2. SEM Images for A layer without diamond (a) for 1000X (b) for 2000X (c) for 3000X (d) for 5000X.

The diamond content in the coatings of AC_0.1 also appears to be considerably lesser than both AB_0.1 and AD_0.1. However, upon close analysis it is revealed that the surface contamination is greater in case of AB_0.1 as compared to AD_0.1, an idea which is corroborated by the presence of a large number of white spots which are most likely to be oxygen and carbon contaminates encapsulated into the coating during the deposition process. The SEM analysis acts as an important tool to identify and ascertain the optimal and suitable conditions for the NI-W-diamond coatings.

Fig. 3. SEM Images for (a) AB_0.1 (b) AC_0.1 (c) AD_0.1.

3.3 EDS analysis

Table 3. Elemental composition of Ni and W in the deposits as reported by the EDS results.

Sample	Ni (At %)	W (At %)
A_ND	95.42	4.58
A	99.06	0.94
B	97.62	2.38
C	78.58	21.42
D	85.56	14.44
AB_0.1	92.97	7.03
AC_0.1	82.87	17.13
AD_0.1	75.75	24.25

The EDS results suggest enhanced Tungsten content in the AD_0.1 layer. This is possibly because D is the direct current which allows for an enhanced duty cycle and thereby and enhanced plating rate leading to a higher co-deposition of tungsten. However, the Tungsten Content decreases drastically when diamond is added to the A layer. This is due to the encapsulation of diamond particles into the coating matrix. The carbon content results in the EDS analysis might also contain carbon contaminates on the coated surface which is why the diamond content by weight in the coatings needs to be investigated by dissolving the coating and the entire substrate in iron and in turn measuring the difference in weight to give the diamond content.

3.4 Relationship between Grain Size and Tungsten content

Table 4 shows a clear relationship between grain size and tungsten content wherein the smaller grain size facilitates an increased concentration of tungsten in most cases. Upon analysis of AD_0.1, A_ 0.1, C and D it's clear that a smaller grain size enables more co-deposition of tungsten in the coated matrix thereby enhancing the tungsten content in the deposits. However, the nature of the current that causes the deposition is also an important factor to determine the tungsten content in the coating which is evident from the fact that the samples AD_0.1 which has D layer as direct current deposited layer has the highest tungsten content among all the coated layers. However, the tungsten content probably follows a limiting value of grain size beyond which tends to violate the aforesaid trend which is why a very reduced tungsten content is observed in case of single layers A, B, C and D and also in case of A_ND layer without diamond. The substantial reduction in grain size for the A single layer without diamond and with diamond can also be credited to the fact that diamond encapsulation into the base matrix tends to reduce the tungsten content. The SEM images as in Fig. 1 and 2 along with the relative data of tungsten content reduction and grain size increment can be used as arguments to prove that diamond has been embedded into the coatings uniformly and not mere weak physical adhesion or mechanical cohesion.

Table 4. Grain Size and Tungsten Content comparison for various conditions.

Layer Conditions	Grain Size(nm)	W (At %)
A_ND	18	4.58
A	23.1	0.94
B	19.1	2.38
C	4.1	21.42
D	5.1	14.44
AB_0.1	16.5	7.03
AC_0.1	8.4	17.13
AD_0.1	7	24.25

4 Conclusion

AD_0.1 has an optimal grain size of 7 nm and the most suitable SEM morphology in terms of uniform distribution of diamond particles, film quality, finer grain size and less pronounced grain boundaries as compared to AB and AC 0.1 min multilayer deposits. There are also almost nil cracks as per SEM morphology for AD_0.1 and AB_0.1 unlike AC_0.1

The incorporation of diamond is meagre, and the distribution also not uniform in the case of AC_0.1. AC_0.1 layer also doesn't have the most suitable grain size, or the tungsten content as compared to AD_0.1layer.

The tungsten content is the maximum in case of AD_0.1 layer and it is the most desirable in terms of both SEM morphology and grain size. However, the actual values of diamond content in the deposits along with various other mechanical properties namely hardness, wear and corrosion resistance need to be figured out before attributing it as the most suitable condition for multilayer. Smaller grain size and suitable layer conditions facilitate larger deposition rate of tungsten in the coatings and an overall increased tungsten content as per EDS results.

References

1. Eliaz, N., T. Sridhar, and E. Gileadi, Synthesis and characterization of nickel tungsten alloys by electrodeposition. Electrochimica Acta, 2005. 50(14): p. 2893-2904.
2. Xiaohe, W. and X.B.-s.H. Zhen-feng, Progress in developing electrodeposition for substitute hard chromium‖. New Technology and New Process, 2009.
3. Nagai, T., K. Hodouchi, and H. Matsubara, Relationship between film composition and microhardness of electrodeposited Ni–W–B films prepared using a citrate–glycinate bath. Surface and Coatings Technology, 2014. 253(0): p. 109-114.
4. Shang, Y.-L., et al., Electrorheological property of M-doped (M=Na, Zr) nano-sized TiO2 particle materials, influence of surface composition and microstructure. Colloids and Surfaces A: Physicochemical and Engineering Aspects, 2008. 325(3): p. 160-165.
5. An, Y.-l., et al., Interfacial structure and mechanical properties of surface iron–nickel alloying layer in pure iron fabricated by surface mechanical attrition alloy treatment. Materials & Design, 2013. 46(0): p. 627-633.
6. Krauss, G., Advanced surface modification of steels. Journal of Heat Treating, 1992. 9(2): p. 81-89.
7. Yuan Xuetao, W.Y., Sun Dongbai, Yu Hongying, Influence of pulse parameters on the microstructure and microhardness of nickel electrodeposits. Surface & Coatings Technology, 2008. 202: p. 1895-1903.
8. El-Sherik, A.M., U. Erb, and J. Page, Microstructural evolution in pulse plated nickel electrodeposits. Surface and Coatings Technology, 1997. 88(1–3): p. 70-78.
9. B.D.Cullity, S.R.S., S.Stock,, ElementsofX-rayDiffraction Fraction. 2001, LONDON: Addison-Wesley.
10. Zemanová, M., et al., Pulse current electrodeposition and corrosion properties of Ni–W alloy coatings. Journal of Applied Electrochemistry, 2011. 41(9): p. 1077-1085.

Pelletization performance of Panzhihua ilmenite concentrate

Bing Song[1,2*], *Ping* Huang[3], *Yong* Ma[4], *Zan* Song[5], and *Zhake* He[4]

[1]Pangang Group Research Institute Co.Ltd. , Panzhihua, 617000, Sichuan, China
[2]State Key Laboratory of Vanadium and Titanium Resources Comprehensive Utilization. Panzhihua, 617000, Sichuan, China
[3]Panxi Science and Technology Innovation Center, Panzhihua University, 617000, Sichuan, China
[4]Pangang Group Vanadium Titania & Resources Co. Ltd., Panzhihua, 617000, Sichuan, China
[5]Metallurgical Center for International Exchange and Cooperation. Beijing, 100711, China

Abstract. The influences of particle size distribution and the binder on the pelletization performance of Panzhihua ilmenite concentrate were studied in this paper. The results showed that the particle size distribution of Panzhihua ilmenite concentrate has the most important on its pelletization. The particle size distribution of ilmenite concentrate was significant changed and the specific surface area increased after grinding. The particle size of 45μm should be controlled less than 55% and 74μm no more than 74% in order to acquire a better performance of pellets. While the same performance of pellets can be obtained when the specific surface area of ilmenite concentrate was 1493cm2/g. The performance of pellets added with an organic binder is better than added bentonite. The pellet's performance could be further improved under the mixed ratio of 1.05% bentonite and 0.45% organic binder.

1 Introduction

The development of titanium industry requires large number of high quality titanium ore. Although natural rutile and titanium mineral were main raw material to titanium industry, the resources decreased worldwide. The low grade titaniferrous ores such as ilmenite should be paid more attention. Titania slag prepared by smelting of ilmenite in an electric arc furnace (EAF) can be served as a superior quality raw materials containing titanium to the pigment or other titanium products industrial processes[1]. Therefore, titania slag is the key to improve the comprehensive utilization level of titanium resources. However, particle size of Panzhihua ilmenite concentrate has become more and more fine. Using these fine ilmenite concentrate directly to smelt, some problems would appear, such as pollution environment, decreasing of the smelting stability and increasing of the smelting cost[2]. Therefore, pelletizing is necessary to using the fine ore[3].

The possible technical routes for treating Panzhihua fine ilmenite concentrate can be classified into two methods, namely, briquetting and pelletizing. Han [3] studied on the effect of mixing time, pressure, binder and rotational speed on compressive strength of

* Corresponding author: sdbingsongbing@163.com

pellets. Ren[4] reported on the effect of grinding time, binder and drying temperature on pelleting of Panzhihua fine ilmenite concentrate. But they did not consider the effect of particle size distribution. There are few studies on pelletization performance of Panzhihua ilmenite concentrate in the literature.

The purpose of this present investigation was to study the effect of particle size distribution and binder on the pelletizing of ilmenite concentrate, while, obtaining a better parameter to improvement pelletization performance of Panzhihua ilmenite concentrate.

2 Experimental

2.1 Raw materials

The chemical composition of ilmenite concentrate is listed in Table 1. Bentonite and the organic binder were employed as a binder and its chemical composition is given in Table 2.

Table 1. Chemical composition of ilmenite concentrate (wt.%).

Ilmenite concentrate	TiO_2	FeO	Fe_2O_3	CaO	MgO	SiO_2	Al_2O_3	S
A	47.80	35.91	4.89	0.87	5.18	3.08	1.24	0.15
B	48.18	35.91	4.89	0.59	5.64	2.72	1.00	0.10
C	46.76	34.95	6.55	0.75	4.86	2.86	2.01	0.27

Table 2. Chemical composition of binder (wt.%).

Binder	C	Na_2O	Fe_2O_3	CaO	MgO	SiO_2	Al_2O_3
Bentonite	0.29	1.48	1.77	2.24	2.90	48.66	12.95
Organic binder	12.30	10.40	1.00	2.69	14.56	9.13	2.78

2.2 Experimental apparatus and methods

A disk pelletizer was employed to acquire pellets. The schematic diagram of pelletizer is shown in Figure 1. The pelletizing flow chart of ilmenite concentrate is shown in Figure 2.

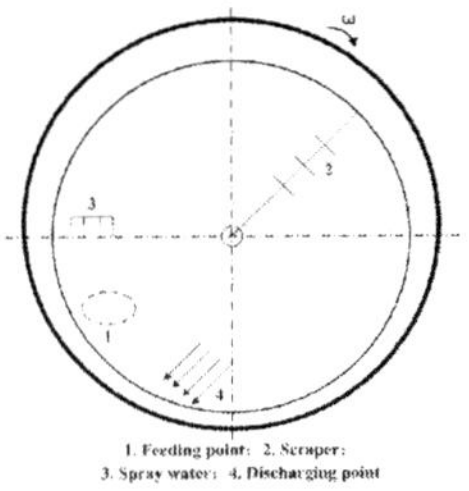

Fig. 1. Schematic diagram of pelletizer.

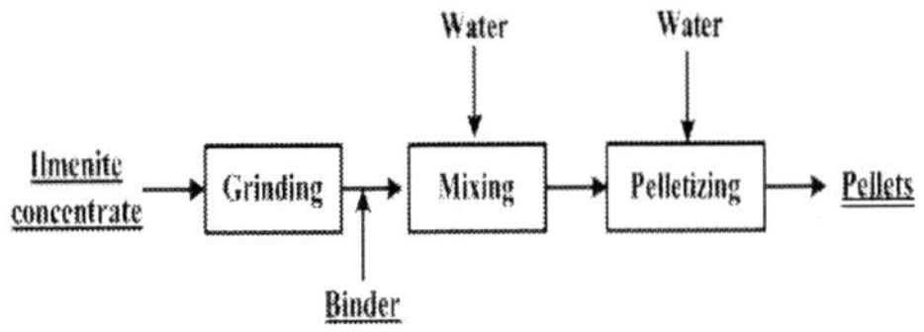

Fig. 2. Flow chart of pelletizing.

50g of sample was screened with 45 μm and 74μm set sieve. Screened sample was dried 2 h at 393 K and then weighed. Scanning electron microscope (SEM, JSM-5600LV, Japan) was employed to study the ilmenite concentrate particle surface morphology. Specific surface area of ilmenite concentrate was determined using liquid nitrogen adsorption method in the automatic surface area analyzer (GEMINI VII 2390, USA).

5000g ilmenite concentrate with a certain amount of binder was pelletized which is 8 to 12mm in diameter. The average value of compressive strength and drop strength was used as the performance criterion of ilmenite concentrate pellets[5,6]. The compressive strengths of pellets was messured using a compression strength tester (HXQT-10D, China). The drop strengths of pellet is the free falling time from 0.5 meters height until the pellets break.The standard indicators of ilmenite concentrate pellets are more than 800g to the compressive strength and more than 6 times to the drop strength according to the previous literature [5].

3 Results and discussion

3.1 Effect of particle size distribution

Ilmenite concentrates D, E, and F is obtained after the grinding of A, B, and C, respectively. Particle size distribution of concentrates used in the tests is shown in Table 3.

The binder content is 1.5% in each mixtures. A better pelletization performance can be acquired when the content of ilmenite concentrate particle size -74 μm more than 85% during the production process of pellets from our previous research. Therefore, ilmenite concentrate E maybe a better raw material to produce a pellet according to the particle size distribution [7]. The -74μm content in ilmenite concentrate A is only 51.84%. The results indicated that the ilmenite concentrate pelleting needs Fine particle size and binder. Compared with ilmenite concentrate A, ilmenite concentrate D is not only particle size significantly fine but also the compressive strength and drop strength increased 416 g and 4.7 times, respectively. It is indicated that the particle size of ilmenite concentrate is reduced can improve the pelletization performance.

Drop strength of pellets was reduced, and the compressive strength of pellets without significantly increased with decreased the particle size from E and F. This result strongly suggests that the smaller ilmenite is not obviously helping to improve performance. Therefore, the strength of pellets does not necessarily significantly improved via to reduce the particle size of ilmenite concentrate. Ilmenite concentrate needs to have a suitable particle size. The results are consistent with those reported in the literature [8].

Table 3. Particle size distribution of ilmenite concentrate/wt.%.

Ilmenite concentrate	Particle size distribution	
	-74μm	-45μm
A	51.84	24.68
B	77.74	43.22
C	54.60	28.54
D	71.36	46.84
E	98.65	85.90
F	76.10	53.90

Table 4. Average performance of pellets.

Ilmenite concentrate	Compression strength/g	Drop strength/times
A	—	—
B	465	5.1
C	683	6.4
D	416	4.7
E	1170	4.8
F	490	5.0

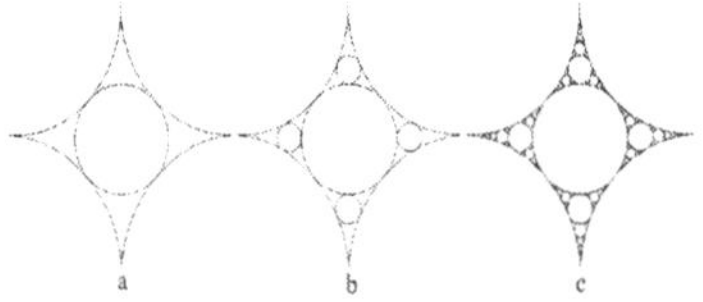

Fig. 3. Arrangement of ilmentie concentrate particle.

Figure 4 shows that the performances of all the pellets which were prepared using single ilmenite concentrate cannot meet the production requirements. Ilmenite concentrate C and E was carried out according to a certain ratio to study the pelletization performance of ilmenite. The ratio of raw material and indexes of pellets are shown in table 5 and table 6, respectively.

The average diameter of the capillary is reduced and the inter-particle binding force is increased, as shown in figure 3. Figure 3 (a) shows that the relatively uniform particle size could increase the porosity. Therefore, a suitable particle size composition is required to improve the strength of ilmenite concentrate pellets [9].

Table 5. Raw material ratio and particle size of distribution of ilmenite concentrate /wt. %.

No	Ilmenite concentrate		Binder /%	Particle size distribution	
	C	E		-74μm	-45μm
1	75	25	1.5	62.28	40.47
2	50	50	1.5	74.09	54.57
3	25	75	1.5	85.9	68.66

Table 6. Average performance of pellets.

No.	Compression strength/g	Drop strength/times
1	789	4.2
2	872	6.4
3	1070	8.4

3.2 Effect of the surface area

Some literature has pointed out that there is a linear relationship between the compressive strength and the specific surface area of pellets. Many researchers also use the specific surface area to measure the particle size of material.

The specific surface area of ilmenite concentrate B after grinding as show in figure 4. Figure 4 demonstrates that the specific surface area of ilmenite concentrate is increased with increasing the content of particle size less than 45μm from 40% to 80%. The particle size of ilmenite concentrate and surface area has a significant positive correlation.

Compressive strength and drop strength of obtained pellets from titanium concentrate B with different specific surface areas is shown in figure 5. Compressive strength and drop strength of pellets is increasing with the increase the specific surface area. It is indicated that the contact surface of particles was increased each other during the pelletization processing if the surface area of ilmenite concentrate larger. The strength of pellets was further increased. The produced pellets can be directly used in industrial production when the specific surface area of ilmenite concentrate reach 1493 cm^2/g. That is the content of particle size less than 45μm is about 70% in ilmenite concentrate.

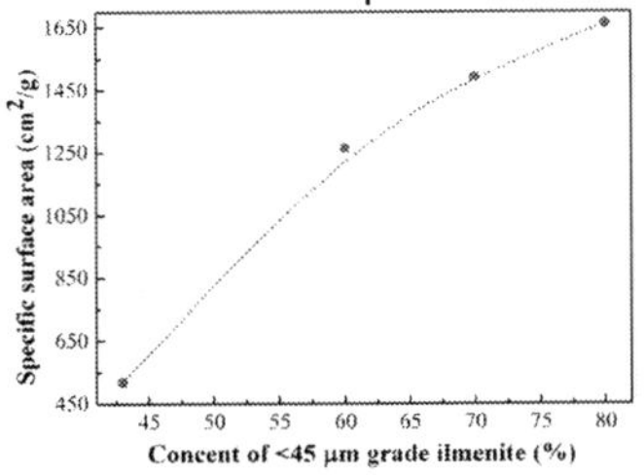

Fig. 4. Specific surface area of ilmenite concentrate B at different particle size.

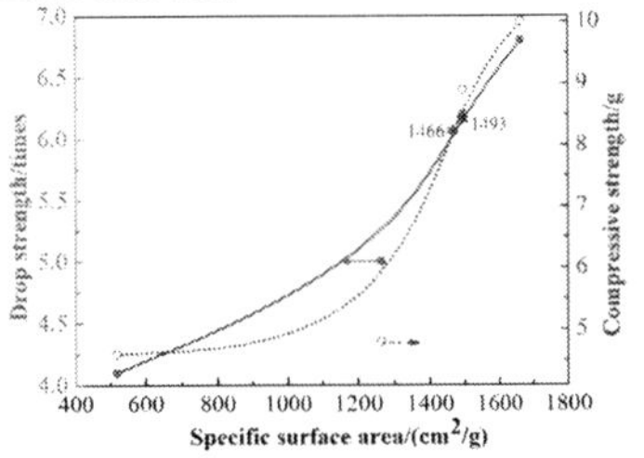

Fig. 5. Indexes of ilmenite concentrate B pellet at different surface area.

Particle microscopic morphology of ilmenite concentrate C is shown in Figure 6. Figure 6 (a) is ilmenite concentrate C; Ilmenite concentrate particles in (a), (b) and (c) are relatively regular in shape and have a dense and smooth surface. The particle shape of ilmenite concentrate is main long-sized in figure 6(C). The contact area of spherical particles is significantly less than long-sizes particles during the pelletization processing. Therefore, the particle size is reduced and the particle shape also changed after grinding of ilmenite concentrate and improved the pelletization performance of ilmenite concentrate.

Ilmenite concentrate which particle size less than 45μm reach 95% is so fine that the fine particles will adsorbed on the surfaces of larger particles to agglomerate. Agglomerate phenomenon of fine particle can be clearly observed in (e) or (f). Specific surface area of ilmenite concentrate will be decreased after agglomerate of particle. Specific surface area is directly determines the contact area between particles and also reflects the adhesion degree.

Fig. 6. Particle surface morphology of ilmenite concentrate C.((a)、(b)、(c) and (d) is 250×; (e) and (f) is 500× and 1000×, respectively).

3.3 Effect of binder

3.3.1 Organic binder

Ratio of mixture of 50% ilmenite concentrate C and E with binders as shown in table 7. Performances of pellets are shown in table 8. Compressive strength of pellets was decreased with increasing content of organic binder. Drop strength of pellets was reduced with decreasing the content of organic binder. Drop strength of pellets and moisture in pellet was increased with increasing organic binder content. Drop strength of $1^{\#}$ and $2^{\#}$ is without significantly difference. The better added content of organic binder could be controlled in 1.2% to meet performances of pellets and consider the cost of organic binder.

Table 7. Ratio of ilmenite and organic binder.

No	Ilmenite concentrate		Organic binder
	C	E	
1	50	50	1.5
2	50	50	1.2
3	50	50	0.9

Table 8. Average performance of pellets.

No	Compression strength/g	Drop strength/times	Moisture /wt.%
$1^{\#}$	872	6.4	9.48
$2^{\#}$	945	5.4	9.41
$3^{\#}$	959	3.2	8.53

3.3.2 Mixed binder

The above results shows that pellets with only organic binder can meet the smelting conditions, but organic binder cost is higher. Although the price of bentonite is low, the performance of pellets with bentonite poor. The mixed binder of organic binder and bentonite was employed to improve the performance of pellets. The ratio of ilmenite concentrate and mixed binder as shown in table 9. The average performance of pellets is shown in table 10.

Compared with table 8 or table 10, the performance of pellets with an amount of mixed binder are better than a single binder. It is indicated that the mixed binder is helping to improve the pelletization performance. The better ratio of mixed binder is 1.05% organic binder and 0.45% bentonite.

Table 9. Ratio of ilmenite concentrate and mixed binder.

No	Ilmenite concentrate		Organic binder
	C	E	
1	C	E	1.5
2	50	50	1.2
3	50	50	0.9

Table 10. Average performance of pellets.

No	Compression strength/g	Drop strength/times	Moisture
1	872	6.4	9.48
2	945	5.4	9.41
3	959	3.2	8.53

4 Conclusions

1. Panzhihua fine ilmenite concentrate can be pelleted through controlled the particle size diatribution or added the binder. The better process parameters were obtained and the index of ilmenite concentrate pellets established.

2. Reduced particle size of ilmenite concentrate will improve pelletization performance during a certain range. Finer ilmenite concentrate mixed with a certain amount of coarser titanium concentrate is more conducive to pelleting. Particle size -45μm content should be controlled at 55%, or -74μm at 74% in mixture of ilmenite concentrate.

3. The mixture binder of bentonite and organic binder can improve the performance of ilmenite concentrate pellets. The better ratio is 1.05% bentonite and 0.45% organic binder.

References

1. LYU X W, Song B, Han K X, Zhang K. Effect of ferrosilicon addition on direct reduction of ilmenite concentrate and Its separation between iron and slag [J]. Journal of Northeastern University (Natural Science), 2015, 36(2):223-226.
2. P, Pistorius and Coetzee C. Physicochemical aspects of titanium slag production and solidification [J]. Metallurgical and Materials Transactions B, 2003, 34 (5): 581~588.
3. Han K X. Study on granulation of fine grained ilmenite concentrate [J]. Panzhihua Sci&Tech Information, 2011,36(4):52-56.
4. Ren C C. Pelletizing experimence of micro-fine grained ilmenite concentrate [J].Express Information of Mining Industry, 2001,(10):296-299.
5. Fu J Y, Jiang T, Zhu D Q. Sintering pelletization [M]. Changsha: Central South University of Technology Press, 1996.
6. P. chindaprasirt, S. Hatanaka and N. Mishima et. al. Effects of binder strength and aggregate size on the compressive strength and void ratio of porous concrete[J]. International Journal of Minerals Metallurgy and Materials.
7. Hu Z Q, Pan J, Zhu D Q, Yue S Y. Raw material characteristics of iron ore concentrate and its influence on balling performance [J]. Sintering and Pelletizing, 2013, 38(4):42-49.
8. Zhang Y X, Rao J T, He M G, Li Y H. Research on the suitable particle size of production pellets about vanadium titanium ore concentrate [J]. Sichuan Metallurgy, 2010,33(5):12-15.
9. Wang C A, Luo L M. The influence of pelletizing quality by using different size distribution of concentrate [J].Journal of Wuhan Institute of Chemical Technology, 2005,27(2):38-41.

MATEC Web of Conferences **277**, 03004 (2019) https://doi.org/10.1051/matecconf/201927703004
JCMME 2018

C^3I cooperative decision system simulation and optimization based on genetic algorithm for surface warship formation

Mai-di Liu, *Ke-wei* Yang, *Qing-qing* Yang, and *Zhi-wei* Yang*

College of Systems Engineering, National University of Defense Technology, Changsha, Hunan, P.R.China

Abstract. The C^3I cooperative decision system is the guarantee for combat capabilities of surface warship formations. At present, research on the C^3I system simulation lays more emphasis on finding structural logic defects by executing the simulation model, lacking the application of intelligent optimization algorithms to optimize parameters in the system. In this paper, the cooperative decision-making process of surface warship formation defense system is studied. Meanwhile, modelling and optimization methods for cooperative decision system are proposed. Based on simulation models built on the ExtendSim platform, this paper optimizes the staffing strategy of decision makers based on the genetic algorithm, to improve the per capita decision efficiency. The optimized staffing strategy meets objectives and requirements. The research in this paper can provide a scientific and objective reference for relevant decision-makers and researchers.

1 Introduction

With the continuous progress of the performance of equipment such as remote cruise missiles and stealth aircrafts, surface warship formation is facing various types of threats.[1] A significant factor to ensure the coordinated operation of equipment on surface warship formation is the efficiency of ship-borne C^3I systems. The C^3I system is a military electronic comprehensive information system, connecting various subsystems such as command, control, communication and intelligence. Although the C^3I system has no direct lethality, it can coordinate killing weapons to achieve maximum combat effectiveness. [2]

To model the C^3I system, both static characteristics and dynamic behaviors should be considered. Numerous scholars used Petri net to simulate and analyze the C^3I system. *Li Dajian*[3] set up a Petri net model of air defense C^3I decision system for the camp and brigade decision organization. *Chen Xingyou*[4] established an air defense C^3I system model based on hierarchical fuzzy colored Petri net. *Zuo Xiaofeng*[5] built a colored Petri net simulation model for the warship formation combat C^3I system, and the system decision delay was analyzed. *Zhao Yanquan*[6] applied the object colored Petri net to analyze the relationship of the composition of ship-borne C^3I system. Researchers mostly focus on

* Corresponding author: zhwyang@nudt.edu.cn

finding logic defects in the C^3I system by executing simulation models, or manual optimizations by modifying the system structure. Current research lacks using heuristic algorithm to search optimal solutions of parameters in C^3I system, which ignores the optimization ability of simulation models.

In this paper, the *ExtendSim* software is used to simulate and analyze the decision-making process of C^3I cooperative decision system of surface warship formation. The genetic algorithm is applied to optimizes the parameters, which improves the efficiency of decision-making.

2 Study on C^3I cooperative decision system of surface warship formation

In order to disperse the combat decision pressure, the cooperative decision system is adopted by surface warship formations.

2.1 Classification rules for threat targets

When the scouting system detects threat enemies, it will report parameters including direction, velocity, quantity and distance to C^3I system to complete the battle decision.

According to parameters of targets, decision-makers obey rules to cooperate and determine the category and the threat level of targets, which helps the defense system complete interceptions.

Table 1. Category classification rules.

Velocity (km/h) / Quantity	≤400	400~800	>800
<8	F	S	B
8~12	B	B	F
>12	B	B	B

Table 2. Threat level classification rules.

Distance (km) / Category	>120	15~120	<15
S	Small	Small	Small
F	Small	Median	Great
B	Median	Great	Great

Note: S means less lethality; F means moderate lethality; B means great lethality.

2.2 The flow of warship formation cooperative combat decision making

Faced with "Saturated Attacks", an effective tactic, warship formations will be threatened by high-density attacks of different types from all directions.[7] In this case, the cooperative combat decision making system is used to relieve the pressure of defense system. The combat area is divided into 12 equal sector areas (r1~r12), meanwhile, 2 primary decision makers (DM1, DM2) and 4 secondary decision makers (DM3, DM4, DM5, DM6) are appointed to be responsible for 12 areas.[8]

R1~r6 are under the jurisdiction of the primary decision maker DM1, in which r1~r3 are assigned to the subordinate secondary decision-maker DM3, and r4~r6 are assigned to the subordinate secondary decision-maker DM4; r7~r12 are under the jurisdiction of the primary decision maker DM2, in which r7~r9 are assigned to the subordinate secondary decision-maker DM5, and r10~r12 are assigned to the subordinate secondary decision-maker DM6.

If the threat is attacking at the jurisdiction boundary (r1, r6, r7, r12) of primary decision makers, DM1 and DM2 will cooperate to complete the decision. For example, if the threat is attacking at r1 (under the jurisdiction of DM1), the decision-making authority will be assigned to DM1, and the proposal authority will be assigned to DM2. DM1 will complete the battle decision by referring to the proposal from DM2, obeying rules as Table 3. This method is also applicable to two secondary decision makers subordinate to a same primary decision maker.

The flow chart of the cooperative combat decision making is as Figure2.

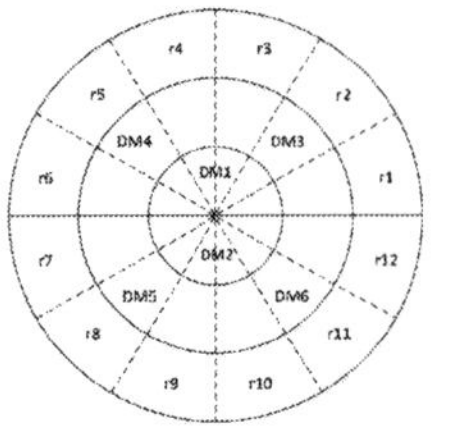

Fig. 1. Division method of certain warship formation combat area.

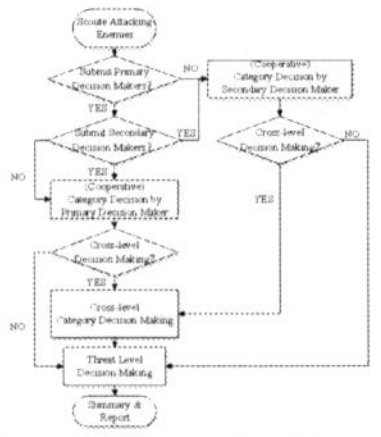

Fig. 2. Flow chart of C^{3}I cooperative combat decision system for warship formation.

Table 3. Decision Rules for Cooperative Decision Making.

Proposal DM Result / Decision Making DM Result	S	F	B
S	S	F	F
F	S	F	B
B	F	F	B

Primary decision makers and secondary decision makers cooperate by the following rules:

If the target distance is shorter than 15km, the primary decision maker is assigned decision-making authority and the secondary decision maker is assigned no authority. If target distance is between 15km and 120km, the primary and second decision maker are assigned proposal authority and decision-making authority respectively. If the distance is longer than 120km, the secondary decision maker is assigned decision-making authority and the primary decision maker is assigned no authority.

3 Research on modelling and optimization method of C^{3}I cooperative decision system for surface warship formation

3.1 Modelling method of C^{3}I cooperative decision system for surface warship formation

This paper mainly focuses on the decision process of the C^{3}I cooperative decision system of surface warship formation.

Hypothetical conditions are as follows:

- The incoming target stream obeys Poisson distribution, whose direction a, velocity v, quantity q and distance d are independent random variables.
- $NumDM_i$ represents the quantity of staffs equipped by the decision maker DM_i.
- The number of attacking targets in time period t_0 is $NumAttacking(t_0)$.
- The discrepancy of decision makers' ability is not taken into consideration.

- The processing time $t(S_i)$ for decision makers to complete the certain decision S_i obeys normal distribution, namely $t(S_i) \sim N(\mu, \sigma^2)$.
- A decision cycle is defined as the process from detecting the targets to summary and report. The number of decision cycles completed in time t_0 by cooperative decision-making system is recorded as $NumDecision(t_0)$.
- In time t_0, the target decision rate $r(t_0)$ of the decision system is defined as $NumDecision(t_0) / NumAttack(t_0)$. If $r(t_0)$ >90%, the cooperative decision system is considered to have a strong threat processing ability.

The objective function is:

$$\max W_p = \frac{NumDecision(t_0)}{t_0 \times \sum_{i=1}^{6} NumDM_i} \quad \text{s.t.} \begin{cases} \dfrac{NumDecision(t_0)}{NumAttack(t_0)} \geq r(t_0), t_0 > 0; \\ NumDecision, NumAttack, NumDM_i \in N_+ \end{cases}$$

3.2 Optimization method for staffing strategy of C³I system based on genetic algorithm

Genetic algorithm is an iterative probability optimization algorithm based on the natural selection principle. The application of genetic algorithm in optimizing staffing strategy of C³I cooperative combat decision system is as follows.

3.2.1 Coding method

In order to facilitate the coding, decoding, cross and mutation on the computer, a binary encoding method is adopted. A chromosome contains 6 segments, each consisting of m genes. Genes in 6 segments represent the number of staffs equipped in 6 decision makers respectively, of which values are integers in the closed interval from 0 to 2^m-1.

3.2.2 Fitness function

To ensure the sufficient influence from the number of staffs, the fitness function is defined as W, taking out the constant t_0 from the denominator of the per capita decision-making efficiency W_p. The fitness function is $W = NumDecision(t_0) / \sum_{i=1}^{6} NumDM_i$.

4 Simulation model and optimization analysis of surface warship formation C³I cooperative decision system based on *ExtendSim*

In this paper, *ExtendSim*, a universal simulation platform based on the C language, is used to build a simulation model for the warship formation C³I cooperative decision system, and optimize the parameters in the system. [9]

4.1 Simulation model building with ExtendSim

The simulation model is divided into 4 main functional modules as follows:

4.1.1 Target stream generation module

Target flow is the premise of the simulation operation. This paper assumes that the target flow obeys Poisson distribution, of which the average interval time is 3 minutes. Other attributes are set randomly.

4.1.2 Decision-making authority allocation module

Decision-making authority allocation is fundamental to ensure orderly cooperation among decision makers. C^3I system assigns authorities to different levels of decision makers according to the attributes of the target. The primary-level authority allocation process is introduced as follows.

The routing item divides targets into 12 paths according to directions: targets from r2, r3, r4, r5 are submitted to DM1 alone; targets from r8, r9, r10, r11 are submitted to DM2 alone; targets from r1, r6, r7, r12 are submitted to both DM1 and DM2 to make cooperative decisions.

4.1.3 Target category cooperative decision module

This module is divided into 3 parts: cooperative decision making of primary decision makers, cooperative decision making of secondary decision makers, and cross-level decision making.

The cooperative process of primary decision makers is as follows. The routing item determines decision-making authorities and proposal authorities according to the target direction. For example, the action of DM1 is divided into 3 cases: independent decision making, accepting proposals from DM2, send proposals to DM2.

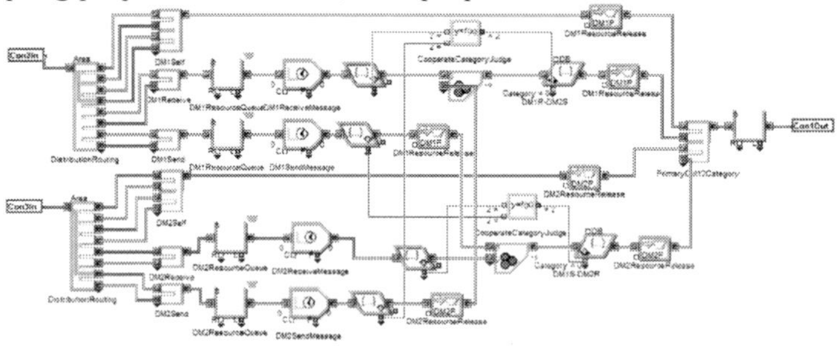

Fig. 3. Cooperative decision of target category module.

4.1.4 Target threat level decision making and summary module

Decision makers determine the threat level with category and distance attributes. A summary of decisions is sent to the defense system of the warship formation, as the end of the simulation.

4.2 Model optimization and analysis

On the basis of the simulation model, this paper attempts to improve the per capita decision-making efficiency and find an optimal solution for the staffing strategy for cooperative decision. The model is optimized by the "Optimizer" item in *ExtendSim*, applying genetic algorithm.

Parameters of resources occupied by each decision are set up. Time consumption of activities follows normal distribution, whose variances are 1 and mean values are as follows.

Table 4. Resource occupied by decision-making activities.

Decision-Making Activities	Time Consumption (Minute)	Human Resource (Person)
1. Authority Allocation for Different Decision Levels	1	/
2. Subdividing Authorities in Each Level	1	/
3. Determine Target Category by Decision Makers Alone	3	2
4. Target Category Decision by Decision Makers in the Same Level (Send Proposals)	2	1
5. Target Category Decision by Decision Makers in the Same Level (Receive Proposals and Make Decisions)	4	2
6. Target Category Decision by Cross-level Decision Makers (Send Proposals)	3	1
7. Target Category Decision by Cross-level Decision Makers (Receive Proposals and Make Decisions)	5	2
8. Determine the Target Threat Level by Decision Makers	3	2
9. Decisions Summary and Report	1	2

The target of optimization is to find an optimal solution for staff strategy to enhance the per capita decision-making efficiency, while ensuring the value of target decision rate.

The optimization-related variables are cloned into the "Optimizer" item to set the objective function in genetic algorithm. The numbers of the staffs are integers in the closed interval from 1 to 15 (on account of the coding method of 4 bit binary), and the simulation time is set to 1440 minutes (24 hours).

The number of running times is set to 1000 times; a single sample runs 5 times at most; the mean of samples is used as the comparison standard; the standard of convergence degree is over 96% (checked after 100 generations). Taking the first round of optimization as an example, the process of optimization costs 19 minutes (140 generations). The optimization curve is shown in Figure 4.

In order to enhance the probability of getting the global optimized solution, the optimization is processed by multiple rounds, of which 10 rounds are shown as Figure 5.

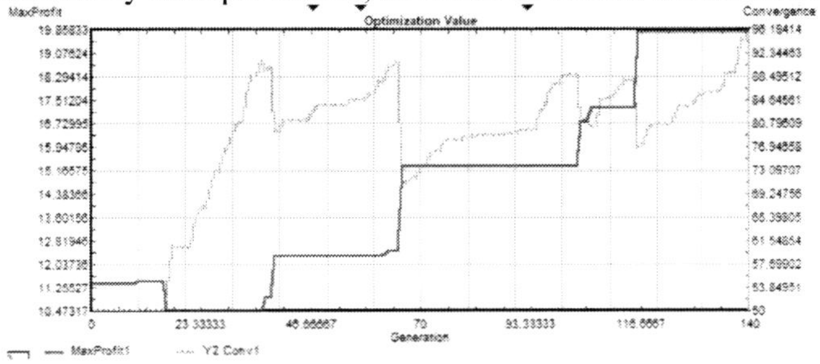

Fig. 4. Process optimization curve. (Note: The green full curve is the objective function, which reaches 19.86. The red imaginary curve is the convergence curve, which reaches 96.2%.)

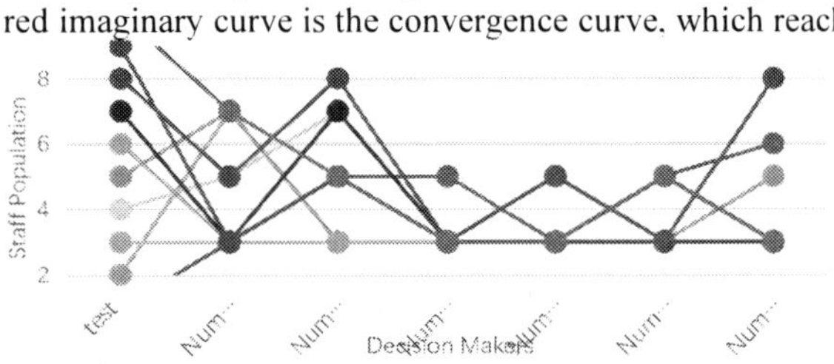

Fig. 5. Results of 10-round optimization.

The result with the maximum objective function value is selected to be the optimized staffing strategy for the C^3I cooperative decision system.

Table 5. Optimized staffing strategy for cooperative decision system.

Decision Maker	DM1	DM2	DM3	DM4	DM5	DM6
Staff Quantity (Person)	3	5	3	5	3	3

The target decision rate of the above staffing strategy is calculated to be 93.9%, which conforms to the minimum constraint of 90%. Consequently, with the optimized staffing strategy, it is believed that the C^3I cooperative decision system has a strong threat processing capability.

5 Future research

Modelling and optimization methods as well as the final result in this paper have scientific and objective reference significance for relevant researchers. Future research can focus on the aspects:

- Priority can be introduced to targets, so that targets of close distance can be treated first.
- The difference of personnel decision-making ability can be considered.
- The parameters of the incoming target flow can be further studied, such as the probability distribution types of frequency, direction, speed, quantity and distance of the target.

References

1. ZHU Chuan-wei, YANG Xing-bao, LU Yong-hong. Research on Anti-Missile System of Warship Formation Network-Centric[J]. Journal of Sichuan Ordnance, 2015(2):12-14
2. YUE Liang, HUANG Zheng, TAN Meng-quan. Methods For Air Defense Operation Mission Reliability Modeling of Warship Formation[J]. Ship Electronic Engineering, 2015(9):123-126
3. LI Da-jian, WANG Feng-shan. Petri Net-based Model Analysis for Air Defense C^3I Decision System[J]. Systems Engineering and Electronics. 2005(27):1600-1602
4. CHEN Xing-you, YUE Xiao-bo, CAO Wei, ZHOU Kai-qing. Modeling and Analysis of Anti-aircraft C^3I System Based on Herarchical Fuzzy Colored Petri Nets[J]. Computer Engineering & Science.2016(38):102-107
5. ZUO Xiao-feng. CPN in Modern Formation of Naval Vessels Combat Teamwork C^3I System Modeling Simulation Application [D]. Xian: Xidian University, 2008
6. ZHAO Yan-quan. Colored Petri Net for Ship-borne C^3I System Based on Objects[J]. Fire Control & Command Control. 2004(2)
7. XIE Yu-peng, WANG Zong-jie, YANG Shi-feng, ZHONG Ling. Research on Anti-saturation Attack Model of Ship Formation to ASM[J]. Fire Control & Command Control. 2014(39):23-26
8. HAN Qiang. Model and Application of C^3I System for Aircraft Carrier Groups Based on Petri Nets[D]. Xian: Xidian University, 2007
9. QIN Tian-bao, WANG Yan-feng. Application Oriented Simulation Modeling and Analysis with ExtendSim[M]. Beijing: Tsing-hua University Press, 2011.

Skeleton based gait recognition for long and baggy clothes

Abrar Alharbi*, *Fahad* Alharbi, and *Eiji* Kamioka

Graduate School of Engineering and Science, Shibaura Institute of Technology, Tokyo, Japan

Abstract. Human gait is a significant biometric feature used for the identification of people by their style of walking. Gait offers recognition from a distance at low resolution while requiring no user interaction. On the other hand, other biometrics are likely to require a certain level of interaction. In this paper, a human gait recognition method is presented to identify people who are wearing long baggy clothes like Thobe and Abaya. Microsoft Kinect sensor is used as a tool to establish a skeleton based gait database. The skeleton joint positions are obtained and used to create five different datasets. Each dataset contained different combination of joints to explore their effectiveness. An evaluation experiment was carried out with 20 walking subjects, each having 25 walking sequences in total. The results achieved good recognition rates up to 97%.

1 Introduction

The employment of modern technology to serve human security is considered one of the most essential services that should be provided to humanity with fewer errors. Due to the technological evolution, the security services are becoming better in accuracy and quality, which has increased its usage in securing critical places such as airports and stations. Security systems based on video cameras have been commonly used because of many advantages and large coverage domain. However, its results are affected by several factors such as lighting conditions, the angle of view, video quality, and the changes in the personal appearance. Therefore, biometric systems such as fingerprint, iris, voice, and facial recognition techniques have been widely used as powerful tools to identify and recognize a person [1]. Gait has been considered as a promising and unique biometric for person identification. Gait recognition is defined as automatic identification of an individual based on the style of walking. Gait recognition systems can be divided into either model-based or model-free approaches. Model-based systems generate the gait signature from modeling and tracking different parts of the body as segments or joint positions over time [2]. The model-free approach extracts gait features directly from the motion or shapes for recognition [3]. Implementing model-free approach is easier than model-based and less computationally expensive. However, the model approach free approach lacks the precision in identifying a subject's gait when silhouettes change due to poor visibility, varied carrying conditions, bulky clothing and occlusions [4]. The goal of

* Corresponding author: mg17002@shibaura-it.ac.jp

this work is to design a method for identifying people who are wearing long and baggy clothes like Thobe and Abaya. Figure 1 shows the Abaya which is a traditional female cloth in Saudi Arabia as well as the Thobe which is the male version cloth. In this study, an experiment was conducted to find out which joints provide reliable information for recognizing the subjects while their body is obscured. The method used in the experiment will be described and the result also will be presented in this paper.

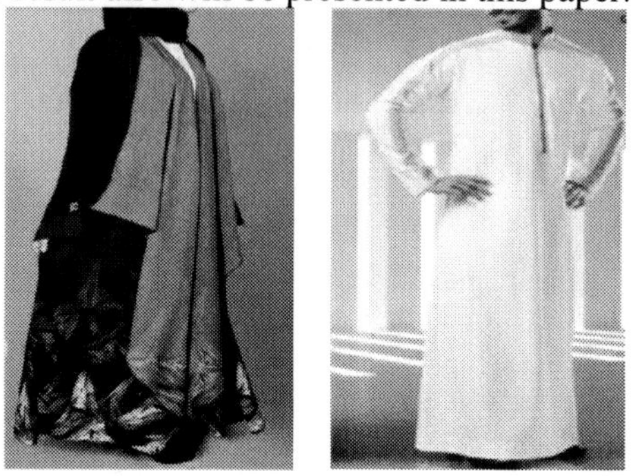

(a) (b)

Fig. 1. An example of long and baggy is shown above. In (a) a traditional female cloth in Saudi Arabia (Abaya) and in (b) Traditional male cloth in Saudi Arabia (Thobe).

2 Related work

Many researchers have used Microsoft Kinect sensor for gait analysis and recognition in the past few years. For example, Ahmed *et al.* [5] presented a method for gait recognition from two sets of dynamic features. The first set is Horizontal Distance Features that is based on the distances between each pair of (Ankles, knees, hands, shoulders). The second set is Vertical Distance Features that provide significant information of human gait extracted from the height of (hand, shoulder, and ankles) to the ground. Their results reach 92% on their own database of 20 walking subjects with 10 walk cycles per subject on average. Dikovski *et al.* [6] evaluated a broad spectrum of geometric features and classifiers. Static body parameters like joint angles and inter-joint distances were aggregated and combined to construct seven different feature datasets. Multilayer perceptron, support vector machine with sequential minimal optimization and J48 algorithms were used for classification on these datasets to find out the role of different types of features and body parts in the recognition process. On their database of 15 walkers where each contributed with about 2 full gait cycles, the 89.8% recognition rate (10-fold cross-validation by MLP) was reached using a descriptor of 71 geometric features. Kumar & Babu [7] proposed an unrestricted gait recognition algorithm which uses trajectory covariance of joint points. The covariance matrices between these joint point trajectories formed the gait model. By computing the minimum dissimilarity measure between the gait models of the training data and the testing data the subject identity was classified. The recognition rate of 97.5% (2-fold cross-validation by 1-NN) is achieved for their dataset of 20 walking subjects, 10 gait samples each. Preis *et al.* [8] defined 13 biometric features, 11 static body features, and two dynamic features. Only four static features of them were sufficient to correctly identify a person with a recognition rate of 91% which was achieved by the Naïve Bayes classifier for their dataset of nine walking subjects. Balazia *et al.* [9] extracted 21 joint angles from 2 signature pose to form the gait pattern descriptor. Subject's identity was classified by the 1-NN classifier. The accuracy of 92.4% is reached on their database of 4188 walk samples of 48 people. Andersson *et al.* [10] used the skeleton joints to extract anthropometric information and calculate lower joint (hips, knees, and ankles) angles as gait kinematic parameters. The mean and standard deviation of the body segment were also used as attributes. Walker identity was classified K-Nearest Neighbor. On their Kinect database of 160 individuals of about five walks each, the method achieves (10-fold cross-validation) 80% recognition rate. Faisal *et al.* [11] introduced a dynamic time warping (DTW) kernel based rank level fusion that takes a collection of joint relative distance (JRD) or joint relative angle (JRA) sequences as parameters and computed a dissimilarity measure between the training samples and the unknown samples. The JRD–JRA fusion achieves a recognition rate of 92 % on their database of 20 walking subjects with three walk cycles per subject.

3 Proposed method

The purpose of this study is to identify people who are wearing long baggy clothes like Thobe and Abaya using gait-based skeletal data from Kinect V2. The main difficulty is derived from extracting the human gait characteristic features to be used for identification, especially when the person body limbs are obscured, and the gait shape is hidden. The proposed gait recognition system is executed in three steps: First, building a database by recording skeleton information using the Kinect sensor. Second, extracting the gait features by detecting the skeleton information and obtaining the joint positions. Finally, training and recognizing using four different classifiers, Support vector machine (SVM), J48 decision tree algorithm, K-Nearest Neighbor (KNN) and Multilayer Perceptron (MLP) artificial neural network

3.1 Data acquisition

There was not any existing database that delivers the skeleton information from Kinect sensor for people wearing long baggy clothes. Therefore, a database for this study was created through an experiment. The environment for recording the skeleton information is as follows: The Kinect sensor was placed parallel to the walking line at the height of 0.26 meters from the ground level and 3m away from the walking line. Twenty different people wearing either Thobe or Abaya joined the experiment. The recordings were done in a lab room at daytime. Each person walked 25 times from right to left in front of the Kinect sensor at their normal speed. All the skeleton information provided by the Kinect sensor were collected from each frame to create the database.

3.2 Feature extraction

The Kinect sensor and its SDK provide a 3D skeleton model of the entire human body [12]. The skeleton consists of 25 joints. The estimated skeleton structure provided by the Kinect and joint names are shown in figure 2. The Kinect provides approximately 30 skeleton frames per second. For each frame, the joint position of each joint is expressed in the X, Y and Z coordinates. In this study, the skeleton joints position in XYZ-coordinates is extracted. However, only the Y-coordinate is used as input feature because the distance between the Kinect sensor and the walking line was fixed as well as the direction of walking which means that the X and Z coordinates are irrelevant. Twenty subjects joined the experiment; each subject had 25 walk sequences which resulted in 500 gait samples obtained for a side view. For each sequence, 32 frames were chosen based on the estimated time to generate one gait cycle on average. The Skeleton joints for each subject were extracted in a column vector of 25 by 32. Recall that, thirty samples were allocated for each subject, which resulted in 800 by 500.

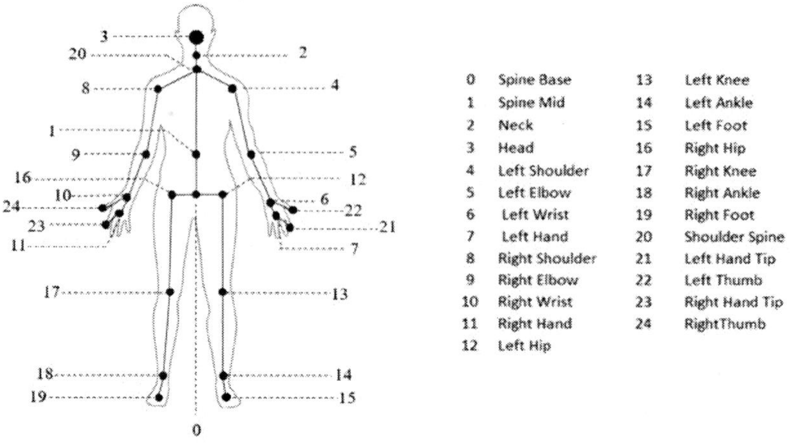

Fig. 2. The skeleton structure of a subject as provided by Microsoft Kinect SDK 2.0.

4 Experimental setup and design

From the skeleton information explained in the previous section, five different datasets were created. The first dataset contains all the skeleton joints positions per walking cycle. The second dataset contains the joint position of (Spine Mid, Neck, Head, Shoulders, Elbows, Wrists, Ankles, Feet, Shoulder Spine). These joints were chosen because they can represent the motion of the human body [13]. However, the hip and knee joints were not considered because their positions are difficult to detect. The third dataset is similar to the second one excluding the joints position on the right side of the body in order to focus on the side facing the Kinect. The fourth data set is constructed from the joints of the upper body part, and the final fifth dataset contained the joints of the lower body part. The final number of joints used in each dataset is shown in table 1.

Table 1. Different datasets used.

Dataset	Joint
1	All joint
2	1,2,3,4,5,6,8.9,10,14,15,18,19,20
3	1,2,3,4,5,6,14,15,20
4	1,2,3,4,5,6,7,8,9,10
5	0,12,13,14,15,16,17,18,19

5 Results and discussion

The performance of the proposed approach was evaluated using a self-constructed Kinect-based side view gait database. In the database, Kinect is used to extract twenty-five different walking sequences for twenty subjects. The achieved results using ten-fold cross-validation in the experiments for each method are shown in table 2. The datasets are listed in the first column of the table, while the remaining columns show the performance of each classifier for every dataset. The MLP and SVM classifiers gave similar results except for the first dataset the SVM achieved better results than the MLP due to its ability to work with high dimensional data. The second and the third datasets achieved the highest recognition rate which showed the effectiveness of the selected joints. However, the third dataset achieved better accuracy than the second one, because few right side joints were detected incorrectly. Finally, the fourth data set achieved better results than the fifth data set which indicated that upper body joints give more information than the lower body joints in the process of gait recognition for long and baggy clothes.

Table 2. Recognition rates from test combinations using ten-fold cross-validation.

Dataset	MLP	SVM	KNN	J48
1	82	90	89	81
2	95	95	93	89
3	97	97	94	83
4	92	91	89	80
5	88	85	84	73

Table 3 shows recognition rates obtained in the Comparison of the proposed method with some of the other studies of gait recognition based on Kinect sensors that have been discussed in the related work section using the Thobe and Abaya database.

Table 3. Comparison According to Recognition Rate.

Method	Recognition Rate
Ahmed [5]	64
Preis [8]	50
Proposed method	95

6 Conclusions and future work

In this paper, a human gait recognition method was introduced. The skeleton joints provided by the Kinect sensor were utilized to achieve proper recognition for the people who are wearing long and baggy clothes like Thobe and Abaya. The skeleton information of each person was detected and used to create five data sets. Based on the experiment on a database of 20 subjects and 25 walk sequences, a high recognition rate of 97% was achieved by using the joints of the spine, head, neck, left shoulder, left elbow, left ankle, and left foot. Therefore, the proposal confirmed the effectiveness of gait recognition for long and baggy clothes. In the future, the dataset will be expanded with more number of subjects and walking scenarios, and advanced techniques like deep learning will be used to get a better recognition accuracy.

References

1. Xiao Q 2007 Technology review-biometrics-technology, application, challenge, and computational intelligence solutions *IEEE Computational Intelligence Magazine* vol **2** no 2 pp 5-25
2. Liu L, Jia W and Zhu Y 2009 Survey of gait recognition *Emerging Intelligent Computing Technology and Applications. With Aspects of Artificial Intelligence* pp 652-659

3. Han J and Bhanu B 2006 Individual recognition using gait energy image *IEEE transactions on pattern analysis and machine intelligence* vol **28** no 2 pp 316-22
4. Wang J, She M, Nahavandi S and Kouzani A 2010 A Review of Vision-Based Gait Recognition Methods for Human Identification *2010 International Conference on Digital Image Computing: Techniques and Applications* pp 320-7
5. Ahmed M, Al-Jawad N and Sabir A 2014 Gait recognition based on Kinect sensor *Proc. SPIE Real-Time Image and Video Processing* vol 9139 pp 1-10
6. Dikovski B Madjarov G and Gjorgjevikj D 2014 Evaluation of different feature sets for gait recognition using skeletal data from Kinect pp 1304 -1308
7. Kumar M S N and Babu R.V 2012 Human gait recognition using depth camera: a covariance-based approach pp 1-6
8. Preis J, et al., 2012 Gait recognition with Kinect *Int. Workshop on Kinect in Pervasive Computing*
9. Balazia M and Plataniotis K 2015 Human gait recognition from motion capture data in signature poses *IET Biometrics* vol **6** no 2 pp 129-37
10. Andersson V, Dutra R and Araujo R 2014 Anthropometric and human gait identification using skeleton data from Kinect sensor pp 60-1
11. Ahmed F, Paul P P and Gavrilova M 2015 DTW based Kernel and Rank Level Fusion for 3D Gait Recognition using Kinect vol 31 no 6 pp 915-24
12. Zhang Z 2012 Microsoft Kinect Sensor and Its Effect *IEEE MultiMedia* vol **19** no 2 pp 4–10
13. Johannson G 1973 Visual perception of biological motion and a model forits analysis *Perception Psychophys* vol 14 pp 201–211

A novel approach for solving multi-objective unit commitment based on decomposition-coordination

Yungui Zhang[1,*], *Huizhao* Li[1] , *Xue* Wang[1], and *Yan* Zeng[2]

[1]State grid Hubei Electric Power Co., Ltd., 430000 Wuhan, China
[2]State grid Wuhan Electric Power Co., Ltd., 430000 Wuhan, China

Abstract. Multi-objective unit commitment (MOUC) considers simultaneously both economic and environmental objectives, then finds the best trade off with respect to these objectives. This paper proposes a novel model for MOUC, and a decomposition coordination approach is presented to solve the model. The MOUC model considers environmental objective by introducing a novel penalty term, and it's a quantized term for preference of environmental objective, which could be a basis for carbon tax makers. The model is solved by a decomposition coordination approach, which decomposes the whole system into subsystems and performs an iterative process. During each iteration step, the tie-line is updated based on the margin price in connected subsystems, then, each subsystem is solved by Lagrangian relaxation (LR), and the result is improved during iterations as shown in case studies. Besides, as LR does not require uploading units' parameters, it protects the privacy of generating companies. Numerical case studies considering different scenarios, conducted using the proposed multi-objective model, are applied to illustrate the robustness as well as the performance of the approach.

1 Introduction

Currently, a great number of scholars have conducted in-depth research on unit commitment(UC) problem and have achieved great performance. UC is a mixed integer, nonconvex, non-linear problem, and with the scale of the system grows, the computation time increases exponentially. The objective of UC is to determine the optimal or near optimal operating schedule of the power system in a feasible time, and the schedule must minimize the operating and commitment cost of a giving forecasted system load, considering the unit and system constraints as well as the tie-line limitation between any pairs of regions.

In [1], a stochastic process is performed to update the multiplier by combining genetic algorithm and LR. In [2], subgradient method is presented, but may suffer from infeasible spinning reserve solution [1]. After 1995, some papers use heuristics method to solve UC problem. Ma [3] firstly applied the genetic algorithm to UC, then, multiple modified

* Corresponding author: 1322404080@qq.com

versions of genetic algorithm were proposed, much of them focused on coding schemes. In [4], a integer-coded genetic algorithm is used to achieve a reduction of chromosome size, and the robustness and computation time are improved. Some papers focus on reducing the dimensions by reducing the number of binary variables or decomposition of the problem. Carrion [5] utilizes less binary variables to model intertemporal constraints, which reduces the computation burden and improves the speed performance. In [4], the uncertainties of the units and transmission lines is presented by a stochastic long-term UC, and the stochastic process is decomposed to long-term subproblems which could be solved by a hybrid method. Nowadays, researchers tend to use optimizers like GUROBI or CPLEX to deal with mixed integer problems, which use branch and bound algorithm, and models based on these optimizers have been proposed [6, 7, 8, 9].

For the above literature about UC, most of the research failed to consider the emission or other environmental indicators, but the thermal cost only. However, researchers have addressed that issue by search of the entire Pareto-optimal front, such as weighted sum, etc. This article presents a novel model and a decomposition-coordination method for solving MOUC problem, which takes environmental and economic indicators into consideration. The whole system is decomposed to subsystems by their geographic region, and the coordinator is the Lagrangian multiplier in subsystems which stands for the margin price. The method performs an iterative process, and achieves a improved result in feasible time considering the tie-line constraints.

2 Model

The MOUC proposed in this study is to optimize the economic and environmental objectives, and there has been many techniques to deal with MOUC problem. The most popular strategy is to modify the objective or constraints in order to find the Pareto optimal front, which is also known as non-dominated solutions, then performs a technique likes Nash [10] to find a single solution that satisfies the subjective preference of scheduling decision-makers. The strategy above requires calculation of UC on the whole system for multiple times, but could be solved in parallel. Though finding Pareto optimal front could consume little time, the strategy usually confirms the final solution by observing the percentage improvement of the two objectives, which could be highly subjective. In this study, two objectives are considered. The first objective is to minimize the overall production cost over the scheduling horizon, which could be expressed as the sum of the start-up cost and the fuel cost. The fuel cost function can be expressed as quadratic form, the coefficients for unit i are a_i, b_i, c_i.

$$\min F_1 = \sum_{t\in\{1\ldots T\}} \sum_{i\in\{1\ldots N\}} u_{i,t}(a_i p_{i,t}^2 + b_i p_{i,t} + c_i) + u_{i,t}(1-u_{i,t})ST_i \quad (1)$$

$u_{i,t}$ is the binary variable corresponding to the status of unit i at time interval t. T is the number of time intervals and N is the number of units. The generation of unit i at time interval t is denoted as $p_{i,t}$, and the start up cost of unit i is denoted as ST_i. The second objective is to minimize the emission, the emission function can be written in form of quadratic function.

$$\min F_2 = \sum_{t\in\{1,\ldots,T\}} \sum_{i\in\{1,\ldots,N\}} u_{i,t}(d_i p_{i,t}^2 + e_i p_i,t + f_i) \quad (2)$$

These objectives must be minimized over a set of constraints.

(i) System power balance

(a) For any giving time intervals, the total power production must equal to the overall loads.

$$D_t = \sum_{i\in\{1,\dots,N\}} u_{i,t} p_{i,t} \tag{3}$$

D_t is the load demand at time interval t.

(b) System reserve constraints During each time interval, sufficient spinning reserve must be available.

$$\sum_{i\in\{1,\dots,N\}} u_{i,t} p_{i,max} > D_t + R_t \tag{4}$$

In our numerical examples. R_t is set to 0.05 times of D_t.

(ii) Unit constraints

$$\begin{cases} u_{i,t} = 1 & T_{i,on} < T_{i,up} \\ u_{i,t} = 0 & T_{i,off} < T_{i,down} \\ u_{i,t} \in \{0,1\} & \text{otherwise} \end{cases} \tag{5}$$

$$p_{i,min} < p_{i,t} < p_{i,max}, -ramp_i < p_{i,t} - p_{i,t-1} < ramp_i, u_{i,t} \in \{0,1\} \tag{6}$$

(iii) Tie-line constraint

$$-P_{max}^{(tie)} \le P_t^{(tie)} \le P_{max}^{(tie)} \tag{7}$$

The whole system is separated to subsystems based on their geographical location, and tie-line between subsystems should not be overloaded. The power flow of tie-line is defined as a sequence $P^{(tie)} = \{p_1^{(tie)}, p_2^{(tie)}, \dots, p_{T-1}^{(tie)}, p_T^{(tie)}\}$, and the tie-line power flow in any time interval should not exceed $P_{max}^{(tie)}$.

Unlike traditional model, we propose a novel multi-objective model based on pre-specified optimization direction. Considering reducing production cost and emission, giving a preference ν, the objective could be formed as following.

$$\min F = F_1 + penalty(\nu) \times F_2, 0 \le \nu \le 1 \tag{8}$$

We regard F as MOUC cost. Note that the penalty is a function of ν, and the fundamental unit must be $/t, representing the unit penalty cost of emission. Hence, multi-objective problem could be reformed as single objective problem. ν is a subjective indicator that stands for the degree to which we attach importance to environmental indicator.

$$\min F' = (1-\nu)F_{1'} + \nu F_{2'}, \quad F_{1'} = \frac{F_1}{F_{1min}}, F_{2'} = \frac{F_2}{F_{2min}} \tag{9}$$

F_{1min} and F_{2min} correspond to the single-objective problem solved by optimizing (1), (2). $F_{*'}$ is the degree of deterioration of the objective with respect to the F_{*min}. Hence, ν stands for our emphasis on the emission. And the penalty function could be written as following.

$$penalty(\nu) = \frac{\nu F_{1min}}{(1-\nu)F_{2min}} \tag{10}$$

The fundamental unit of F_{1min} is \$, while F_{2min} is t, hence \$/t is the fundamental unit of $penalty(v)$, representing the penalty cost of unit emission. Further, giving the subjective emphasis on the environmental indicator, the $penalty(v)$ could be a basis for carbon tax makers.

3 Methodology

We propose a method by updating tie-line based on the margin price in connected subsystems. The basic hypothesis is that for giving subsystems, in any time intervals, the total MOUC cost would be decreased, if a little part of load is transferred from high margin price subsystem to the other. Note that the margin price here is the MOUC cost of the next kwh of energy. The optimal solution is obtained when the difference of margin price between subsystems is zero. We illustrate the approach with two subsystems, and the proposed methodology for MOUC is shown below.

Step1: Calculate penalty cost by (10), then formulate the original MOUC to single-objective problem (8).

Step2: Initialize tie-line based on the total capability and local load in each subsystem.

Step3: Update virtual load in each subsystem based on tie-line.

Step4: Solve local UC problem in each subsystem by LR based on virtual load.

Step5: Update tie-line based on Lagrangian multipliers in subsystems calculated in step4.

Step6: If $|\lambda_t^A - \lambda_t^B| < \theta$ in all time intervals or max iteration steps is met, then ends, otherwise updates tie-line and returns to step3.

3.1 Initialize tie-line

A good initialization strategy of tie-line could lead to a improved result when performing the following procedures. The initialization strategy follows three rules.

(i) The tie-line constraint (7) should be met.

(ii) Subsystems with low margin price generate more than they did.

(iii) The total generation level in each subsystem depends on the total capability in each subsystem.

$$\frac{\text{Capability}_\text{A}}{\text{Capability}_\text{B}} = \frac{Generation_\text{A}}{Generation_\text{B}} \tag{11}$$

The priority of these conditions is 1, 2, and 3.

3.2 Update virtual load

The virtual load in subsystem is not the real load in the subsystem, but virtual load with consideration of the tie-line. The sum of virtual loads in each subsystems is still equal to the sum of real loads. Firstly, we define the direction of tie-line. Then, we update the virtual load based on the real local load and tie-line.

$$VL_t^A = D_t^A + p_t^{(tie)}, VL_t^B = D_t^B - p_t^{(tie)} \tag{12}$$

VL_t^* is the virtual load in area * during time interval \$t\$, while D_t^* is the real load in area * during time interval t. During the iteration process, we calculate the virtual load by (12), then, solve local MOUC with consideration of constraints in Section 2. The load demand in (3) is modified to virtual load.

3.3 Calculate local MOUC

We apply LR to each local MOUC problems. The local MOUC problem is generated by the giving tie-line as discussed above. LR decomposes the coupling problem into multiple N independent subproblems, where N is the number of units. The original problem is modified by introducing Lagrangian multiplier λ_t with constraint (3) and μ_t with constraint (4). And the objective could take the new form.

$$\min L = F' + \sum_{t\in\{1,\ldots,T\}} \lambda_t (D_t - \sum_{i\in\{1,\ldots,N\}} u_{i,t} p_{i,t}) + \sum_{t\in\{1,\ldots,T\}} \mu_t (D_t + R_t - \sum_{i\in\{1,\ldots,N\}} u_{i,t} p_{i,\max}) \tag{13}$$

By looking into the dual problem and interchange the order of the summation, the original problem could be modified to a two layers problem. The lower layer solves N independent subproblems corresponding to every single unit and the units' parameters are not necessary to be uploaded. These subproblems could be solved locally, which protects the privacy of generating companies. The lower layer problems could be solved by dynamic programing [11], and the solution is delivered to the upper layer problem. The upper layer solves a linear optimization problem. Then, calculate the gap between the dual solution and primal problem. If the gap exceeds a threshold, updates Lagrangian multilier and calculates the lower layer problem, or LR is terminated. The Lagrangian multiplier has an economic interpretation corresponding to the price of the next kwh of energy during each time interval, which means marginal price at the time interval. Further, we use the interpretation for the next steps.

3.4 Update tie-line

The tie-line is updated based on the hypothesis that for any time intervals, if a little part of load is transferred from high margin price area to the other, the total MOUC cost in the whole system could be decreased, and the amount of change signifies the step size with direction defined by the difference between subsystems. Hence, the rules of updating tie-line is summarized as follows.

(i) For any giving time interval, the change of tie-line should decrease the total MOUC cost in the whole system. As the Lagrangian multiplier has been gotten in previous step, which stands for margin price [12], the subsystem with high margin price should generate less power, and tie-line could be updated as follows.

$$\begin{aligned} &P^{(tie)} = P^{(tie)} + \Delta P^{(tie)}, \\ &\Delta P^{(tie)} = \arg\min_{\Delta P^{(tie)}} \lambda^A (\Delta P^{(tie)})^T - \lambda^B (\Delta P^{(tie)})^T \\ &s.t. \quad -Step \le \Delta P^{(tie)} \le Step \end{aligned} \tag{14}$$

Note that we have defined the direction of tie-line in (12), the objective term in (14) can guarantee to meet the load constraint in (3). Solving (14) is equivalent to the following procedure.

$$\begin{aligned} &P^{(tie)} = P^{(tie)} + direction \cdot E \cdot Step \\ &direction = [direction_1, direction_2, \ldots, direction_T], Step = [Step_1, Step_2, \ldots, Step_T] \end{aligned} \tag{15}$$

E is unit matrix with T columns, and direction is determined by the difference between margin price of the two subsystems.

$$direction_t = I(\lambda_t^B - \lambda_t^A), I(x) = \begin{cases} 1 & , x > 0 \\ 0 & , x \le 0 \end{cases} \tag{16}$$

(ii) Giving the direction, $Step$ must be chosen to update the tie-line. $Step$ could neither be too big nor too small. A big $Step$ may cause the total MOUC cost increased, because the direction is a decent direction only if the $Step$ is not too big in any time interval. Further, a small $Step$ may lead to the total MOUC cost to be almost invariant. Here, $Step$ is chosen empirically, based on the following rules.

(a) The $Step$ would be big, if the difference of margin price between the two subsystems is significant.

(b) The $Step$ shall not only meet the tie-line constraint, but also ensure that it does not exceed a certain percentage of the local load of the two subsystems to which tie-line is connected.

$$0 \le Step_t \le \min(P_{\max^{(tie)}}, \delta \times \min(D_t^A, D_t^B)) \tag{17}$$

δ represents the percentage of load allowed to be changed. Hence, the algorithm of choosing the $Step$ could be stated as follows.

Algorithm 1 Find the *Step*

1: **for** each $i \in \{1, \ldots, T\}$ **do**
2:     **if** $\max(\lambda_t^A, \lambda_t^B) < min\_price$ **then**
3:         break
4:     **end if**
5:     $Step_t = \frac{|\lambda_t^A - \lambda_t^B|}{\max(\lambda_t^A, \lambda_t^B)} \times \delta \times \min(D_t^A, D_t^B)$
6:     **if** $|p_t^{(tie)} + Step_t \times direction_t| > P_{\max}^{(tie)}$ **then**
7:         $Step_t = I(p_t^{(tie)} + Step_t \times direction_t) \times P_{\max}^{(tie)}$
8:     **end if**
9: **end for**
10: **return** $Step = [Step_1, Step_2, ..., Step_T]$

3.5 Stop criterion

The basic criterion is stated in step 6. As the derivative function of (8) is a linear function, and most of the time, the derivative function is positive number if p is positive or $p \ge p_{min}$. The derivative function of (8) could be written as follows.

$$\frac{d^2F}{dp^2} = 2u(a + penalty \times d) > 0 \tag{18}$$

When p is increased, margin price could also be increased. Hence, as subsystem with high margin price generate less as stated in previous section, the margin price in both subsystems could be closed with each other as the iteration process moving on.

If the difference of margin price in both subsystems is lower than threshold θ or max iteration steps is met, the process is terminated.

4 Numerical examples

The algorithm was coded in JAVA, and have been performed on a server with double Intel Xeon E5-2697 processors and 256GB of RAM under Linux operating system.

A 46 units case is applied to illustrate the performance of the proposed algorithm, which contains two subsystems. The subsystem A has 10 units while B has 36. The characteristics of units are referenced from [13, 14]. The algorithm is compared with the operating approach in Nordic countries (ANC) [15].

We consider emission and economy as of equal importance, hence the preference ν is set to 0.5. And the δ in (18) is set to 0.25. The penalty term calculated with ν is 47.88\$/t, which had been edited to (8) to form the new objective.

The tie-line obtained during iterations are shown in Fig.1. As the iteration process moving on, tie-line is updated according to the difference of margin price between subsystems. As a result, the difference of margin price between subsystems are decreased as shown in Fig.2, which is represented as the area between curves. Note that change of tie-line in any time interval will make a difference in margin price at the whole 24-h horizon.

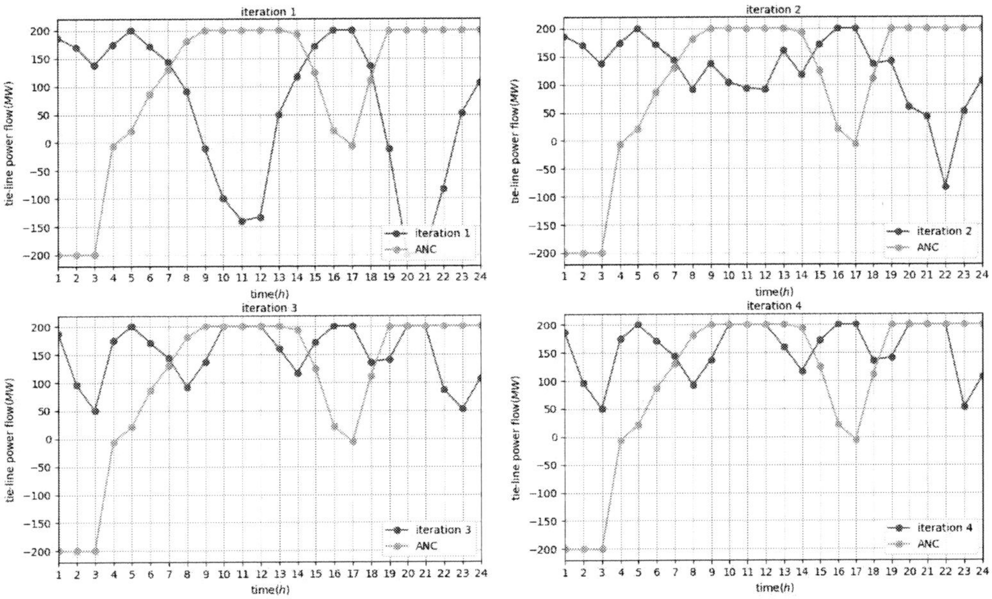

Fig. 1. Tie-line during iteration process.

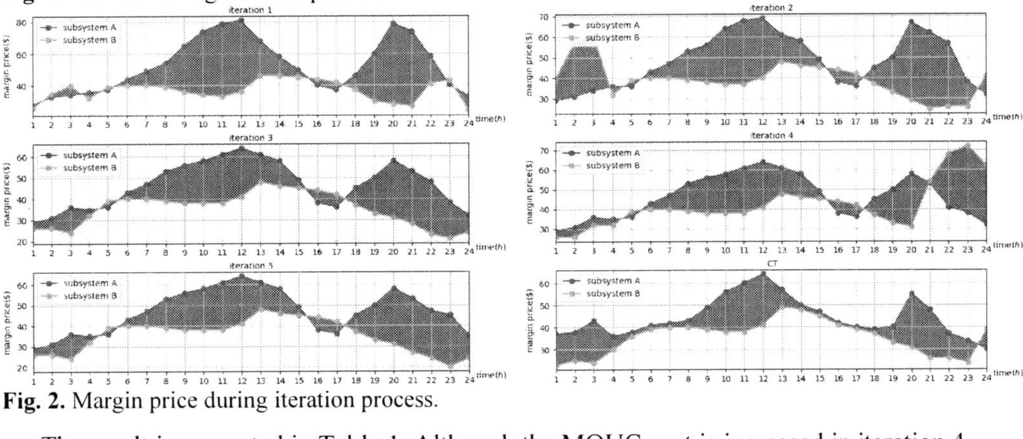

Fig. 2. Margin price during iteration process.

The result is presented in Table 1. Although the MOUC cost is increased in iteration 4, the overall trend is decreasing, and the final MOUC cost is lower than that obtained from ANC.

Table 1. MOUC cost.

	Iteration 1	Iteration 2	Iteration 3
result($)	3611406.85	3326772.53	2939905.71
	Iteration 4	Iteration 5	ANC
result($)	3106927.73	2929619.11	3034115.20

5 Conclusion

This paper proposes a novel MOUC model and a decomposition coordination method to solve it considering tie-line constraint, in order to minimize the production cost and emission. Giving the subjective preference for emission reduction, a penalty is calculated and the multi-objective problem could be revised to single objective. Besides, the penalty could be a basis for carbon tax makers. Then, we calculate the revised problem by an iteration process. During the iteration process, the margin price in connected subsystems is regarded as the coordinator to update tie-line, and we realize exact step control by considering the margin price, total load, etc. Each subsystem is solved separately and locally by LR. Test cases indicate that our method could obtain a improved result than ANC. Besides, solving local MOUC problem by LR avoids the need for uploading units' parameters, which protects the data privacy of generating companies. For any giving two subsystems, if subsystem with high margin price generate less while the other generate more, the total cost would be decreased. However, the step size is chosen by a subjective approach, and we have not found the optimal step. Further, we could apply neural networks to find the underlying relationship among units, load, tie-line as well as step size in the future.

References

1. Chuan Ping Cheng, Chih Wen Liu, and Chun Chang Liu. Unit commitment by lagrangian relaxation and genetic algorithms. IEEE Transactions on Power Systems, 15(2):707–714, 2000.
2. Arthur I Cohen and S. H Wan. A method for solving the fuel constrained unit commitment problem. IEEE Transactions on Power Systems, PER-7(8):39–39, 1987.
3. X. Ma, A. A. El-Keib, R. E. Smith, and H. Ma. A genetic algorithm based approach to thermal unit commitment of electric power systems. Electric Power Systems Research, 34(1):29–36, 1995.
4. Ioannis G Damousis, Anastasios G Bakirtzis, and Petros S Dokopoulos. A solution to the unit-commitment problem using integer-coded genetic algorithm. IEEE Transactions on Power Systems, 19(2):1165–1172, 2004.
5. M. Carrion and J. M. Arroyo. A computationally efficient mixed-integer linear formulation for the thermal unit commitment problem. IEEE Transactions on Power Systems, 21(3):1371–1378, 2006.
6. Xuejiao Lei, Xiaohong Guan, and Qiaozhu Zhai. Constructing Valid Inequalities by Analytical Feasibility Conditions on Unit Commitment with Transmission Constraints. IEEE Transactions on Power Systems, 31(5):3484–3494, 2016.
7. Kai Pan, Yongpei Guan, Jean Paul Watson, and Jianhui Wang. Strengthened MILP Formulation for Certain Gas Turbine Unit Commitment Problems. IEEE Transactions on Power Systems, 31(2):1440–1448, 2016.
8. Raca Todosijevic, Marko Mladenovic, Said Hanafi, Nenad Mladenovic, and Igor Crevits. Adaptive general variable neighborhood search heuristics for solving the unit

commitment problem. International Journal of Electrical Power & Energy Systems, 78:873–883, 2016.

9. Chunheng Wang and Yong Fu. Fully parallel stochastic security-constrained unit commitment. IEEE Transactions on Power Systems, 31(5):3561–3571, 2016.
10. Susanne Albers, Stefan Eilts, Eyal Even-Dar, Yishay Mansour, and Liam Roditty. On nash equilibria for a network creation game. Acm Transactions on Economics& Computation, 2(1):1–27, 2014.
11. Dimitri P Bertsekas, Dimitri P Bertsekas, Dimitri P Bertsekas, and Dimitri P Bertsekas. Dynamic programming and optimal control, volume 1. Athena Scientific Belmont, MA, 1995.
12. S. Virmani, E. C. Adrian, K. Imhof, and S. Mukherjee. Implementation of a lagrangian relaxation based unit commitment problem. IEEE Transactions on Power Systems, 4(4):1373–1380, 1989.
13. CHEN Xing-ying ZHANG Xiao-hua, ZHAO Jin-quan. Unit commitment by enhanced adaptive lagrangian relaxation. Proceedings of the CSEE, 30(22):71–76, 2010.
14. Haiti Ma and SM Shahidehpour. Unit commitment with transmission security and voltage constraints. IEEE transactions on power systems, 14(2):757–764, 1999.
15. Rickard Nilsson. organisation, history, products and roles in the deregulated Nordic Market, 2007.

MATEC Web of Conferences **277**, 03007 (2019) https://doi.org/10.1051/matecconf/201927703007
JCMME 2018

Helix grounding electrode with good grounding performance

Zhanlong Zhang, *Yihua* Dan*, *Daojun* Mei, *Jing* Zou, and *Peiyu* Jiang

State Key Laboratory of Power Transmission Equipment & System Security and New Technology(Chongqing University), Chongqing, 400030, China

Abstract. Grounding electrode is used for discharging current and ensure the safe and stable operation of electrical equipment. Grounding performance of grounding electrode will degradation with limited installation space. This paper proposes helix grounding electrode and builds theoretically and simulating model on the helix grounding electrode. Analyzing grounding surface potential and step voltage distribution, grounding resistance and surface current density distribution. Results show helix grounding electrode can reduce grounding resistance, improve grounding surface potential distribution and the discharging current distribution.

1 Introduction

Grounding device buried in the soil is used to ensure the safe and stable operation of various electrical equipment. Grounding electrode is the core part of the grounding device, the current discharging performance of grounding electrode is an important indicator of grounding performance. Horizontal grounding electrode and vertical grounding electrode are the common grounding ways in the most project. Both grounding electrodes are affected by terrain easily, such as some location closed to the rock or river, the length and depth of the grounding electrode will be limited and causing the poor current discharging performance, grounding resistance increasing, and the operation can't meet the thermal stability requirements power accidents. Horizontal grounding electrode and the downlead are easy to be corroded in the soil. Especially the junction between the grounding electrode the downlead because of end-affect caused by non-uniform discharging current distribution. Corrosive breakage of the grounding electrode can cause serious power accidents.

Domestic and foreign scholars have made remarkable achievements in the research on the dispersion of grounding devices at present. Professor Chen Ci-xuan realized the calculation of the leakage current distribution by the grounding device through zero-order interpolation of the boundary element and the first-order interpolation of the boundary element, and the sub-section closed to the intersection is subdivided into much smaller two, more accurate current density distribution can be obtained [1]. To improve the analysis accuracy of the grounding device's current discharging characteristics, Wen Xishan established a mathematical model of vertically stratified soil structure for calculating the

* Corresponding author: microdanyihua@outlook.com

grounding grids of large-scale water conservancy and hydropower plants [2]. These grounding grids are located in horizontal and vertical stratified soil with complex resistivity of soil. Water, underwater media and the soil resistivity across both sides of the straits are very different, a mathematical model that vertically divides the three-layer soil structure need to be established. The formula of ground current distribution is deduced by Laplace equation, which solves the problem of numerical calculation of vertical three-layer soil grounding parameters. Yuan Tao proposed "needle" grounding electrode from the concept of improving the impact current discharging efficiency of the grounding electrode and unevenness of the discharging impact current, effectively improve the resistance reduction characteristics of grounding electrode [3-5].

The above studies mainly focus on present grounding electrode, most scholars seldom studied the discharging current distribution characteristics of the other structure grounding electrode. Based on discharging current maximum concept, this paper proposed a helix grounding electrode. Theoretical calculation method for space current discharging characteristics, the surface current density, the end effect and the optimal buried depth range of the helix grounding electrode and ground surface potential distribution are analyzed.

2 Theoretical analysis for helix discharging current

Calculation and analysis of grounding electrode current discharging characteristics, widely used numerical methods such as the boundary element method, finite element method, but the above methods are too complicated. In order to simplify the computation, this paper mainly analyzes the low-frequency steady-state grounding characteristics of helix grounding electrode, uses electrostatic fields equivalent the current field in the soil and discharge the charge distribution on the surface of the grounding electrode to indicate current discharging performance of the grounding electrode. Analysis of the discharging current distribution of helix grounding electrode by moment method

Assuming that the helix grounding electrode is in an infinitely uniform soil and establishing the calculation model as shown in figure 1.

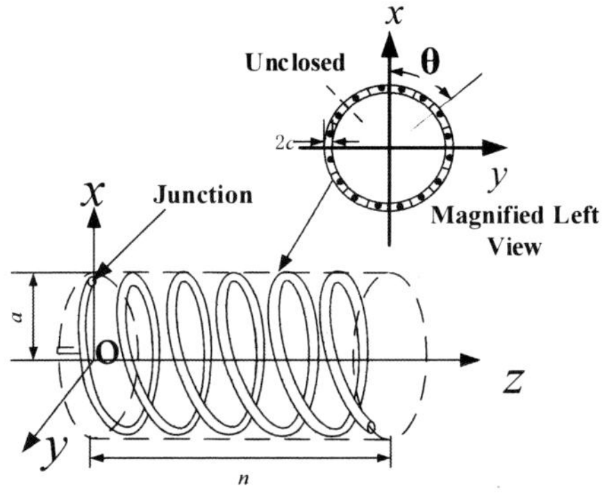

Fig. 1. Calculation model of helix grounding electrode.

As shown in figure 1, the major radius of helix grounding electrode is a, the minor radius of helix grounding electrode is c, the number of turns is n. The equal length of helix grounding electrode is $2\pi an \gg 2c$, The distribution of surface charge density $\sigma(\boldsymbol{r}')$ of a helix grounding electrode can be regarded as the linear charge density $\tau(\boldsymbol{r}')$ along the spiral line of the grounding electrode. According to the analog charge method, the potential equation of the grounding electrode and the equivalent spiral line equation can be established as below.

$$\int_S \tau(\boldsymbol{r})\left(4\pi\varepsilon_0 \left|\boldsymbol{r}-\boldsymbol{r}'\right|\right)^{-1} ds = \varphi \tag{1}$$

$$\begin{cases} x = a\cos\theta \\ y = a\sin\theta \\ z = b\theta \end{cases} \tag{2}$$

S is the equal length of the helix grounding electrode, $\left|\boldsymbol{r}-\boldsymbol{r}'\right|$ is the distance between the field point and the source point. Based on the idea of micro-element, the continuous line charge source $\boldsymbol{\tau}(\boldsymbol{r})$ within the one turn spiral grounding electrode is approximately divided into m discrete point charge sources $\boldsymbol{\tau}_i$ The point charge source $\boldsymbol{\tau}(\boldsymbol{r})$ is discretized to τ_i by the pulse function $N_i(\theta)$ on θ .

$$\boldsymbol{\tau}(\boldsymbol{r}) = \sum_{i=1}^{N} N_i(\theta_i)\tau_i \tag{3}$$

Establishing discrete point charge source integral equation by bringing equation (2) (3) into equation (1).

$$\int_{\Delta\theta} N_i(\theta_i)\left|\boldsymbol{r}-\boldsymbol{r}'\right|^{-1} d\theta \sum_{i=1}^{N}\tau_i = 4\pi\varepsilon_0\varphi\left(a^2+b^2\right)^{-1/2} \tag{4}$$

According to the moment method, using Dirac function as the weight function and establishing a point matching equation.

$$\sum_{i=1}^{N}\tau_i \int_0^{\theta_{\max}} \delta(\boldsymbol{r}-\boldsymbol{r}_j')d\theta \int_{\Delta\theta} N_i(\theta_i)\left|\boldsymbol{r}-\boldsymbol{r}'\right|^{-1} d\theta = 4\pi\varepsilon_0\varphi\left(a^2+b^2\right)^{-1/2}\int_0^{\theta_{\max}} \delta(\boldsymbol{r}-\boldsymbol{r}_j')d\theta \tag{5}$$

Equation (5) is changed into matrix form.

$$\mathbf{K}\boldsymbol{\cdot}\boldsymbol{\tau} = \mathbf{C} \tag{6}$$

$\boldsymbol{\tau}$ is a column of to-be-solved vectors consisting of N point charge source τ_i as we can see from equation (5). Grounding electrode is usually made of good conductor such as copper or galvanized steel to ensure the effect of current discharging. This paper focus on the current discharging characteristic of grounding electrode in power frequency and steady state condition, potential φ of grounding electrode can be approximated as a constant.

Ground potential φcan be measured by the fall of potential method, C is a known vector. Charge distribution of the grounding electrode can be given by solving the coefficient matrix K. According to the equation (5) and equation (6), the equation can be obtained.

$$\mathbf{K} = \int_{\Delta\theta} N_i(\theta)\left|\boldsymbol{r}_j'-\boldsymbol{r}_i'\right|^{-1} d\theta \tag{7}$$

1) $j \neq i$, matching point and the source point are not in the same segment

In order to select the coordinates of the matching point on the surface of the grounding electrode, and the coordinates of the points of the charged points of the discrete grounding electrode, the equation (2) is taken into the equation of the coefficient matrix. $\boldsymbol{r}_j'$ is the selected matching point coordinates of the grounding electrode surface, $\boldsymbol{r}_i'$ is the selected source point coordinates of the grounding electrode surface, element k_{ji} in the matrix $\mathbf{K}$ can be solved through equation (2) and equation (7).

$$k_{ji} = \int_{\theta_i-\frac{\Delta\theta}{2}}^{\theta_i+\frac{\Delta\theta}{2}} \left\{b^2(\theta_j-\theta)^2+2a^2\left[1\text{-}\cos(\theta_j-\theta)\right]\right\}^{-1/2} d\theta \tag{8}$$

2) $j=i$, matching point and the source point are in the same segment

A much smaller segment of a helix grounding electrode with a circular section is equivalent to a cylinder to ensure a higher calculation accuracy, and a model is established as shown in figure 2.

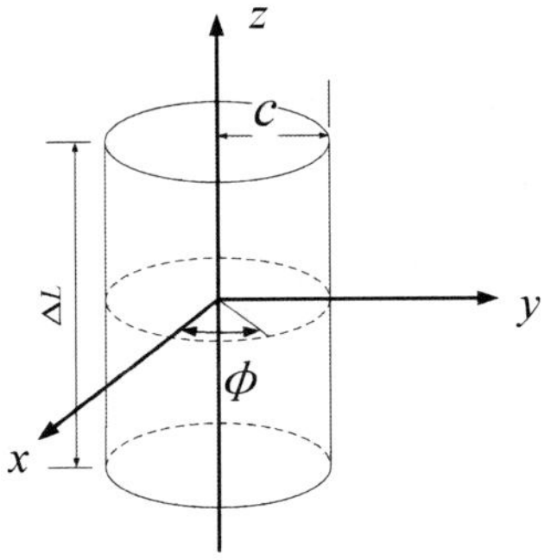

Fig. 2. Calculation model of coefficient.

As shown in figure 2, with the source of the segment ***i*** as the origin coordinates, the matching segment ***j*** is selected on the surface of the grounding electrode. Surface charge density σ of the grounding electrode need to be solved first, then solving the surface discharging current. The micro segment area of the grounding electrode as below.

$$\Delta A = 2\pi \cdot c \cdot \Delta L \tag{9}$$

Surface charge density σ.

$$\sigma = \tau_i (\Delta L)(\Delta A^{-1}) = \tau_i (2\pi c)^{-1} \tag{10}$$

Potential in centred point I_i caused by surface current can be obtained through integrating the entire surface area A when the surface charge density of the grounding electrode is known.

$$\begin{aligned}\phi_j &= (4\pi\varepsilon)^{-1} \int_{\Delta S} \sigma \cdot \vec{R}^{-1} dS = (4\pi\varepsilon)^{-1} \cdot \int_0^{2\pi} \int_{-\frac{\Delta L}{2}}^{\frac{\Delta L}{2}} c \cdot \tau_i \cdot (2\pi c)^{-1} \cdot \left(c^2 + z'^2\right)^{-1/2} dz' d\varphi \\ &= (4\pi\varepsilon)^{-1} \cdot \tau_i \cdot 2 \cdot \ln\left\{c^{-1} \cdot \left[\Delta L/2 + \left(c^2 + (\Delta L/2)^2\right)^{1/2}\right]\right\}\end{aligned} \tag{11}$$

Whereas $\Delta L = \Delta\theta * a$, simplified equation (11) as below.

$$\tau_i \cdot 2\left(a^2 + b^2\right)^{-1/2} \cdot \ln\left\{c^{-1} \cdot \left[\Delta L/2 + \left(c^2 + (\Delta L/2)^2\right)^{1/2}\right]\right\} = 4\pi\varepsilon\phi_j \left(a^2 + b^2\right)^{-1/2} \tag{12}$$

Comparing equation (5) and equation (12).

$$k_{ji} = 2c \cdot \left(a^2 + b^2\right)^{-1/2} \cdot \ln\left\{c^{-1} \cdot \left[\Delta L/2 + \left(c^2 + (\Delta L/2)^2\right)^{1/2}\right]\right\} \tag{13}$$

Solving the coefficient matrix $\mathbf{K}$ through equation (8) and equation (13), point charge τ_i distribution on helix grounding electrode surface can be solved. To solve the charge distribution of helix grounding electrode surface, each discrete segment τ_i is normalized $\tau_i / |\boldsymbol{\tau}|$, the charge distribution ratio at each point on the helix grounding electrode surface can be obtained. The current is the charge passing some section of the conductor in a unit time, so the discharging current distribution of the helix grounding electrode can be approximated by the distribution of charge distribution.

3 Analysis of the discharging current on helix grounding electrode

Current discharging characteristic is one of the most important factors to measure grounding performance of grounding electrode. From the aspects of surface potential distribution, step voltage, grounding resistance, end effects, surface current density distribution, the length and buried depth of the helix grounding electrode, the current discharging characteristic of the helix grounding electrode is comprehensively analyzed.

3.1 Analysis of the discharging current on helix grounding electrode

Frequency of the grounding current is usually power frequency, and the grounding current field can be approximated as constant electric field, so the 1A DC is taken as power source in this simulation. Building a simulation model as shown in figure 3, Assuming that the turns of the helix grounding electrode is 20, the minor radius is 0.02m, the axial pitch is 0.5m, the major radius 0.5m, and installing grounding electrodes along the *x*-axis. To compare the current discharging characteristic of helix grounding electrode and horizontal grounding electrode. Establishing horizontal grounding electrode model with the same axial installation length (10 m) and 0.02 m cross-section radius to comparative analysis, buried depth is 0.8m, soil resistivity is $200\Omega\cdot M$, radius of the equivalent model of the infinite soil is 50m.

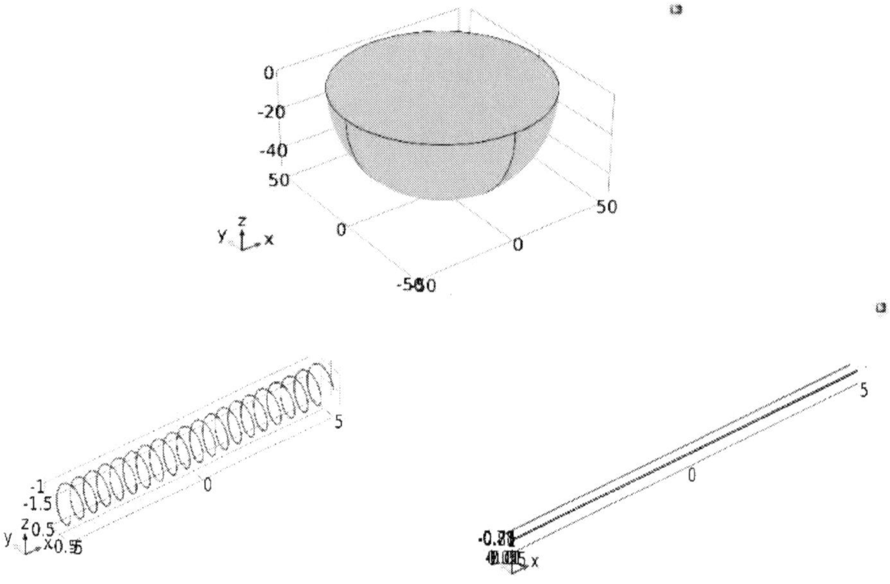

Fig. 3. Simulation Model Schematic.

3.2 Ground surface potential distribution and step voltage

Electric field will be generated in the soil surrounding the grounding electrode because the current is injected into the grounding electrode. The ground surface potential distribution above the grounding electrode is as shown below.

 https://doi.org/10.1051/matecconf/201927703007

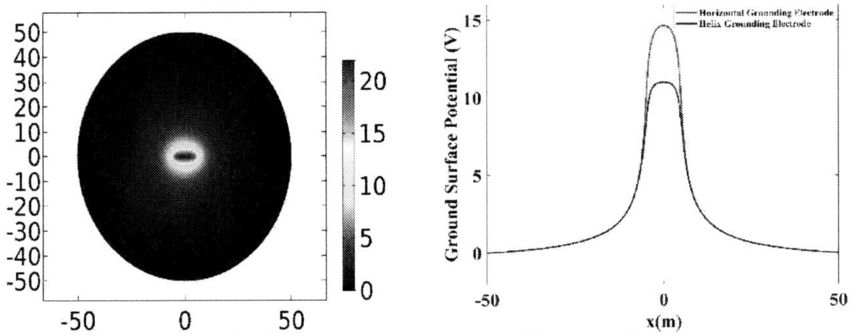

(a) Surface (b) x-axis direction

Fig. 4. Surface Potential Distribution.

Figure 4(a) shows that the surface potential above the helix grounding electrode appears an elliptical potential distribution pattern. Helix grounding electrode reduces the potential soil area effectively and electromagnetic interference caused by grounding electrode for surrounding environment. Figure 3(b) shows the maximum potential on the ground surface is 29.6V in buried horizontal grounding electrode condition, the maximum potential on the ground surface is 21.9V in buried helix grounding electrode condition. The maximum difference between the within the two inflection points of the curve is 26.01%, the distribution are almost the same outside the two inflection points of the curve. Results indicates helix grounding electrode effectively reduces the step voltage around the inflection point of the curve.

The soil resistivity in actual projects is easily affected by the external environment. To analyse the current discharging performance of helix grounding electrode more comprehensively, the resistivity to (step length is $50\Omega{\cdot}M$) of soil are selected. Ground surface potential distribution is as follows in the different soil resistivity condition.

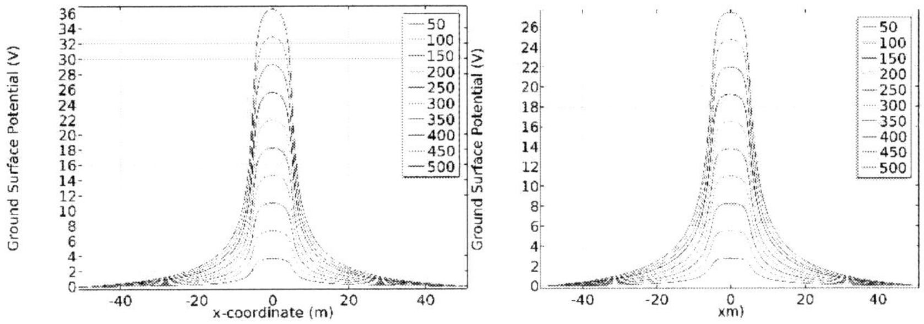

(a) Horizontal Grounding Electrode (b) Helix Grounding Electrode

Fig. 5. Ground Surface Potential Distribution in Different Resistivity Soil.

Figure 5 shows that as the resistivity of the soil decreases, the maximum potential on the ground surface of helix grounding electrode and the horizontal grounding electrode both show a decreasing trend, and step voltage decreases at the inflection point of the curve (both ends of the grounding electrode), ground surface area affected by the grounding electrode is also gradually reduced. The maximum ground potential rise of the helix grounding electrode is lower than about 25% of the horizontal grounding electrode at the same resistivity soil. The step voltage is significantly improved. Results indicate that the helix grounding electrode can achieve a better ground potential distribution in any resistivity soil.

3.3 Grounding Resistance

Grounding resistance is an important indicator to measure the current discharging performance of helix grounding electrode. Power source is selected as 1A DC in this simulation, grounding electrode can be considered as a good conductor in the low frequency steady state field, the potential on grounding electrode surface is equal. So the potential value on grounding electrode surface means grounding resistance, the potential distribution on grounding electrode surface as shown below.

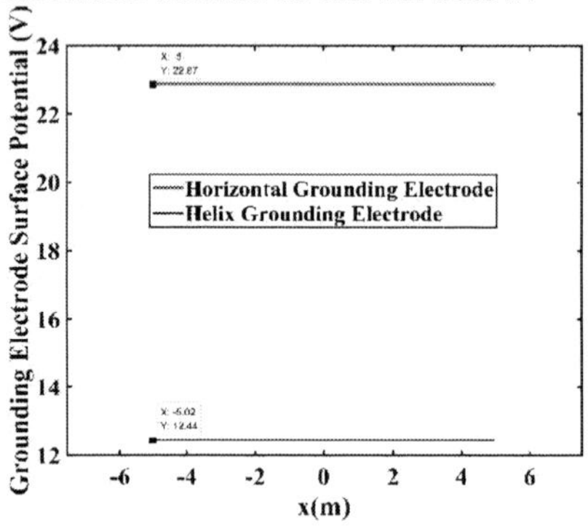

Fig. 6. Potential Distribution on Grounding Electrode Surface.

Figure 6 shows that the grounding resistance of the helix grounding electrode, and the resistance of the helix grounding electrode is in the same axial installation length. Grounding resistance of the helix grounding electrode was reduced by 45.6% compared to the same axial installation length of the horizontal grounding electrode. When the installation space is limited, helix grounding electrode maximizes the current discharging performance, greatly reducing the grounding resistance and improving the grounding performance.

To study the grounding resistance of grounding electrode in different resistivity soil, to soil resistivity (step length) is chose to analyze the ground performance, potential distribution on grounding electrode surface in different soil resistivity as shown below.

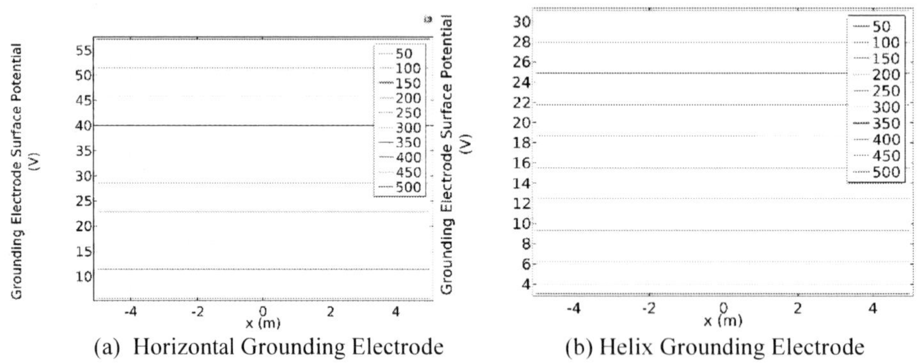

Fig. 7. Potential Distribution on Grounding Electrode Surface in Different Soil Resistivity.

Figure 7 shows that grounding resistance of the grounding electrode tends to increase with the soil resistivity increasing. At the same resistivity, the grounding resistance of the helix grounding electrode is consistently lower than about 46% of the grounding resistance of the horizontal grounding electrode. On the one hand, helix grounding electrode has better current discharging characteristics and smaller grounding resistance in any resistivity soil, On the other hand, helix grounding electrode can ensure a relatively lower touch voltage and the safety of the operator when fault current flow into grounding electrode.

3.4 Discharging current distribution on helix grounding electrode surface

Further study on the discharging current distribution on the helix grounding electrode surface, the discharging current distribution on the helix grounding electrode surface is as follow.

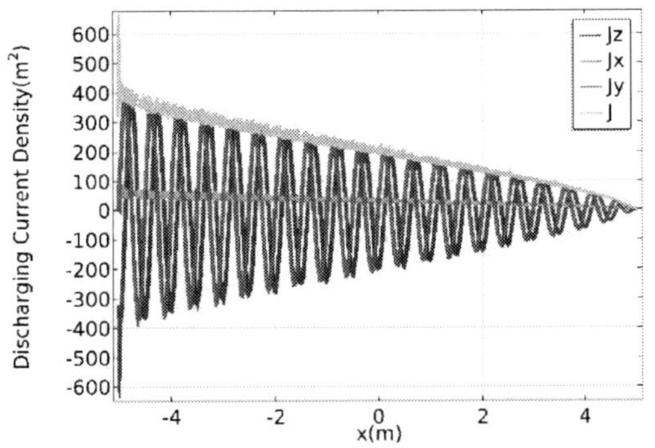

Fig. 8. Discharging Current Density Distribution on Helix Grounding electrode.

Figure 8 shows the y-axis component and the z-axis component of discharging current density on helix grounding electrode exhibit a decaying sinusoidal function distribution. With the structure of the helix grounding electrode, it can be seen that the effective current discharging length of the grounding electrode is increased under the same axial installation space. The linear attenuation of the discharging current density on the helix grounding electrode surface indicate that the helix grounding electrode has a good current discharging performance.

4 Conclusion

This paper proposes a helix grounding electrode based on the concept of maximizes the grounding current discharging. The following conclusions are drawn through extensive analysis.

1) Compared with the horizontal grounding electrode with the same axial installation length, the helix grounding electrode can effectively improve the surface potential distribution, reduce the step voltage and reduce the grounding resistance by about 45.6% in any resistivity soil.

2) Helix grounding electrode increases the contact area with the soil effectively and ensure a better discharging current distribution on the helix grounding electrode surface. The grounding performance of helix grounding electrode has obvious advantage over the horizontal grounding electrode.

In summary, the helix grounding electrode has good current discharging characteristics and grounding performance in low-frequency steady-state electrical field. So helix grounding electrode has far-reaching engineering significance with installation space consideration.

Acknowledgments

This work was supported by the National Natural Science Foundation of China (51577017).

References

1. Chen Ci-xuan. Application of Boundary Element Method in Ground Calculation—Calculation of lightning protection grounding resistance [J]. High Voltage Engineering, 1983(2):15-25.
2. Wen Xishan, Dai Chuanyou, Fangyu. Computation of Current Field in a Vertical Three-layer Earth Structure[J], Journal of Wuhan University (Engineering Edition), 1997(1):53-56.
3. Tao Y. Analysis of Grounding Resistance Reduction Effect Based on Enhancing Impulse Current Leakage Efficiency[J]. Transactions of China Electrotechnical Society, 2012, 27(11):278-284.
4. Zhu Bing. Structure Parameter optimization of Grounding Device With Needle—Shaped Conductors Based on Electric Field Distribution in Soil[J], Power System Technology, 2015, 39(10):2907-2914.
5. Li Jingli. Simulation and Experiment Study on Resistance-Reducing Mechanism ofGrounding Device With Spicules[J]. Power System Technology, 2013, 37(1):211-217.

MATEC Web of Conferences **277**, 03008 (2019) https://doi.org/10.1051/matecconf/201927703008
JCMME 2018

Replacement of several single function generator protection relay at Badak LNG

Iqbal Nusya Perdana*

Lead Electrical Engineer, Technical Department, Badak LNG, Indonesia

Abstract. Badak LNG is a company that operates the largest LNG Plant in Indonesia with maximum production capacity of 22.5 MTPA. The company operates twelve (12) steam turbine generators and one gas turbine generator and one diesel engine generator to supply electricity for process, utility, storage, loading and plant support facility areas, which constitutes the criticality of power generator equipment to keep the reliability of production process. Generator protection relay is one of the equipment that monitors and protects the generator from fault resulting from either internal or external disturbance, preventing the generator from serious damages. Before 2015, Badak LNG used single function relays to protect the generators. The generators are protected by stator differential protection (87), current unbalance (46), loss of excitation (40), reverse power (32), time overcurrent with voltage restraint (51V), and time overcurrent ground (51G). These single function relays wield more disadvantages compared to the newer technology held by the digital relays. Since 2015, Badak LNG have started to replace the single function generator protection relays with the digital type ones.

1 Introduction

Badak LNG (PTB) is a natural gas liquefaction plant situated in Bontang, East Kalimantan, Indonesia. Operating since 1977, PTB has eight process trains, namely Train A to H, with a maximum sustainable capacity of 22.1 MTPA (Million Tonnes per Annum). Badak LNG is comprised of three main areas which are: Process Trains, Storage and Loading, and Utilities. Utilities area covers several systems that supports the LNG production process, such as: electricity supply, steam supply, cooling water system supply, instrument air, nitrogen system and water treatment system.

In terms of electricity supply as can be viewed in figure 1, Badak LNG has fourteen (14) generators to supply power electricity to its system. These generators are equipped with several protections to prevent damages subject to generator and its prime mover from occurrences of failures. However, the vast majority of the generator protections in Badak LNG still utilizes electromechanical relay which only has single purposive function. Over times, these electromechanical relays suffer from problems due to reliability and obsolescence issues.

* Corresponding author: iqbalnusya@badaklng.co.id

Facing this issue, Badak LNG had an intention to start replacing the existing technology (electromechanical relays) with the latest technology which possesses several advantages, some of which are more complete protection, diagnostic capabilities for optimization, and other operational advantages e.g. sequence of events, present values, and communication capabilities. Since 2015, Badak LNG have started to replace these electromechanical relays with the digital type ones. Upgrading protection technology was the best solution to increase the reliability of generators in Badak LNG plant.

2 Generator protection

Generator is a machine that converts mechanical energy into electrical energy. Generator works based on the principle of Faraday's law of electromagnetic induction. In Badak LNG, generator plays an important role as the backbone of utility systems in producing electricity. Consequently, generator becomes a critical equipment for the dependence of LNG production on electric power supply and therefore, it shall be protected from internal or external faults and operated properly to ensure the sustainability of electricity.

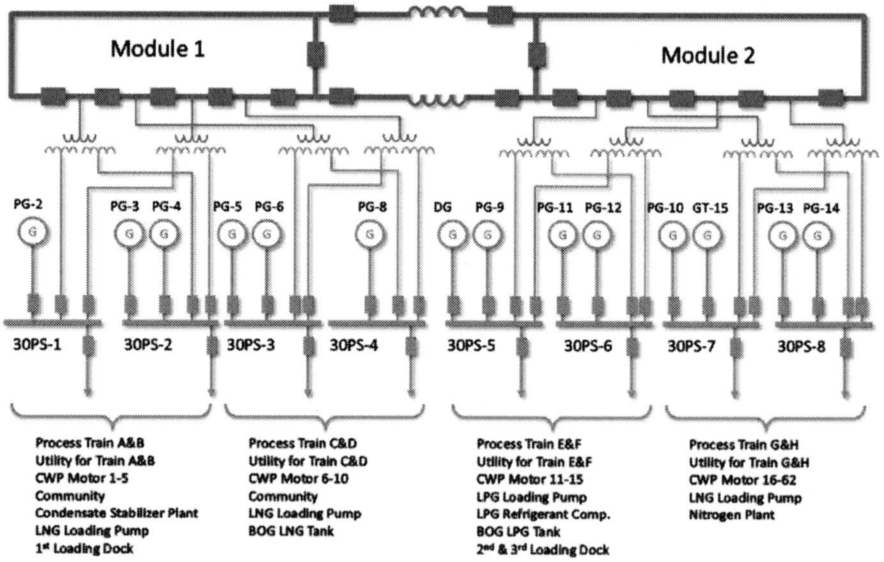

Fig. 1. Overall Single Line Diagram Badak LNG.

Generator has to be protected not only from electrical faults (stator and rotor faults) and mechanical problems, but also from adverse system interaction if it is going out of step with the rest of system, loss of field winding, etc. Under certain situations e.g. internal faults, the generator has to be isolated quickly, while other problems might only require alarms to alert the operator.

3 Replacement relay

Generator protection in Badak LNG utilizes single function electromechanical relays. These electromechanical relays, as can be viewed figure 2, have been suffering from several problems due to reliability and obsolescence issues, see table 1. Manufacturer has stated that their product will be obsolete starting in 2019 which translates to no further spare-part related support will be available by then. On account of this issue, Badak LNG has to start to do replacement program immediately. Newer technology, which is digital protection as can be viewed in figure 3, offers more advantages compared to the electromechanical relays, some of which are: more complete protection, diagnostic improvement, life extension with minimum effort, and operational advantages e.g. events recorder and log sheet.

Table 1. Status Generator Protection in Badak LNG.

No	Protection	ANSI	Model	Status
1.	Stator Differential Protection	87	CFD	Obsolete in 2019
2.	Current Balance	46	SGC	Obsolete in 2019
3.	Loss of Excitation	40	CEH	Obsolete in 2019
4.	Anti Motoring	32	GGP	Obsolete in 2019
5.	Overcurrent Voltage Restrain	51V	IFCV	Obsolete in 2019
6.	Stator Ground Fault	51GN	IAC	Obsolete

Nowadays, technology of protective relay has used microprocessor and microcontroller and well known as numerical or digital relay. These relays offer several unique advantages. These include more complete protection, diagnostic capabilities, life extension with minimal implementation and operational advantages of sequence of events, present values and communication capabilities.

With these all advantageous from digital relay, operation and maintenance are able to maintain the performance reliability of generator easily. In order to ensure new digital relay has same setting with the existing relay, current-time characteristic is needed to be examined between digital relay and existing relay.

Fig. 2. Electromechanical relay for generator protection.

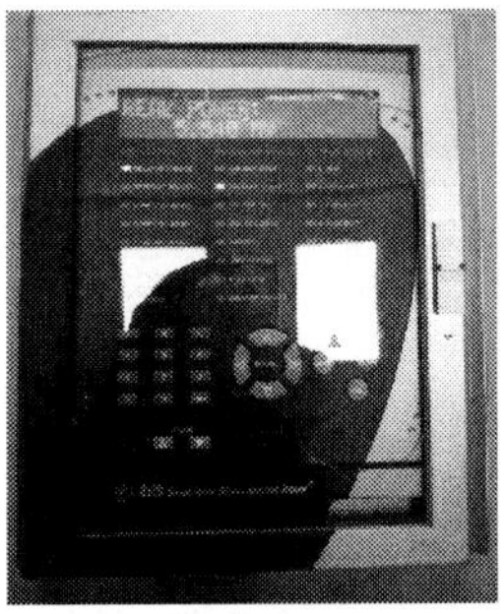

Fig. 3. New digital relay protection for generator.

3.1 Stator differential protection (87G)

Stator Differential Protection protects generator windings against internal faults e.g. phase-to-phase and three phase-to-ground faults. This type of fault is very serious due to the ability of high current to flow and produce significant amount of damages to the winding if it is allowed to last. This protection system requires two identical current transformers which are mounted on both sides of the protection zone. The secondary terminals of the current transformers are connected in stars, and their end terminals are connected through the pilot wire as can be viewed in figure 4.

The existing stator differential protection is electromechanical relay with single slop setting of 100% percentage slope and minimum pick-up of 0.2A on secondary side. This relay is adjusted at the factory and the setting cannot be changed. Based on IEEE Std 242 Protection and Coordination of Industrial and Commercial Power System [1], the percentage slope characteristic can vary from 5% at low values of through current up to 50% or more at high values of through current. This characteristic result in a relay that is sensitive to internal faults and insensitive to CT error currents during severe external faults. Since the new digital relay could provide double slope protection,as can be viewed in figure 5, a new setting must be considered. Referring to the study, the optimum value of percentage slope setting is 25% (slope 1) at low and 95% (slope 2) at high. The minimum pick-up current setting of new relay having range of 0.05 CT secondary or 0.25A (Using CT 1000:5).

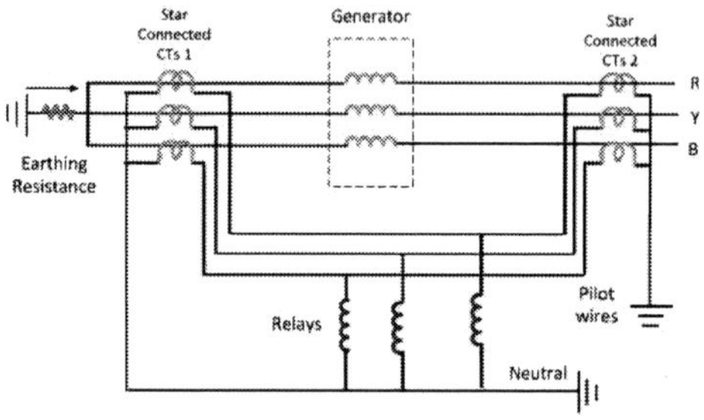

Fig. 4. Stator differential protection for generator.

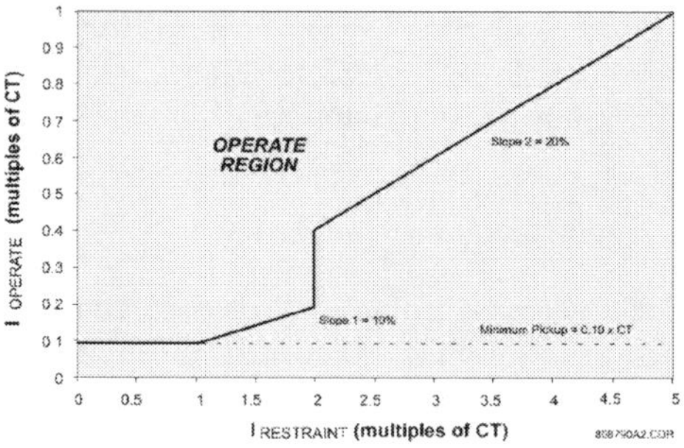

Fig. 5. Differential elements on new digital relay. [2]

The differential element for phase A will operate when

$$I_{operate} > k \times I_{restraint} \tag{1}$$

where the following hold

$$I_{operate} = \overline{I_A} - \overline{I_a} = operate\ current \tag{2}$$

$$I_{restraint} = \frac{|I_A| + |I_a|}{2} = restraint\ \ current \tag{3}$$

k = characteristic slope of the differential element in percent

$$k = Slope1\ if\ I_{restraint} < 2 \times CT;\ \ k = Slope2\ if\ I_{restraint} \geq 2 \times CT \tag{4}$$

I_A = phase current measured at the output CT
I_a = phase current measured at the neutral end CT

3.2 Current unbalance protection (46)

There are numbers of system conditions that may cause a lack of symmetry of three-phase current in a generator. The most common causes are system asymmetries, unbalanced loads, unbalanced system faults, and open phases. These system conditions produce negative-phase-sequence components of current which induces a double-frequency current in the surface of the rotor, the retaining rings, the slot wedges, and to a smaller degree, in the field winding. These rotor currents may cause high and possibly dangerous temperatures in a very short time.

A generator shall be capable of withstanding, without injury, the effects of a continuous current unbalance corresponding to a negative-sequence current I_2 of the following values on table 2, providing the rated kVA is not exceeded and the maximum current does not exceed 105% of rated current in any phase as shown as shown in figure 6.

Table 2. Constant Withstand Generator Limits. [3]

Type of Generator	Permissible I_2 (percent)
Salient Pole	
With connected amortisseur windings	10
With non-connected amortisseur windings	5
Cylindrical Rotor	
Indirectly Cooled	10
Directly Cooled - to 960 MVA	8
961 to 1200 MVA	6
1201 to 1500 MVA	5

The ability of a generator to accommodate unbalanced currents is specified in datasheet of generator in terms of negative sequence current I_2. The negative sequence current is expressed as a percentage of rated stator current. The continuous I_2 capability of a generator and the short-time capability of a generator specified in terms I^2t. The existing setting is minimum pick-up at 4% FLA and K=20. The existing generator short-time capability is I^2t=K=30. Therefore, K setting must be under 30. The existing setting of K=20 and minimum pick-up of 4% is already sufficient protection of the generator. Therefore, the setting for the digital relay is also going to be set at 4%, and K=20.

3.3 Loss of excitation protection (40)

Loss of field or loss of excitation results in loss of synchronism between rotor flux and stator flux. Under loss of excitation condition, the synchronous machine will operate as an induction machine at higher speed and draws more reactive power from the grid which results in the flow of slip frequency currents in the rotor body as well as severe torque oscillations in the rotor shaft. The rotor is not designed to sustain such currents or to withstand the high alternating torques which results in rotor overheating, coupling slippage and even rotor failure. This is where loss of excitation protection comes in handy.

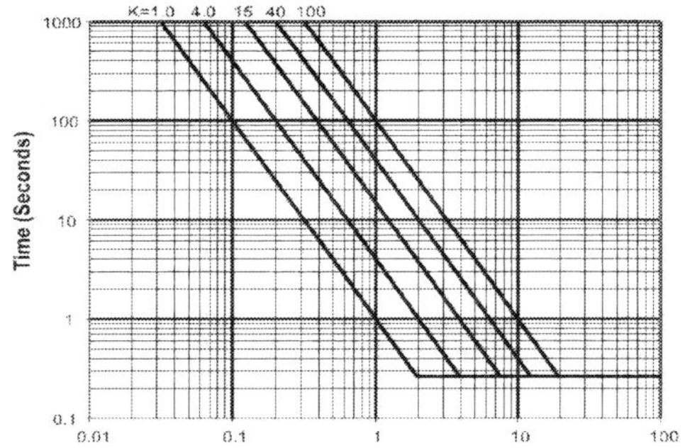

Fig. 6. Negative sequence current / rated current.

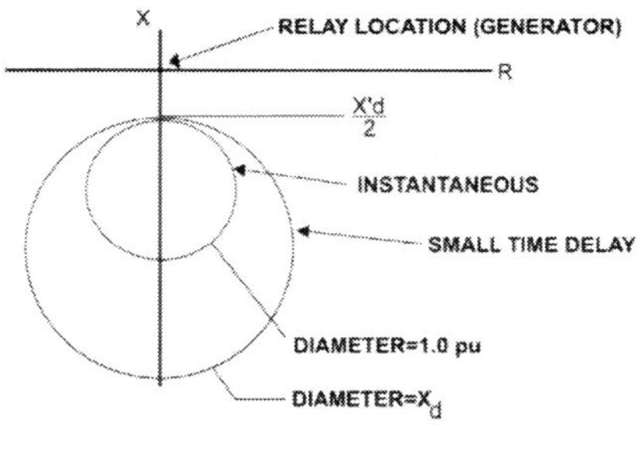

Fig. 7. Loss of excitation using impedance relay.

When impedance relaying is used to sense loss of excitation, the trip zone is typically marked by a mho circle centered about the X axis, with an offset from the R axis by X'd/2 as shown in figure 7[4]. Two zones are used on occasions: a high speed zone and a time delayed zone. To protect the generator from loss of excitation occurrences, some parameters of generator as list on table 3 shall be considered:

Table 3. Generator Parameter.

No	Parameter	Value
1.	X'd	0.23 pu
2.	Xd	1.65 pu
3.	Base MVA	15.625 MVA
4.	Base kV	13.8
5.	CT Ratio	1,000 / 5 = 200 / 1
6.	PT Ratio	14,400 / 120 = 120 / 1

Then the setting can be calculated as follow:

$$Z_{base(\sec)} = \frac{(kV\,base)^2}{MVA\,base} \times \frac{CT\,Ratio}{PT\,Ratio} \quad (5)$$

$$Z_{base(\sec)} = \frac{(13.8)^2}{15.625} \times \frac{200}{120} = 20.31\ ohm \quad (6)$$

$$X'_d = Z_{base(\sec)} \times X'_{d(\,perunit)} = 20.31 \times 0.23 = 4.67\,ohm(\sec)$$

$$X_d = Z_{base(\sec)} \times X_{d(\,perunit)} = 20.31 \times 1.65 = 33.52\,ohm(\sec) \quad (7)$$

From this calculation, the relay setting can be calculated as follows:

Offset tap settings $= \frac{X'_d}{2} = \frac{4.67}{2} = 2.33\,ohm$, use higher tap setting 2.5 ohm

Circle Diameter $= X_d = 33.52\,ohm$, use lower tap setting 33.5 ohm

3.4 Anti motoring or reverse power protection (32)

Anti Motoring or Reverse Power Protection protects the generator and its prime mover (turbine) from drawing out power from the grid (motoring). Motoring of a generator will occur when turbine output is reduced to less than its no-load losses while the generator is still on-line. In this condition, the generator will operate as a synchronous motor and drive the turbine. The generator will not be harmed by synchronous motoring, but in contrast a steam turbine can be harmed by overheating during synchronous motoring if it lasted long enough. The motoring of the turbine output can be detected by reverse power protection.

The maximum motoring power of steam turbine is 3% of rated power (see table. 4), therefore the motoring load will be 3% x 12.5 MW = 375 kW. This number equal with secondary current of 0.078A. While for the minimum motoring power of steam turbine is 0.5% of rated power so the motoring load will be 0.5% x 12.5 MW = 62.5 kW or equal with 0.013A in secondary current. Therefore for setting will be within range 0.5% - 3% or 0.5 - 3 per unit. Sufficient time delay should be provided to prevent undesired operation on transient power reversals such as those occurring during synchronizing or system disturbances. A setting of 2% of rated power is recommended with time delay of 5 seconds.

Table 4. Generator Maximum Motoring Power IEEE Std 242-2001 [1].

Type of Prime Mover	Maximum Motoring Power of Unit Rating
Gas Turbine	50%
4 Cycle Diesel	25%
Steam Turbine	0.5% - 3%

3.5 Overcurrent voltage restrain protection (51V)

Generator overcurrent protection is provided by voltage restraint overcurrent protection relay (51V). The voltage restraint overcurrent relay must be set higher than generator decrement curve and lower than generator damage curve at 100 % voltage restraint.

In the voltage-restrained relay, the pick-up current varies as a function of voltage applied to the relay. In one type of relay with zero voltage restraint, the current pick-up is 25% of the pick-up setting with 100% voltage restraint as can be viewed in figure 8. On units that have a short, short-circuit time constant, the 51V voltage-restrained overcurrent relay should be used.

The pick-up current at full restraint voltage is set at 150% of Full Load Amps. The pick-up at no restraint must be lower than (Current FLA/Xd), based on IEEE recommendation. The existing protection setting is set at pick-up current of 1,000 A (5A secondary) and time dial 10.

FLA generator = 654 A

Tap value $= I_{FL} \times \frac{1.5}{CT\ Ratio} = \frac{654 \times 1.5}{1000 \div 5} = 4.905A \approx 5A$, $I_{FaultL} = \frac{I_{FL}}{X_d} = \frac{654}{1.65} = 396A$.

At 0% voltage restraint, the relay will pick-up at: $I_{pickup} = 5 \times 0.25 \times \frac{1000}{5} = 250A$

Since the fault current I_{fault} is higher than the pick-up current during 0% voltage restraint, the existing setting is adequate to protect generator during fault since the pick-up value is lower than the fault current (FLA/Xd).

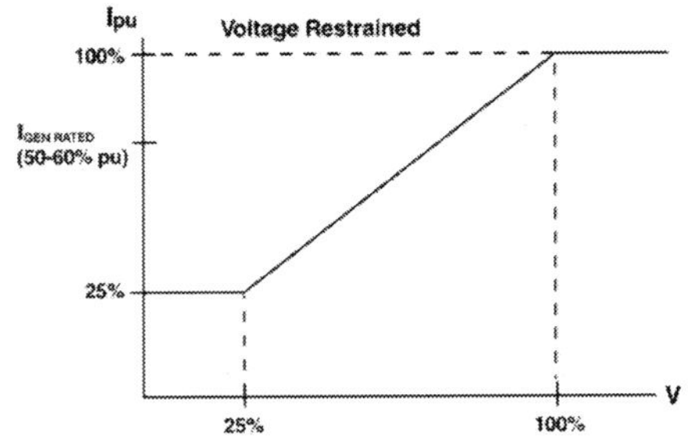

Fig. 8. Voltage-restrained time-overcurrent phase fault protection.

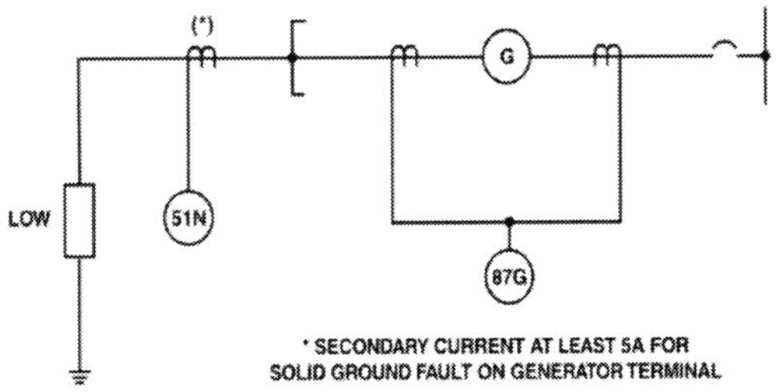

Fig. 9. Ground fault relaying for generator low impedance grounding.

3.6 Stator ground fault protection (51GN)

Normally the generator stator neutral operates at a potential close to ground. If a faulty phase winding is connected to ground, the normally low neutral voltage could rise as high as the line-to-neutral voltage depending on the fault location. Although a single ground fault will not necessarily cause immediate damage, the presence of one fault increases the probability of a second one. A second fault, even if detected by differential relay, may cause some serious damages.

If 87G relaying is provided, relay 51GN backs up the 87G, as well as the external relays. If an 87G is not provided or is not sufficiently sensitive for ground faults, the 51N will serve as the primary protection for the generator as shown in figure 9. The advantage of the 87G is that there is no delay to coordinate with external protection; however, a delay is required for the 51N [5].

The existing grounding system of generator is low impedance grounding with an 80Ω Neutral Ground Resistance (NGR), having with the capacity of 8000 V, 100 A for 10 seconds. There is a linear correlation between the current identified by ground fault relay and the ground fault position in the stator winding. The existing setting for stator ground fault is 20 A on primary side. It means that the relay will protect 80% of the generator winding.

4 Cost and benefit analysis

The investment cost on table 5 is acquired from equipment price (new numerical relay) and man power for installation, testing and commissioning.

Table 5. Total Investment Cost.

No	Description	Price / Unit	Quantity	Total Cost	Remark
1.	Generator Protection	US$ 9,200	14	US$ 128,800	Refer to GE SR 489
2.	Material Accessories	US$ 500	14	US$ 7,000	
3.	Installation Cost	US$ 1,000	14	US$ 14,000	
4.	Service Engineer for Testing and Commissioning	US$ 2,000	14	US$ 28,000	
Total Investment Cost				**US$ 177,800**	

While loss of production cost are derived from information on table 6 which shown the production loss due to nuisance trip or inability of generator protection to clear the fault in past time.

Table 6. Total Loss of Production Caused by Generator Protection Case.

No	Date	Equipment	Outage Production	Outage LNG (m^3)
1.	12 Oct 1988	31-PG-5	0.2 train days (D)	3,456
2.	26 Aug 1996	31-PG-4	0.65 train days (A&B)	11,232
3.	22 Jul 1997	31-PG-2	1.39 train days (A)	23,018
4.	7 Mar 2001	31-PG-8	0.32 train days (C)	5,530
5.	24 Jul 2002	31-PG-8	0.18 train days (C)	3,067
6.	13 May 2014	31-PG-10	-	-
Actual LNG Loss Production				**US$ 9,793,080**
Total Average LNG Loss Production / Year (30 Years)				**US$ 326,436**
Remark: Assumption • Liquid Density = 453 kg/m^3 • Higher Heating Value = 1,107 BTU/SCF • 1 m^3 LNG is equal with 23.50 MMBTU • LNG Price = US$9/MMBTU				

By definition, reliability function represents the probability that generator will have no nuisance trip at time "t" could be defined as

$$R(t) = e^{-\lambda t} \tag{8}$$

where R(t) = Reliability Function of generator, λ = failure rate of generator protection.

Since 1988 or for 30 years, Badak LNG has been suffering 6 times of nuisance trip. It means the failure rate λ = 6 / 30 years = 0.20 / year.

Therefore Reliability per years of generator system in Badak LNG per year

$R(t) = e^{-\lambda t} = e^{(-0.2)(1)} = 0.819$

And then, the probability that the generator fail because of nuisance trip or inability to clear the fault.

$$F(t) = 1 - R(t) \tag{9}$$

F(t) = probability of failure = 1 – 0.819 = 0.181

And the cost of consequences for not replace relay protection.

$$E(t) = F(t) \times C \tag{10}$$

E(t) = cost of consequences.

C = consequences, the amount of US$ losses average in per year.

E(t) = 0.181 x US$ 326,436 / year = US$ 59,085 / year.

With several assumptions basis data such as: project lifetime about five years and cost of capital Badak LNG about 10.18% so this replacement will give positive NPV (Net Present Value) started in fourth year (payback period in fourth year). From this result it can concluded that this project is justified from economical evaluation perspective.

5 Conclusion

Badak LNG had to start replacing the existing electromechanical relays with digital technology which has several advantages.New digital relay must have same setting with the previous to keep selectivity and proper function of machine. Replacement will give benefit to Badak LNG because it gives positive NPV (Net Present Value) from cost and benefit analysis point of view.

Acknowledgments

The author would like to thanks Mr. Rahmat Safruddin and Mr. Feri Sulistyo Nugroho for encouraging the author to write this paper and many thanks to all Maintenance and Operation Crew for making this project upgrading implemented.

References

1. IEEE Std 242-2001 Recommended Practice for Protection of Industrial and Commercial Power System
2. General Electric 2013 *489 Generator Management Relay Instruction Manual*
3. IEEE Std C.37.102-2006 IEEE Guide for AC Generator Protection
4. Rockfeller G 1994 *Generator Protection Application Guide* (Basler Electric)
5. IEEE Std C.37.101-1993 IEEE Guide for Generator Ground Protection

MATEC Web of Conferences **277**, 03009 (2019) https://doi.org/10.1051/matecconf/201927703009
JCMME 2018

Comparison of harmonic limits and evaluation of the international standards

Zijiang WANG[1,*], *Qionglin* LI[2], *Yuzheng* TANG[2], *Shuming* LIU[2], and *Shuangyin* DAI[2]

[1]Industrial Technology Research Institute of Zhengzhou University, P.R. China
[2]State Grid Electric Power Research Institute of Henan Electric Power Company, P.R. China

Abstract. The paper takes the Chinese standard GB/T 14549-1993, the British Engineering Recommendation G5/4-1, the Institute of Electrical and Electronics Engineers IEEE Std 519-2014 and the part of IEC 61000-3 series standard as an example. Then summarize the difference of harmonic limits, the different processing methods of harmonic impedance, and the different summation law of each standard. Compare the harmonic limit calculation and evaluation methods of Chinese GB/T 14549-1993, British G5/4-1 and IEEE Std 519-2014 standard in detail, summarize the characteristics of the three standards in disturbance emission evaluation and introduce examples to verify the conclusion. Finally, summarize the characteristics and differences of mentioned harmonic standards.

1 Introduction

With the large use of power electronic switching devices, which generate non-power frequency harmonic voltages during operation [1], harmonic voltage components and resulting harmonic current in the power grid have been increasing rapidly. These harmonic currents flow into the busbars and affect the normal use and life of other powered devices connected to the busbars. Therefore, in order to minimize the impact of harmonics on the grid and customers, and to ensure the quality of power supply, countries have established harmonic limit standards according to the actual situation of the national grid and customers. Different standards use different ideas and methods. This paper compares and summarizes the difference of harmonic limits, the different processing methods of harmonic impedance, the different summation laws of Chinese standard GB/T 14549-1993, the British Engineering Recommendation G5/4-1, the American Engineers Association IEEE Std 519-2014 and the part of IEC 61000-3 series standards. Then introduce an instance to compare the calculation and evaluation methods of Chinese GB/T 14549-1993, British G5/4-1 and IEEE Std 519-2014 standard.

2 The comparison of harmonic limits

* Corresponding author: wzj23@126.com

This part is mainly to compare the specific value of harmonic limits from different standards. At first, there is a concept of compatibility level and planning level. The compatibility level refers to the maximum level that power grid can be operated normally, above which there will be an impact. The planning level refers to the specific value set by the operator. It is the planning goal level, which ensures the stable operation of the power grid. Therefore, the planning level is generally stricter than compatibility level. Among the standards mentioned in this paper, IEC 61000-3 series standards give the indicative values of planning level but the evaluation of disturbance emission is mainly based on the compatibility level [2]. The evaluation of the other three standards is based on the planning level.

As for the THD (Total Harmonic Distortion) limits, Chinese standard GB/T 14549-1993 and British G5/4-1 are basically same. The THD limits for 35kV and below the grid of IEEE Std 519 is relatively loose. For 35kV~69kV grid, IEC 61000-3 series standards are stricter. For grid above 69kV, IEEE Std 519 is stricter than the other three standards, except for 110kV~161kV grid, Chinese standard GB/T 14549 is stricter.

As for the limits of harmonic voltage, IEC 61000-3 series standards and British G5/4-1 give the harmonic voltage in three categories which are odd harmonics non-multiple of 3, odd harmonics multiple of 3, and even harmonics. In a comparison of the two standards, the harmonic voltage limits of British G5/4-1 is stricter than IEC 61000-3 series standards under same voltage level. For the 35kV~145kV grid, the limits of these two standards are basically the same. For a grid of higher voltage level like 275kV and 400kV, the British standards stricter.

3 Different processing methods of harmonic impedance

The processing methods of harmonic impedance can be divided into two categories, the first is simplified method applied by Chinese standards GB/T 14549, British G5/4-1 and IEC 61000-3 series standards, which considers that the harmonic impedance is approximately equal to the harmonic order multiplied by the fundamental impedance as:

$$Z_h = hZ_1 \tag{1}$$

The other method is applied by IEEE Std 519-2014. This method can bypass the direct use of harmonic impedance [3]. The concept of short-circuit ratio (SCR) is introduced in this standard, which refers to the ratio of the available short-circuit current to the load current. The harmonic current limit is translated into a relationship between the harmonic voltage limit and the SCR. The process to find harmonic impedance is bypassed. The harmonic current limit is given by the total demand distortion (TDD), which refers to the ratio of the harmonic current to the maximum demand current. It can be calculated as:

$$\frac{I_h}{I_L} = \frac{\dot{U}_h \cdot U_N}{Z_{sh} I_L} \approx \frac{\dot{U}_h \cdot U_N}{hZ_s I_L} = \frac{\dot{U}_h}{h} \times \frac{I_{SC}}{I_L} = \frac{\dot{U}_h}{h} \times SCR \tag{2}$$

In equation (2), I_h is the harmonic current in amperes (RMS) at point of public coupling (PCC), I_L is the maximum demand current in amperes (RMS) at PCC, U_h is the harmonic voltage limit in per-unit value based on the rated phase voltage, Z_{sh} is the harmonic impedance of order h at PCC, Z_s is the fundamental impedance converted from the short-circuit capacity. As we can see from Equation (2), the harmonic current limit can be calculated directly by the SCR. Furthermore, the IEEE Std 519-2014 standard determines the harmonic current limit by assuming that the harmonic impedance is the order h multiplied by the short-circuit impedance, ignoring the effects of capacitance and load. In this way, for high order harmonics, the harmonic impedance is larger so that the harmonic current limit is lower, and the margin is larger.

4 The comparison of summation laws

When assessing the harmonic level in the power grid, it is necessary to consider both the background harmonics generated by the upper-level power grid to the lower-level power grid and the harmonics generated by the harmonic sources of this level. Different hypothesis are adopted when calculate the summation of different components. This part summarizes the harmonic summation laws by taking IEC 61000-3-6, British G5/4-1 and GB/T 14549-1993 standards as examples.

IEC 61000-3-6 gives a general summation law of harmonic voltage and harmonic current as [4]:

$$U_h = \sqrt[\alpha]{\sum_i U_{hi}{}^{\alpha}} \tag{3}$$

The summation law of harmonic current is the same as harmonic voltage. U_h is the magnitude of the resulting harmonic voltage (order h), U_{hi} is the magnitude of the various individual emission levels (order h) to be combined, α is an exponent as table 1. The exponent α considers that the harmonic emission co-ordination mainly refers to 95 % non-exceeding probability values, the aggregated global emission of different installations, the high variety of high order harmonic and so on [4].

Table 1. Summation Exponents for Harmonics (Indicative Values).

Harmonic order	α
h<5	1
5≤h≤10	1.4
h>10	2

British G5/4-1 adopts different summation laws for different order harmonic. For harmonic order up to and including 5th, for all the triplens and for the harmonic order having the highest measured value, add the different components arithmetically as equation (4). U_{hp} is summated voltage distortion, U_{hm} is background harmonic value, U_{hc} is the total emission from the customer's new load [5].

$$U_{hp} = U_{hm} + U_{hc} \tag{4}$$

For all other harmonics and to allow for diversity, the vector sum with 90° phase angle is adopted as:

$$U_{hp} = \sqrt{U_{hm}^2 + U_{hc}^2} \tag{5}$$

The summation law adopted by the Chinese standard GB / T 14549-1993 mainly uses the cosine theorem, as [6]:

$$C_h{}^2 = A_h{}^2 + B_h{}^2 + 2A_h B_h \cos\theta_h \tag{6}$$

C_h is summated value, A_h is the background component from the upper-level grid, B_h is the component generated in this level, θ_h is the phase angle of A_h and B_h. In the actual power grid, considering the randomness of the harmonic phase and the statistical results of the power flow calculation, and the harmonic sources are generally close to each other, the summation is usually calculated as:

$$C_h{}^2 = A_h{}^2 + B_h{}^2 + A_h B_h \tag{7}$$

5 The comparison of limits calculation and disturbance emission evaluation

This part compares the limits calculation and evaluation of GB/T 14549, British G5/4-1 and IEEE Std 519-2014 standards and introduce an example for analysis.

5.1 Theoretical comparison of limits calculation and evaluation

At present, most of the electrical equipment is connected to the low-voltage power grid. Therefore, the Chinese standard GB/T 14549 prioritizes the THD and the harmonic voltage limit of the low-voltage power grid. Then the limits of the other voltage levels are based on this. So that most of the electrical equipment can be ensured to operate normally. Considering the influence of stator winding heating on the life of AC induction motor, the influence of capacitor overvoltage and overcurrent on its life, and the influence of harmonic distortion on the computer, the Chinese standard GB/T 14549 determines that the THD limit of the 0.38kV grid is 5%. Then, according to the transmission relationship between the different voltage levels, the THD limits of other voltage levels are determined as shown in table 2.

Table 2. Harmonic Voltage Limits of Power System of *GB/T 14549-1993.*

Voltage Level/kV	THD/%	Harmonic Voltage/%	
		Odd Order	Even Order
0.38	5.0	4.0	2.0
6, 10	4.0	3.2	1.6
35, 66	3.0	2.4	1.2
110	2.0	1.6	0.8

Since the actual measurement results show that the harmonic of phase voltage is higher than the line voltage, the harmonic voltage limit calculation is based on the phase voltage. The harmonic voltage limits of table 2 are the result after the summation of background component from the upper-level grid and the component generated in this level. equation (7) can be transferred into equation (8) to calculate the harmonic voltage limits of harmonic sources in each voltage level.

$$B_h = \sqrt{C_h^2 - 0.75A_h^2} - 0.5A_h \tag{8}$$

Taking the influence of network component impedance such as transformers and resonance amplification into account, the transmission coefficient of the Chinese standard GB/T 14549-1993 to transfer harmonic voltage from the upper-level grid to the lower-level grid is 0.8. At the same time, for the harmonic components transferred to 110kV grid from 220kV grid, THD is approximately 0.8%, odd harmonic voltage is approximately 0.6%, even harmonic voltage is approximately 0.3%. Then we can get the harmonic voltage limits of harmonic sources in each voltage level as table 3.

Table 3. Harmonic Voltage Limits of Harmonic Sources in Different Levels of *GB/T 14549-1993.*

Voltage Level/kV	THD/%	Harmonic Voltage/%	
		Odd Order	Even Order
0.38	2.6	2.0	1.0
6, 10	2.2	1.8	0.9
35, 66	1.9	1.5	0.7
110	1.5	1.2	0.6

Equation (1) can be calculated as:

$$Z_h = h\frac{(1000U_N)^2}{10^6 S_k} = \frac{hU_N^{\ 2}}{S_k} \tag{9}$$

S_k is the reference short-circuit capacity in MVA, U_N is the nominal voltage of the grid in kV. The h order harmonic voltage (HRU_h%) can be calculated as:

$$HRU_h = \frac{U_h}{U_1}\times 100\% = \frac{I_h Z_h}{1000U_N/\sqrt{3}}\times 100\% = \frac{\sqrt{3}I_h Z_h}{10U_N}\% \tag{10}$$

Combine equation (9) and equation (10), we can get:

$$I_h = \frac{10S_k \cdot HRU_h}{\sqrt{3}U_N \cdot h} \tag{11}$$

According to the equation (11) and table 3, the total harmonic voltage limits of each voltage level grid can be calculated. Then we can get the harmonic voltage limits for each customer as [6]:

$$I_{hi} = I_h\left(S_i / S_t\right)^{1/\alpha} \tag{12}$$

S_i is the capacity of the customer i, S_t is the sum of the capacity allocations of all installations at PCC, α is Phase Superposition Coefficient as table 4.

Table 4. The Value of Phase Superposition Coefficient α.

h	3	5	7	11	13	9，>13，Even Order
α	1.1	1.2	1.4	1.8	1.9	2.0

When the actual short-circuit capacity S_{ki} isn't totally the same as the reference short-circuit capacity S_k, the harmonic current limits can be transferred as [6]:

$$I_h' = I_h \cdot S_k' / S_k \tag{13}$$

The evaluation process is divided into three stages in the British G5/4-1 standard. Stage 1 is simplified evaluation for the 230/400V grid. When the connections of equipment comply with the relevant limits and other conditions, they can be connected to the grid. Stage 2 and stage 3 are the core of the evaluation process. The idea of evaluation is different with Chinese standard GB/T 14549. Chinese standard requests customer to inform the capacity they need before the limits are given. But in British standard, connections of customer's equipment which comply with emission limits may be without a voltage evaluation provided that the levels of network distortion are less than 75% of the planning levels. If the connections don't comply with the emission limits, further harmonic voltage prediction calculations of equation (4) and (5) are needed. Different from the idea that the Chinese standard calculates the harmonic limit of each customer according to the customer's capacity and the total limit, G5/4-1 summates the customer's harmonic voltage with the background harmonic voltage and compares it with the overall limit. The specific calculation method can be transformed roughly from equation (11). What is different is that the British G5/4-1 considers the harmonics generated by the low-order parallel resonance that may occur in this frequency range, introducing the empirical coefficient k as:

$$HRU_h = \frac{I_h kh\sqrt{3}U_N}{10S_k}\% \tag{14}$$

For the 400V grid, k is 1 when h≤7, k is 0.5 when h>7. For 6.6kV, 11kV, 20kV and 22kV grid, k is 2 when h≤8, k is 1 when h>8.

In comparison with the limits of the Chinese standard and the British G5/4-1, the Chinese standard is relatively stricter. The total harmonic planning limits are divided according to the customer's capacity in Chinese standard GB/T 14549-1993. So that there will be a lot of margins when customers control their emission equal to or below the limits. But it might be difficult for new customers when connecting to the grid if they don't inform the system operator before the limits are set. The main advantage of the British G5/4-1 is its flexibility. There are no certain harmonic limits for customers to connect to the grid. It can be adjusted according to the actual situation of the grid and the needs of customers. The disadvantage is also obvious that there will be more emission space for earlier connection but nearly no more space for later connection. It is not fair for customers who connect to the grid later.

The IEEE Std 519-2014 simplifies the harmonic voltage limits and THD limits only for different voltage levels [6]. However, the harmonic current limits of IEEE Std 519-2014 are more detailed. In different voltage levels, the standard firstly calculates the ratio of the short-circuits current to the maximum demand current to get SCR of customers. Then calculate the TDD of customers using equation (2) and compare the TDD with the limit according to the different interval of SCR. The main advantage to use TDD as the harmonic current limit is that it can prevent that in some cases the THD value is large but the actual amplitude of the harmonic current is small. Because sometimes little amplitude of harmonic current may account for a large proportion in which case will not impact the power system.

5.2 Comparative analysis of examples

Taking the Guangzhou and Zhuhai Electrified Railway Guanyao Traction Station as an example, according to the PSCAD/EMTDC simulation results, it is analyzed whether it complies with the limits of three harmonic standards in section 5.1. The Guangzhou and Zhuhai Railway Guanyao Traction Station uses a 110kV/27.5kV three-phase V/V wiring transformer with an impedance voltage of 8.4% and with twice the overload capacity. The recent 95% probability maximum load of the Guanyao Traction Station is 26454kVA. Considering the situation of the power grid near the Guanyao Traction Station and the power supply capacity of the 220kV transformer, the power station that can be considered near the Guanyao Traction Station are 110kV Taobu Station and 110kV Leping Station. Next, the harmonic levels of the two power supply schemes are analyzed and evaluated. The THD results of Taobu Station and Leping Station obtained from the simulation results are shown in table 5.

Table 5. The THD of PCC of *Guanyao Traction Station* and THD Limits.

Power Station	THD (%)			GB/T 14549 THD Limits (%)	G5/4-1 and IEEE Std 519 THD Limits (%)
	Phase A	Phase B	Phase C		
Taobu	0.512	1.008	0.496	2	2.5
Leping	0.546	1.074	0.527		

It is shown from table 5 that the THD of the two power supply schemes of the Guanyao Station does not exceed the three standards' limits. According to the simulation results, the harmonic current can be obtained as shown in table 6.

Table 6. The Harmonic Current of *Guanyao Traction Station* (A).

Power Station	3rd Harmonic Current			5th Harmonic Current			7th Harmonic Current		
	Phase A	Phase B	Phase C	Phase A	Phase B	Phase C	Phase A	Phase B	Phase C
Taobu	8.380	16.446	8.066	5.116	10.064	4.948	5.041	9.914	4.873
Leping	7.854	15.410	7.546	5.009	9.851	7.082	4.952	9.737	4.785

The capacity data of PCC is shown in table 7. The short-circuit capacity of the 110kV busbar of the Taobu Station is 2050MVA, and the short-circuit capacity of the 110kV busbar of Leping Station is 1804MVA.

Table 7. The Capacity of Power Station Connected to the PCC of 110kV Grid.

Power Station	Taobu	Leping	Longkou	Beiqu	Gulao	Sancun
Capacity (MVA)	2*63	2*50	40	2*40	31.5+40	2*50

5.2.1 The evaluation according to Chinese standard GB/T 14549-1993

The odd harmonic voltage limit of the 110kV grid of Chinese standard is 1.6%, the even harmonic voltage limit is 0.8%. Combined with data in Table 5, some of them are above the limit. The harmonic current limits of the 110kV grid PCC of Chinese standard are shown as table 8.

Table 8. The Harmonic Current Limits of PCC of *GB/T 14549.*

Voltage Level (kV)	Reference Short-Circuit Capacity (MVA)	The Harmonic Order and Limits (A)		
		3	5	7
110	750	9.6	9.6	6.8

According to equation (13) and table 8, the PCC harmonic current limits of Guanyao Traction Station can be calculated as:

Table 9. The PCC Harmonic Current Limits of Guanyao Traction Station (A) of *GB/T 14549.*

Power Station	3rd Harmonic Current Limit	5th Harmonic Current Limit	7th Harmonic Current Limit
Taobu	26.24	26.24	18.59
Leping	23.09	23.09	16.36

According to the data of table 9 and table 7, the harmonic current limits of the single customer can be calculated by equation (12) and table 4 as:

Table 10. The Harmonic Current Limits of Single Customer of *GB/T 14549.*

Power Station	3rd Harmonic Current Limit	5th Harmonic Current Limit	7th Harmonic Current Limit
Taobu	7.264	8.085	6.777
Leping	5.181	5.868	5.056

By comparing the data of table 6 and table 10, according to the Chinese standard GB/T 14549 limit, there are situations that the two power supply schemes' harmonic current exceeds the limits.

5.2.2 The evaluation according to british G5/4-1

According to British G5/4-1, with the data of table 6 and equation (14), the harmonic voltage generated by the harmonic source in the 110kV voltage level can be calculated as:

Table 11. The Harmonic Voltage Generated by Harmonic Source in 110kV voltage level of *G5/4-1*.

Power Station	3rd Harmonic Voltage Limit%			5th Harmonic Voltage Limit%			7th Harmonic Voltage Limit%		
	Phase A	Phase B	Phase C	Phase A	Phase B	Phase C	Phase A	Phase B	Phase C
Taobu	0.639	1.253	0.615	0.650	1.278	0.628	0.896	1.763	0.867
Leping	0.599	1.174	0.575	0.636	1.251	0.899	0.881	1.732	0.851

Assume that the background THD of the 110kV grid is 0.8%, and the background odd harmonic voltage is 0.6%. According to the summation law given in British G5/4-1, as shown in equation (4) and equation (5), summate the data of Table 11 with the background harmonic. Then the summated harmonic voltage is table 12:

Table 12. The Harmonic Voltage of 110kV Grid of *G5/4-1*.

Power Station	3rd Harmonic Voltage Limit%			5th Harmonic Voltage Limit%			7th Harmonic Voltage Limit%		
	Phase A	Phase B	Phase C	Phase A	Phase B	Phase C	Phase A	Phase B	Phase C
Taobu	1.239	1.853	1.215	1.250	1.878	1.228	1.079	1.862	1.054
Leping	1.199	1.774	1.175	1.236	1.851	1.499	1.066	1.833	1.041

The British G5/4-1 standard stipulates that for the 20kV ~145kV grid, the harmonic voltage limit of 3rd, 5th, 7th harmonic is 2%. It can be seen from the data in table 11 that the harmonic voltage of the Taobu and Leping station meet the requirements of G5/4-1.

5.2.3 The evaluation according to IEEE Std 519-2014

According to the SCR method mentioned in Section 3, Taobu Station can be used as an example to get the SCR as:

$$SCR = \frac{I_{SC}}{I_L} = \frac{S_{SC}/\sqrt{3}U_N}{S_L/\sqrt{3}U_N} = \frac{2050 \times 1000/110\sqrt{3}}{26454/110\sqrt{3}} = 77.493 \tag{15}$$

The SCR of Taobu Station ranges from 50 to 100. According to the limit of IEEE Std 519-2014, for the grid between 69kV and 161kV, the ratio of the 3rd to 10th harmonic currents must not exceed 5%, calculate its specific harmonic current limit as:

$$I_h = I_L \times 5\% = \frac{26454}{110\sqrt{3}} \times 5\% = 6.9425A \tag{16}$$

The harmonic current can be obtained as table 13:

Table 13. The Harmonic Current Limits of Single Customer of *IEEE Std 519-2014*.

Power Station	3rd Harmonic Current Limit	5th Harmonic Current Limit	7th Harmonic Current Limit
Taobu	6.9425	6.9425	6.9425
Leping	6.9425	6.9425	6.9425

In comparison with table 6 and table 13, there are harmonic currents which exceed the limits of IEEE Std 519-2014.

5.2.4 Comparison of evaluation results

After the evaluation of three standards for two power supply schemes of Guanyao Traction Station, it can be seen that these two schemes only meet the limits of British G5/4-1 standard. From the analysis of this example, the limits of the Chinese standard GB/T 14549 are directly dependent on the customer's capacity, and the limits for the small-capacity customer are stricter. The limits of IEEE Std 519 are not directly dependent on the customer's capacity but is determined by the SCR interval [7]. The limits are the same if the SCR of different customers is in the same interval. The British G5/4-1 standard is relatively more flexible. The limits of G5/4-1 are depended by the background harmonic level. In this paper, under the circumstance of an average background harmonic level, G5/4-1 allows more harmonic emission than the other two standards.

6 Summary

Overall, the concept and evaluation method of the IEC 61000-3 series standards is the basis for other standards [8]. IEC 61000-3 series standards focus on discussing the EMC compatibility of grid [4],[9]-[11]. IEEE Std 519-2014 offers a new method which introduces SCR to solve the difficulties that harmonic impedance is hard to estimate. So that it is much easier to calculate the limits of harmonic current. At the same time, IEEE Std 519-2014 uses TDD for the harmonic current limits, which can avoid the situation that small harmonic current causes high THD. British G5/4-1 standard is based on the IEC 61000-3 series standards, but the limits are relatively higher. And G5/4-1 takes the operation of the actual power grid into account. It is more flexible for distortion sources to connect to the grid. Especially for customers with large capacity, it will be much easier for them to connect to the grid. Chinese standard GB/T 14549-1993 is relatively strict. It divides the total harmonic voltage limits of the whole grid into several parts according to the capacity of each customer and distributes the limit to every customer. So the total harmonic level will not exceed the total level of the whole grid, but it is not flexible for new customers when connecting to the grid.

References

1. Haixue LIN Introduction to the Foreign Standards of Harmonic Current Limits *Distribution & Utilization* **2008(05)**:1-8
2. Qing MO, Haoliang SUN Basis and Derivation of Harmonic Current Limits in IEC Standards for Power Equipments *Power System Technology* **2000(04)**:67-70
3. Haixue LIN Introducing Current Limits in US Harmonic Standard IEEE Std. 519 *Distribution & Utilization* **2012,29(05)**:11-14
4. Electromagnetic compatibility (EMC) - Part 3-6: Limits -- Assessment of emission limits for the connection of distorting installations to MV, HV and EHV power systems (IEC 61000-3-6:2008)
5. Engineering Recommendation G5/4-1, Planning Levels For Harmonic Voltage Distortion And The Connection Of Non-Linear Equipment To Transmission Systems And Distribution Networks in The United Kingdom, Issue 4 Amendment 1-October 2005.
6. Chinese Standards, Quality of electric energy supply Harmonic in public supply network (GB/T 14549-1993)

7. IEEE Recommended Practice and Requirements for Harmonic Control in Electric Power Systems (IEEE Std 519-2014)
8. Haixue LIN On Difference Between National Standard and IEC 61000 Standards of EMC in Harmonic *Power System Technology* **1999(05)**:64-67
9. Electromagnetic compatibility (EMC) - Part 3-2: Limits - Limits for harmonic current emissions (equipment input current ≤ 16 A per phase) (IEC 61000-3-2:2014)
10. Electromagnetic compatibility (EMC) - Part 3-12: Limits - Limits for harmonic currents produced by equipment connected to public low-voltage systems with input current >16 A and ≤ 75 A per phase (IEC 61000-3-12:2011)
11. Electromagnetic compatibility (EMC) - Part 3-4: Limits - Limitation of emission of harmonic currents in low-voltage power supply systems for equipment with rated current greater than 16 A (IEC 61000-3-4:1998)

MATEC Web of Conferences **277**, 03010 (2019) https://doi.org/10.1051/matecconf/201927703010
JCMME 2018

Optimum threshold of group formation in multi-agents

Kodai Shinkawa, and *Isamu* Shioya*

Faculty of Science and Engineering, Hosei University, Kajino-cho 3-7-2, Koganei-shi, Tokyo, Japan

Abstract. This paper presents a simple multi-agent model of group formation, while the behaviour is complicated. The statistical characteristics of the formation are suddenly changed at some states. Their agents are carrying characteristic vectors, meet each other in non-dimensional free spaces without any restrictions randomly, and the agents make groups in the spaces if their characteristics are similar, that is, they have high similarities. Actually, for a given threshold on similarities, when the characteristic vectors between two agents are similar in the threshold, the agents join into one. They are not only for agents but also for groups, i.e. two groups can become into one if they are similar, and we repeat it. On the other hand, making groups decrease a satisfaction on groups. In this paper, we show, for a given threshold, there is not only an optimal threshold to maximize the satisfactions among groups which satisfy the threshold, but also the other one to minimize the satisfactions. The thresholds to maximize satisfactions are experimentally approximated by some function on the size of characteristic vectors rather than the number of agents.

1 Introduction

To discover a new model is essential that the model is to be simple. Our model is of multiple agents, and each agent is quite simple. And, also the interactions among agents are simple, but the behavior sometimes becomes complex. In this paper, we present a new model for multi-agents to form groups, their agents meet each other at random in free moving spaces autonomously and combine the groups into a new group if the characteristic vectors are similar or compatible (an extended similarity). We suppose each agent carrying a characteristic vector that are properties represented by numerical vectors, and assume that when two persons meet each other in the free spaces and also their similarities are higher than a certain threshold, they form a new group. That is, if an agent is close to any other one in the degrees of similarities, two agents form a new group. It applied to not only agents but also groups, i.e. two agents/groups are combined into new one. In this paper, we first define a similarity, that is a compatibility by inner product between two characteristic vectors of agents, and extend this definition to the similarity between two groups consisting groups, that is a set of agents. However, forming group might reduce the satisfactions of the groups by getting the size of groups larger, because each group consists of similar agents,

* Corresponding author: shioyai@hosei.ac.jp

but each agent is different more or less in general. For an example, suppose A is similar to B, B is similar to C, and they form an agent group. Then, A is always similar to C? When an agent group becomes larger, the satisfaction of the agent becomes lower. Proving a threshold is a lower limit of "ideal height", which allows to form groups. When the threshold value becomes lower, the ideal height also becomes lower. Conversely, the ideal height becomes higher when the threshold becomes higher. Forming groups is to become lower on the satisfaction of groups. Then, it is difficult to form a group if the ideal is too either high or low. This paper presents an interesting result for forming groups, and there is not only a threshold value to maximize the amount of the satisfactions of the groups, but also the other threshold to minimize the amount of satisfactions. By choosing two threshold values appropriately, we show that the amount of the fitness among the group can become the highest, and can also become lowest. It says that two kinds of optimal thresholds we have.

The nearest work to ours is [1, 2] for looking another kind of partner in the context of two kinds of agents as same as in our free moving spaces, and two agents become partners each other. The partners are keeping finite time before disassociating, where the compatibility of two agents is defined as well as symmetric, and the goal is to maximize the duration keeping the pairs. [3] is discussion on the real world, not so small, on the conference community of DBLP. It uses the techniques of decision trees, time series and diffusion, they are trying to find some incentive to join the conference community so that to present their papers to the community is not only to related the number of friends, but also how those friends connected to other friends within or out of the community. Even, they are close to their research topics of interest. However, the community as well as others is complicated for obtaining explicit results, how many friends are interrelated. The Schelling segregation model [4] is not only a simple model to produce a segregation among groups as well as our work, also the results are also simple. But, these behaviors are sometimes complicated as [5, 6]. Our discussions can be applied to our real world as collective dynamics [7] and scaling features [7, 8] as we have to try in the next. These are on the discussions in colored free spaces rather than complete free spaces.

This paper is organized as following. We define our model and present an example in the following section. In Section 3, we show our experiments and our discussions. We present our conclusions in the final section.

2 Proposed model of group formation

Consider N agents $A_1, A_2, \ldots, A_N$ of which each has a fixed feature *S*-dimensional vector $\mathbf{x}_i$, and they are moving at random in a moving free space. Then, the agents form groups consisting of some agents if their agents meet at random in the space, their feature vectors match or the similarity value is higher than a certain threshold, i.e. we consider a model in which agents with high compatibility between agents form groups.

Suppose two agents A_i and A_j ($i \neq j$), and two have the feature vector $\mathbf{x}_i$ and $\mathbf{x}_j$, respectively. For the simplicity, assume every elements of their vectors are either 1 or -1. Then we define a compatibility $C'_{i,j}$ between A_i and A_j as well as in [1, 2]

$$C'_{i,j} = (\mathbf{x}_i \cdot \mathbf{x}_j + S)/2,$$

where · denotes an inner product between two vectors, and S is the number of charcteristic vectors. Then, we have $C'_{i,j} = C'_{j,i}$ and $0 \leq C'_{i,j} \leq S$, i, j=1,..,N. Also, $C'_{i,j}$ follow a binomial distribution Bin(S,1/2). The higher similarity $\mathbf{x}_i$ and $\mathbf{x}_j$ becomes the higher compatibility $C'_{i,j}$, and vice versa.

We suppose all the agents belong to one of the groups $G_1, \ldots, G_N$, and they are exclusive. Further, we extend the degree of conformity between agents to the degree of conformity between groups.

$$C_{i,j}=\frac{\sum_{k\in G_i,l\in G_j,k<l} C'_{k,l}}{|G_i|\,|G_j|}$$

$C_{i,j}$ denote a compatibility between G_i and G_j, and it is the amount of the compatibilities of all the agents belonging to their groups.

Similar to $C'_{i,j}$, we have a symmetric $C_{i,j} = C_{j,i}$ and $0 \le C_{i,j} \le S$. Two groups G_i and G_j meet in free spaces, and when their compatibility $C_{i,j}$ are higher than a threshold τ $(0 \le \tau \le S)$, two groups united into one group, and a new group is generated. We note that when $C_{i,j} = \tau$, two groups G_i and G_j into one with even probability 1/2.

A satisfaction S_i of a group G_i is defined by the sum of inner products as:

$$S_i = \sum_{k,l\in G_i,k\neq l} x_k \cdot x_l$$

We assume that the satisfaction of G_i is zero, if G_i are a singleton. S_i are the sum of inner products of two agents among groups, and the sum of the satisfactions for every groups by:

$$S_{sum} = \sum_{i=1}^{M} S_i.$$

By varying the threshold τ, we can find the threshold τ_{op} to maximize the S_{sum}. Also, we can find the other optimal threshold to minimize the satisfactions.

Example 1 (table 1):

Consider the five agents A_1, A_2, A_3, A_4 and A_5, whose vectors x_i have the same dimension $S = 4$, and we define the vectors as

$\mathbf{x}_1 = [1,-1,1,1]$,
$\mathbf{x}_2 = [-1, 1, 1,-1]$,
$\mathbf{x}_3 = [1, -1, 1,1]$,
$\mathbf{x}_4 = [1, 1, 1,-1]$, and
$\mathbf{x}_5 = [1, 1, -1,1]$.

Table 1. C'_{ij} in Example 1.

C'_{ij}	1	2	3	4	5
1		1	4	2	2
2	1		1	3	1
3	4	1		2	2
4	2	3	2		2
5	2	1	2	2	

The compatibilities can be computed, for an example, $C'_{1,2}$ is as

$$C'_{1,2} = \frac{x_1 \cdot x_2 + 4}{2} = 1.$$

At the initial agent setting, each agent belongs to a group having only its own group as follows:

$G_1 = \{1\}$, $G_2 = \{2\}$, $G_3 = \{3\}$, $G_4 = \{4\}$, $G_5 = \{5\}$.

Set the threshold $\tau = 2.5$, and suppose G_1 and G_3 meets at random, and merge into one by $C_{1,3} = C'_{1,3} = 4 > \tau$ and the following groups are formed.

$G_1 = \{1, 3\}$, $G_2 = \{2\}$, $G_3 = \{4\}$, $G_4 = \{5\}$.

Also, if G_2 and G_3 encounter and $C_{2,3} = C'_{2,4} = 3 > \tau$, G_2. Then, the running groups are:

$G_1 = \{1, 3\}$, $G_2 = \{2,4\}$, $G_3 = \{5\}$,

The compatibilities between the groups are as:

$C_{1,2} = \frac{C'_{1,2}+C'_{1,4}+C'_{3,2}+C'_{3,4}}{2\times 2} = 1.5 < \tau$,

$C_{1,3} = \frac{C'_{1,5}+C'_{3,5}}{2\times 1} = 2.0 < \tau$, and

$C_{2,3} = \frac{C'_{2,5}+C'_{4,5}}{2\times 1} = 1.5 < \tau$.

Forming groups are inactive, anymore. The compatibilities of their groups are:
$S_1 = \mathbf{x}_1 \cdot \mathbf{x}_3 + \boldsymbol{x}_3 \cdot \mathbf{x}_1 = 8.0$,
$S_2 = \mathbf{x}_2 \cdot \mathbf{x}_4 + \boldsymbol{x}_4 \cdot \mathbf{x}_2 = 6.0$, and
$S_3 = 0$.

The amount S_{sum} of the compatibility of the agents is:
$S_{sum} = S_1 + S_2 + S_3 = 14.0$

The final group composition obtained may differ depending on the order to encounter each other in the free space. In this research, the group encounters randomly, and it does not depend on group size. But, we will see in the following section that the thresholds to maximize the satisfactions do not depend on the number of agents.

3 Experiments and the results

We examined the formation of groups by agents and computed the thresholds to maximize S_{sum}. We present the following algorithm:

(1) Provide thirty agents, i.e. N=30. Each element of all the feature vectors is either 1 or -1 with even probability.

(2) Two groups meet in the free space randomly and are unified into one if the compatibility of G_i and G_j is greater than the threshold $\boldsymbol{\tau}$. If the threshold is equal to $\boldsymbol{\tau}$, they are unified with even probability.

(3) Repeat (2), until group formation is impossible.

(4) Finally, we show the amount S_{sum} of satisfiability.

We examined 1,000 times for each threshold $\boldsymbol{\tau}$, and their average values are shown in figure 1 and 2. The figures show that there is the optimum threshold $\boldsymbol{\tau}_{op}$ so that the thresholds to maximize the compatibilities S_{sum} can be experimentally approximated by the following expression:

$$\boldsymbol{\tau}_{op} = \frac{S}{2} + \frac{\sqrt{S}}{3}.$$

We present a comparison of the experimental values and the approximated expression values shown in figure 3. Although it looks like a straight line, the values of the second term become relatively smaller as $\boldsymbol{\tau}$ is larger. On the other hand, the thresholds to minimize the S_{sum} can be find in the figures, and also the values are independent to the number of agents, and the values, the optimum thresholds, are low fifty.

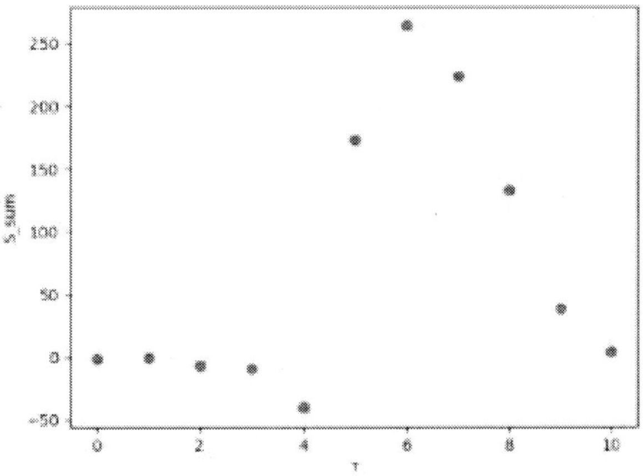

Fig. 1. $\boldsymbol{\tau}$ vs. S_{sum}. The experiment of the case N=30, S=10.

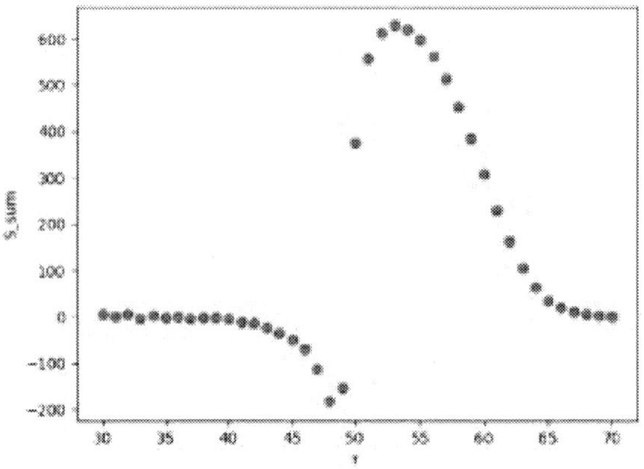

Fig. 2. τ vs. S_{sum}. The experiment of the case N=30, S=100.

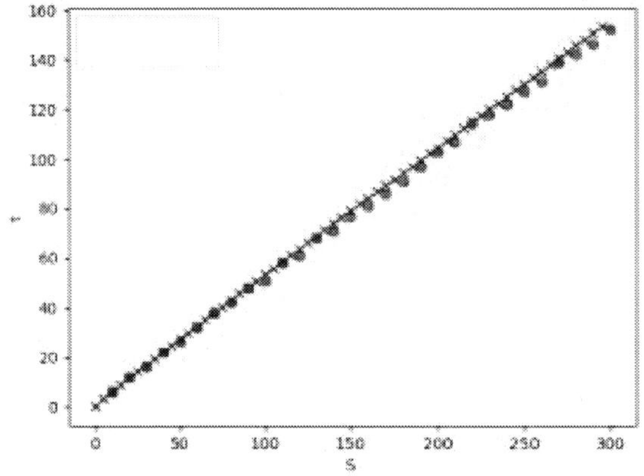

Fig. 3. S vs. τ(maximized). The experimental values (red dot) and approximated values (×).

We have presented the group formation of agents in free spaces. For every two agents encounter each other in the free space, and join into a new group if two are similar. The compatibilities $C'_{i,j}$ to represent the similarities follow binomial distribution Bin(S,1/2), so that the average is S/2, and the variance is S/4 along normal distributions. If two agent similarities is greater than a given threshold $\boldsymbol{\tau}$, they form new groups. So, our model simple and also no restrictions to apply on real world. To become the size of the groups larger decrease the satisfactions of groups lower. So, we expect to derive an optimum threshold and also to find the theoretical values S_{sum}. Our experiments show that our discussions are true. The similar approach can be applied to the threshold which the satisfactions make worst.

4 Conclusions

This paper experimentally showed that, when agents are carrying feature vectors and they meet at random in moving spaces forming groups, the optimum thresholds exist.

We have discussed the group formation of agents in free spaces. For every two agents meet each other in the free space, and join into a new group if two are similar. To become the size of groups larger decreases to become the satisfactions of groups lower. So, we expect to derive an optimum threshold and also to find the theoretical values S_{sum}. Our experiments show that our discussions are true. Similarly, the threshold to minimize the compatibilities can be found. Choosing the worst threshold will not only form a group, but it also become at the worst level of satisfaction.

For our future works, since our experiment was conducted on a simple network that the agent groups randomly meets in free spaces, we will perform on more complex networks as in [7, 8]. Furthermore, consider the model in which the probability of meeting depends on the size of the groups, and we will derive the theoretical expression of S_{sum}.

References

1. David M.D. Smith and Neil F. Johnson,Pair Formation within Multi-Agent Populations, lanl.arXiv.org, arXiv:physics/0604142v1, 2006.
2. David M.D. Smith and Neil F. Johnson,Pair Formation within Multi-Agent Populations, Physica A: Statistical Mechanics and its Applications, 363, 1, pp. 151-158, 2006.
3. Lars Backstrom, Dan Huttenlocher, Jon Kleinberg and Xiangyang Lan, Group formation in large social networks: membership, growth, and evolution, KDD '06 Proceedings of the 12th ACM SIGKDD international conference on Knowledge discovery and data mining, pp. 44-54, 2006.
4. Thomas Schelling, Dynamic models of segregation, Journal of Mathematical Sociology, 1, pp. 143-186, 1971.
5. S. Strogatz, Nonlinear Dynamics and Chaos, Perseus Books, 2000.
6. Y. Kuznetsov, Elements of Applied Bifurcation Theory, Springer, 1998.
7. D. J. Watts and S. H. Strogatz, Collective dynamics of ‘small-world’ networks, Nature, 393, pp. 440-442, 1998.
8. A. L. Barabási and R. Albert, Emergence of scaling in random networks, Science, vol. 286, pp. 509-512, 1999.

Determination of temperature dependences of Young's modulus and internal friction of fuel cladding by resonance method

A A Moshev, *S P* Martynenko, *S S* Martynenko, *E M* Kudryavtsev, *A N* Ableyev, *A N* Tokarev, *R A* Panasenko, *I I* Rodko, *A I* Maksimkin*, *Y O* Zhukovskii, and *E G* Belendryasova

National Research Nuclear University MEPhI (Moscow Engineering Physics Institute), 115409 Moscow, Russia

Abstract. We study elastic characteristics and internal friction of fuel claddings to improve computer codes for VVER-1000 fuel rods. We analytically described elastic characteristics of cladding material and obtained coefficient of the form of the first longitudinal frequency numerically. We described new measuring module for automatic acquisition data. We've established temperature dependences of Young's modulus and internal friction via high-temperature facility and developed electronic module and noted maximum of these characteristics at the temperature 1160 K. It can be explained by the destruction of the texture in the material of claddings.

1 Introduction

The development of new types of nuclear fuel involves the study of their physical and mechanical properties [1]. One of the most important parameters required for fuel rod codes is the temperature dependence of the elastic characteristics (Young's modulus, shear modulus, Poisson's ratio) of the fuel rod cladding material in the operating temperature range. This paper describes the resonance method [2, 3], ultrasonic facility and experimental results on the temperature dependences of Young's modulus and internal friction of fuel rod claddings from zirconium alloys. The measurements were carried out via special measuring module developed in NRNU MEPhI. A distinctive feature of the module is a uniform amplitude- frequency characteristic of the measuring channel in the frequency interval 20 — 200 kHz with its high sensitivity and noise immunity, which allows measurements of elastic characteristics and internal friction of samples in a wide range of their *q*-factor. In this case, the sample is clamped between the pointed ends of the acoustic waveguides, one of which provides the transfer of vibrations from exciting piezoelectric transducer to the sample, and the second — from the sample to receiving transducer [4–7].

The facility allows to measure the temperature dependence of the characteristics of elasticity and internal friction of a wide class of products of different sizes, including composites, fuel cladding, powder metallurgy products, etc. up to their melting

* Corresponding author: aimaksimkin@mephi.ru

temperatures, as well as to investigate the features of structural and phase transformations in them, the kinetics of sintering of powder metallurgy products, to carry out flaw detection of serial products, to assess the quality of composites, including the adhesion of components, etc.

Our paper is structured as follows: in the section 2 we give basic theory states of our measurements. Experimental facility is described in the section 3. The results of experiments are presented in the section 4. In conclusion 5 we summarize important achievements.

2 Theoretical basis of measurements

The principle of operation of the facility is based on the relationship of the resonance characteristics of the samples (resonance frequencies f_i and Q_i, where i is the number of the resonance peak) with their elastic and inelastic characteristics.

The elasticity characteristics of isotropic materials are: Young's modulus E, shear modulus G, bulk modulus of elasticity B, Poisson's ratio v. Of these four characteristics for isotropic materials, two are independent, since there are coupling equations:

$$G = \frac{E}{2(1+\nu)} \quad (1)$$

$$B = \frac{E}{3(1-2\nu)} \quad (2)$$

In General, the resonance spectrum of an arbitrary shape product is a function of several parameters. The frequencies f_i of natural oscillations of any isotropic defect-free body with characteristic size d can be represented as:

$$f_i = P_i(d,v)C_L \quad (3)$$

where $P_i(d, v)$ is the coefficient of the form of oscillations at the i-th frequency, which depends on the geometry of the product and its Poisson's ratio v. C_L is compressional velocity in the product:

$$C_L = \sqrt{\frac{E}{\rho}} \quad (4)$$

where E — Young's modulus, ρ — density of material.

For products of the cylindrical form made of one material (for example, segments of claddings of fuel elements), the coefficient of the form can be represented as:

$$P_i(d,v) = \frac{1}{2}hk_i \quad (5)$$

where h is the height of the sample; k_i is the coefficient of the i-th mode of oscillations, depending on the ratio of height to the diameter of the segment.

For standard segments of claddings 30 mm long, the Young's modulus E is:

$$E = 4h^2 f_{com}^2 k_{com} \rho \quad (6)$$

And shearmodulus:

$$G = 4h^2 f_{tor}^2 \rho \quad (7)$$

where f_{com} — the resonant frequency of the first longitudinal form of vibrations of the segment, k_{com} — the coefficient of the form of the first longitudinal frequency, close to one, f_{tor} — the resonant frequency of the first torsional form of vibrations of the fuel element segment.

The internal friction is determined by the formula:

$$Q_i^{-1} = \frac{\Delta f_i^{0.7}}{f_i} \tag{8}$$

where $\Delta f_i^{0.7}$ is the width of the resonance peak at the height of 0.7 of the maximum amplitude.

The coefficient k_{com} is calculated using the finite element method (FEM) in ANSYS for different lengths of fuel cladding segments. The calculation results are shown in figure 1.

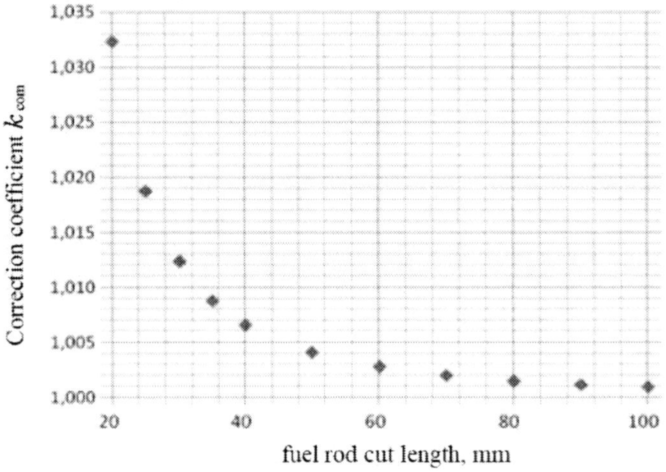

Fig. 1. The dependence of the correction coefficient k_{com} on the length of the fuel cladding segment.

3 Experimental facility

The block diagram of the experimental sample of a spectroscopic high-temperature facility for measuring the elastic characteristics and internal friction of the samples is shown in figure 2. The facility consists of two parts in the form of stainless steel pipes with a flange connection— a fixed 5 with an ultrasonic measuring module and a movable one with a heater 14. The segment of the cladding or the test sample 1 is clamped between two stainless steel acoustic waveguides, one of which 2 provides transfer of ultrasonic vibrations from exciting piezoelectric transducer to the segment of cladding, and the second 3 — from the cladding to receiving piezoelectric transducer. After that, the movable part is moved in the horizontal direction and is connected fixed via a flange gasket 4. Dampers 6 are used to optimize the amplitude-frequency characteristics of the measurement channel. At opposite ends of acoustic waveguides there are transducers — one 8 for exciting vibrations and 7 for receiving them. The electrical leads from the piezoelectric transducers and the chromium-aluminum thermocouple 10 pass through the rubber seal 9. To generate sinusoidal signals in a given frequency range, register an informative signal and communicate with the control computer 13, the interface block 12 is used. The system of vacuuming and filling the facility with inert gas includes a nozzle 11, flexible hoses and a system 16, which provides an inert environment in the facility. The power supply of the heater 15 heats the nichrome heater 14 and is controlled by the computer 13 through the interface unit 12. The main technical characteristics of the module are given in table 1.Measurements are carried out automatically using the PC.

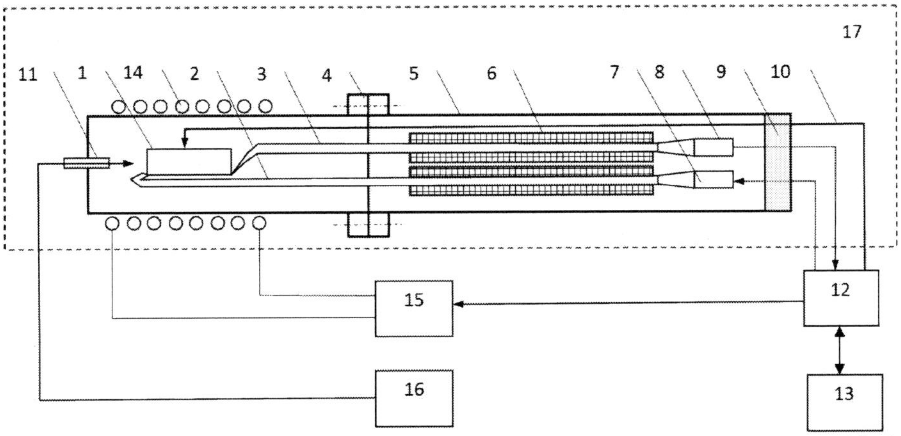

Fig. 2. Block diagram of the facility: 1 – short segment of a cladding; 2 – acoustic waveguide for the transfer of vibrations from the excitation piezoelectric transducer to the sample; 3 – acoustic waveguide for the transfer of vibrations from the sample to receiving piezoelectric transducer; 4 – flange gasket; 5 – case; 6 – damper; 7 – exciting the piezoelectric transducer; 8 – receiving piezoelectric transducer; 9 – flange gasket; 10 – chromel-aluminum thermocouple; 11 – nozzle of filling inert gas; 12 – the interface block; 13 – control computer; 14 – heater; 15– power supply to the heater; 16 – filling system with an inert gas; 17 – hot chamber.

Table 1. Main technical characteristics of the facility.

Temperature measurement range	300–1500 K
Environment	Vacuum, inert
Sample mounting method	point
Measurement error of young's and shear modules, not more than %	1
The range of internal friction of samples	10^{-4} –0,2
Measurement error of internal friction, not more than %	5
The working range of frequencies	20–200 kHz
Unevenness of the frequency response of the measuring path, dB	6
Material of acoustic waveguides	Stainless steel
The material of the transducer	PZT–19

4 Experimental results

Via the developed facility, measurements of Young's modules, shear and internal friction of a series of fuel rod segments from zirconium alloys of promising compositions were carried out. In figures 3(a) and 3(b), the results obtained for the VVER standard cladding are presented in the form of temperature dependence graphs. It can be seen from the graphs that at the first heating of the samples (lower curve) at a temperature of 1160 K there is a characteristic change in Young's modulus and internal friction (the so-called peak of internal friction), indicating the phase transition in the samples from α-Zr with a hexagonal lattice (a = 0.3231 nm, C = 0.5146 nm) to β-Zr with a cubic lattice (a = 0.361 nm). In the subsequent cooling of samples from high temperatures, the young modulus increases in the temperature range of 300 — 1100 K by 10-20 %, which can be explained by the destruction of the existing texture in the cladding after the passage of phase transformations.

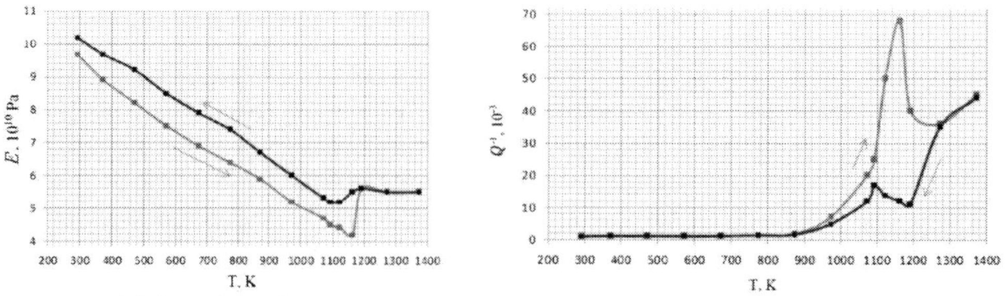

(a)Young's modulus (b) Internal friction

Fig. 3. Temperature dependences of characteristics of standard cladding of VVER fuel rod on heating (blue curve) and cooling (red curve).

5 Conclusion

We developed ultrasonic module of high-temperature facility allows to measure the temperature dependence of elastic characteristics and internal friction of standard fuel rod claddings. Obtained characteristics behave non-monotonously in narrow temperature range 1100 — 1200 K and at the temperature 1160 there is a maximum — internal friction peak. Such phenomenon can be explained via the destruction of material texture.

We also found correction coefficient of the form the first longitudinal frequency via finite element method (see Figure 1).

The results obtained during the presented experimental work will become the fundamental for developing the system of integrated digital twins of fuel rods. The framework establishes the relationship between thermal-physical and neutron design codes with a change in the technical state of the fuel rods, determined by the methods presented.

Acknowledgments

The research was supported by the Ministry of science and higher education of the Russian Federation, the Agreement №14.578.21.0258, a unique identifier of the project RFMEFI57817X0258.

References

1. Reshetnikov F G, Bibilashvili Y K, Golovnin I S, Gorsky V V, Kazennov Y I, Menshikova T S, Nikulina A V and Romaneev V V 1995 *Development, production and operation of fuel elements of power reactors* vol 1 (Moscow: Energoatomizdat)
2. Baranov V M 1989 *Acoustic measurements in nuclear energy* (Moscow: Energoatomizdat)
3. Baranov V M, Karasevich A M, Kudryavtsev E M and et al 1999 *Acoustic measurements in nuclear energy*(Moscow: Energoatomizdat)
4. Maximkin A I, Kryukov I S, Ableev A N, Berestov A V and Rodko I I 2018 *ICONE 2018* vol 3 (London, United Kingdom: American Society of Mechanical Engineers)
5. Martynenko S P, Martynenko S S, Ableev A N, Panasenko R A, Anufriev B F and Maksimkin A I 2016 *ICCICCT* 2015 (Kumaracoil, India: Institute of Electrical and Electronics Engineers) pp 668–671 [6]Martynenko S S, Ableev A N, Martynenko S P, Panasenko R A, Berestov A V and Rodko I I 2016 *ICCICCT* 2015 (Kumaracoil,

India: Institute of Electrical and Electronics Engineers) pp 672–676 [7]Panasenko R A, Martynenko S P, Martynenko A S, Ableev A N, Kudryavtsev E M, Frolov E S and

6. Varyatchenko E P 2016 *ICCICCT* 2015 (Kumaracoil, India: Institute of Electrical and Electronics Engineers) pp 677–681

Effect of different curing conditions on compressive strength and durability of kenaf fibre reinforced blended cementitious composites

R. Ahmad, *R.* Hamid*, and *S. A.* Osman

Smart and Sustainable Township Research Centre, Faculty of Engineering and Built Environment, Universiti Kebangsaan Malaysia, 43600 UKM Bangi, Selangor, Malaysia

Abstract. This paper investigates the effect of normal curing (NC), air curing (AC), and burlap curing (BC) under different curing periods on the mechanical strength and durability of kenaf fibre reinforced blended cementitious composites (KFRBCC) with thermally activated alum sludge ash (AASA). The aim is to determine the most efficient condition and period for curing KFRBCC and to assess the effect of accelerated ageing on strength and durability of KFRBCC after wet/dry cycles. Meanwhile, the microstructure of these mixes is observed via scanning electron microscopy (SEM). The KFRBCC is designed to achieve strength beyond 50 MPa after 28 days (d) of curing by adding 2% treated kenaf fibre (KF) and by replacing Ordinary Portland Cement (OPC) with AASA. The findings suggest that compressive strength and durability of KFRBCC with 40% AASA cured under BC after age of 28 d are enhanced. The addition of treated KF with AASA have helped in limiting the reduction in the compressive strength and durability of the KFRBCC, particularly under prolonged curing period.

1 Introduction

The most influential factors on mechanical and durability of cementitious composites is a satisfactory curing condition for adequate moisture content and temperature to assist the hydration of OPC and to attained higher strength [1]. The importance of curing is critical in the mixtures contains pozzolanic materials, or are otherwise intended to meet specific performance requirements [2]. Curing for concrete containing pozzolanic material takes at least 4 weeks due to pozzolanic reaction occurs at the mature age of more than 28 d which enhance the durability [3]. Compressive strength should also be characterize along with durability to identify the satisfactory performance of composite and to withstand the damaging effects of environmental factors under severe service conditions. Several researchers working on improvement of concrete strength and durability by incorporation of pozzolans in mixes [4-5]. The use of pozzolans in concrete results in reduction of early

* Corresponding author: roszilah@ukm.edu.my

strength and increase in late strength which require high replacement contents of pozzolans [6-7] and AASA also shows this behaviour [8].

The alum sludge (AS) are classified as a Class N pozzolans while AASA is similar to thermally activated kaolin after calcination at 800 °C [9] and Tantawy [10] approved that calcination at 800 °C is adequate to improve the pozzolanic properties of AS. The strength of composites improves with the presence of AASA as partial replacement of OPC, but the optimum replacement to accelerated long term strength enhancement and durability has been reported between 10 to 30% [11-12]. The replacement of AASA are in percentage by wt. of the total binder material. This study will investigate the replacement level of AASA with treated KF, strength and durability improvement of KFRBCC under different curing conditions. Benli et al. [13] investigated the effects on the mechanical properties of FA self-compacting mortars in WC and AC at different curing periods. The findings indicate that mortar cured in WC gave highest compressive strength for 180 d. According to Amjadi et al. [14], the mechanical strength of mortar improved at extended curing period and the pozzolans added into the mixes has help to accelerate the pozzolanic reaction at later ages.

The objective of the research was to obtain the required strength and durability during the life cycle of KFRBCC for future applications in construction in which offering a good alternative for recycling AS. Determination of the appropriate method of curing is one of the base for producing a composite with the strength and durability in the long term.

2 Experimental procedures

2.1 Materials

Type I OPC complying with ASTM C150 and AASA was used to produce KFRBCC with randomly dispersed short KF. The chemical composition are shown in Table. 1. AS was oven dried at 105±5 °C for 24 hours (h) and later the dried AS was calcined in furnace oven at heating rate of 5 °C/min at 800 °C for 2 h to achieve the thermal steady state. Then, the AASA was allowed to cool gradually to the room temperature. AASA was ground in steel ball mill until the fineness reach the specification recommended by ASTM C618-17 and sieved using 45 μm sieve to increase the pozzolanic reaction.

Table 1. Chemical composition of ordinary Type 1 Portland cement and AASA.

Chemical composition (%)	SiO2	Al2O3	Fe2O3	CaO	MgO	MgO	Na2O	K2O	Loss of Ignition	Specific gravity	Fineness	Median Particle Size	Pozzolanic Activity Index (7 d)	
											m2/kg	μm	7 d	28 d
OPC	20.18	5.23	3.34	64.4	1.8	0.07	0.44	2.17	3.15	338	16.9	100	100	20.18
AASA	47	41.94	4.86	0.41	0.4	0.09	0.99	2.64	2.53	11.6	10.1	84	93	47

10 mm length KF were treated with 6% w/w $NaHCO_3$ alkali solution and immersed for 72 h. After the treatment, KF were thoroughly washed with running water to remove absorbed alkali from the fibre surface. The washed KF were allowed to dry at room temperature for 24 h and then oven dried for thermal treatment at 70 °C±5°C for 24 h to release lignin and increase the crystallinity index (CrI) of the KF before added into the mixes. Superplasticizer (SP) was used to control the flow of mortar. 4.5% of low viscosity

liquid Darex Super 20 is the maximum dose for mixes but the doses use are vary to comply flow rate.

2.2 Mix proportion and preparation of specimens

Eight mixes were cast with 2% volume fraction of KF by volume of total mix design (constant for all mixes) into standard cube 100 mm with cement/sand (s/c) ratio of 1:2. Optimum water/cement (w/c) ratio was obtained at 0.4 (constant for all mixes) after three different ratio of 0.3, 0.4, and 0.45 were used for trial mix. The amount of AASA (20, 30, and 40%) are added into mixture for partial replacement of OPC. The mix proportions of the composites are given in Table 2. The composite with 100% OPC and cured in NC was used as control specimen.

Table 2. Mortar mix proportion.

Index	**w/b ratio**	**Cement**	**Sand**	**AASA**	**Water**	**SP**	**Fiber**	**Curing Conditions**		
		(g)	(g)	(g)	(g)	(%)	(%)	NC	AC	BC
OPC1	0.30	192	384	0	58	0	0	/		
OPCR2	0.40	151	378	0	76	3.5	2	/		
AASA20	0.40	161	378	28	76	4.5	2	/	/	/
AASA30	0.40	151	378	38	76	4.5	2	/	/	/
AASA40	0.40	132	378	57	76	4.5	2	/	/	/

2.3 Curing conditions of test specimens

The specimens were demoulding after 24 h and then exposed to the selected curing condition. The average temperature recorded in the laboratory was 27± 2 °C with relative humidity, RH 85±5%. The cementitious mortar specimens were removed from curing conditions 24 h before testing and the maximum curing age was 180 d for all conditions due to slower hydration rate than OPC. For NC, the curing conditions involving specimen immersed in water outside the laboratory and for AC, the specimen were exposed to the ambient air in the laboratory. The specimens were covered with wet burlap outside the laboratory for BC. The burlap was immersed in water every 2 d to control the moisture content. Three specimens were tested after 28 d of curing for wet/dry cycle whilst for compressive strength, after 180 d of curing.

3 Test procedures

3.1 Compressive strength and wet/dry cycles

The average of three samples were tested for compressive strength test following ASTM C109 and the reported compressive strengths were determined at various ages of 28, 56, 90 and 180 d. For wet/dry cycles, specimens was covered with a piece of plastic and left in laboratory. After 24 hours, the specimen is removed from the mould and soaked in tap water for 15 h and left to dried for 1 h in laboratory. Next, the specimen is exposed to a temperature of 70 ± 5 °C for 25 h in the ventilated oven and left to cool down gradually to room temperature in the oven with the door slightly open to prevent excessive cooling. Each wet/dry procedure represent a cycle within 24 hours and performed by 10, 20, 30, and 60 cycles, respectively.

4 Result and discussion

4.1 Effects of curing conditions on the compressive strength development of mortar

Compressive strength development results of KFRBCC for different curing conditions are presented in Fig. 1(a-c) and when OPC was replaced by 20% AASA, the strength of specimens is increased up to 30.6% compared to control specimen (OPC1 and OPCR1). The incorporation of 40% AASA by wt. of OPC increased the mechanical strength of cementitious composite about 34.3 % after 28 d due to smaller particle size of AASA and higher surface contact between fibres and mortar. The strength continued to increase by an average rate of 10% for each specimen up to the age of 180 d and the rate after 28 d are more significant for the strength development. This maybe owing to the higher pozzolanic reaction of AASA at later age compared to early age.

The specimens cured by BC provide the highest compressive strength of 20.3%, attributed to the suitable curing conditions compared to NC and AC. The increase in strength also related to hot weather with high humidity, which increased the rate of cement hydration during the early stage of curing. It could be observed from all of the curing conditions, BC produced composite specimen that met the targeted compressive strength of 50 MPa and NC has the lowest compressive strength of 47.32 MPa, while AC recorded the strength of 47.74 MPa at 180 d respectively. It was observed that the strength increases with increase in AASA content (20 to 40%) and curing periods for all types of curing conditions and reaches maximum 50.34 MPa for 180 d in 40% replacement of AASA with OPC. Obviously, AASA and treated KF can be effectively used to improve the strengths of KFRBCC with BC at a later age. In addition, the 28 d compressive strengths of KFRBCC were found to be higher than mixture containing OPC alone. As for mixture containing AASA, similar results was reported and suggesting that AASA is more effective in increasing the strengths of KFRBCC. This shows that the replacement of OPC with AASA has the effect of filling with fibre dispersed randomly. This proved that curing conditions is one of the most important parameters to improved microstructure by improving the water holding capacity of composites in which curing involves maintaining sufficient moisture content to the last stage of the placement of the composite.

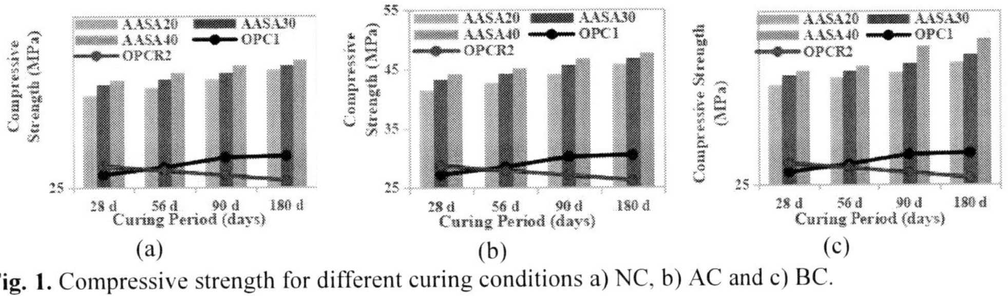

Fig. 1. Compressive strength for different curing conditions a) NC, b) AC and c) BC.

4.2 Effect of curing conditions on residual compressive strength of concrete mixes subjected to wet/dry cycling

The effect of wet/dry cycling on the residual compressive strength for specimens cured for 28 d in NC, AC, and BC are shown in Fig. 2. After 10 wet/dry cycles (30 d), all specimens showed reduction in the compressive strength. This result point to the mineralization of KF where fibre became slightly brittle due to some partial degradation of treated KF in cement matrix which deteriorate the fibre–matrix bonding. However, the incorporation of AASA into the mixtures gave significant enhancement in the residual compressive strength of KFRBCC. These test revealed after 20 wet/dry cycles (60 d), the compressive strength of AASA40 in NC, AC, and BC continue to dropped 43.1, 33.7, and 21 % of the initial compressive strength at 56 d. Specimens in BC curing show lower reduction compared to others curing conditions. Moreover, the compressive strength of KFRBCC with 40% AASA in NC, AC, and BC reduced by 73.8, 68.3 and 65.7% compared to their values at 180 d after 60 wet/dry cycles (180 d).

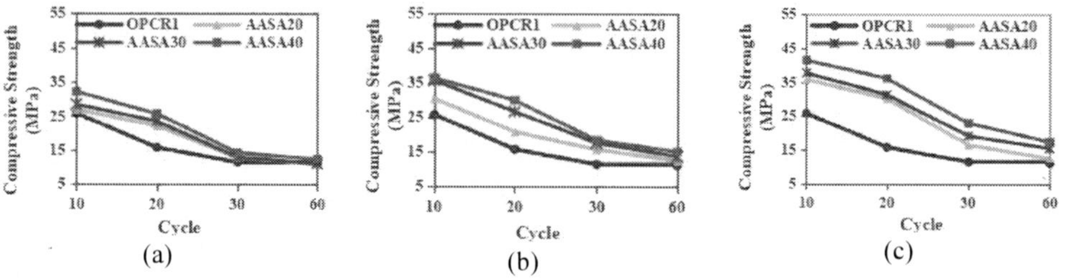

Fig. 2. Residual compressive strength prior to wet/dry cycle a) NC, b) AC and c) BC

4.3 Microstructure analysis of the composites using BC of AASA40

SEM investigations demonstrated different variations in the morphology of OPC and AASA mixtures with treated KF after tested with designated regimes. Small fragments of AASA40 cured in BC were taken after compressive test was used for microstructure analysis. Fig. 3 shows the SEM of all specimens after compressive strength test at designated ages. The pozzolanic reaction of AASA and filler effect have contributed to a denser microstructure and enhanced the strength particularly at 180 d of curing. From Fig. 3(a1-a4), all specimens shows that void and fluid structure are almost negligible. The chemical composition of the AASA used in this investigation suggests that almost negligible hydraulic activity (0.41 % of CaO content) and a relatively high pozzolanic activity because of its 47 % of SiO_2(Table 1). Therefore, the effect of using AASA on the properties of the composites will be more noticeable at later ages. Fig. 3(b1-b4) shows SEM images obtain from elevated temperatures test. There is no significant changes in the specimens when temperature increase up to 250 °C. However, noticeable cracks appeared when the temperature reached 500 °C, and responsible for a decrease in compressive strength. The specimens was severely damaged as a result of fibres melting but the presence of AASA contributes to the prevention of crack extension and improving the durability of cement composites.

Fig. 3(c1-c4) shows images obtain from wet/dry cycles with regards to the effect of AASA on the degradation of treated KF in a cement matrix. After 30 wet/dry ycles, the the treated KF suffer slightly degradation in the cement matrix but the fibre still manage to bridging a micro-cracks. However, with addition of AASA has helped to mitigate the degradation and resulted in high level of intactness with the embedded treated KF. Partial replacement of 40 wt.% AASA verified that cement modification reduce the degradation and improving the durability of treated KF. This confirms that AASA in cement modification had accelerated high pozzolanic activity and gave more significant role than its physical filling.

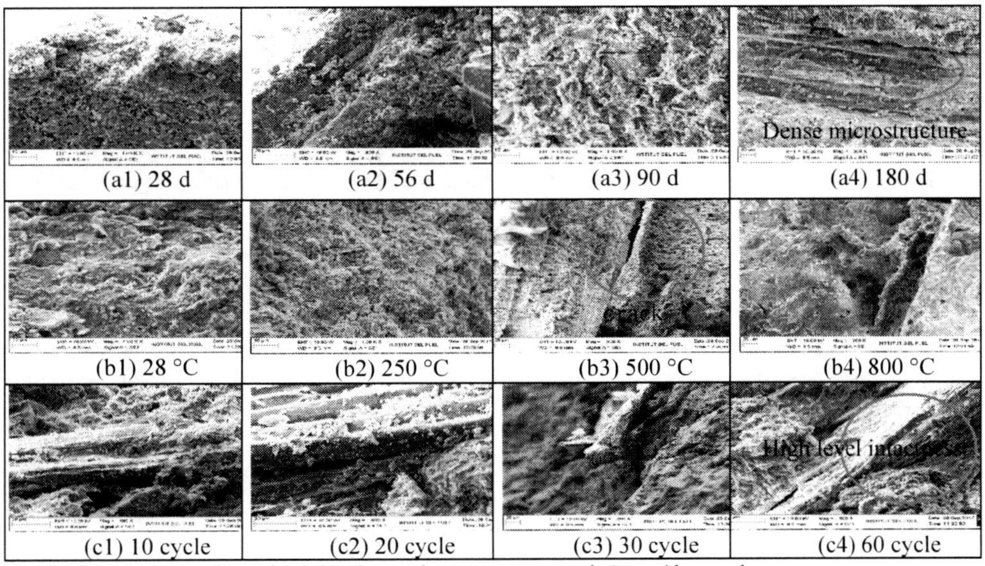

Fig. 3. SEM images of a) initial, b) elevated temperature, and c) wet/dry cycle.

5 Conclusions

The influences of treated KF, partially replacement of OPC with AASA, curing conditions and duration of exposure on accelerated ageing conditions on compressive strength of KBRCC are determined. The results indicate that compressive strength can increase with BC and longer exposure duration. Meanwhile, study on compressive strength indicates that for BC there were increases in strength up to 60 % for 180 d curing period. The results also show that strength loss in KFRBCC depends on cement fineness and the regime of curing. Curing conditions has an important effect to reduce or increase the compressive strength at prolong curing periods and durability of composites after wet/dry cycle. However, the limitation of the study is rely on AASA contents which indicate that 40 % wt. was the maximum contents for OPC replacement. The mixes will require more liquid if AASA were added beyond maximum contents and will effect the workability. In general, the strength loss was slightly lesser for AASA specimens than OPC specimens. Test results indicate that the effect of curing conditions on the durability was much lower than on the mechanical properties, and strength should not be the single control indicator while considering the effect of curing. Apart from that, the technology of internal curing and different types of pozzolans material (nanosilica, nanoclay, metakaolin, and etc.) should be taken into consideration for further research to indicate the positive impacts of aspects of KFRBCC performance.

Acknowledgment

The authors acknowledge the institution for financial supports under Fundamental Research Grant Scheme (FRGS/1/2016/TK06/UKM/02/2) and AP-2015-011.

References

1. G. Espinoza-Hijazin and M. Lopez, "Extending internal curing to concrete mixtures with W/C higher than 0.42," *Constr. Build. Mater.*, vol. 25, no. 3, pp. 1236–1242, 2011.

2. M. Radlinski and J. Olek, "Effects of Curing Conditions on Properties of Ternary (Ordinary Portland Cement / Fly Ash / Silica Fume) Concrete," no. 112, pp. 49–58, 2015.
3. A. S. Al-Gahtani, "Effect of curing methods on the properties of plain and blended cement concretes," *Constr. Build. Mater.*, vol. 24, no. 3, pp. 308–314, 2010.
4. E. R. Grist, K. a. Paine, a. Heath, J. Norman, and H. Pinder, "Structural and durability properties of hydraulic lime–pozzolan concretes," *Cem. Concr. Compos.*, vol. 62, pp. 212–223, 2015.
5. G. K. Al-chaar, M. Alkadi, and P. G. Asteris, "Natural Pozzolan as a Partial Substitute for Cement in Concrete," *open Constr. Build. Technol. J.*, vol. 7, pp. 33–42, 2013.
6. S. Haruehansapong, T. Pulngern, and S. Chucheepsakul, "Effect of the particle size of nanosilica on the compressive strength and the optimum replacement content of cement mortar containing nano-SiO2," *Constr. Build. Mater.*, vol. 50, pp. 471–477, 2014.
7. N. Y. Mostafa, Q. Mohsen, S. A. S. El-Hemaly, S. A. El-Korashy, and P. W. Brown, "High replacements of reactive pozzolan in blended cements: Microstructure and mechanical properties," *Cem. Concr. Compos.*, vol. 32, no. 5, pp. 386–391, 2010.
8. H. M. Owaid, R. Hamid, and M. R. Taha, "Strength - Ultrasonic Pulse Velocity Relationship of Thermally Activated Alum Sludge Multiple Blended High Performance Concretes," *Key Eng. Mater.*, vol. 594–595, pp. 521–526, 2013.
9. H. M. Owaid, R. Hamid, and M. R. R. Taha, "Influence of thermally activated alum sludge ash on the engineering properties of multiple-blended binders concretes," *Constr. Build. Mater.*, vol. 61, pp. 216–229, Jun. 2014.
10. M. A. Tantawy, "Characterization and pozzolanic properties of calcined alum sludge," *Mater. Res. Bull.*, vol. 61, pp. 415–421, 2015.
11. M. Frias, R. Vigil De La Villa, I. De Soto, R. Garcia, and T. A. Baloa, "Influence of activated drinking-water treatment waste on binary cement-based composite behavior: Characterization and properties," *Compos. Part B Eng.*, vol. 60, pp. 14–20, 2014.
12. A. Yagüe, S. Valls, E. Vázquez, and F. Albareda, "Durability of concrete with addition of dry sludge from waste water treatment plants," *Cem. Concr. Res.*, vol. 35, no. 6, pp. 1064–1073, Jun. 2005.
13. A. Benli, M. Karataş, and Y. Bakir, "An experimental study of different curing regimes on the mechanical properties and sorptivity of self-compacting mortars with fly ash and silica fume," *Constr. Build. Mater.*, vol. 144, no. July, pp. 552–562, 2017.
14. R. Amjadi, M. Monazami, E. Mohseni, H. Azar Balgouri, and M. M. Ranjbar, "Effectiveness of different recycled materials in self-compacting mortar," *Eur. J. Environ. Civ. Eng.*, vol. 21, no. 12, pp. 1485–1501, 2017.

A review of causes of variation order in the construction of terrace housing projects in Malaysia

Noraziah Mohammad*, and *Zabidi* Hamzah

Department of Architecture, Faculty of Engineering and Built Environment, National University of Malaysia (UKM), Bangi, Selangor, Malaysia

Abstract. Variation order refers to any changes from the scope, plan, specification or contract document in the construction project. This changes of work is a common problem in construction projects in Malaysia. It is inevitable in any construction project and this problem can become more serious when variation order causes work progress to be impaired as a results it increased the construction time and cost that affects the construction industry in Malaysia. This paper critically review the causes of variation order in the construction industry on the double-storey terrace housing project in Malaysia. The discussions began with the construction industry in Malaysia and then related parties involved in the construction industry. The extensive local and abroad literature review on the definitions of variation order and analysis of past studies conducted on variation order and the previous study of the prediction model of variation. The extensive literature on variation order in the construction of building project may help to improve the performance in term of time, cost and quality of the construction of building projects in Malaysia.

1 Introduction

Variation order refers to any changes to the scope of work, planning, and specification or contract documents in a construction project. There are many interpretations of variation order from the literature reviews. [1] states that in a tender and building contract, the term variation is any changes to the works under the supervision of an architect as the contract administrator on behalf of the client or employer. On the other hand, [2] interprets the variation in works as modifications to the scope and schedule agreed. Generally, it is changes or any modification to the contract awarded to a contractor as a guide to the client or client's representative. This include variations to the planning, specification, or any of the contract documents. According to [3] and [4] from the construction context, variation order is also known as variations. Any variation instruction is an official document that is

* Corresponding author: noraziahmohammad@ukm.edu.my

used to modify the original contract given to the contractor by the client or client's reptresentative and become part of the project documents. Increase in variation orders in a project may resulted in additional time and cost. When variation order is issued, contractors are inclined to charge at a higher rate for the work changes, thus affecting the client's budget. The change of work is a common problem in construction projects in Malaysia. This change of work is inevitable in any construction project and this problem can become more serious when variation order causes work progress to be impaired as a results of the increased in the construction time and cost that affects the construction industry in Malaysia.

2 Research Objectives

The main objectives of this research are as follows:

2.1 Comprehensive literature review relating to variation order in the construction industry carried out by previous researcher.
2.2 Identifying the causes of variation order in the construction project.
2.3 Identifying the parties contributing to the variation order.

3. Literature Review

Even though the project planning is properly implemented, the possibility of changes happening cannot be dismissed due to various factors [3]. There are various causes why changes of variation order taking place in a construction project, whether identifiable or not.

3.1 The causes of variation works

From the literature review, it is found that there are 18 main causes of the variation works identified by most researches ([5],[6],[2],[7],[3],[8],[9],[10],[11],[12], [13], [14], [15],[16] and [17]). The causes of variation order are described as follows:

Changes of scope by client: Changes to planning or the project scope is one of the main causes in the construction project [18]. This is due to improper planning at the project definition or due to lack of collaboration by the owner in the design phase [9]. [19], in his research identified the effects of variation works in the sport facilities in Egypt and found that additional scope of work by the client is one of the factor that give a big impact during the project implementation.

Owner's financial problems: The owner's financial problems can effect the quality and progress of the project [9] and [3]. This problem often brings changes to the schedule and work specification and subsequently the construction quality.[11] have studied the causes and effects of variation works in road construction projects in Iran and found that variation works are caused by changes in the scope of work by the employer, mistakes and outdated design as wll as finacial problem of the owner.

Inadequate project objectives: Inadequate project objectives can cause variation in the construction projects. According to [20], it limits the designers with the appropriate design that can result in variation works in the later stages of the construction process.

Replacement of materials or procedures: Replacement of materials or work procedure can cause changes during the construction phase. Replacement of work procedure also involve changes to the work method) [21].

Impediment to prompt decision-making process: Prompt decision-making is an important factor in the success of a project [11]. Failure to make a quick decision can cause the delay and the needs for work changes due to raise in cost.

Obstinate nature of owner: A building project is the outcome of work collaboration of professionals that are involve in the various phases of a construction project [9].The owner's attitude with individual opinion without considering the professional involve can cause the changes in work in the last stage of a project.

Change in specifications by the owner: Changes in specification by owners are common phenomena in construction projects due to inadequate project objectives [3]. If this change is on the client's requirement, this leads to a change in the construction phase

Change in design by the consultant:A change in design for improvement by the consultant is a norm in contemporary professional practice [9]. Changes in design are frequent in projects where construction starts before the design is finalized [2]. Such changes can affect the project in various ways depending on the timing of the change.

Errors and omissions in design: Errors and omissions in design are a significant cause of project delays [9]. Dependent upon the timing of the errors in the project, delays and variations may occur [13], work changes are a key issue in the construction industry in Ghana and are largely due to client design changes.

Conflicts among contract documents: Conflict between contract documents can result in misinterpretation of the actual requirement of a project [22]. It is essential that contract documents are clear and precise. Inadequate information in contract documents may result in delays in project completion or cause changes in costs.

Technology changes: Changes in technology could potentially result in variations on a project. . Project planning should be flexible for accommodating new beneficial variations [23]. Emergence of new technology can affect construction methods and processes.

Value engineering: Value engineering should ideally be carried out during the design phase [24]. Value engineering carried out during the construction phase can become an extremely costly exercise and may result in variations.

Lack of coordination: A lack of coordination between parties may cause major variations with adverse impact on the project [9], leading to dissatisfaction of the owner, reconstruction work and work changes.

Design complexity: Complex designs require unique skills and construction methods [9]. Complexity affects the flow of construction activities, whereas simpler and linear construction works are relatively easy to handle [2]. This would suggest that the more complex a construction design is, the more there is a chance that variations may occur.

Inadequate working drawing details: To convey a complete concept of the project design, the working drawings must be clear and concise [25]. Insufficient working drawing details can result in misinterpretation of the actual requirements for the project [6], causing variations in the project.

Consultants' lack of required data: A lack of data can result in misinterpretation of the actual requirements of a project [26]. If there is not enough data available to the consultant then the design is based more on the consultants' own perception of the requirements of the project. Should these perceptions be wrong, then variations will occur.

Poor knowledge of available materials and equipment: Knowledge of available materials and equipment is an important factor for developing a comprehensive design [25]. If the consultant has a poor knowledge of available materials or equipment that can be used in the construction process, variations are more likely to occur during the construction phase.

Ambiguous design details: A clearer design tends to be comprehended more readily. Ambiguity in design may cause misinterpretations and the need for rework leading to delay and increased cost. [27] studied factors influencing thirty-one high-rise buildings in

Indonesia and found that design changes were one of the most important factors that led to the increase in time and cost. Summary of the major causes of variation order based on the 18 factors identified by the results of the literature review are shown in table 1.

3.2 Parties that contribute to variation order

The major contributors to variation order have been grouped under four categories which is client, consultant, contractor and other parties [8], [28], [29] and [30]. The connection between the parties who contribute to the variation order will be discussed as below:

Variations order caused by owner: The causes of change in road construction in Iran are due to changes in scope by employers, errors and omissions in design and financial problems by owners [11] and most of the causes of change of work are change of design by clients and details of work drawings are not complete [13].

Variations order caused by consultant : [31] recommends consultants to understand the scope and project requirements. In general, the role of the consultant is to advise clients on technical matters, laws and financial aspects. The following are the variation order associated with the consultants which is design mistakes and changes, the failure to determine the best alternative design for the project in the feasibility study [32].

Variations order caused by contractor: [33] suggesting for each project, the parties concerned should be concerned with possible problems. It is the contractor's responsibility to advise the consultant to issue a change of working instructions when there is a technical problem in the project.

Variations order caused by others: Conditions beyond the control of other parties that are not in the contract, may cause changes in the work. Among them are weather conditions, security, changes in government regulations, changes in economic conditions, socio-cultural factors and unforeseen problems [28]. The summary of the parties contributing to the variations order identified as a result of the literature review is shown below table 2.

Table 1. Summary of literature review of the causes of variation order.

CAUSES OF VARIATION ORDER AND AUTHOR	Arain et al. (2004)	Arain et al. (2006)	Keane at al. (2010)	Alnuaimi et al. (2010)	Pourrostam et al. (2011)	Ismail et al. (2012)	Amiruddin et al. (2012)	Rashid et al. (2012)	Asamaoh & Offei-Nyako	Odura et al. (2013)	Memon et al. (2014)	Dickson et al. (2014)	Rayan et al.(2014)	Gokulkarthi &	Majid et al. (2015)	Masalam (2015)	Uttam (2015)	Hanif (2016)
Changes of scope by client	√	√	√	√	√	√	√	√	√	√	√	√	√	√	√	√	√	√
Owner's financial problems		√	√			√	√	√	√		√			√	√	√		√
Inadequate project objectives	√	√	√				√		√		√			√			√	√
Replacement of materials or procedures		√	√		√	√	√						√	√			√	√
Impediment to prompt decision-	√	√	√	√			√		√		√					√	√	

making process																		
Obstinate nature of owner	√	√	√		√		√	√			√						√	
Change in specifications by the owner.	√	√	√				√		√		√							√
Change in design by the consultant	√	√	√	√			√	√			√	√	√	√	√		√	
Errors and omissions in design	√	√	√	√		√	√	√				√	√	√				√
Conflicts among contract documents		√	√				√		√		√	√		√	√	√		√
Technology changes		√	√		√		√		√					√		√		
Value engineering	√	√	√			√	√	√						√				
Lack of coordination	√	√	√				√		√			√			√	√		
Design complexity	√	√	√				√		√		√							√
Inadequate working drawing details	√	√	√				√		√		√					√	√	
Poor knowledge of available materials and equipment	√	√	√						√					√				
Consultants' lack of required data	√	√	√					√							√	√	√	√
Insufficient details in the contract documents	√	√	√				√	√		√						√	√	

Table 2. Summary of literature review of parties contributing to variations order.

PARTIES CONTRIBUTING TO VARIATION ORDER AND AUTHOR	Thomas & Napolitan	O'Brien (1998)	Ibbs et al. (1998)	Wang (2000)	Gray & Hughes (2001)	Mokhtar et al. 2000)	Arain (2002)	Arain et al.(2004)	Arain & Pheng (2006)	Acharya et al. (2006)	Uyun (2007)	Keane et al. (2010)	Alnuaimi (2010)	Asamaoh dan Offei-Nyako (2013)	Dickson et al. (2014)	Desai (2015)
Owner	√	√	√	√	√		√	√	√		√	√	√	√	√	√
Consultant	√	√	√	√		√		√	√	√		√	√		√	√
Contractor	√	√	√	√			√	√	√						√	√

Others	√	√		√				√	√						√	√

4 Conclusion

The literature review shows that major causes of the variation orders are due to changes of scope by client. Thus, the main contributer to the variation order are from the client. The finding from the literature studies can be used as reference in reducing the variation order in the construction of terrace housing projects in Malaysia.

Acknowledgment

The authors would like to thank the Research Incentive Grant research fund (GGP-2017-082) under the Center for Research and Instrumentation Management, National University of Malaysia (UKM) who have been supportive of this study.

References

1. Harbans Singh K.S. 2003. Enginerring and Construction Contracts Management. Post-Commencement Practise. Ed. ke-2. Malaysia: Penerbit LexisNexis, A Devision of Reed Elsevier (Singpore) Pte Ltd.: 423-555.
2. Fisk, E.R. 2013. Construction Project Administration, edisi kesepuluh. Prentice Hall, Upper Saddle River, New Jersey, United Stated.
3. O'Brien J.J. 1998. Construction Change Orders. McGraw-Hill, New York.
4. Arain, F.M., & Pheng, L.S. (2005). The potential effects of variation orders on institutional buildings projects. Facilities. 23(11/12): 496-510.
5. Thomas, H.R., & Napolitan,C.L. 1995. Quantitave effects of construction changes on labor productivity. The Journal of Construction Engineering and Management-ASCE. September, 121(3): 290-296.
6. Clough R.H., & Sears G.A 1994. Construction Contracting, edisi keenam. John Wiley and Sons Inc., New York.
7. Mokhtar, A., Bedard, C., & Fazio, P. 2000. and scheduling of interrelated design changes. J. Archit. Eng. ASCE, 6(2): 66-75.
8. Gray, C, & W. Hughes. 2001. Building Design Management. Butterworth Heinemann, Oxford, UK.
9. Arain, F.M., Assaf, A, S., & Low, S.P. 2004. Causes of discrepancies between design and construction. Architectural Science Review. 47(3): 237-249.
10. Pourrostam, T., Ismail, A., & Mansournejad, M. 2011. Identification and Evaluation of Causes and Effects of Change Orders in Building Construction Projects. Applied Mechanics and Materials 94-96: 2261-2264.

11. Ismail, A., Pourrostam, T., Soleymanzadeh, A., & Ghouyounchizad, M. 2012. Factors causing variation orders and their effects in roadway construction projects. Research Journal of Applied Sciences, Engineering and Technology. 4(23): 4969-4972.

12. Rashid, K.A. 2002. Construction Procurement in Malaysia, Process and Systems, Constraints and Strategies. Gombak: IIUM Press. First Edit.: 1-191.

13. Asamaoh, R.O., & Offei-Nyako, K. 2013. Variation Determinants in Building Construction: Ghanaian Professionals Perspective. Korea Institute of construction Engineering and Management (KICEM), Journal of Construction Engineering and Project Management 3(4): 20-25.

14. Memon, A.H., Rahman, I,A, & Abul Hasan, M.F. 2014. Significant causes and effects of variation orders in construction projects. Research Journal of Applied Sciences, Engineering and Technology. 7(21): 4494-4502

15. Gokulkarthi, M.,& Gowrishankar, K.S. 2015. A Study On Impacts Of Change Order In Construction Projects. International Journal of Science and Engineering Research (IJ0SER) 3(4): 3221 5687-3221 5693

16. Msallam, M., Abojaradeh, M., Jrew, B., & Zak, I. 2015. Controlling Of Variation Orders in Highway Projects in Jordan. Journal of Engineering and Architecture 3(2): 95-104

17. Hanif, H., Khurshid, M.B., Nauman, S., & Lindhard, S. 2016. Impact of variation orders on time and cost in mega hydropower projects of Pakistan. Journal of Construction in Developing Countries. Penerbit Universiti Sains Malaysia.: Pg. 1-20

18. (CII) Construction Industry Institute. 1990b. Scope definition and control. Publication 6-2, CII, The University of Texas at Austin, Austin, Tex

19. Rashid, I. A., Mohamed, A. E-M., & Mohammed, E. A-H. 2012. The Impact of Change Orders on construction projects Sports Facilities Case Study. Journal of American Science. 8(8): 628-631

20. Ibbs, W. & Allen, W. E. 1995. Quantitative impacts of project change. Source Document 108, Construction Industry Institute, University of Texas at Austin, Austin, Tex

21. Chappell, D., & Willis, A. 2005. The Architect in practice. Edisi kesembilan, Blackwell Science, Oxford, U.K.

22. (CII) Construction Industry Institute. 1986. Constructability: A Primer. Construction Industry Institute, The University of Texas at Austin, TX

23. (CII) Construction Industry Institute. 1994b. Project change management. University of Texas at Austin, Austin, Tex

24. Dell'Isola, A. J. 1982. Value engineering in the construction industry. 3rd Ed., Van Nostrand Reinhold, New York

25. Geok, O. S. 2002. Causes and improvement for quality problems in design and build projects. Unpublished B.Sc. thesis, National University of Singapore, Singapore.

26. Arain, F.M. 2002. Design-construction interface dissonances. Thesis, King Fahd University of Petroleum and Minerals, Dhahran, Saudi Arabia

27. Kaming, P.F., Olomolaiye, P.O., Holt, G.D., & Harris, F.C. 1997. Factors influencing construction time and cost overruns on high rise projects in Indonesia. Construction Management and Economics. 15(1):83-94

28. Arain, F.M., & Pheng, L.S. 2006. Developers' views of potential causes of variation orders for institutional buildings in Singapore. Architectural Science Review. 49(1): 59-74.

29. Dickson, O.D., Gerryshom, M., & Wanyona, G. 2014. Factors Contributing to Variations Construction Projects in Kenya. International Journal of Social Science and Enterpreneurship. 1(13):1-11

30. Desai, J.N., Pitroda, J., & Bhavsar, J.J. 2015. A Review on Change Order and Assessing Causes Affecting Change Order in Construction. Journal of International Academic Research for Multidisciplinary. 2(12) January 2015: 152-162

31. Acharya, N.K., Lee, Y.D., & Im, H.M. 2006. Design Errors: Tragic for the Clients. Journal of Construction Research. 7(1/2): 177-190.

32. Al-nuaimi, A.S., Taha, R., Al Mohsin, M., & Al-Harthi, A.S. 2010. Causes, effects, benefits and remedies of change orders on public construction project in Oman. Journal of Construction Engineering and Management. 136(5): 615–622.

33. Levy, S.M. 2002. Project Management in Construction, 4th Ed. Columbus: McGraw-Hill

Effect analysis of grouting reinforcement ring considering fluid solid coupling

Heng Zhou[1,*], *Xiaochen* Wang[1], *Haojie* Liu[1], *Yanan* Liu[1], *Chenyang* Ma[1], and *Kexian* Li[2]

[1]College of civil engineering and water conservancy of Shandong University, Jinan, China, 250061
[2]Qingdao metro line one line limited company, Qingdao, Shandong, 266000, China

Abstract. Geological hazards such as instability of surrounding rock and leakage of water are easily occurred in subway tunnels crossing water rich sand layers. Based on the principle of fluid solid coupling in porous media, this paper studies the plastic zone distribution of surrounding rock and reinforcement ring, vertical settlement of tunnel vault and water seepage of tunnel before and after grouting reinforcement for water rich sand layer. Considering the compressibility of rock mass, the relationship between porosity, permeability and volume strain is further deduced and simplified. A set of numerical calculation formula is set up to study the change of physical parameters of the water rich sand layer before and after grouting to determine the grouting effect and select the thickness of the best grouting reinforcement ring. The results show that the deformation and water permeability of the tunnel decrease with the increase of the thickness of the grouting reinforcement ring. The thickness of the grouting reinforcement ring is the most reasonable when the thickness of the reinforcing ring reaches a certain value, and the thickness of the grouting reinforcement ring is designed to be 5~6m. The research results have been successfully applied in the grouting project of Qingdao Metro stone elderly bathing beach. The reliability of the numerical simulation is verified by comparing the data from the field monitoring and measurement, providing reference for the related projects.

1 Introduction

With the rapid development of construction, municipal, water conservancy and transportation projects, more and more projects need to be constructed in extremely complex strata, which leads to a large number of underground disaster prevention and control problems, and grouting[1-3] reinforcement is a traditional method of underground engineering disaster prevention and control. The full-section curtain grouting is often used to reinforce the water-rich strata in subway tunnels. However, due to the different rock formations in different areas, the properties of the rock and soil are complicated and changeable, so it is difficult to put forward a practical grouting design scheme.

About the mechanism of grouting reinforcement ring and the stability of surrounding rock. Wang Gang[4] optimized the parameters selection of grouting reinforcement ring for typical water rich stratum tunnels. Chen Jiankuang[5] analyzed the process of pore water

* Corresponding author: 1192582720@qq.com

pressure dissipation in grouting reinforcement ring. Study on fluid structure interaction in tunnel excavation. Li Lu et al.[6] established a fluid solid coupling model describing the diffusion of viscous slurry in porous media.Ye Yuanxin et al.[7] systematically summarized and introduced the coupling characteristics of rock seepage. The numerical simulation of fluid structure interaction is considered. Ju Minghui[8] studied the pore fluid flow due to tectonic deformation by numerical simulation. Li Shuchen et al.[9] established a quantitative index to evaluate the stability of surrounding rock by solving the safety factor of surrounding rock stability. A large number of scholars have studied the effect of grouting reinforcement ring. But they often simplify the tunnel into a circular section and consider the influence of the lining system on the surrounding rock. In fact, tunnels are generally designed to be eggshell or horseshoe shaped. When curtain grouting is used to control bad strata, the reinforcement ring should ensure excavation safety before the lining is formed. It is more practical to consider only the effect of grouting reinforcement ring on surrounding rock. Furthermore, the coupling effect of stress field, displacement field and seepage field should be considered in the study of grouting reinforcement circle.

This paper takes the underground horizontal channel grouting project of Qingdao stone old bathing station as an example. The relationship among porosity, permeability and volumetric strain is further simplified by using the Kozeni-Kalman equation, and a set of numerical formulas for fluid-solid coupling is established. With the help of numerical simulation software, the variation characteristics and interaction of stress field, displacement field and seepage field of surrounding rock before and after grouting are studied, and the influence of thickness and permeability of reinforcement ring on stability of surrounding rock and water inflow of tunnel is further studied. The grouting effect is discussed and the parameter design of reinforcement ring is optimized to provide reference for other similar projects.

2 Engineering survey

2.1 An overview of the station

Qingdao Metro stone elderly bathing station is a station on Qingdao Metro Line 2. The station crossing water rich sand layer. The grade of strata is grade VI, water content is high, and self stability ability is very poor. Influenced by the Wanggezhuang fault, mylonite, sandy clastic rock and massive clastic rock are developed in the tectonic fracture zone. Site grouting construction is as shown in figure 1.

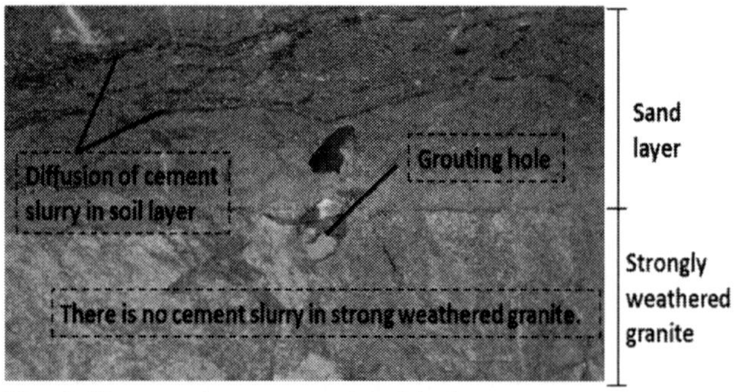

Fig. 1. Site grouting construction drawing.

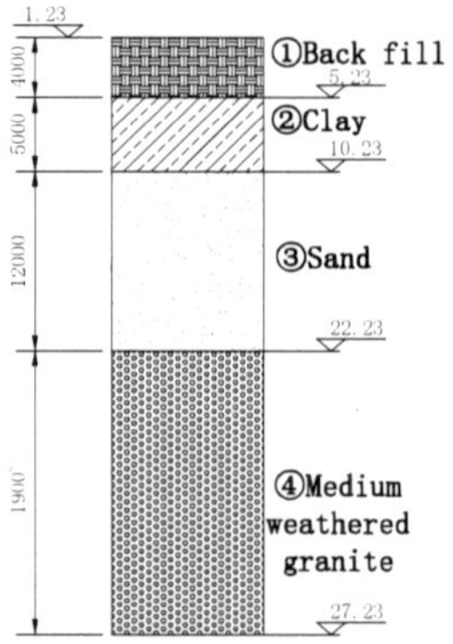

Fig. 2. Soil layer distribution map.

Table 1. layer parameter table.

Type of surrounding rock	bulk modulus K (MPa)	shear modulus G (MPa)	angle of friction φ (°)	cohesive strength C (MPa)	unit weight γ (kN·m^{-3})
Back fill	6	2.22	9	0.2	18.5
Clay	12	4.62	18	0.275	19.7
Sand	25	9.68	3.00	0.1	25
Medium weathered granite	50	22	45	12	41

2.2 Parameters of surrounding rock

According to the geological investigation report of stone elderly bathing station of Qingdao Metro Line 2 and the Engineering Geology Manual, the distribution of underground rock and soil in the study section is shown in figure 2. The physical and mechanical parameters of the engineering section obtained from the relevant documents are shown in table 1. The ultimate aim of the grouting reinforcement in this project is to form a grouting stone in the direction of 40m in the direction of the excavation face along the direction of the tunnel, so as to prevent the instability of the surrounding rock excavation caused by the tunnel crossing the water rich sand layer.

3 Model establishment

The underground passageway of Qingdao stone elderly bathing station is the engineering background of numerical simulation. The coupling effect between stress field, displacement field and seepage field of surrounding rock during grouting treatment is considered when the corresponding finite element analysis model is established. The coupling system consisting of water rich sand layer, grouting reinforcement area and surrounding rock is simulated.

3.1 Constitutive equation

Fluid solid interaction has a significant effect on the grouting reinforcement process in the water rich formation. In order to simplify the problem, the surrounding rock can be regarded as homogeneous, isotropic elastic porous media. The governing equations of fluid solid coupling are as follows.

(1) the motion equation of groundwater seepage.

The groundwater seepage in the water rich stratum belongs to the passive non steady state problem. Its continuity equation is:

$$\frac{\partial(\rho\phi)}{\partial t}+\nabla\cdot(\rho V)=0 \tag{1}$$

Considering the fluid solid coupling[10-11], the equation of state of surrounding rock:

$$\rho_s=\rho_{s0}\left[1+\frac{p-p_0}{E}-\frac{\sigma'-\sigma'_0}{(1-\phi)E}\right] \tag{2}$$

where p is ground water pressure(Pa),and p_0 is initial groundwater pressure(Pa). E is modulus of elasticity(Pa). ϕ is porosity of rock and soil. ρ_s is density of surrounding rock(kg/m^3),and ρ_{s0} is initial density of surrounding rock(kg/m^3). σ' is effective stress of surrounding rock, and σ_0' is effective stress of initial surrounding rock.

(2)The control equation of the stress field of rock and soil [12].

Considering the effect of hydrostatic pressure p on the basis of elastic relation, the constitutive equation is written as follows:

$$\varepsilon_{ij} = \frac{1+\nu}{E}\sigma_{ij} - \frac{\nu}{E}\sigma_{kk}\delta_{ij} + \frac{\alpha}{3K'}(p-p_0)\delta_{ij} \tag{3}$$

$$\varepsilon_{kk} = \frac{1-2\nu}{E}\sigma_{kk} + \frac{\alpha}{3K'}(p-p_0)\delta_{ij} \tag{4}$$

where σ_{ij} , σ_{kk} is stress of rock and soil mass(Pa). ν is poisson ratio. ε_{ij} , ε_{kk} is strain of surrounding rock. α is $Biot$ modulus. K' is volume modulus of surrounding rock. δ_{ij} is kronecker symbols in tensor form.

(3)Permeability[13-14] of surrounding rock.

Considering the fluid solid coupling, the relationship between porosity and volumetric strain can be expressed as:

$$\phi = \frac{\phi_0 + \varepsilon_v + (1-\phi_0)\Delta p / K_s}{1+\varepsilon_v} \tag{5}$$

where ε_v is volumetric strain of surrounding rock, and $\varepsilon_v = \frac{\partial u}{\partial x} + \frac{\partial v}{\partial y} + \frac{\partial w}{\partial z}$. ϕ_0 is initial porosity. Δp is variation of hydrostatic pressure(Pa). K_s is model parameter.

According to the Calman equation[15],the volume change of surrounding rock is simplified as the pore volume change of surrounding rock. After further simplification, the permeability of surrounding rock satisfies the equation:

$$K = K_0 \frac{1}{1+\varepsilon_v}\left(1+\frac{\varepsilon_v}{\phi_0}\right)^3 \tag{6}$$

where K_0 is initial permeability.

3.2 Constitutive equation

The stone elderly bathing station of Qingdao Metro passes through the water-rich sand layer, and the risk of grouting reinforcement is very great. Therefore, it is necessary to evaluate the stability of surrounding rock of this stratum. In this paper, the finite element[15] software COMSOL Multiphysics is used to simulate the coupling between porous media and solid mechanics. Combined with the possible instability of surrounding rock and groundwater seepage inflow, the numerical analysis and calculation with strong pertinence are carried out.

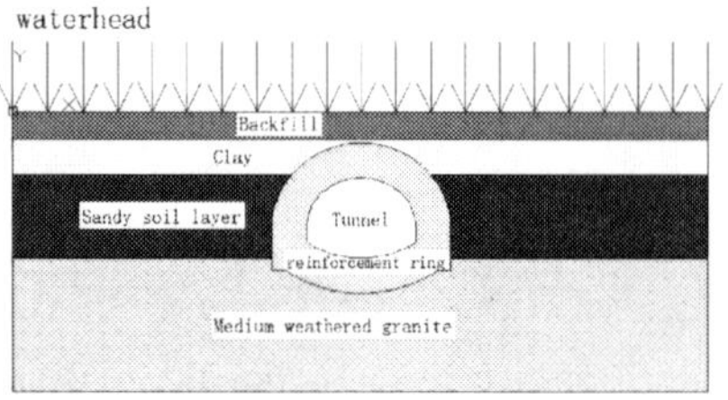

Fig. 3. Numerical simulation model.

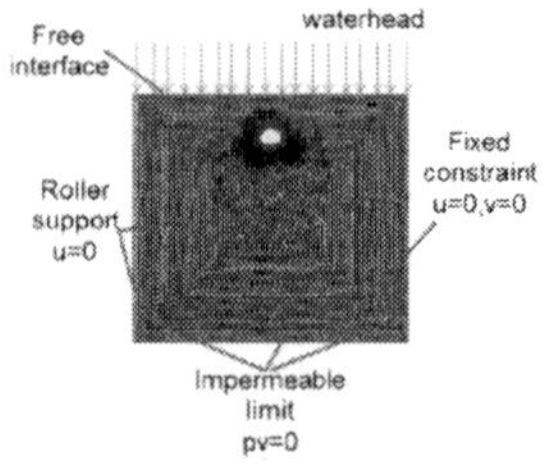

Fig. 4. Model boundary condition.

The numerical model is composed of surrounding rock and grouting reinforcement circle. The rock and soil are saturated with water. The research problem is steady state. The calculation model is shown in figure 3. The study is simplified to plane strain problem, and two dimensional finite element model is established as shown in figure 4.

Referring to the data of related projects, the physical and mechanical parameters of the reinforcing ring are determined by sampling experiment of the reinforcing ring after grouting in water-rich sand layer, as shown in table 2.

Table 2. Calculation parameters of numerical model.

Material name	Density (10^3 kg/m^3)	Modulus of elasticity (MPa)	Poisson ratio	internal friction angle (°)	Cohesive force (MPa)	Permeability (m^2)	porosity
Grouting reinforcement ring	3.00	2000	0.32	25	2.15	5×10^{-12}	0.01

4 Result analysis

In this simulation, the mechanical parameters of water-rich sand layer before and after grouting are brought into the numerical simulation to calculate. The effect of grouting reinforcement ring is analyzed, and the risk of tunnel excavation is determined.

4.1 Analysis of effect of grouting reinforcement ring

The settlement of vault caused by tunnel excavation before and after grouting is shown in figure 5 and figure 6. The figure shows that the closer the settlement is to the part of the tunnel, the more obvious the effect is. The maximum settlement is at the center of the tunnel, which is 100 meters away from the initial section. Comparing with figure 5 and figure 6, the settlement of tunnel vault decreased from 0.24M to 0.019m after grouting reinforcement circle was formed. It can be seen that the grouting reinforcement forms a compact stone around the tunnel, enhances the elastic modulus, porosity and other physical and mechanical parameters of the water-rich sand layer, and the grouting reinforcement ring has obvious effect on limiting the tunnel vault settlement.

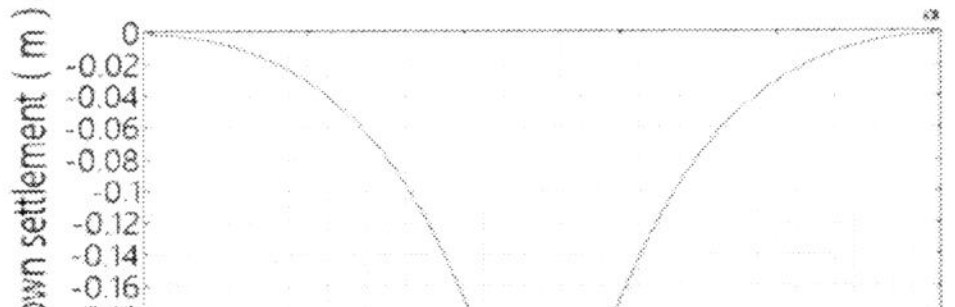

Fig.5. Settlement of tunnel vault before grouting reinforcement.

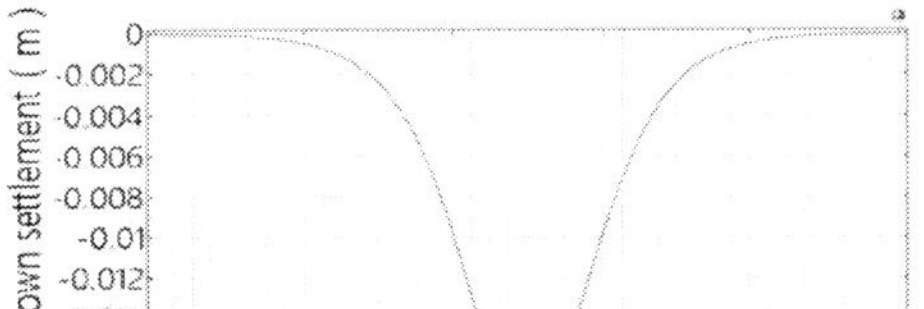

Fig. 6. Settlement of tunnel vault after grouting reinforcement.

The plastic deformation around the tunnel before and after grouting is shown in figure 7 and figure 8. From the figure7, it can be seen that a large number of plastic zones appear at the arch foot and waist of the tunnel before the grouting reinforcement circle is formed, and the plastic zones almost run through the left and right. At this time, the rock and soil mass basically loses the ability to bear the stress and deformation. According to figure 8, after forming the grouting reinforcement ring, the plastic zone is formed only in the minimum area of the arch foot of the tunnel. There is no tendency to pass through the reinforcing ring, which ensures that the reinforcing ring will not be unstable due to plastic deformation. The cohesion and internal friction angle of grouting reinforcement ring are greatly increased compared with before reinforcement, which greatly limits the plastic deformation and failure of surrounding rock. The effect of grouting reinforcement ring on the plastic deformation of surrounding rock in tunnel excavation is obvious.

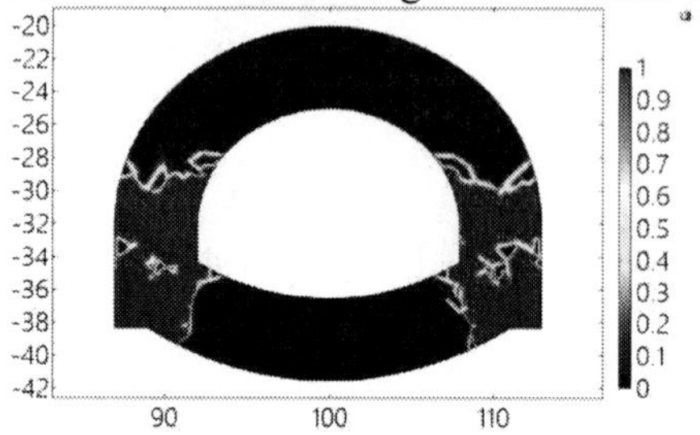

Fig. 7. Plastic deformation around the surrounding tunnel.

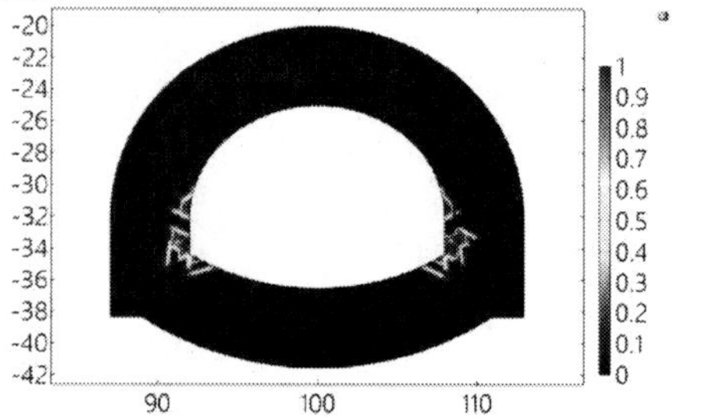

Fig. 8. Plastic deformation of reinforced ring after reinforcement.

The seepage velocity before and after grouting is shown in figure 9 and figure10. After the tunnel is excavated, the groundwater flows into the tunnel under the action of Darcy's law. Excessive infiltration of tunnel will form a permeable channel around the tunnel and destroy the stability of surrounding rock. As shown in figure 9, the arrow indicates the size of the water head. The soil around the tunnel is a water-rich sand layer with strong permeability and high porosity. Due to the gap formed by the tunnel excavation, the groundwater flows from the outside of the tunnel to the interior. The seepage effect of groundwater is obvious, which is extremely unfavorable to tunnel excavation. From figure 10, we can see that after the formation of the grouting reinforcement ring, the seepage velocity of the ground water decreases. This is because the physical and mechanical parameters of the water-rich sand layer around the tunnel are improved, and the permeability coefficient and porosity are greatly reduced. The formation of grouting reinforcement ring has obvious effect on controlling the leakage of water in the tunnel.

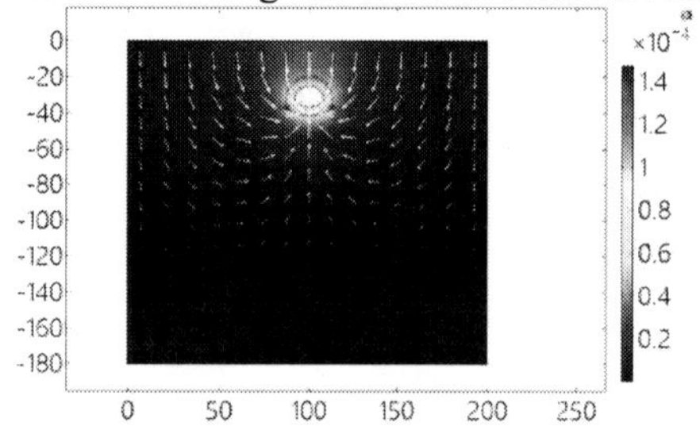

Fig. 9. Seepage velocity distribution before reinforcement.

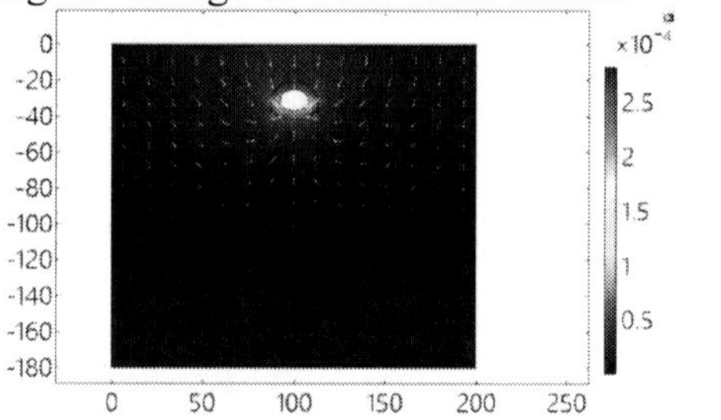

Fig.10. The velocity of seepage flow after reinforcement.

4.2 Optimization of parameters of grouting reinforcement ring

From the above analysis, we can see that the grouting reinforcement ring has the function of improving the water rich sand layer. It can be seen that grouting reinforcement plays an obvious role in the tunnel crossing the poor water bearing stratum. However, the above analysis only takes into account the condition that the thickness of the grouting reinforcement is 5m. Obviously, in the case of the same grouting material and the same grouting method, the thickness of the grouting reinforcement ring plays a decisive role in the effect of the grouting reinforcement.

The above analysis shows that the settlement of tunnel vault and the horizontal displacement of tunnel are mainly determined by the elastic modulus of grouting reinforcement ring, and the amount of water gushing is mainly controlled by the permeability coefficient of grouting reinforcement ring. Therefore, when the parameters of grouting reinforcement ring are optimized, the variation of tunnel deformation and seepage with the thickness of grouting reinforcement ring under different elastic modulus and different permeability coefficient are considered respectively.

4.2.1 When the elastic modulus of reinforcement ring is different, the settlement and offset values of the tunnel vary with the thickness of the reinforcing ring

The thickness of the reinforcing ring is set by 16 parameters from 1~16m. M is taken as the ratio of elastic modulus(E_g)and surrounding rock elastic modulus(E_s)of grouting reinforcement ring. The elastic modulus E_s =100Mpa and m of the surrounding rock take 6 different values, as shown in table 3.

Table 3. Elastic modulus value table of reinforcement

Thickness of reinforcement ring(m)	$m=E_g/E_s$	Elastic modulus of reinforcement ring E_g(MPa)
1~16	1	100
	5	500
	10	1000
	20	2000
	40	4000
	80	8000

Table 4. Penetration rate table ring of reinforcement ring

Thickness of reinforcement ring (m)	$k=k_s/k_g$	Permeability of reinforcement ring $k_g(m^2)$
1~16	1	1×10^{-11}
	5	2×10^{-12}
	10	1×10^{-12}
	20	5×10^{-13}
	50	2×10^{-13}
	100	1×10^{-14}

The numerical simulation results are shown in figure 11. The vertical settlement of tunnel vault before grouting have serious safety risks. After grouting, the settlement values of the tunnel are greatly reduced. It can be seen from the diagram that under the same thickness of grouting reinforcement ring, the settlement value of tunnel decrease with the increase of elastic modulus of grouting reinforcement ring. But when the elastic modulus of the reinforcement ring increases to a certain extent, the thickness of the reinforcement ring has little effect on the settlement of the tunnel. Under the same elastic modulus of grouting reinforcement ring, the settlement of tunnel decreases with the increase of the thickness of reinforcement ring. But when the thickness of the reinforcing ring is greater than 5m, the vertical settlement value of the tunnel changes gently.

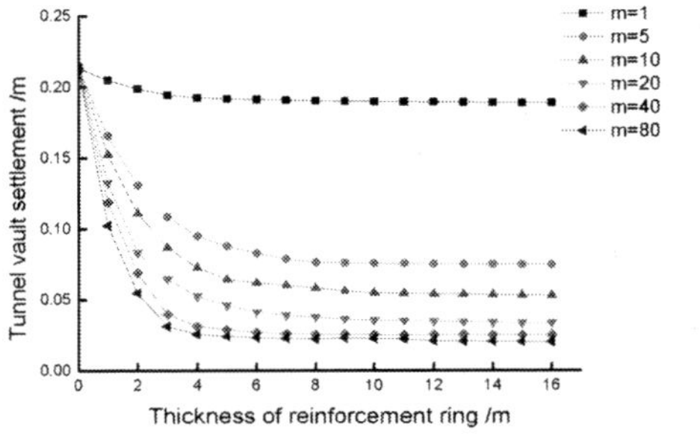

Fig. 11. Change chart of vault settlement.

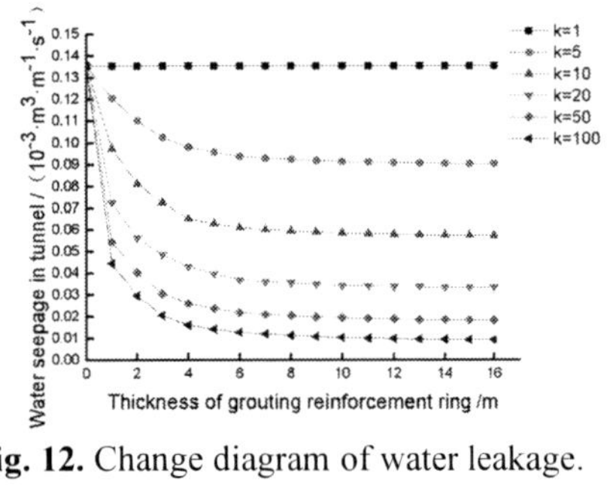

Fig. 12. Change diagram of water leakage.

4.2.2 The permeability of the tunnel varies with the thickness of the reinforcing ring when the permeability of the reinforcement ring is different

Thickness of reinforcing ring is set from 1 m to 16 m, and n is taken as the ratio of permeability of surrounding rock (ks) to that of grouting reinforcing ring (kg). The permeability of the surrounding rock is $k_s=1\times10^{-11}m^2$, and n takes 6 different values, as shown in table 4.

The numerical simulation results are shown in figure 12. Before grouting, the leakage of the tunnel is 1.35×10^{-4} m^3/ (m/s). The permissible seepage is set to 5×10^{-5} m^3/ (m/s). The permeability of the tunnel before grouting is far greater than the allowable value, and there is a serious safety hazard. When the grouting reinforcement ring is formed, the leakage amount of the tunnel will be greatly reduced. It can be seen from the diagram that when the thickness of grouting reinforcement ring is unchanged, the leakage of tunnel decreases obviously with the decrease of the leakage rate of reinforcement ring. When the permeability of reinforcement ring is reduced to a certain extent, the change of seepage volume tends to be gentle. When the permeability of the reinforcing ring is lower than that of the surrounding rock 1/50, the leakage quantity of the tunnel will not change any more. When the grouting reinforcement circle leakage remains unchanged, the tunnel leakage decreases significantly with the increase of reinforcement circle thickness, but when the reinforcement circle thickness is greater than 6 m, the tunnel leakage changes gentle.

4.2.3 The permeability of the tunnel varies with the thickness of the reinforcing ring when the permeability of the reinforcement ring is different

The actual project is to reinforce grouting according to the thickness of the selected reinforcement ring 5m.The numerical simulation software is used to simulate the working condition. The maximum vertical settlement of the tunnel vault is 0.0192m, and the tunnel leakage is $2.369\times10^{-5}m^3$/ (m/s) by integrating the tunnel profile with seepage velocity. The maximum settlement value and leakage volume of the tunnel are calculated as safety allowable values. The maximum vertical settlement of the tunnel vault measured in situ is 0.0213m and the measured tunnel leakage is 2.293 m^3/ (m/day), that is, $2.654\times10^{-5}m^3$/ (m/s). The measured data show that the effect of grouting reinforcement is remarkable, and the settlement value and leakage amount all reach the control standard range. The results show that the maximum settlement of the simulated tunnel is 9.85% less than the measured value and the water inflow of the simulated tunnel is 10.73% less than the measured value. The data obtained from the numerical simulation are basically consistent with the measured data, which verifies the reliability of the numerical simulation.

5 Conclusion

(1) Based on the fluid-solid coupling principle of grout in porous media with small pores, a fluid-solid coupling numerical simulation model is established considering the coupling changes of physical and mechanical parameters such as volume strain, porosity, permeability and density of surrounding rock mass in stress field, displacement field and strain field. The vertical settlement of tunnel vault and water leakage caused by tunnel excavation are numerically simulated.

(2) Through the numerical simulation analysis before and after the grouting reinforcement circle formation, the grouting reinforcement circle formation significantly reduces the tunnel vault vertical settlement and tunnel leakage. This is because the elastic modulus, density, cohesion and internal friction angle of grouting stone have been greatly improved compared with water-rich sand layer. At the same time, the grouting reinforcement circle is more compact, and the porosity and permeability decrease significantly compared with the water rich sand layer.

(3) When the elastic modulus of grouting reinforcement ring is invariable, the vertical settlement of tunnel vault decrease with the increase of grouting reinforcement ring thickness; when the thickness of grouting reinforcement ring is invariable, the vertical displacement decrease with the increase of grouting reinforcement ring elastic modulus. When the permeability of grouting reinforcement ring remains unchanged, the leakage of tunnel decreases with the increase of the thickness of grouting reinforcement ring; when the thickness of grouting reinforcement ring remains unchanged, the leakage decreases with the increase of the permeability of grouting reinforcement ring.

(4) Considering the theoretical feasibility and economic applicability, the optimum thickness of grouting reinforcement ring is 5~6m. It is applied to the grouting reinforcement project in stone elderly bathing station of Qingdao Metro. The maximum settlement, displacement and leakage of the tunnel measured on site are in good agreement with the results of numerical simulation, which verifies the reliability of numerical simulation and provides reference for similar projects in the future.

References

1. Liu R Study on diffusion and plugging mechanism of quick setting cement based slurry in underground dynamic water grouting and its application .2012 *China. D. Shandong University.*
2. Sui W, Liu J, Hu W, et al Experimental investigation on sealing efficiency of chemical grouting in rock fracture with flowing water 2015 *China. J. Tunnelling and Underground Space Technology incorporating Trenchless Technology Research.* 50(1):239-49.
3. Zhang Q S, Han W W, Li S C, et al Comprehensive grouting control for water inrush from limestone breccia fracture zone 2012 *China. J. Chinese Journal of Rock Mechanics and Engineering.*31(12) :2412-19.
4. Wang G Study on parameters and stability of curtain grouting reinforcement ring for water rich stratum in tunnel. 2014 *China. D. Shandong University.*
5. Chen J G, Yuan J Y, Yuan H Q, et al Research and analysis of water inrush and grouting reinforcement ring in Xiang Jia tunnel 2014 *China. J. Journal of highway.*39(04) :52-56.
6. Li L, Cheng P D, Zhong B C, et al Fluid solid coupling analysis of viscous slurry diffusion in porous medium with small pores 2011 *China. J. Research and development of hydrodynamics.* 26(02):209-16.
7. Ye X Y and Liu G Y Study on seepage stress coupling characteristics of rock 2005 *China. J. Chinese Journal of Rock Mechanics and Engineering.* (14):2518-25.
8. Ju M H Numerical simulation of nonlinear dynamic coupling process of mineralization in Dachang ore field, Guangxi 2011 *China. D. Central South University.*
9. Li S C, Li S C and Xu B S Minimum safety factor method for stability analysis of tunnel surrounding rock 2007 *China. J. Rock and Soil Mechanics.* 28(03):549-53.
10. Ding W T and Xu W J Study on the multiphase fluid-solid interaction in granular materials based on an LBM-DEM coupled method 2018 *Switzerland. J. Powder Technology.*335(15):301-14.
11. Cui W, Gawecka K A, Potts D M, et al Numerical analysis of coupled thermo-hydraulic problems in geotechnical engineering 2016 *J. Geomechanics for Energy and the Environment.*6:22-34.

12. Tang L S, Sang H T, Song J, et al Mechanical model for failure modes of rock and soil under compression 2016 *J. Transactions of Nonferrous Metals Society of China.* 26(10):2711-23.

13. Zhao C, Hobbs B E, Muhlhaus H B, et al Numerical modelling of double diffusion driven reactive flow transport in deformable fluid-saturated porous media with particular consideration of temperature-dependent chemical reaction rates 2000 *J. Engineering Computations.* 17:367-85.

14. He H Z and Wang X H Study on parameters of grouting reinforcement ring for subsea tunnel construction 2008 *China. J. Highway traffic technology.* (05) :99-102.

15. Monforte L,Arroyo M,Carbonell J M, et al Coupled effective stress analysis of insertion problems in geotechnics with the Particle Finite Element Method 2018 *England. J. Computers and Geotechnics.* 101:114-29

Study on the actual case of deep excavation in sands and pebble with plenty water

Feng Wang[1,*], *Yuexin* Gao [1], *Xun* Zhou[1], *Yan* Lv [2], and *Xin* Xu[3]

[1]Zhongjiao Luqiao North Engineering Co, Ltd, Beijing 100024, China
[2]College of Construction Engineering, Jilin University, Changchun 130026, China
[3]College of Exploration and Surveying Engineering, Changchun Institute of Technology, Changchun 130021, China

Abstract. Based on the deep excavation of Fengxihe Station of Chengdu Metro Line 17, the groundwater treatment plan for the deep sandy pebble stratum in Wenjiang district of Chengdu, the hole forming process of the retaining pile and the design of the retaining structure and the on-site monitoring data were analyzed and studied. Through systematic research on key technologies, the design and construction optimization schemes are proposed, and the relevant information of on-site information construction is enriched, which provides reference and reference for the design and construction of station excavations in similar strata.

1 Introduction

A prominent problem in the construction of the subway is the design and construction of the deep excavation. How to rationally design the excavation support and construction plan, which is economical, reasonable, safe and reliable, has become one of the important research topics in the geotechnical engineering field. The subway excavation deep excavation project is the most risky stage in the subway civil construction process. Especially when there are important buildings and pipelines around the excavation, the large ground displacement and settlement will affect its normal use and excavation safety[1]. The row pile and internal support structure are a kind of support form widely used in deep excavations in China, and have been widely used in the deep excavation engineering of metro stations in Chengdu area.

The water-rich sandy pebble stratum in Chengdu has the characteristics of loose structure, high content of pebbles and boulders, no cementation, large permeability coefficient, poor self-stability, high strength of single stone, and skeleton function of pebble in the stratum. At present, there is still little domestic engineering experience for the excavation and retaining structure design of excavations under such stratum conditions[2]. Therefore, how to ensure the rational design, safety and reliability of the excavation of Chengdu sandy pebble stratum station has become an urgent problem for engineers.

2 Project overview

* Corresponding author: 1419575643@qq.com

Fengxihe Station is the fourth station from the Yiyuan Park Station on the first phase of Line 17, and is interchanged with Fengxihe Station on the 4th line. The station is located on the west side of the intersection of Fengxi Avenue and Nanxuan Avenue, and lay along the east-west direction of Fengxi Avenue.

The excavation of Fengxihe Station of Chengdu Metro has the characteristics of the general pit of the subway station, namely "deep" and "big". The station is a three-story, two-column, three-span island station with a total length of 216.2m and a standard section width of 22.7. m, both ends of the section are shield sections, the depth of the excavation is about 28.7~30.0m, and the bottom plate of the station is located in the dense pebble soil layer. The station groundwater is mainly the pore type diving in the Quaternary sandy pebble stratum.

Intensive buildings around the station. The main surrounding environment is: the 9th detachment of the armed police on the northeast side of the station (brick 5/7, shallow foundation) is about 10m away from the main body of the station; the Rongxing Garden Community on the north side of the station (brick 3/4, shallow foundation), distance The main body of the station is 5.6m; the south side of the station is Xiangyang Mingyuan (brick 2~6, shallow foundation). As shown in figure 1.

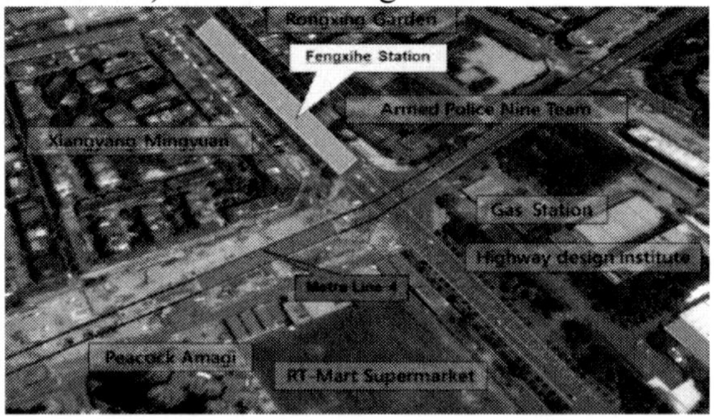

Fig. 1. Location map of Fengxihe Station.

3 Engineering geological conditions

Although Chengdu has successfully built Metro Line 1 Phase 1, Line 2 Phase 1, Line 2 West Extension Line and Line 3 Phase 1, Line 4 Phase 1, Line 4 Phase 2 and Line 7, but according to the preliminary engineering geological and hydrogeological data, the first phase of the Metro Line 17 project in the Wenjiang section is located at the maximum pebble content of more than 75%, the size of the boulders is 20 ~ 70cm, and the uniaxial compressive strength of the pebble It may exceed 132 MPa in large particle size, high strength, water-rich sandy pebble stratum. Compared with the central urban area, this sand-pebble stratum has the characteristics of large diameter of boulders, high content of boulders, and poor density of pebble in some sections. As shown in figure 2 and figure 3.

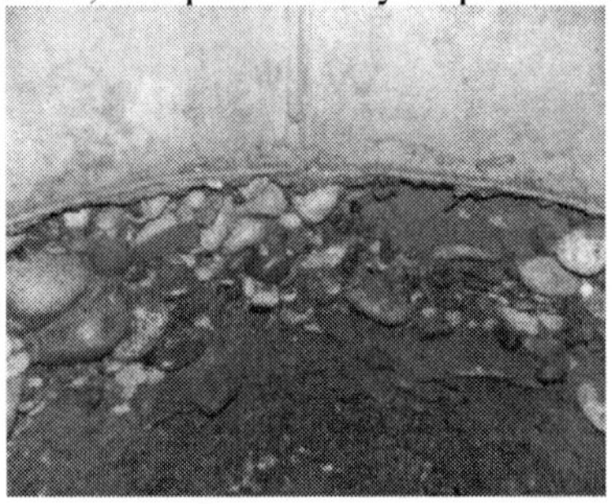

Fig. 2. Great quantity drifting stone.

Fig. 3. Drifting stone 40cm or more.

The main geotechnical engineering features and hydrological characteristics of this project are described in detail in the following table 1:

Table 1. Engineering characteristics and hydrological characteristics of rock and soil layers statistical table.

Layer number	Name	Engineering characteristics	Hydrological characteristics	Uniformity
<1-2>	Artificially filled soil	Mottled, slightly wet, loose, layer thickness of about 1.0 ~ 3.4m.	A small amount of water, water-rich poor, micro-permeable ability.	More miscellaneous ingredients and usually contain pebble, sand and containing a partial construction waste, concrete, poor uniformity
<3-8-1>	Slightly dense pebble soil	Gray, grayish yellow, slightly dense, moist-saturated, filled with medium fine sand. The layer thickness is 2.0 to 9.7 m.	Generally the amount of water, good water-rich, high water permeability.	Pebble content of about 60% to 65%, for medium and fine sand filling, poor sorting
<3-8-2>	Medium dense pebble soil	Gray, grayish yellow, medium dense, saturated, filled with medium fine sand, layer thickness 4.7 ~ 10.8m.	Generally the amount of water, good water-rich, high water permeability.	Pebble content of about 65% to 75%, for medium and fine sand filling, poor sorting
<3-8-3>	Compact pebble soil	Gray, blue-gray, brown-yellow, dense, saturated, filled with medium fine sand.	Generally the amount of water, good water-rich, high water permeability.	Pebble content of 75%, for medium and fine sand filling, poor sorting

4 Groundwater treatment plan

The groundwater of the site is mainly Quaternary pore diving, and the main aquifer is sandy pebble stratum, which is a moderately complex site for precipitation projects. The groundwater static level of the site is about 3.80~7.30m. The stable water level elevation is 532.65~541.25m. During the flood season, the water table may rise.

4.1 Excavation dewatering design

The main parameters of this excavation dewatering project in the following table 2:

Table 2. Fengxihe Station excavation dewatering parameters summary table.

Fengxihe Station Excavation Dewatering Parameters	Precipitation influence radius R: 1815m;
	Excavation calculation radius: 40.85m;
	Excavation water inflow Q: 22094m3
	Single well water output q: 720m3;
	Number of precipitation wells n: 35 eyes;
	Precipitation well spacing D: 15m;
	Precipitation well depth HW: 42.5m/40m;

The main precipitation well parameters of this project are shown in figure 4:

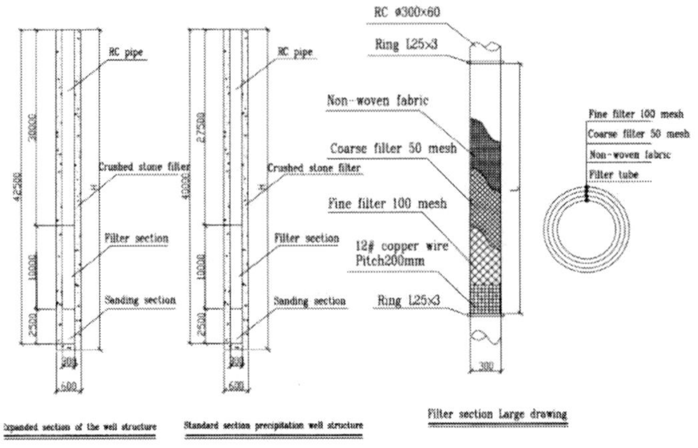

Fig. 4. Structure of the precipitation well.

4.2 Precipitation impact analysis

Excavation precipitation is one of the main factors causing settlement deformation of the surface and surrounding buildings. The buildings around the excavation of Fengxihe Station are densely packed, the underground pipelines are densely covered, the construction site is narrow, and the ground subsidence of the pits will seriously affect the safety of surrounding buildings and underground municipal pipelines. Considering the stratification of the soil, the stratification sum method is used to calculate the settlement caused by the precipitation of the pipe well according to the effective stress increment of the soil. Taking the soil micro-unit of the i-th layer soil, the settlement formula of the i-th layer soil is as follows:

- $\mathrm{L} = \xi \sum_{i=1}^{n} \frac{\Delta P \times H}{E}$,where:
- L is the final settlement, mm;
- δ is the empirical coefficient for calculating the compressive deformation of the soil layer;
- ΔP is the additional stress of the soil layer, kPa;
- $H\ i$ is the thickness of the i-th layer soil, mm;
- $E\ i$ is the i-th layer The compressive modulus of the soil, kPa.

From this, it is calculated that the surface subsidence caused by precipitation at Fengxihe Station is 1~3 mm. Through on-site verification, the surface subsidence caused by construction precipitation in Chengdu sandy pebble stratum is relatively small.

5 Enclosure structure design and construction plan

5.1 Row pile and inner support design

In combination with the geological and hydrogeological conditions of the first phase of Chengdu Metro Line 1 and the requirements for surrounding environmental protection, the selected supporting structure should meet the feasibility of construction, and should also meet the restrictions on the horizontal displacement of the excavationand the surface settlement of the station environment. Finally, it is determined that the envelope structure of Fengxihe Station is a cast-in-place pile. The row pile support form of the deep excavation mainly includes two types: pile and anchor and steel pipe inner support. Bolt support can provide a relatively open working environment for excavation excavation, but due to its relatively poor reliability, the deformation of the excavation is not easy to control, it is restricted by underground pipelines, and easy to have a certain impact on the surrounding pipelines and buildings[3]. In view of the above reasons, the Century Avenue Station does not use bolt support, and the main excavation support is recommended to adopt the plan of row pile and internal support. The site construction photos are shown in figure 5 and figure 6.

Fig. 5. Rotary drilling rig construction.

Fig. 6. Row pile and inner support structure.

The excavation depth of the excavation of the metro station is 28.7~30.0m, and the standard section width is 23.8m. The retaining pile is made of ⌀ 1200@1800 size rotary drilling bored pile, the pile length is 34m, and the buried depth of the excavation is or so 6m below. The crown is set at the top of the pile, and the Φ8@150×150 steel mesh is used between the piles. The reinforcement mesh is reinforced by Φ16@600, and the pile is sprayed with 150mm thick C20 concrete. The support structure is provided with five supports vertically. The first support is 800×1000 concrete support, the horizontal spacing is 1.0m, and the second to fifth support is supported by steel pipe with diameter Φ609×16 (t=16mm). The end is supported on the steel cofferdam of the two sides of the retaining pile, the horizontal distance of the steel support is 3m, and the vertical distance of each steel support from the top surface of the first concrete support is shown in figure 7.

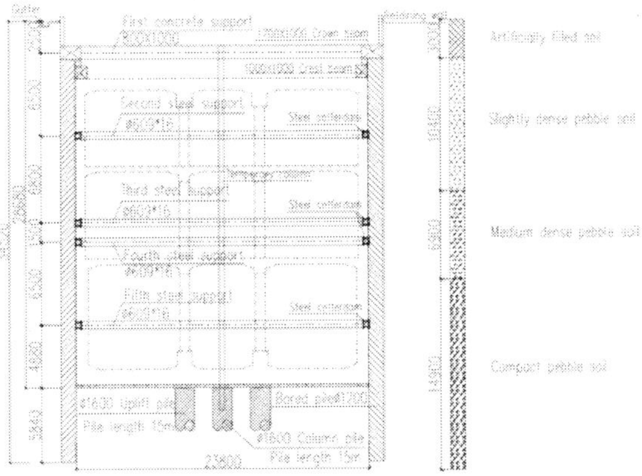

Fig. 7. Sectional view of the excavation retaining structure of the subway station.

5.2 Selection of hole forming technology for retaining piles

By comparing the hole forming process and investigating the hole forming process of the civil excavation and bridge pile foundation near the station, the project is planned to use a rotary drilling rig to form a hole. At the same time, the adaptability of the rotary drilling rig to the soil layer is poor, and the particle size exceeds 0.3 m. The pebble is difficult to drill, and it is difficult to apply to the larger particle size of the pebble stratum or boulders. The following measures are mainly taken during construction[4]:

①The sandy pebble stratum has a large wear on the drill bit, and the drill bit reserve should be increased;

② Large-diameter drifting stone affects the construction of rotary drilling rig, and the ordinary drill bit should be replaced with a spiral drill bit, and the larger-sized boulders must be broken;

③In addition to the expansive soil mud, a certain amount of yellow mud should be stored, if necessary, added to the mud, thereby increasing the mud viscosity, increasing the volume of the mud pool, and increasing the mud reserve to prevent the sandy pebble stratum from leaking and collapsing.

6 On-site monitoring results

During the construction of the station excavation, the construction unit monitored and measured the horizontal displacement of the bored pile top, the inclination of the pile body, the axial force of the steel support, the surface settlement around the excavation, and the groundwater level, and adjusted according to the on-site monitoring results , optimization, to achieve information design[5]. The pit depth of the station is about 28.7~30.0m, and the deformation control protection level is one level. The displacement requirements are:

1) Groundwater level control standard: 2000mm; alarm value: 1600mm.

2) Settlement control standard for surrounding buildings: 20mm; alarm value: 16mm.

Figure 8 shows the groundwater level monitoring points SW03~SW11 within one month that a plot of the distance of the water level from the excavation face, Before the earth excavation, the inner and outer sides of the excavation began to dewater. During the excavation of the excavation, the groundwater level gradually decreased with the increase of the excavation depth, but it remained at 1~2m below the excavation surface of the excavation, which will prevent excessive water head drop .The impact on the surrounding environment is reduced to a minimum. Figure 9 is the cumulative settlement curve of the surrounding building settlement monitoring points JZ1-1~JZ2-12 within one month. The closer to the excavation, the greater the impact. However, they are all controlled within a reasonable range and have little effect on the surrounding buildings and the deformation tends to be stable before the next working condition begins[6].

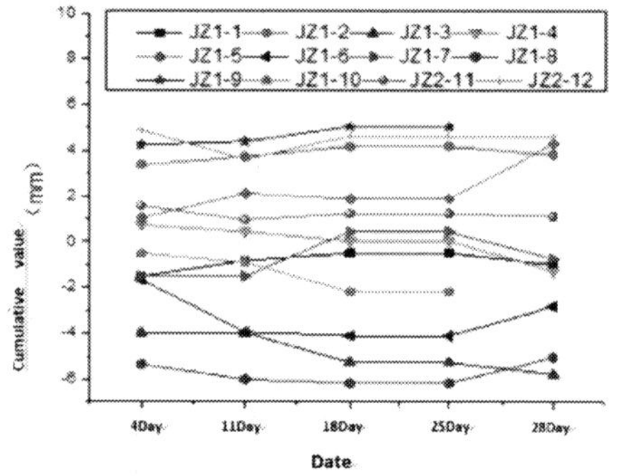

Fig.8. SW03～SW11 water level from the excavation surface distance curve.

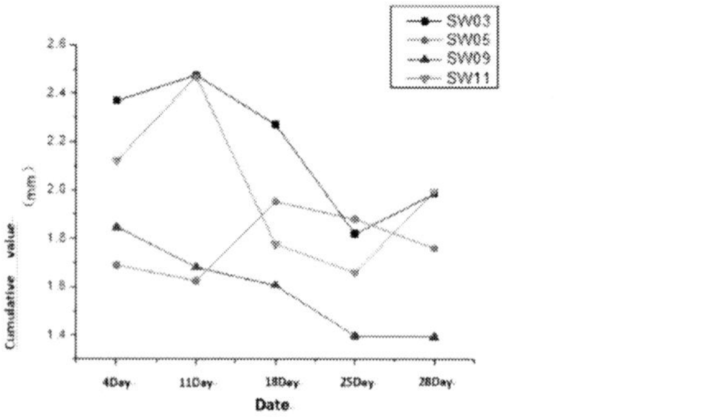

Fig.9. JZ1-1～ JZ2-12 cumulative settlement curve.

7 Conclusion

Through theoretical calculation and actual monitoring results, the design and treatment measures for the excavation of Fengxihe Station in Chengdu metro are safe, feasible and economical. The research on the formation of sandy pebble stratum precipitation measures and hole forming process will provide important technical guarantee for the station design of sand pebble stratum[7]. By comparing the monitoring measurement data with the theoretical analysis results, it is found that the deformation trend of the enclosure structure and the supporting axial force are basically in line with the design expectations, indicating that the design method is reasonable and correct. The main conclusions are as follows:

①Treatment of groundwater in Chengdu water-rich sandy pebble stratum is an effective treatment method for using excavation outer tube well precipitation. According to precipitation settlement analysis and construction monitoring, when fine particles are controlled within the required range, due to the skeletal effect of the sandy pebble stratum, the surface subsidence caused by precipitation is relatively small, and the resulting subsidence value is basically 2 to 5 mm.

②The hole-forming experience of bored piles in Chengdu's water-rich sandy pebble stratum is the process of forming the hole of the station retaining pile. Steel casings shall be used in the top miscellaneous fill soil and plain fill soil sections to ensure the verticality of the enclosure structure and the stability of the top soil. Mud wall is used in the hole forming process, and the mud concentration is increased at the same time. In addition to the expansive soil mud, a certain amount of yellow mud is stored. If necessary, it is added into the mud to increase the mud viscosity, increase the volume of the mud pool, and increase the mud reserve to prevent The sandy pebble stratum leaks and collapses.

References

1. Sun Shuangxiang. Research on the Support Scheme of Deep Excavations of the Provincial Wenlian Station of Chengdu Metro Line 4 [D]. Southwest Jiaotong University, 2012.
2. Jiang Huachun. Stability and Impermeability Analysis of Water-rich Gravel Stratum Retaining Structure [D]. Central South University, 2012.
3. Wu Bing. Study on The Force Law of Row Pile Support Structure in Deep Excavation of Metro Station in Chengdu Area [D]. Southwest Jiaotong University, 2011.
4. Zhou Song. Research on Deformation and Internal Force of Retaining Structure of Deep-supported Piles in Deep Excavation [D]. Chengdu University of Technology, 2017.
5. Cao Haijing. Research on Monitoring Index of Sand Pebble Excavation Deformation in Chengdu Subway [D]. China Academy of Railway Sciences, 2015.
6. Chen Shengjin. Research on Deformation Law of Deep Excavation in Sand and Gravel Stratum of Chunxi Road Station in Chengdu Metro [D]. Chengdu University of Technology, 2011.
7. Liu Zhiyi. Deformation Study and Design Optimization of Excavation Support Structure of Chengdu Metro Station [D]. Southwest Jiaotong University, 2011.

MATEC Web of Conferences **277**, 03016 (2019) https://doi.org/10.1051/matecconf/201927703016
JCMME 2018

Exploration to the construction pattern of Wetland Park---Taking Haizhu Wetland Park as an example

Yan Lv, and *Shao Ping* Guan*

School of Design,South China University of Technology,Higher Education Mega Centre, Panyu District, Guangzhou , China

Abstract. China has been increasingly attaching importance to the construction of wetland parks.But at present, the design methods adopted by most of the domestic wetland parks constructed are sameness, with a short construction period, neither taking into account of the land usage and surrounding environment of the land, nor in-depth land design and mostly the study is from the angle of comprehensive planning or focus on the design of some special projects (such as plant arrangement). In this way, the ecological restoration of the wetland and the minimal intervention in the design of the wetland park cannot be truly realized, nor integration of regional characteristics, resulting in low ecological, landscape and economic value of the constructed wetland park. Some wetland parks have been abandoned once again due to their isolated habitat. This paper aims to probe the construction pattern of wetland park in China. Through the study on the typical cases of wetland parks at home and abroad and taking Haizhu Wetland Park as a major example to reflect on the construction and development of wetland park in China.

1 Introduction

Nowadays,the construction and conversation of wetlands have been generally paid much attention. In recent years, an increasing number of wetlands have been constructed in China. As an important ecological remediation means, the construction of wetland park is the optimal path to conserve wetlands and give play to the multi-purpose function of wetlands. But how can we truly realize the maximization of wetland protection and wetland utility?

1.1 The status quo of wetland park in China

The construction of wetland park, as a significant means of wetland conservation and ecological restoration, has been developed rapidly in China in more than a decade, which has become hot spots especially in recent years. A large number of wetland parks have been built across the country. By the end of 2017, there have been 898 national wetland parks (including pilot projects) in the whole country, a total of 258 were officially opened after

* Corresponding author: shpguan@scut.edu.cn

the acceptance inspection, of which, the well-known practice cases include Xixi Wetland Park and Poyang Lake Wetland Park. The construction trend of wetland park embodies the increasing attention to wetland protection in China, yet the construction of wetland park in China is still at the exploratory stage, so a lot of problems are still existed in its concrete construction process.

1.2 There are general issues in the construction practice of wetland park

The primary and and essential goal for the construction of the wetland park is to protect the wetland habitat, restore ecosystem, etc., and basically, all the practice will also claim like that. However, for most of the wetland parks, these problems have not been settled. And they have not achieved the goal of protecting biodiversity and carrying out ecological restoration through planning and design or technology application with positive attitude and effective achievements[1], which deviates from the original intention of the construction. The problems in construction are generally reflected in the following aspects:

- It is not well disposed for the disturbance of human activities to wetland organisms: most of the parks do not really solve the interference of human on birds and other wetland habitat animals, so as to realize the minimum intervene to the organism in the block;
- The science popularization education is insufficient: the science popularization education for the conservation of biodiversity in the wetland park is merely reflected in some exhibition halls, which does not enable people to truly establish the awareness of organism conservation;
- Landscape has been attached importance in the construction: For most of the wetland park construction, it focuses too much on creating landscape value but ignores the original intention of building wetland park;
- The design is predominantly sameness: mostly vanity projects, without on the basis of regional features;
- After completion, it cannot give economic benefits to the the plot around: most of the domestic wetland parks are routinist in construction, and they are neglected in management after completion, and it fails to promote the surrounding environment and the economic value through the construction of the wetland park.

All these points result in the weak construction significance of most existing wetland parks.

1.3 The research on wetland parks in China is less

The theoretical basis for guiding the ecological restoration and construction and development of wetland parks in China is not adequate.At present, there are no laws and regulations on wetland protection in China. The domestic "Management Measure of Wetland Park", "Planning Guideline of Wetland Park", "Construction Standard of Wetland Park" and "Check and Acceptance Method of National Wetland Park" are all development-guided in wetland park from the planning level[2], while the "Guidelines for Urban Wetland Park Design" only gives the major principles of wetland park planning and design, and fail to elaborate the construction means and other details of wetland park. There are few theory books on wetland research in China.

1.4 Research significance and objectives

The construction trend of the wetland park, the situation with the universal problem that is insufficient synchronization with is weak in study and extensive in practice, all indicate the importance of promoting wetland park research. The construction of wetland park should be integrated with ecology, science popularization, leisure and culture.

2 Examples of excellent wetland parks

The wetland park can be divided into urban wetland park, suburban wetland park and exurb wetland park according to the relationship between the site selection of the park and the city[3]. The geographical position of the urban wetland park is relatively superior, which also causes the relatively large pedestrian volume involved in the urban wetland park. It requires more attention to coordinate with the surrounding plot, and the economic value that it can bring and the impact on the city are higher than other wetland parks. The construction of urban wetland park has more factors to be considered than other ones, The wetland park can be divided into urban wetland park, suburban wetland park and exurb wetland park according to the relationship between the site selection of the park and the city. The geographical position of the urban wetland park is relatively superior, which also causes the relatively large pedestrian volume involved in the urban wetland park.

In the construction of urban wetland park, there are many good cases, and the construction of London Wetland Park and Hong Kong Wetland Park is very worthy of reference.

2.1 London Wetland Park

London Wetland Park is built in metropolitan center, just 4.8km from downtown London.

On the basis of the habitat and hydrological characteristics of species, the park is set to be relatively independent for each area, taking into account the different environmental requirements of breeding and reproduction of various species. As a public place, London Wetland Park is undoubtedly open to the public. In order to realize the purpose of wetland park construction, it is inevitable to consider the harmonious coexistence of "people" and the creatures in the plot.

How can visitors observe wildlife at close range without disturbing the rest of the creatures? The solution to this problem in London Wetland Park is to disperse stream of people and design the streamline sequence for people stream. Definite dynamic and static partitions are set up in the park to separate the stream of people from the visitor center, undoubtedly, the small-sized people flow has relatively low disturbing force on the environment. Tourists visit through various winding paths and are guided to the static and dynamic zones in the park successively. Through the dual experience in dynamic and static zones, tourists can comprehend the value of wetlands and cultivate their awareness of protecting biodiversity subjectively[4].

2.2 Hong Kong Wetland Park

The design of Hong Kong Wetland Park is similar to that of London Wetland Park, which is also a way to gather and disperse people to reduce the interference of people on the block. Beyond that, in the design of Hong Kong Wetland Park, the design idea of environmental protection is highlighted. In the processing of building and other details, environmental protection materials are used in large, which greatly improves the service efficiency of energy[5]. Noise isolation design is also ubiquitous in the park, and all of these practices achieve the minimum interference from visitors to the block.

2.3 Case summary

The construction of the above two wetland parks embodies the respect for nature and humanity in the design, and the point of strength for the design is on ecological remediation. Meanwhile, the construction of the two parks not only lift the environmental quality and ecological restoration of the land within the plot, but also significantly boost the environmental quality of the surrounding area. By reselling the surrounding site, the real estate developers promote the land value and achieve win-win results.

3 The probe for the construction mode of wetland park---taking Haizhu Wetland Park as an example

3.1 Basic information of Haizhu Wetland Park

Haizhu Wetland Park, located in Guangzhou, is the largest wetland park in the central urban area of megacities in China, which is honoured as "Guangzhou Wetland". Haizhu Wetland Park is very advantageous in geographical location, lied on the new central axis of Guangzhou. The unique geographical position of Haizhu Wetland Park also signifies that itself can create economic value for the region and even the whole city. However, the improvement of this economic value cannot be achieved at the cost of damaging the internal environment of the wetland park, nor can it occupy the land within the wetland park to build profitable architectures.

Haizhu park has a lot for reference in the process of construction. Superior geographical location results in the pedestrian volume into the wetland park is ineluctably large. Therefore, more emphasis on should be laid on the principle of minimum interference to the natural background of wetlands through design.

In the construction process of Haizhu Wetland Park, in line with the principle of minimum interference to the natural background of the wetland, the people filament lines that can minimize disturbances to the land parcel are set up. For example, the design of the path avoids the main habitat of the organism[6], and adopts the approach of aerial trails in areas with high ecological sensitivity, as well as the approach of structures, such as building bird watching house to accelerate people's understanding of wetland organisms. All these approaches, from the design of optimized paths and functional zoning, taking the path of static and dynamic zoning as sequential progression, allow visitors to establish the awareness of protecting species subjectively between static and dynamic.

3.2 Optimization measures for Haizhu Wetland Park

In China, the construction of wetland park is still in the exploration stage, and the designers concern more on ecological restoration and landscape value, but overlook that the wetland park also needs the integration of local culture.

Haizhu Wetland Park has taken notice of blending in local features on the basis of reconstructing its ecological background, such as the entrance memorial archway, which shows Lingnan characteristics, and the pavilions in the wetland park are also adopted Lingnan architectural style. Yet, in addition to this, Haizhu Wetland Park still lacks the expression of Lingnan culture.

As the wetland park should pay more attention to ecological restoration and put emphasis and foundation on ecological restoration, it is not recommended to build a large number of structures for the purpose of constructing regional features. However, the regional features can be reflected from the facilities of landscape. Through the reconstruction of landscape facilities, the wetland park is integrated in local characteristics on the basis of ecological restoration. For example, the construction of trash cans promoting Lingnan culture and the laying of floor tiles with Lingnan culture will make the tourists tour in them, with interest and pursuit. It can publicize Lingnan culture in details, thus the Haizhu Wetland Park is blended in local characteristics. Moreover, besides reflecting Lingnan cultural characteristics from the plant configuration, it can also be incarnated from the unique Sangji fish pond in Lingnan. Setting Sangji fish pond as an experience area can not only play the role of popular science and leisure, but also achieve the role of cultural publicity.

3.3 Wetland park construction promoting park city interaction

Because of the special geographical location of Haizhu Wetland Park, it means that Haizhu Wetland Park cannot only meet the ecological restoration, and the impact of it on the economic value of surrounding plots is also huge. On the foundation of wetland background repair and wetland protection, local features can be integrated to increase landscape benefits, accordingly, to attract more tourists and improve the surrounding environment quality, making the value of the Haizhu Wetland Park plot itself and surrounding plots enhance. As a result, the real estate developers will invest more energy and capital to create more value for the whole city, and meanwhile, the government will truly realize the economic value of the wetland park as well. The protection of wetlands is not only to awaken public awareness of conservation, but the government also plays an important role. In this way, the government will have more"impetus"to spend mind and energy on the construction of the wetland park and intensify efforts to promote ecological conservation. The construction of the wetland park can further accelerate the interaction between the park and the city, and embody more significant ecological and social benefits in ecological restoration and urban image promotion.

4 Conclusion

The construction of wetland park can intensively embody the ecological， economic and social benefits. In addition to restoring the ecology and maintaining the regional ecological balance, the environment of the region can be improved so as to boost the economic value of the surrounding land, and science popularization education can be conducted to the society as well. Therefore, the construction of the wetland park is particularly crucial. Majority wetland park construction is not based on local characteristics, and most of the wetland parks built are vanity projects. The government should also strengthen the control over the construction of wetland parks, and build wetland parks that can truly regulate regional climate and balance regional ecology, so as to maximize the benefits of wetland parks. This can make the constructed wetland park not only elevate the environmental quality of the plot, but also enhance the economic value of the surrounding area, to achieve better interaction between the park and the city.

Acknowledgments

The work is upported by "the Fundamental Research Funds for the Central Universities" (project number: XMS17).

References

1. Yurui Liu,Research contents and problems of Urban Wetland Parks.*Jiangxi agriculture* **17**(2017)

2. Yinyu Liang.Study on Design of Wetland Park in Pearl River Delta.*South China University of Technology,Guangzhou(*2015)
3. Shengyong Wang, Xiaoyan Wang and Yanbo Sun,Understanding and discussion of Wetland Park classification..*Journal of Shandong Forestry Science and Technology* **4**(2007) p95-96
4. Jinghua Bu and Yang Wang.Function Mode and Design Conception of London Wetland Center.*Huazhong Architecture* **2**(2015) p103-105
5. Ruifang Lin,Concepts and Principles of Planning for the Hong Kong Wetland Park.*Wetland Science and Management* **1**(2006) p51-54
6. chnical Approach to Ecological Restoration of Guangzhou Haizhu National Wetland Park.*Wetland Science and Management* **12**(2016)

Overall buckling analysis on different formed circular steel tubular columns under axial compression

Xi Feng Yan*

Dept. of Architecture and Building Engineering, Kanagawa Univ., Kanagawa 2218686, Japan

Abstract. This paper presents a numerical finite element model (FEM) investigation on the overall buckling behaviour of hot-rolled (HR), submerged arc welded (SAW) and high-frequency welded (HFW) steel circular columns under axial compression. Three dimensional FEM of circular hollow sections were developed using shell elements considering material nonlinearities, geometric imperfections and residual stress. The established FEM was used to simulate experimental studies conducted by past researchers. Good agreement has been found between numerical analysis and past researchers results, which has validated the reasonability of the FEM to carry out further investigation. Based on the validated FEM, numerical analysis incorporating 180 numerical generated HR, SAW and HFW steel circular columns with various section sizes and slenderness were carried out. The numerical analysis results were compared with the existing column design curves in Chinese, European and American codes. The numerical results showed that the design resistances for hot-rolled and welded steel circular columns calculated based on design curve a in both GB 50017-2003 and Eurocode 3 and the design formula in ANSI/AISC 360-2016 should be accepted. In addition, to further improve the design efficiency, new column design curves for hot-rolled and welded steel circular columns were recommended based on the expressions in GB 50017-2003 and Eurocode 3.

1 Introduction

Modern steel structures are not only developing to large-span, super-tall and heavy-load, but also can adapt strong earthquakes and hostile environments, such as storm, flood and atmospheric corrosion, etc.. Steel tubular structure is a new energy saving and environmental protection architectural structure, famed as the green building in 21st century. Circular steel tubular structures are aesthetically attractive, reduce the amount of surface exposed to corrosion, and in some instances offer certain advantages in structural capacity. According to the manufacturing technique, circular steel tubes are divided into hot-rolled (HR) and welded tubes [1].

There have been a few experimental investigations reported on the overall buckling behaviour of different shaped steel circular columns, with the tests having focused mainly

* Corresponding author: ensaihou@gmail.com

on specimens fabricated by SAW. Shi et al.[2] tested 24 SAW steel circular columns, and Zhao [3] investigated the section capacity and local buckling limits of 1350 MPa SAW circular tubes under compression through testing 12 SAW stub circular columns. Martin et al.[4] researched local buckling behaviour of 5 bare SAW thin-walled steel circular columns and 5 filled concrete SAW thin-walled steel circular columns under axial compression, in addition, 10 bare SAW steel circular columns' local buckling behaviour under eccentric compression was also tested. 15 HR circular specimens were tested by Georgios [5] and presented the theoretical models about plastic mechanism according to the experiments of cold-formed steel circular tubes subjected to pure bending. Recent tubular researches also extend to the new form of tubular cross sections-elliptical hollow sections [6].

However, the overall buckling capacity closely relates to the initial geometric imperfections and residual stress in the cross-section [1]. Different forming methods produce different residual stress and initial geometric imperfections [2,3]. It is considered that the design rules as presently merely reflect the member capacity for hot-rolled and welded columns under axial compression. Therefore, it is necessary to investigate the design methods for SAW, HFW and HR steel circular columns, respectively.

This paper presents a numerical finite element model (FEM) investigation on the overall buckling behaviour of hot-rolled (HR), submerged arc welded (SAW) and high-frequency welded (HFW) circular steel tubular columns under axial compression. Based on a large of parameter analysis results, and comparisons with column design curves in the Chinese code GB 50017-2003 [7], Eurocode 3 [8] and ANSI/AISC 360-2016 [9], column design curves selection guidance for all three kinds of circular steel columns is proposed, furthermore, new column curves for 345 MPa hot-rolled and welded circular steel tubular columns in GB 50017-2003 and Eurocode 3 are recommended.

2 Verification of finite element analysis model

The FEM need to be verified against previously done experiments for reliability and accuracy. For this purpose, experimental study performed by Shi, et al. [2,10] has been taken into consideration. For the initial bending, the out-of-straightness of columns is measured in the length direction, defined as the maximum deviation from the straight lines connecting the points at the centroid of both column ends in the quartile following the length direction, and denoted by v_1, v_2, v_3. As well as, the eccentricity of the load is measured in the both column ends, and donated by e_{0t} (top) and e_{0b} (bottom). A comparison between the results of the numerical analysis and the experiment (Shi, et al.) is illustrated in table 1.

Table 1. Comparison between experimental and FEA results.

Specimen	$N_{u,FEA}$(kN)	$N_{u,Exp}$(kN)	$\Delta=N_{u,FEA}/N_{u,Exp}$	Average of Δ
D420-20-1	2055.59	2094.82	0.9813	0.9756
D420-20-2	2139.83	2173.08	0.9847	
D420-20-3	2092.58	2178.10	0.9607	
D420-30-1	2195.71	2075.09	1.0581	1.0171
D420-30-2	2116.22	2094.09	1.0106	
D420-30-3	2006.61	2042.43	0.9825	
D420-40-1	1974.79	2027.24	0.9741	1.0121
D420-40-2	2154.00	2099.50	1.0260	
D420-40-3	2099.87	2026.38	1.0363	
D420-50-1	2024.89	2037.92	0.9936	

D420-50-2	1963.22	1997.43	0.9829	
D420-50-3	2216.66	2012.03	1.1017	
D420-60-1	1985.26	1956.83	1.0145	
D420-60-2	1908.99	1974.01	0.9671	0.9818
D420-60-3	1888.51	1959.65	0.9637	
Average				1.0025
Standard Deviation				0.0394

Note: $N_{u,FEA}$, $N_{u,Exp}$ -- the ultimate axial bearing capacity of FEA and experimental results of Shi, et al., respectively.

From table 1, we can have the common cognition that FEA results are agreement with the results of experiment of Shi et al.. The average ratio of FEM with experiment (Shi, et al.) is about 1.0025, and the standard deviation is about 0.0394. Hence, the FEM established in this investigation is validated to be accurate and reliable to simulate the overall buckling behaviour of circular steel tubular columns.

3 Parametric analysis and design curves selection

One purpose of this study is to obtained the overall buckling capacities of 345 MPa HR, SAW and HFW circular steel columns under axial compression. According to the steel tubular design code GB/T 17395-2008 [11] and JG/T 381-2012 [12], 180 specimens with 3 different manufacturing techniques, 6 cross sections and the principle slenderness λ_n ranging from 0.4 to 2.2 are designed. Specific details are shown in the table 2. The initial geometric imperfection was applied by updating the geometry based on the first-order eigenvalue buckling model. For overall geometric imperfections, 1‰ of the length of the specimens are adopted in all specimens as overall geometric imperfection amplitude, and not more than 10.0mm [13]. Residual stress mainly is influenced by the craftsmanship and molding method of specimens. There are three different forming methods in this study, which include HR, SAW and HFW. Yan et al. [14] measured the residual stress about these three types steel circular columns and have put forward the residual stress distribution models, respectively.

Table 2. Dimensions of specimens.

Specimens	Outside diameter D(mm)	Thickness t(mm)	Ratios of radius to thickness	Principle slenderness λ_n
SAW-1	114	6	19.00	0.4, 0.6, 0.8, 1.0, 1.2, 1.4, 1.6, 1.8, 2.0, 2.2
SAW-2	273	8	34.13	
SAW-3	325	8	40.63	
SAW-4	406	8	50.75	
SAW-5	508	8	63.50	
SAW-6	610	8	76.25	
HFW-1	114	6	19.00	0.4, 0.6, 0.8, 1.0, 1.2, 1.4, 1.6, 1.8, 2.0, 2.2
HFW-2	273	8	34.13	
HFW-3	325	8	40.63	
HFW-4	406	8	50.75	
HFW-5	508	8	63.50	
HFW-6	610	8	76.25	
HR-1	114	6	19.00	0.4, 0.6, 0.8, 1.0, 1.2, 1.4, 1.6, 1.8, 2.0, 2.2
HR-2	273	8	34.13	
HR-3	325	8	40.63	
HR-4	406	9	45.11	
HR-5	508	9	56.44	
HR-6	610	9	67.78	

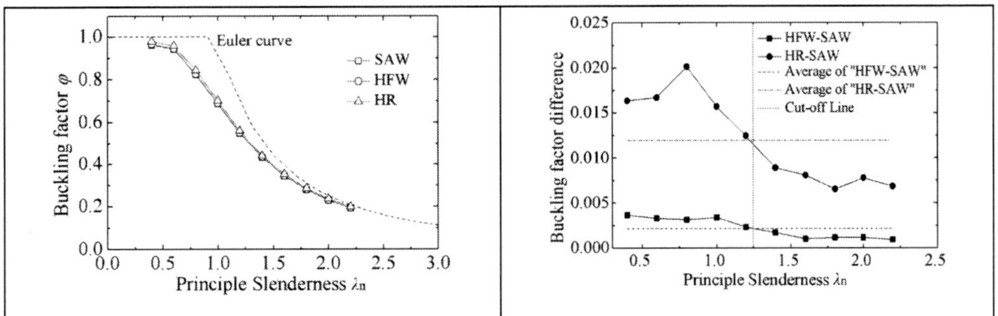

(a) Column curves (b) Subtractions

Note: SAW means submerged arc welded; HFW means high-frequency welded; HR means hot-rolled.

Fig. 1. Comparison of three different types of columns.

The FEA results of buckling factor for SAW, HFW and HR circular steel columns are shown in figure 1. According to the figure 1, the buckling factors of SAW and HFW circular steel columns are adjacent, as well as, the buckling factors of HFW circular steel columns are slightly higher than SAW circular steel columns, the buckling factors of HR circular steel columns are higher than two welded circular steel columns visibly. This suggested that the overall buckling performance of HR circular steel columns is the best; HFW circular steel columns take second place, and SAW circular steel columns are the worst. During the manufacturing technique, welding technology produces more heat than hot rolling method, besides, SAW generates higher heat relative to HFW. On account of different manufacturing techniques, there are different heat influences in the cross sections, then, different residual stress magnitude and different residual stress distribution are produced in the cross sections, and finally, there are different overall buckling behaviours for different formed circular steel columns.

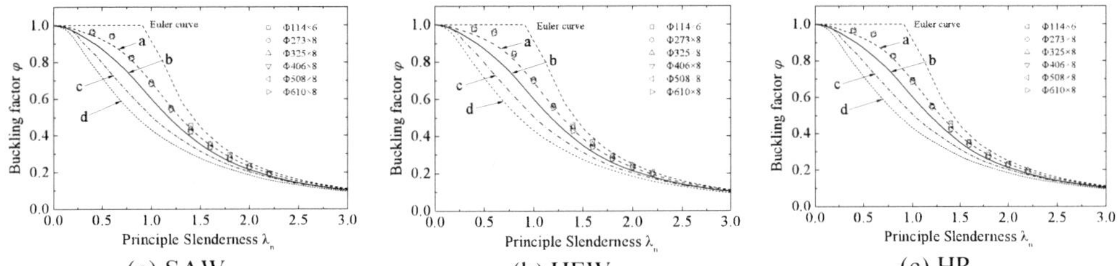

(a) SAW (b) HFW (c) HR

Fig. 2. Comparison of FEA results with GB 50017-2003.

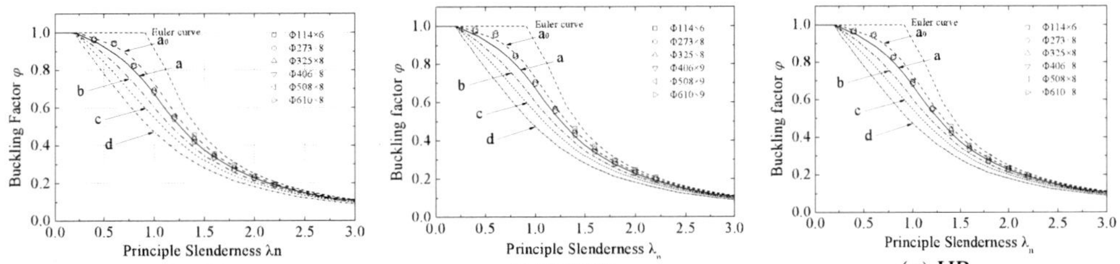

(a) SAW (b) HFW (c) HR

Fig. 3. Comparison of FEA results with Eurocode 3.

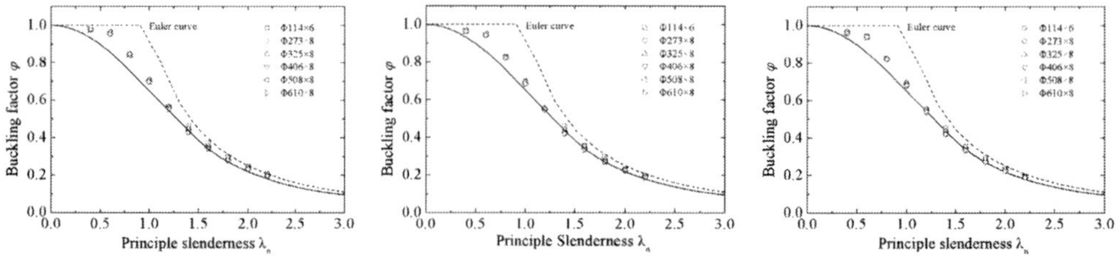

(a) SAW (b) HFW (b) HR

Fig. 4. Comparison of FEA results with ANSI/AISC 360-2016.

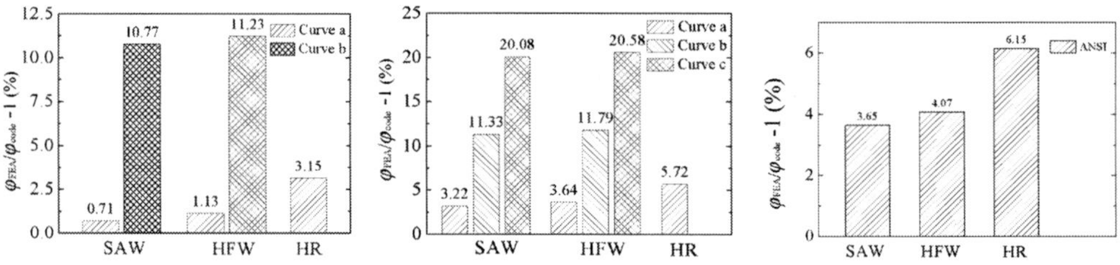

(a) GB 50017-2003 (b) Eurocode 3 (c) ANSI/AISC 360-2016

Fig. 5. Comparison of FEA results with the current design codes.

When principle slenderness $\lambda_n \leq 1.25$, φ_{HFW}-φ_{SAW} and φ_{HR}-φ_{SAW} exceed the average, respectively, manifests that residual stress of manufacturing technique has larger influence for circular steel columns when $\lambda_n \leq 1.25$. According to Eurocode 3, when $\lambda_n \leq \sqrt{2}$, the axial compressed members are in elastic-plastic failure buckling mode, reversely, in elastic failure buckling mode. According to ANSI/AISC 360-2016 code, the column curve in ANSI/AISC 360-2016 code is reasonable for all three kinds of 345MPa steel circular columns when $\lambda_n \geq 1.2$, conversely rather safe when $\lambda_n < 1.2$. But on the whole the column curve in ANSI/AISC 360-2016 is recommended. The FEA results demonstrate that manufacturing technique has major influence on elastic-plastic failure buckling capacity, minor effect on elastic failure buckling capacity.

For figure 5 (a) and figure 2, the column curve a is reasonable for 345MPa HR circular steel columns, but the column curve b is not reasonable and rather safe for 345MPa SAW and HFW circular steel tubular columns. Therefore, the column curve a in GB 50017-2003 is suggested to design SAW, HFW and HR circular steel tubular columns.

For figure 5 (b) and figure 3, it is reasonable to choose curve a for designing HR circular steel tubular columns, however, it is rather safe to select curve c for designing SAW and HFW circular steel columns. Therefore, the column curve a in Eurocode 3 is suggested to design SAW, HFW and HR circular steel columns.

For figure 5 (c) and figure 4, the column curve in ANSI/AISC 360-2016 is recommended to design SAW, HFW and HR circular steel columns.

4 Proposed columns design curves

The expressions for the column curves in both GB 50017-2003 and Eurocode 3 are derived from the Perry equation, and herein are applied to determine the new column design curves. The coefficients in the expressions (i.e. the imperfection factors) are obtained through a non-leaner fitting of the results of the parametric analysis. The fitting process was based on the average level of all parametric analysis results, which is adopted in current design codes. From the section 3, there is negligible distinction of the overall buckling behaviour for SAW and HFW steel circular columns.

Therefore, 345 MPa circular steel columns are designed by two column curves (hot-rolled and welded), and the results of hot-rolled circular steel tubular columns for the imperfection factor α_1, α_2 and α_3 based on the expression in GB 50017-2003 is α_1=0.086, α_2=0.975 and α_3=0.134, respectively. The results of welded circular steel tubular columns for the imperfection factor α_1, α_2 and α_3 based on the expression in GB 50017-2003 is α_1=0.101, α_2=0.969 and α_3=0.165, respectively, the new column curves are plotted with the results of the parametric analysis in figures 6. (a) and (b). Based on the expression in Eurocode 3, the fitting result about hot-rolled and welded circular steel tubular columns for the imperfection factor is 0.138 and 0.164, respectively, and the corresponding new curve is plotted in figures 6. (c) and (d), respectively.

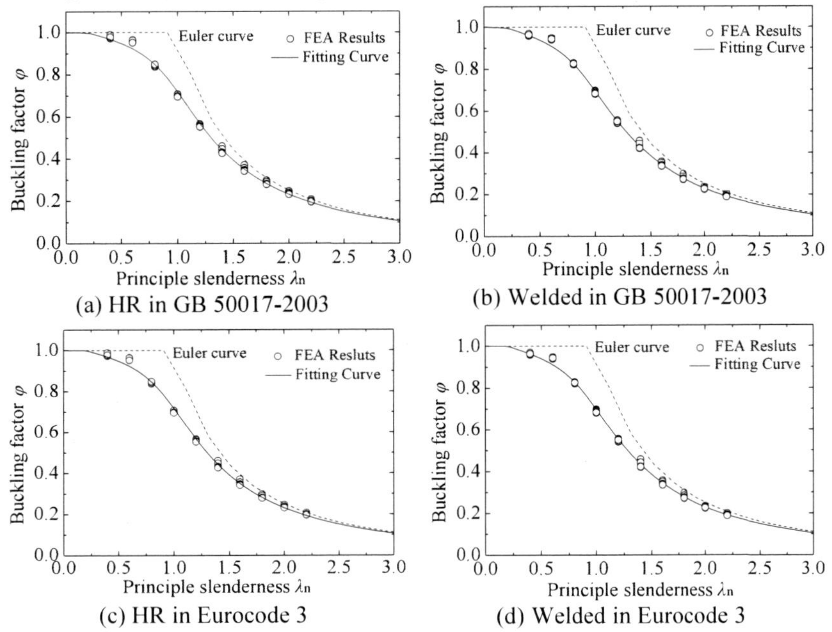

(a) HR in GB 50017-2003
(b) Welded in GB 50017-2003
(c) HR in Eurocode 3
(d) Welded in Eurocode 3

Fig. 6. Proposed curves in GB 50017-2003 and Eurocode 3.

5 Conclusion

This paper presents a numerical finite element model (FEM) investigation on the overall buckling behaviour of hot-rolled (HR), submerged arc welded (SAW) and high-frequency welded (HFW) circular steel tubular columns under axial compression. Based on comparisons of FEA results and the current design codes, the main conclusions are as follows:

(1) Aiming at the overall buckling behavior of 345 MPa HR, SAW and HFW circular steel tubular columns, SAW is the worst, HR is the best, and HFW is between SAW and HR. When principle slenderness $\lambda_n \leq 1.25$, manufacturing technique has a bigger influence in the overall buckling behavior than $\lambda_n > 1.25$. Therefore, the manufacturing technique mainly influences the overall buckling performance in elastic-plastic failure stage;

(2) Compared with the experimental results, the FEM established in this investigation is validated to be accurate and reliable to simulate the overall buckling behavior of steel circular columns;

(3) Based on comparisons of the FEA results and the current design codes, the column curve a in GB 50017-2003 and Eurocode 3 is suggested to design SAW, HFW and HR

circular steel tubular columns.

(4) New column curves are proposed in this paper to predict the buckling behavior of circular steel tubular columns. The fitting imperfection factors based on the expression in GB 50017-2003 are α_1=0.086, α_2=0.975, α_3=0.134 and α_1=0.101, α_2=0.969, α_3=0.165 for HR and welded circular steel columns, respectively, and the corresponding factor based on expression in Eurocode 3 is α=0.138 and α=0.164 for HR and welded circular steel columns, respectively.

References

1. Kulak, Geoffrey L. Tubular members-large and small. Engineering Structures 1996; 18(10): 745-751.
2. Shi G, Jiang X, Zhou WJ. Experimental study on column buckling of 420MPa high strength steel welded circular tubes. Journal of Constructional Steel Research 2014; 100: 71-81.
3. Zhao XL. Section capacity of very high strength (VHS) circular tubes under compression [J]. Thin-Walled Structures 2000; 37(3): 223-240.
4. O'Shea, Martin D, Bridge, Russell Q. Local buckling of Thin-walled Circular Steel Sections With or Without Internal Restraint [J]. Journal of Constructional Steel Research 1997; 41(2-3): 137-157.
5. Elchalakani M, Zhao XL, Grzebieta RH. Plastic mechanism analysis of circular tubes under pure bending. International Journal of Mechanical Sciences 2002; 44(6): 1117-1143.
6. Chan, T. M., and Gardner, L. Flexural Buckling of Elliptical Hollow Section Columns [J]. Journal of Structural Engineering ASCE 2009; 135(5): 543-557.
7. Code for design of steel structures. GB 50017-2003; China; 2003.
8. Eurcode 3. Design of steel structures. Part 1-1: General rules and rules for Buildings. BS EN 1993-1-1; 2005.
9. ANSI/AISC 360-16. Specification for Structure Steel Buildings. Chicago: AISC, 2016.
10. Shi G, Jiang X, Zhou WJ, et al. Experimental study on the buckling behavior of Q420 high strength steel welded tubes under axial compression and design methods. Engineering Mechanics 2015; 32(02):64-73.
11. GB/T 17395 — 2008. Dimensions, shapes, masses and tolerances of seamless steel tubes. Beijing: China Standards Press; 2008.
12. JG/T 381—2012. Cold-formed welded circular tubes for building structures. 2012.
13. GB 50205—2001. Code for acceptance of construction quality of steel structures.2001.
14. Yang JF, Yan XF, Liu HF et al.. Detection and Analysis on Residual Stress Distribution Induced by the Processing Method of Circular Tubes. Engineering Mechanics 2017; 9:202-210.

Building energy performance assessment for existing commercial buildings with heritage values

James Pow Chew Wong*

RMIT University, Melbourne, Australia

Abstract. The building sector accounts for 19% of Australia's energy consumption and around 23% of the greenhouse gas emissions. In 2005, commercial buildings in Australia were responsible for 10% or 26 megatonnes (Mt) of the nation's greenhouse gas emissions. They are two effective ways to reduce the building sector's emissions, namely constructing new energy efficient buildings or retrofitting existing buildings. Due to the life expectancy of existing building stock the greatest reduction in emissions before 2030 could be achieved through retrofits. Building performance assessment and evaluation uses predictive modelling method to investigate and evaluate building performance in energy demands, thermal occupant comfort, indoor air quality, building retrofit options and many other related environmental and wellbeing aspects. This paper discussed the use of building performance simulation software to investigate the impacts of various retrofit options have onto energy performance of existing commercial buildings with heritage values in Australia. The research explores the use of thermal and airflow modelling in assisting understanding of building energy performance with various retrofit options applied to existing commercial buildings. The finding highlights the capabilities of building performance simulation software in decision making for building performance investigations.

1 Introduction

The building sector contributed to around 19% of Australia's energy consumption and around 23% of the greenhouse gas (GHG) emissions [1]. Commercial buildings in Australia contributed to around 10% or 26 megatonnes (Mt) of the nation's GHG emissions in 2005. The Centre for International Economics has noted that energy end use for the commercial sector has nearly tripled since the 1970s [2].

A number of publications highlight the need for more comprehensive information with respect to existing non-domestic building stock typology, distribution and energy use in order to inform future policy formulation [3,4]. Research has shown that energy consumption in commercial buildings is complex due to the wide range of building fabrics and mechanical systems employed [1, 5, 6, 7].

* Corresponding author: james.wong@rmit.edu.au

Energy modelling was used to explore the operational energy use of the case study buildings in their existing state (base case) and in case when energy efficiency measures through selected retrofits were implemented. The findings of this study may therefore inform a better understanding of the challenges involved in analysing the energy demand of commercial buildings with heritage value. This paper investigates the energy performance for two case study buildings with selected retrofits' impacts. The paper also attempts to identify lessons learnt and strategies that could be used to improve building simulation results in the future.

2 Research investigation and modelling methods

Two case study buildings with heritage value were selected from buildings constructed between the 1900s and the 1930s. The first case study is the Old Attorney-General's Building in Adelaide, South Australia. The building was built in the 1900s and consists of an original stone and brick structure with covered verandas, as well as a more modern addition towards the north. The building consists of two-storey with a gross floor area of 1195m2. The second case study, the AMP Building in Townsville, Queensland, is a five-storey building, built in the early 1930's in a classical style. Still displaying its original and unchanged design, it is an important landmark in one of Townville's main streets. Table 1 summaries the characteristics of the two case study buildings. The two case study buildings are selected based on their representativeness of the heritage building type in their regions and periods, the climatic conditions the buildings are located (with Adelaide located in a warm temperate climate and Townsville located in a climate with high humidity during summer and warm winter), and the main building walls' constructions which represented the common heritage building types in Australia.

Table 1. Case study buildings' characteristics.

Case Study	Era	Gross Floor Area, m^2	Net Lettable Area, m^2	Number of Storeys	Roof	Glazing	Walls	Skylight
Old Attorney-General's Building, Adelaide	1900s	1195	894	2	Pitch and flat	Single clear float	Bluestone brick and brick cavity	Single glazed
AMP Building, Townsville	1930s	1289	915	5	Pitch and flat	Single clear float	Uninsulated brick cavity	None

The research used a whole building dynamic simulation software called IESVE for Engineers by Integrated Environmental Solutions [8] to predict the annual electricity and gas usage for the case study buildings. The software takes into account the weather data for the building's geographical location, its three-dimensional geometry, the thermal performance of the building envelope, the type and efficiencies of the heating and cooling systems, hot water systems as well as internal heat gains due to appliances, lightings and occupants.

2.1 Data collection

The building stakeholders were contacted to obtain and confirm key information on the case study buildings' geometry, layouts, construction materials, heating and cooling systems, fuels used, etc. Comprehensive information such as building plans, building operational and usage patterns, fixed appliances used, heating and cooling systems installed was obtained. A Quantity Surveyor consultant was also engaged to determine the construction types and the nature of the materials used when these information are not available. The accuracy of the input information is critical as the accuracy of building energy simulation predictions is largely dependent on it.

2.2 3-D modelling

ModelIT, which is a module of the IESVE for Engineers package, allows the operator to construct three-dimensional models of buildings. Virtual models are usually constructed floor-by-floor using this module. Figures 1 and 2 shows the case study buildings as modelled in the IESVE for Engineers software. Building specific construction elements such as external walls and floors were constructed using the software's building materials database according to the specifications provided by the Quantity Surveyor using the module. In Australia, solar heat gain coefficient (SHGC) is used in the glazing industry rather than Shading Coefficient (SC). SC may be converted to SHGC by multiplying the value by 0.87 [9].

Fig. 1. Model image of the Old Attorney-General's Building in Adelaide.

Fig. 2. Model image of the AMP Building in Townsville.

While the software allowed the customisation of operational profiles, the usage profiles for occupancy, lighting and equipment for the modelling were based on the schedules of the NABERS Energy-Guide to Energy Estimation for Computer Simulations due to some of the building specific information was not available [10]. The default occupant density is 1 occupant per 15m^2 of office space and a sensible heat gain of 75W/person. The values of internal heat gains due to equipment, occupancy and lighting were subsequently adjusted in the input of the modelling in order to reflect the actual conditions in the case study buildings.

The type and efficiency of the HVAC systems used in the case study buildings were assumed based on research carried out in similar types of buildings due to lack of information of the buildings' operational systems. It was assumed that spaces were conditioned by an inefficient central HVAC fan coil system based on the heritage nature of the case study buildings. The set point for space heat is controlled at 19^{0}C and it was assumed natural gas was used. The cooling set point is controlled at 23^{0}C and assumed electricity was used at the central air-conditioning plant. The temperature setting of 19^{0}C for heating and 23^{0}C for cooling were considered a reasonable assumption as they fall within the temperature range of 18^{0}C to 26^{0}C, which is a standard parameter used to model air-conditioned reference buildings according to the Building Code of Australia Section

JV3 [11]. The IESVE for Engineers software assumed the temperature sensor for its HVAC systems at the centre of the space/room for simulation purposes. Constant efficiencies were assumed for both the heating and cooling systems. External climatic data was modelled by the software using real data from the nearest weather station.

SunCast and MacroFlo are additional modules of the IESVE for Engineers software, which are able to take account of the effects of solar shading and of natural ventilation in the dynamic simulation. The SunCast module manages the impact of the shading provided by neighbouring buildings and building integrated shading on the energy demand of the building. SunCast considers the date, time, orientation, as well as the site latitude and longitude. Daylighting effects on lighting use and subsequent impacts were not included in the simulations. The effects of trees and their shadows were not considered.

After all parameters had been assigned to the building model, the dynamic thermal simulation was run using ApacheSim module. A simulation period of 365 days with a simulation time step of 10 minutes as recommended by the IESVE software guidelines was used [8], and with a preconditioning period of 10 days.

2.3 Building retrofit interventions

One of the goals of the study was to identify and review a range of suitable design solutions that could be used to reduce the environmental impacts of heritage commercial buildings. A range of intervention possibilities were considered, including adding insulation, refurbishing existing windows, installing secondary or double glazing, applying ultraviolet (UV) film, upgrading heating, ventilation, and air conditioning (HVAC) systems and lighting systems, etc.

There are a total of 16 intervention strategies applied to the case study buildings for the project but this paper only report on the top five (5) interventions that have the most significant impacts on the energy performance of the case study buildings. The top five interventions applied to the case study buildings for investigations are chiller upgrade, changing cooling set point to 26^0C, lighting upgrade, switching off equipment after office hours, opening windows and increase cooling set point to 26^0C. As the Old Attorney-General's Building is not possible to implement opening windows as an intervention option, therefore intervention in changing the cooling set point to 26^0C is selected.

3 Results and discussions

The Old Attorney General's Building in Adelaide is assumed to be conditioned by a central HVAC fan coil system. Space heating is assumed to be provided by a central boiler fuelled by natural gas. Cooling is assumed to be provided by a central air-conditioning plant fuelled by electricity. Hot water is assumed to be provided by an electric system. The AMP Building is assumed to be conditioned by a central HVAC fan coil system. Space heating and hot water is assumed to be provided by a central boiler fuelled by natural gas. Cooling is assumed to be provided by a central air-conditioning plant fuelled by electricity.

Figures 3 and 4 below show the IESVE for Engineers operational energy simulation results for the buildings. Both electricity and lighting usage were found to be account for a quarter each of the building's energy consumption, with equipment constitute a third of the energy demand for the Old Attorney-General's building. Electricity usage constitutes almost half of the building's energy consumption, followed by lighting and equipment which constitute around 22% of the energy demand for the AMP building.

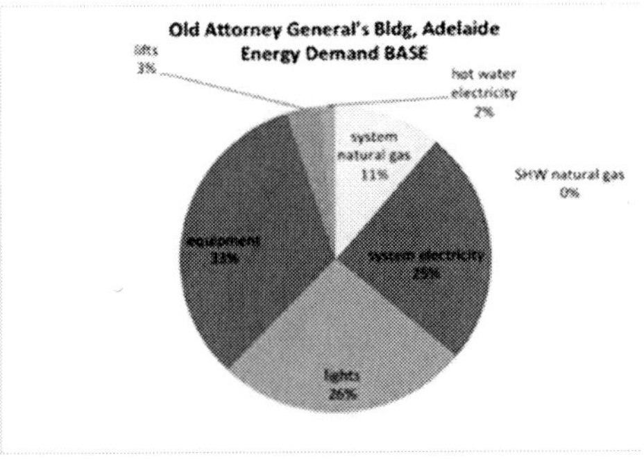

Fig. 3. Operational energy simulation results for Old Attorney-General's Building.

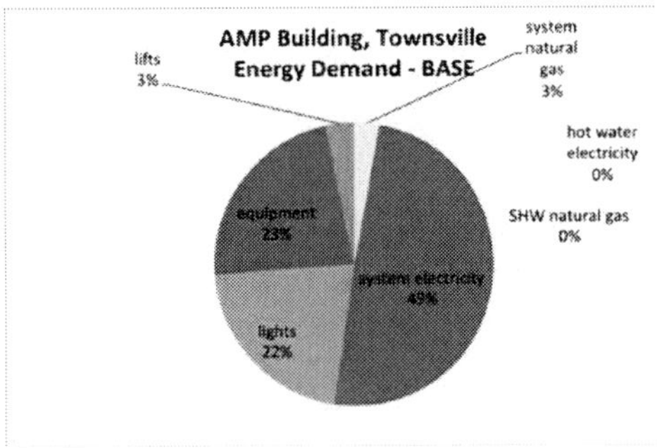

Fig. 4. Operational energy simulation results for AMP Building.

For the case study buildings considered, life cycle global warming impact was predominantly caused by cooling (25%-49%), lighting (22%-26%) and equipment use (23%-33%). Other operational elements of heating (3%-11%) and lifts (3%) represented smaller portions of the life cycle impact. Collectively, this result shows that intervention initiatives that target cooling, lighting and equipment are likely to generate the most significant results.

For the chiller upgrade intervention, information on the efficiencies of the HVAC systems is not provided. Therefore educated estimates were taken for these values. Both buildings were assumed to be equipped with a central fan coil system and cooled by electricity. In this intervention the efficiency of the chiller system is assumed to be increased from a Nominal EER (Energy Efficiency Ratio) of 3.13 to between 4.38 and 5.50.

In the changing cooling set point to 26^{0}C intervention, the cooling set point was assumed to be set at 23^{0}C under the base case. In this intervention, it was examined how the energy use would change if the cooling started operating at a temperature which was three degrees higher, at 26^{0}C.

For the lighting upgrade intervention, both buildings were assumed to be equipped with T8 light fittings in the base cases. In the lighting intervention, it was assumed that all lamps would be replaced by the more energy efficient T5 lighting.

For the switching off equipment after office hours intervention, while the number of computers and their sensible gain values remain the same in this intervention as in the base case scenarios, the variation profile of the computers has been changed. In the base case, the National Australian Built Environment Rating System (NABERS) Appliances and Equipment profile was used.

In the opening windows and increase cooling set point to 26^{0}C intervention, according to the information provided, in both buildings, the windows are kept closed throughout the year. This intervention examines the effect of natural ventilation on the energy demand of the buildings. In this intervention, the opening of half of the windows between 7:00 h and 21:00 h on weekdays if the room air temperature is between 23^{0}C and 26^{0}C. In addition, the chiller's set point is increased to 26^{0}C. In other words, the occupants of the building are assumed to open the windows when the room reaches 23^{0}C, and they are assumed to close them again, when the inside temperature reaches 26^{0}C. Only then will the air conditioning switch on. Instead of setting the comfort level at 23^{0}C, as in the base case, the comfort level is shifted to 26^{0}C.

Table 2. Thermal modelling results for the case study buildings.

		APACHESIM RESULTS					
	Net Lettable Area (m2)	**BASE**	**Chiller upgrade**	**Cooling set point 26^{0}C**	**Lighting upgrade**	**Switch off equipment after hours**	**Opening windows and increase set point to 26^{0}C**
Natural Gas Use (MJ/m2.year)							
AMP, Townsville	915	Not connected to gas					
Old Attorney General's Buildings, Adelaide	894	**93**	93	93	112	113	122
Electricity Use (kWh/m2.year)							
AMP, Townsville	915	**249**	208	220	223	215	219
Old Attorney General's Buildings, Adelaide	894	**206**	192	191	182	168	194

Table 2 above shows the thermal modelling results for the base case and the top interventions for the two case study buildings. Gas usage was mainly higher for most of the interventions for Old Attorney-General's Building but electricity usage saw saving up to 16%.

4 Conclusion

Building operational requirements were found to be driven predominantly by cooling, lighting and equipment use. The large contribution of operation was explored further by implementing building interventions that sought to reduce impacts. Operational heating and cooling (system natural gas and system electricity) contributed the most to the total operation energy usage for the case study buildings. Equipment and lighting contributed around a quarter each to the total annual operational energy usage. The case study buildings performed quite satisfactory in terms of operational energy usage compared to the average commercial buildings in Australia.

A number of building interventions were explored, each of which was intended to reduce building energy use in operation. The results demonstrated that significant improvements in building life cycle global warming impacts could be achieved. For the case study buildings considered, when applied in isolation, high impact outcomes tended to be associated with building equipment use, lighting and mechanical systems. Occupant related interventions, such as switching off equipment after hours and on weekends was particularly effective. Efficiency upgrades such as upgrading lighting systems and the building chiller were also effective.

Acknowledgments

The author would like to thank the Department of Planning and Community Development of the State of Victoria, the Heritage Chairs and Officials of Australia and New Zealand (HCOANZ) and the research team at RMIT University for assisting and supporting this research project and subsequent publication of the findings.

References

1. DCCEE 2010 *The Pathway to 2020 for Low-Energy, Low-Carbon Buildings in Australia: Indicative Stringency Study* (Canberra: Department of Climate Change & Energy Efficiency)
2. Centre for International Economics 2007 *Capitalising on the building sector's potential to lessen the costs of a broad based GHG emissions cut* (Canberra: Centre for International Economics)
3. Bruhns H 2008 Identifying determinants of energy use in the UK nondomestic stock *Proc. 5th Int. Conf. on Improving Energy Efficiency in Commercial Buildings (Frankfurt)* (Joint Research Centre, Institute for Energy, European Commission) pp 389-400
4. Lam J C 2000 Energy analysis of commercial buildings in subtropical climates *Building and Environment* **35** pp 19-26
5. DEWHA 2008 *Mandatory Disclosure of Commercial Building Energy Efficiency: Concept Report* (Canberra: Department of the Environment Water Heritage and the Arts)
6. Kofoworola O F and Gheewala S H 2009 Life cycle energy assessment of a typical office building in Thailand *Energy and Buildings* **41** pp 1076-1083
7. Haase M, Grini C and Wigenstad T 2010 Measurement and simulation of a commercial building in Norway *Proc. 6th Int. Conf. on Improving Energy Efficiency in Commercial Buildings (Frankfurt)* (Luxembourg: Joint Research Centre, Institute for Energy, European Commission) pp 416-429
8. IES 2010. ApacheSim User Guide. IES Virtual Environment 6.4, Integrated Environmental Solutions Limited.
9. ASHRAE Handbook Fundamentals 2009. American Society of Heating, Refrigerating and Air- Conditioning Engineers, Inc.
10. Department of Environment, Climate Change and Water 2009 *NABERS Energy Guide to Building Energy Estimation* (Sydney: Department of Environment, Climate Change & Water)
11. ABCB 2010 *Handbook on BCA Section J – Assessment and Verification of an Alternative Solution* (Australian Government and States and Territories of Australia)

Study on joint detection of fabricated concrete shear wall based on percussion method

Bo Song[1,2], *Jian* Li [1,2,*], and *Man sheng* Wang[1,3]

[1]School of Civil and Environment Engineering, University of Science and Technology Beijing, Beijing 100083, China
[2]Earthquake Research on Earthquake Traffic Engineering in Strong Earthquake Area Beijing International Science and Technology Cooperation Base, Beijing 100083, China
[3]Beijing institute of housing and urban-rural construction science and technology, Beijing 100021, China

Abstract. The knock detection method is one of the non-destructive testing methods for assembly prefabricated components. The knocker is used to knock on the concrete prefabricated components to be tested. The internal conditions of the tested components can be judged by analyzing the time-domain and frequency-domain diagrams of the knocker. In this paper, the impact of the change of the knock point on the detection is studied by using the knock detection experiment. It is found that the location of the knock point has a great influence on the detection. With the change of the position of the knock point, the peak frequency in the spectrum obtained by the detection changes correspondingly. According to the frequency information corresponding to the peak value in the spectrum, the depth of the void to the knock surface can be calculated.

1 Introduction

Assembled concrete building structure is more and more recognized by the public because of its advantages of high efficiency, high assembly speed and fast construction speed[1]. Compared with the traditional building method, the construction efficiency of assembly building can be increased by 80%, and the waste rate of resources can be reduced by 70%. However, due to the defects of honeycomb, hemp surface or reserved holes which are not fully filled in the fabrication or field connection engineering of prefabricated concrete buildings, the existence of these defects will seriously affect the bearing capacity and durability of components[2]. Using effective methods to detect the nature, location and scope of fabricated concrete defects, and taking corresponding measures is an important part of Engineering construction. At present, in the non-destructive testing of connection defects of concrete components, there are mainly three methods: infrared detection, ultrasonic detection and knock detection[3]. Infrared detection technology[4] absorbs the infrared radiation signal of the continuous radiation of concrete by infrared camera and converts the signal processing into the distribution of temperature field in the range of concrete to judge the internal defects and losses of concrete components, but infrared

* Corresponding author:154388870@qq.com

detection technology is effective in detecting defects in deeper parts of concrete components. The fruit is bad. Ultrasound detection technology[5] has high sensitivity for the detection of the structure interior. Ultrasound can penetrate solid components and detect their interior. By judging the information of wave velocity and amplitude, we can deduce whether there are defects in concrete members, but the ultrasonic testing method is suitable for non-metallic components, and has little effect on concrete members with steel tubes reserved inside. In this paper, the knock method is used to detect the internal defects of concrete, and the factors affecting the accuracy of the knock method are explored.

2 Tapping detection principle

The experiment of detecting the internal vacancy of concrete by percussion method is carried out by using small steel balls lagging behind the surface of concrete blocks as vibration source and concrete blocks to be tested as propagation medium. Three kinds of waves are produced after the test block is knocked behind the ball: P wave parallel to the propagation direction, S wave perpendicular to the propagation direction, and Rayleigh wave propagating along the concrete surface, namely R wave.

Because of the characteristics of the fastest propagation speed and the strongest penetration ability of P-wave, the main waveform used in the detection of concrete blocks by percussion method is P-wave. The surface displacement caused by the stress wave produced by percussion is detected and recorded by the sensor to form a voltage-time signal, i.e. waveform. The waveform contains the frequency information of multiple one-wave reflection between the impact surface and the defect and each interface. The propagation path of P wave is shown in the figure.1 The displacement amplitude of concrete surface caused by P wave near the impact point is the largest, while that caused by S wave is the smallest, which is basically zero and negligible.

The measured time history curve is filtered and converted to FFT to obtain the spectrum curve for further analysis and summary.

(a) P-wave and S-wave waveform schematic diagram (b) Schematic diagram of R wave

Fig. 1. Waveform schematic diagram generated by knocking method.

3 Experiment

3.1 Instrument and model

In the experiment, the acceleration time history curve at the impact point was measured, and the frequency spectrum curve was obtained by frequency spectrum transformation (FFT), so as to judge whether the concrete block is defective or not. The instrument used in the experiment is WS-5921N network data acquisition instrument produced by Beijing Spectrum, which can quickly carry out spectrum analysis and display the refined power spectrum with cosine window in real time. It has the characteristics of fast processing speed, good automation and high measurement accuracy.

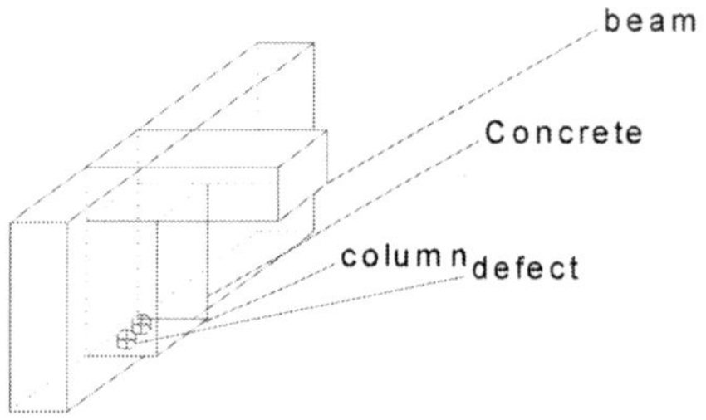

Fig. 2. Assembly structure defect diagram.

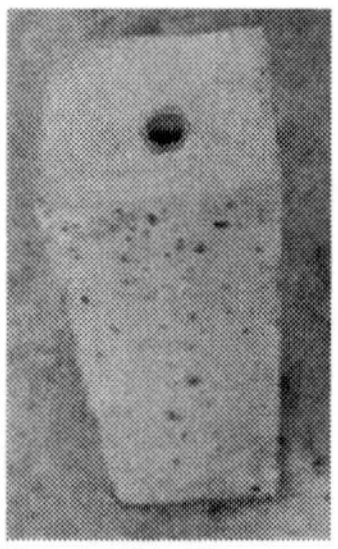

Fig. 3. Test block diagram of percussion detection.

In engineering, the grouting sleeve of assembly building structural components should be tested by joint grouting. as shown in Figure 2. In the actual testing process, the experimental model is simplified to reserve holes in concrete blocks with steel tube sleeves, as shown in Figure 3. Concrete blocks are made of cement, sand, stone and other raw materials in a certain proportion. Among them, cement is 325# Portland cement. The grain size of sand is about 0.35-0.5 mm, the grain size of stone is about 5-20 mm, and the mass ratio of cement, sand and stone is 1:2.89. Complete concrete blocks and defective concrete blocks with reserved voids are made respectively. Numbers 1, 2 and 3 (as shown in the figure 4) are as follows. (b) is a concrete block with reserved holes. The holes run through the cross section of the model, and the diameter of the holes is 20 mm.

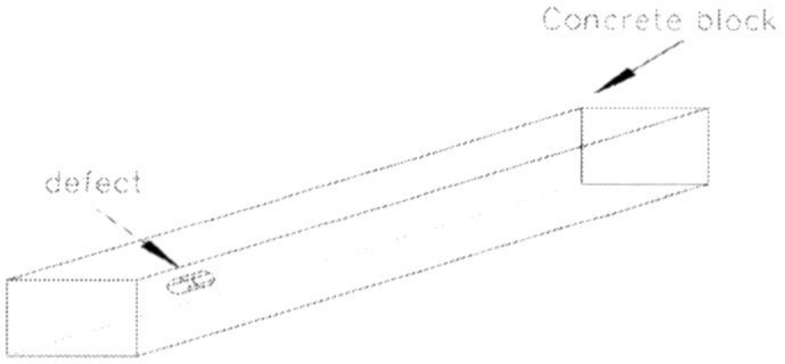

(a) Circular concrete defect blocks

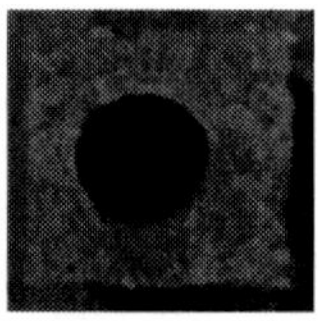

(b)concrete block with circular defect

Fig. 4. Defected concrete sketch.

3.2 Experimental process and method.

The three experimental models are all tested in a single channel. According to the relevant regulations of "Technical Regulations for Testing Concrete Compressive Strength by Rebound Method JGJT 23-2011"[6], the experiment should be carried out at room temperature (5~35) °C, and the hammering ball is fixed to the same height. Falling, the tapping value should take the average of the stable results of three consecutive strokes. Fixed the steel balls to the same height and fell freely. The average value of the stability results of three consecutive knocks should be taken. In order to reflect the real situation of the defects of the test blocks, eight knocking points are evenly arranged on the test surface of each concrete block, in a word arrangement, and the distance between the two adjacent points is 15 mm.. Raw materials and matching as shown in the table 1.

Table 1. Mix ratio.

material	cement	water	Sand	Stone
proportion	1.75	1	3.96	5.05

Draw the line of the concrete specimen that has been demolded, arrange the number of measuring points as shown in the figure, test according to the numbering sequence, and record the data. The picture shows the actual test site map. The obtained test data is subjected to spectrum conversion processing, and after processing, combined with the calculation formula (1), the defect of the concrete can be obtained.

$$Vp = 2fT / c \tag{1}$$

In the expression

Vp— Known wave speed test values for precast concrete members

f— Concrete member thickness reflection frequency

T—Concrete member thickness

Fig. 5. Survey point layout.

The red color in the figure 5 is 0cm at the time of detection, and the distance between each detection point is 1.5cm.

4 Results and analysis of influencing factors

4.1 The influence of test point distance on the measurement

There are many uncertain factors affecting the connection of the fabricated concrete members in the actual engineering. Since the specific location of the defects is not known before the detection, the position of the receiver and the position of the defect at the time of the tapping are caused.

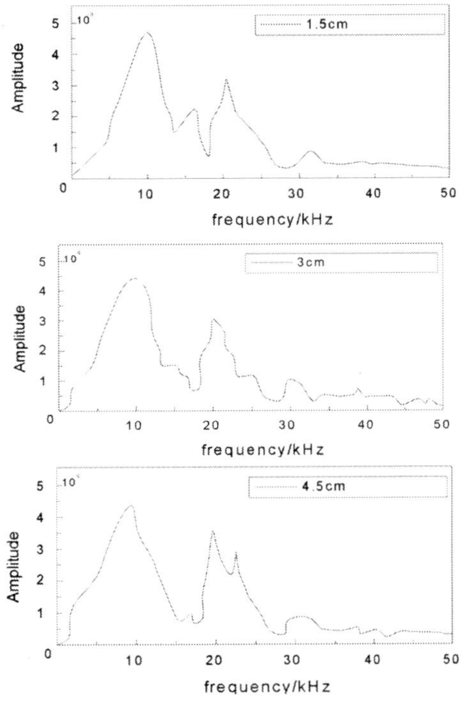

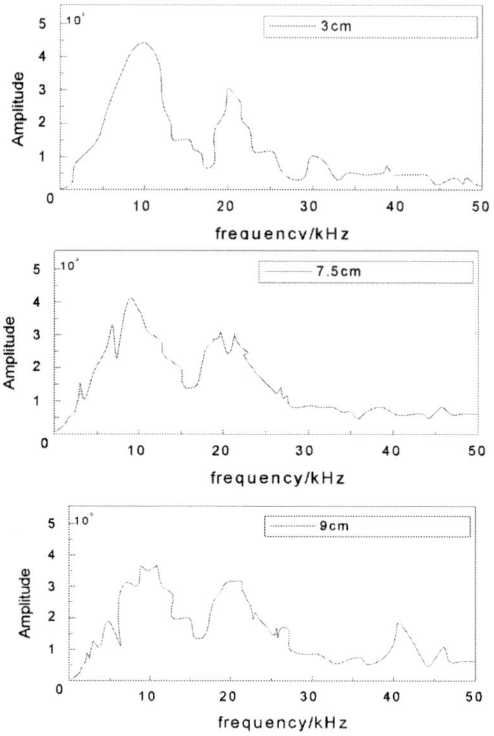

Fig. 6. Spectral map of different tapping points.

As can be seen from the above figure 6, there are 8 measuring points in the range of 1.5 cm to 9 cm in the location of the offset defect at the percussion point. With the offset of the percussion point, the amplitude curves show different spectrograms. According to the spectrum analysis after testing, the signal percussion point is more accurate between 3cm and 6cm. There are two peaks, which are consistent with the actual situation of the test block. The reason may be that the P wave tested has been reflected many times in the tested block, and the internal interference and attenuation have occurred to it. After experimenting the influence of the measured point offset on the percussion test, it is found that when the percussion test is carried out, the signal percussion points should be arranged in the range of 3cm-6cm.

According to the P wave propagation law and formula in concrete blocks, the defect detection depth is calculated when the percussion point is offset as shown in the figure 7. It can be seen that with the offset of the received signal point, the detection error rate reaches the minimum at the offset value of 4.5 cm. Within this range, the detection error increases gradually with the decrease of the offset distance, while in the range larger than this value, the detection error increases gradually with the increase of the offset value. It can be inferred that the signal percussion point should be placed near the range of 4.5 cm.

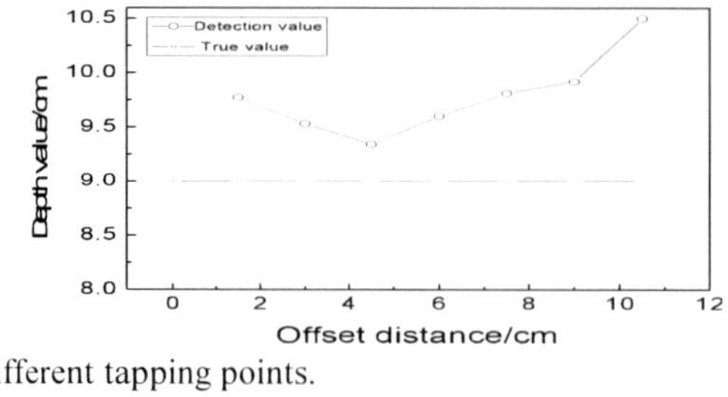

Fig. 7. Spectral map of different tapping points.

5 Conclusion

The thickness of concrete blocks and the depth of internal defects can be calculated by different flat peaks in the spectrum. Through experiments, it can be concluded that the actual test thickness is slightly higher than the actual thickness of concrete blocks.

The different locations of measuring points will have some influence on the test results. With the horizontal displacement of measuring points from the defect position increasing, the error rate of the test results gradually increased. When the measuring points are located within 2-3 cm from the defect, the test results are more accurate. When the measuring points are located outside 3 cm, the influence on the test results is greater.

References

1. Shao Haidong. Experimental study on testing curve of compressive strength of pumped concrete in Ningxia by rebound method [D]. Xi'an University of Architectural Science and Technology, 2012
2. Luanjian, Experimental Study on Quality Inspection of Prestressed Pipeline Grouting [D], Central South University of Forestry Science and Technology, 2011.
3. Huang Jianxin, Application of Shock-echo Method in Nondestructive Testing of Concrete Structures [D], Hehai University, 2006
4. Cheng, C.C. and M. Sansalone, Determining the minimum crack width that can be detected using the impact-echo method Part 1: Experimental study[J]. Materials & Structures, 1995. 28(2): p. 74-82.
5. Men Jinjie et al., experimental study on acoustic emission testing parameters and stress characteristics of reinforced concrete beams [J].Journal of Xi'an University of Architectural Science and Technology (Natural Science Edition), 2015 (06): 793-798.
6. Technical Regulations for Testing Concrete Compressive Strength by Rebound Method JGJT 23-2011"

AUTHOR INDEX

AUTHOR INDEX